Food Science: Sensory Evaluation Techniques

Food Science: Sensory Evaluation Techniques

Edited by **Dorothy Green**

SYRAWOOD
PUBLISHING HOUSE
New York

Published by Syrawood Publishing House,
750 Third Avenue, 9th Floor,
New York, NY 10017, USA
www.syrawoodpublishinghouse.com

Food Science: Sensory Evaluation Techniques
Edited by Dorothy Green

International Standard Book Number: 978-1-68286-310-7 (Hardback)

Printed in China.

Contents

Preface

Sensory evaluation plays a significant role in the food industry. Human senses affect decision making process of end consumers. Thus, it is important to understand them. This book presents the essential concepts in the field of sensory analysis and provides some incredible insights into consumer behavior and response. It discusses in detail the effective testing and affective testing methods for sensory evaluation. It also elucidates the concept of perception. This book will prove extremely beneficial to students, researchers and professionals engaged in this field.

All of the data presented henceforth, was collaborated in the wake of recent advancements in the field. The aim of this book is to present the diversified developments from across the globe in a comprehensible manner. The opinions expressed in each chapter belong solely to the contributing authors. Their interpretations of the topics are the integral part of this book, which I have carefully compiled for a better understanding of the readers.

At the end, I would like to thank all those who dedicated their time and efforts for the successful completion of this book. I also wish to convey my gratitude towards my friends and family who supported me at every step.

Editor

Preliminary studies on the effect of shea kernel size on shea butter quality

A. Abedin, A. H. Abubakari, Quainoo A. K., Chimsah F. A., Nyarko G., Mahunu G .K., Abagale F. K. and Addai G.

Faculty of Agriculture, University for Development Studies, Tamale, Ghana.

The study was carried out between March, 2012 and January, 2013 at the University for Development Studies, Nyankpala campus, Ghana. The objectives of the study were to characterize farmer shea nut varieties to determine the physical and chemical characteristics of the shea nut butter and determine appropriate shea nut quality for the domestic market. The shea nuts were characterized as T1 (small nuts), T2 (medium nuts), T3 (big nuts) and T4 (very big nuts). These nuts were arranged in a completely randomized design and replicated three times for shea nut butter extraction and assessed for free fatty acid (FFA), moisture content and sensory quality. The study reveals strong correlation between shea nut size and butter quality with bigger shea nuts recording lower FFA and moisture content hence higher butter quality. There were differences in colour and texture of the shea butter extracted from the different shea nut sizes. It is recommended that, further research be carried out at different geographical locations to assess the effect of shea nut size on butter quality.

Key words: Shea kernel, free fatty acid, moisture content, sensory quality.

INTRODUCTION

Shea nut is known as '*Kpihi*', '*Nyuuni*' and '*Kyuuma*' among the Dagomba, Frafra and Wala people of Ghana, respectively. It is obtained from the fruit of shea tree *(Vitellaria paradoxa),* which exists in the wild and grows in an uncultivated state in most parts of Africa. Shea fruit is made up of a green epicarp, a fleshy mesocarp (pulp) and a relatively hard shell (endocarp) which encloses the shea kernel (embryo). The kernel, according to Axtell et al. (1993) contains about 60% edible fat (shea butter) and the residual product, from which the butter is extracted (shea cake), is an excellent ingredient for livestock feed production. Shea butter is as good as table oil because of its high nutritive value. It is widely used locally for curing leprosy and other ailments. It also has various industrial uses that include edible oil, soap making, cosmetics, pharmaceuticals, lubricants and paints. According to Russo and Etherington (2001), shea products such as solid fat (butter or stearin) and the liquid oil (olein) are ideal for use as raw materials for cooking oils, margarines, cosmetics, soaps, detergents and candles. However, they have found their primary market as a substitute for cocoa butter in the chocolate and confectionary industry.

According to the American shea Butter Institute (ASBI, 2004), 100% pure natural shea butter is an all-natural vitamin A cream. Shea butter has been shown to be a superb moisturizer, with exceptional skin healing properties. ASBI (2004) compiled a list of skin conditions where 100% shea butter has been proved to be effective in healing. These skin conditions are dry skin, skin rash, skin peeling after tanning, blemishes and wrinkles, itching skin, sunburn, shaving cream for a smooth silky shave, small skin wounds, skin cracks and tough or rough skin (on feet). Others are cold weather, frost bites, stretch marks prevention during pregnancy, insect bites, health skin, muscle fatigue, aches and tension, skin allergies such as poison ivy or poison oaks, eczema, dermatitis and skin damage from heat. It is because of these unique healing properties that the shea tree is called karite tree which means tree of life (ASBI, 2004). It was reported by ASBI (2004) that shea butter has unparalleled moisturizing property and this is due to several natural moisturizers present in it. It was also discovered by ASBI (2004) that the moisturizers in shea butter are the same moisturizers produced by the sebaceous glands in the skin.

In the same development, ASBI (2004) reported that the positive biochemical and physiological effects that shea butter has on skin injuries make it ideal for wound healing. Reports from many users of shea butter have shown that this product promotes and accelerates wound healing. Furthermore, vitamin E in shea butter is helpful to the skin and such benefits could be accomplished by increasing the microcirculation to the skin. This eventually results in increased blood supply to and from the skin. Also, vitamin E in shea butter may serve as an anti-free radical agent thereby preventing the deleterious effects of sun and environmental exposure.

Determination of free fatty acid of shea nut butter is important especially for the screening purposes of large quantities of samples during processing and marketing (Quainoo et al., 2012).

MATERIALS AND METHODS

The study was carried out from March, 2012 to January, 2013 at the Spanish Laboratory, University for Development Studies, Nyankpala. Nyanpkala is located between latitude 09° 25.925 N and longitude 01° 00,420 W. The area is in the Tolon District of the Northern Region of Ghana. The Agro-ecological zone of the area is Guinea Savanna. Shea nut samples were collected in the Zoolanyili community located in the Tolon district near Nyankpala. They were collected by farmers based on their own characterization into the various nut sizes from trees that bare fruits of different sizes. The farmer varieties were re-classified into four size groups for the purpose of this study. The four size groups are shown in Table 1 and Plates 1 to 4, respectively.

Treatments comprising four farmer varieties were used to carry out the study. The four treatments were arranged in a completely randomized design and replicated three times for laboratory analysis for FFA's and moisture content.

Table 1. Size ranges of nut varieties.

Treatment no.	Treatment description	Kernel length (cm)
(T1)	Small nuts	2.45 – 2.60
(T2)	Medium sized nuts	2.65 – 2.81
(T3)	Big nuts	2.95 – 3.00
(T4)	Very big nuts	3.25 – 3.45

Plate 1. Small nuts.

Plate 2. Medium sized nuts.

Snowball sampling was used to select respondents for sensory quality survey on butter colour and texture. Forty (40) women were selected from the Tamale Metropolis for the sensory quality survey on color, texture, taste and aroma.

Shea butter made from the different size categories (Plate 1a and b) were used for the study.

Plate 3. Big nuts.

Plate 4. Very big nuts.

Butter extraction procedure

Until recently, almost all shea butter was extracted by use of small-scale extraction methods. In Ghana, the method of extraction used locally is the same throughout the north. The collection and extraction of the shea nut were carried out based on the following procedure: collection of the shea nuts from the field, drying of the shea nuts, extraction of the kernels from the shells for further drying, washing of the dry kernels and crushing of the kernels into pieces to ensure roasting.

The roasted kernels were cooled and milled into paste using a grinding mill. The paste was then mixed thoroughly with cold and hot water until a cream coloured crude butter separated and floated on the surface of the water. The crude butter on top was collected, washed in clean water and then boiled to separate the oil from the impurities in the cream crude butter. The oil was put in clean containers and cooled whilst steering until it solidified into the familiar solid shea butter.

Laboratory determination of FFA

1% Phenolphthalein in 95% ethanol; 0.1 N potassium hydroxide; Standard solvent: Mix 25 mL ether, 25 mL 95% alcohol and 1 mL of 1% phenolphthalein solution and neutralize with N/10 alkali.

Procedure

1-10 g of oil or melted fat was dissolved in 50 ml of the standard solvent in a 250 mL conical flask. Few drops of phenolphthalein was added to the oil in the standard solvent and titrated against 0.1 N potassium hydroxide. The content was shaken until a pink colour persisted for fifteen seconds.

Calculation

$$\text{Acid Value (mg KOH/g)} = \frac{V_1 - V_0 \times N \times 56.1}{W}$$

V_1 = Titrate value of sample; V_0 = titrate value for blank solution; N = normality of titration; W = weight of sample in grams; the free fatty acid (as oleic) is estimated by dividing the acid value by 1.99.

Laboratory determination of moisture content

The digital moisture content analyzer (Sartorius, MA 45C-000230V1, June 2003) was used for the estimation of moisture content. Sample of the butter was placed on a sample holder and placed inside the analyzer. The machine was tarred and set to record the moisture content of the butter. The machine also recorded the weight of the sample, the dry matter content and the moisture dry matter ratio.

Analysis of data

Microsoft excel and genstat was used to analyze the data and standard error of difference (SED) to compare the means.

RESULTS

Effect of shea nut size on FFA content of shea butter

Figure 1 shows the levels of FFA (expressed as % oleic acid) in shea butter extracted from shea kernels. There were significant differences in the mean FFA values of butter extracted from the shea nuts of the various sizes used in this study. Shea butter extracted from big nuts recorded the lowest mean FFA value of 1.133 (Figure 1).

The Ghana Standards Authority criteria indicated that butter extracted from big, medium and small nuts with mean FFA values of 1.133, 2.227 and 2.420 respectively were grade two while very big nuts with mean FFA value of 3.637 was grade three (Table 2).

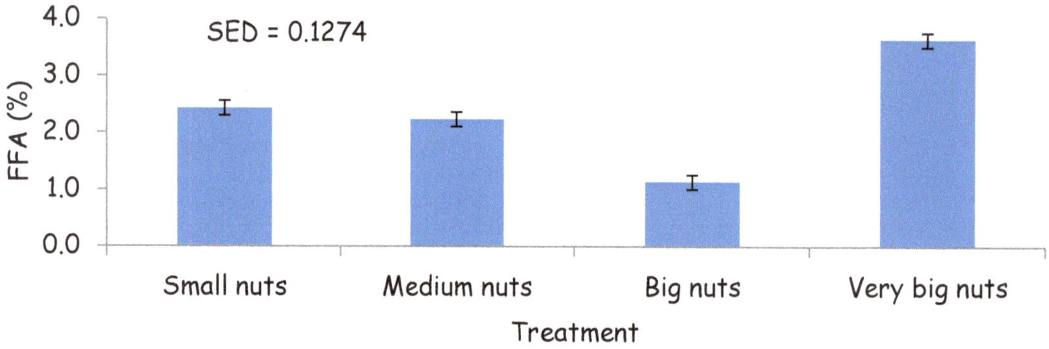

Figure 1. Effect of shea nut size on free fatty acid content of shea butter.

Table 2. Ghana Standards Authority grading of FFA values of shea butter.

Treatment	Grade 1 (1)	Grade 2 (>1.0-3.0)	Grade 3 (>3.0-8.0)
Small nuts	-	2.420	-
Medium nuts	-	2.227	-
Big nuts	-	1.133	-
Very big nuts	-	-	3.637

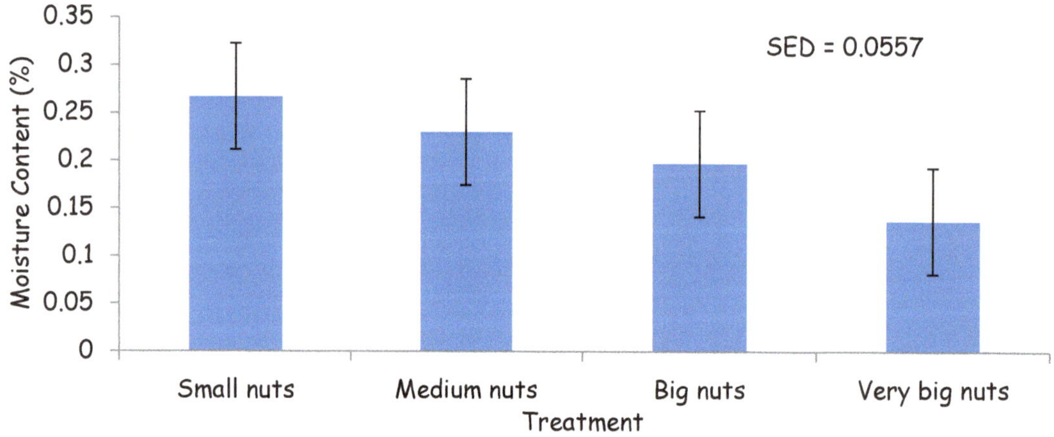

Figure 2. Effect of shea nut size on moisture contents of butter.

Effect of nut size on moisture content of shea butter

The result of the mean moisture contents of shea butter extracted from the different shea nuts is presented in Figure 2.

The results indicate that there were no significant differences in the mean moisture content values of shea butter extracted from the shea nuts of different sizes. Butter extracted from very big shea nuts has the least amount of moisture with a mean moisture content of 0.137% with small shea nuts recording the highest moisture content of 0.267% (Figure 2).

The Ghana Standards Authority criteria indicated that shea butter extracted from very big and big shea nuts with mean moisture contents of 0.137 and 0.197% respectively, were grade 2 while butter from medium and small shea nuts with mean moisture contents of 0.230 and 0.267% respectively, were grade 3 (Table 3).

Sensory qualities of shea nut butter

The sensory qualities of shea nut butter studied were color, texture, taste and aroma. These qualities were

Table 3. Ghana Standards Authority grading of moisture content of shea butter.

Treatment	Grade 1 (0.05%)	Grade 2 (>0.05-0.2%)	Grade 3 (>0.2-2.0%)
Small nuts	-	-	0.267
Medium nuts	-	-	0.230
Big nuts	-	0.197	-
Very big nuts	-	0.137	-

Table 4. Sensory qualities of shea nut butter as assessed by forty respondents.

Treatment	No. of respondents who agreed on purity of butter colour	No. of respondents who agreed on the smoothness of butter texture
Small nuts	16	18
Medium nuts	12	11
Big nuts	8	7
Very big nuts	4	4

observed based on indigenous knowledge. Individual participants did not find any distinction in the taste and aroma of the shea butter extracted from the different shea nut sizes. There were differences in colour and texture of the shea butter extracted from the different shea nut sizes (Table 4).

Table 4 indicated that for color, sixteen people observed shea butter extracted from small nuts to be the whitest, followed by butter from medium nuts with twelve people, then butter from big nuts with eight people and finally, butter from very big nuts with four people observing it to be the whitest.

For texture, 1 people observed shea butter extracted from small nuts to be the smoothest, followed by butter from medium nuts with eleven people, then followed by butter from big nuts with seven people and four people observing butter from very big nuts to be the smoothest.

DISCUSSION

Free fatty acid (FFA)

Shea nut butter extracted from the big nuts recorded the lowest mean FFA value with shea nuts from the very big nuts recording the highest mean FFA value. The lower mean FFA value for the shea nut butter extracted from big shea nuts may be attributed to their lower moisture contents.

The study revealed that, generally the bigger the shea nut kernels, the lower the FFA values and the higher the quality according to Ghana Standards Authority GSA (Figure 1). The only exception was the very big nuts which had very high FFA content. This indicates the need for further classification and categorization of nuts in repeated experiments to arrive at a conclusive relationship between the nut size and the FFA content

Shea nut butter quality is high when its free fatty acid content is low. Free fatty acid content is naturally low in fresh nuts, but it increases rapidly through hydrolysis under poor storage conditions. Hydrolysis occurs through the lipolytic activity of the fruit lipase and micro-organisms and halted by heating by reducing the moisture content to levels lower than 8%.

Results obtained by grading the four different shea nut sizes revealed that shea nut butter from big nuts, medium nuts, and small nuts were graded as high quality butter (grade 2) and can meet the requirement of the food industry (confectionary, chocolate, edible oil or as basis for margarines) based on the GSA grading criteria. However, shea nut butter from very big nuts was graded as low quality butter (grade 3) and may serve the needs of the soap making industries or can be refined for direct consumption (Table 2).

Moisture content

Shea nut butter extracted from the very big nuts recorded the lowest moisture content with shea nut butter from the small nuts recording the highest moisture content. The results show that, the bigger the nuts, the lower the moisture content and this may be attributed to the fact that, bigger nuts have a larger surface area and are therefore subject to more rapid drying than smaller nuts (Figure 2).

Results obtained from grading the four different shea nut sizes based on GSA grading criteria on moisture content revealed that, shea nut butter extracted from very big nuts and big nuts were graded as high quality butter (grade 2), and may serve the needs of the food industry (confectionary, chocolate, edible oil or as basis for margarines). Whiles butter from medium nuts and small nuts were graded as low quality butter (grade 3) and may serve the needs of the soap-making industry or may be refined for direct consumption (Table 3).

Furthermore, shea nut butter with low or no moisture content stores better than one with high moisture content. This is because shea nut butter with higher moisture content is prone to spoilage within one year of storage (Yidana, 2007).

Also, high moisture content in plant fats and oils usually leads to increased microbial load and lipid oxidation resulting in rancidity (Hee, 2011).

Sensory qualities

The colour of the shea nut butter samples after extraction maintained a bright colour with slight differences among them. Sensory analysis conducted by Akingbala et al. (2006) found that unrefined shea nut butter gained lower scores than refined shea nut butter based on color. However, refining procedure also caused the loss of minor but valuable components such as unsaponifiable fraction with medicinal properties (Tasan and Demirci, 2005; Moharram et al., 2006; Van Hoed et al., 2006).

Texture of the extracted shea nut butter was similar to commercially available body butter or vaseline. Since the shea nut butter samples were all solid at room temperature, the first feeling was to take a small portion from the bulk shea nut butter using the fingertip but it was not as smooth as creams or lotion. However, once applied on the skin, all the samples were melted and showed nice creamy texture when spread over the back of the hand.

The results obtained from the study however indicated that for shea nut butter colour, butter extracted from small nuts was the brightest, followed by shea nut butter from medium, big nuts and very big nuts in that order respectively. The texture of the shea nut butter from the small nuts was the smoothest, followed by medium, big and very big nuts respectively in that order. The variations in colour and texture of the shea nut butter extracted from the different shea nut sizes may be probably due to some varying responses of the nut sizes to practices in the extraction process especially washing and roasting of the shea nut kernels.

The quality criteria usually used by buyers at the market level to assess shea nut butter for purchases are color and texture since the two characteristics of shea nut butter may be easily assessed visually and by feeling. Bright solidified shea nut butter with smooth texture is usually preferred by buyers.

Conclusion and recommendation

The research was carried out to assess the chemical and physical characteristics of shea kernel samples obtained from the Zoolanyili community in the Tolon District of Northern Region. The usefulness of shea nut butter is based on its chemical composition. However, the market value of the shea nut butter is determined by both the physical appearance and the chemical composition. The results of the FFA and moisture content test carried out indicated that nut size have a significant effect on the quality of the extracted shea nut butter.

It is recommended that, further research be carried out at different geographical locations so as to provide more thorough understanding of the effects of shea nut kernel size on the quality of shea nut butter.

Conflict of interests

The authors did not declare any conflict of interest.

REFERENCES

Akingbala J, Falade K, Adebesi E, Baccus-Taylor G, Lambert I (2006). Effect of Processing Conditions on Yield, Physical and Chemical Properties of Shea Butter. West Indian J. Eng. 29: 73-80.

ASBI (2004). 21 Reasons to Use Shea Butter. The American Shea butter institute, Columbus, GA 31908-7906, USA.

Axtell B, Kocken E, Sandhu R (1993). Oil processing. United Nations Development Fund for Women (UNIFEM), Food Cycle Technology Source Books. Intermediate Technology Publ. Ltd, London

Hee SH (2011). Quality characteristics of West African shea butter (*Vitellaria paradoxa*) and approaches to extend shelf-life. MSc thesis, State University of New Jersey, 121pp.

Moharram H, Ray J, Ozbas S, Juliani H, Simon JE (2006). Shea Butter, Chemistry, Quality and New Market Potentials. In: C.T. Ho, M. Wang and S. Sang (eds). Herbs: Challenges in Chemistry and Biology. American Chemical Society Symposium Series 925. American Chemical Society, Washington, D.C. USA.

Quainoo AK, Nyarko G, Davrieux F, Piombo G, Bouvet JM, Yidana JA, Abubakari AH, Mahunu GK, Abagale FK, Chimsah FA (2012). Determination of biochemical composition of shea (*Vitellaria paradoxa*) nut using near infrared spectroscopy (NIRS) and gas chromatography. IJBPAS 1(2):84-98.

Russo L, Ethrington T (2001). Non-wood News. An information Bulletin on Non-wood Forest products, 8: 38-39.

Tasan M, Demirci M (2005). Total and Individual Tocopherol Contents of Sunflower Oil at Different Steps of Refining. Eur. Food Res. Technol. 220:251-254.

Van Hoed V, Depaemelaere G, Villa Ayala J, Santiwattana P, Verhé R, De Greyt W (2006). Influence of chemical refining on the major and minor components of rice bran oil. J. Am. Oil Chem. Soc. 83: 315-321.

Yidana JA (2007). Draft report on JICA Trial Project. Submitted to JICA Office, Accra.

A survey on traditional fish smoking in Lagos State, Nigeria

Adeyeye, S. A. O[1] , Oyewole, O. B.[1], Obadina, A. O[1], Omemu, A. M.[2], Oyedele, H. A.[1] and Adeogun, S. O.[3]

[1]Department of Food Science and Technology, Federal University of Agriculture, Abeokuta, Ogun State, Nigeria.
[2]Department of Hospitality and Tourism Management, Federal University of Agriculture, Abeokuta, Ogun State, Nigeria.
[3]Department of Agricultural Administration, Federal University of Agriculture, Abeokuta, Ogun State, Nigeria.

This study examined the socio-economic characteristics, technologies and hygiene of the processors in 20 different 'fish' processing centres in fishing communities in Lagos State. A total of 200 questionnaires were administered through purposive sampling method at 10 respondents/processing centre. Data were collected through field observation and administration of structured questionnaire. Analytical technique used was descriptive statistics. Results reveal that most of the households were relatively poor, using age and educational level of the processors and availability of household amenities as proxies for socioe-conomic status. Majority of the processors (55.5%) were old women, 51.5% had primary school education while 38% had post-primary school education. The study shows that majority (98.0%) of the processors practiced manual operations while 2.0% practiced mechanical operation. Every processor used eviscerating, washing, filleting and de-scaling and 99.0% of the processors operated on non-concrete floor while 1.0% were on mould floor. Majority (98%) used firewood as normal smoking fuel while only 0.5% used charcoal. Majority (77.5%) of the processors used full drum as smoker, 2.0% used half drum while 19.0% used mud oven and the rest (1.5%) used charcoal oven and 92.0% of the processors were from urban communities while 8.0% of them were from rural communities. Very few processors (1.0%) used disinfectants for their processing facilities and environment. The study concludes that fish smoking makes an important contribution to household food and financial security in all processing centres.

Key words: Fish, hygiene, processors, smoking, socio-economic, survey.

INTRODUCTION

Fish is an important dietary component of people all around the world and represents a relatively cheap and accessible source of high quality protein for poorer households (Ikutegbe and Sikoki, 2014). Global production of fish, mussels and crab in 2010 was almost 60 million tonnes, a figure which includes production in marine waters, brackish water and freshwater. Aquaculture production is now about three quarters of that from ocean fish and seafood caught in the wild. In 2011 this amounted to 78.9 million tones, 15% of which was cured in one or another way (Ikutegbe and Sikoki, 2014; FAO, 2013). One third of the cured fish was smoked and about 20% of the smoked fish goes into international trade (Clucas and Ward, 1996). No other food industry has shown such growth as aquaculture in recent decades. Between 1970 and 2008 annual production worldwide increased by an average of 8.4%; much more than poultry farming and egg production, which have the second highest growth rates after aquaculture (da Silva, 2002; da Silva e t al., 2008; Abolagba and Melle, 2008).

In Nigeria, fish production through aquaculture has risen steadily from a few hundred kilograms in the 1950s to over 45,000 metric tonnes in 2004 (FAO, 2007). Today, aquaculture is the fastest growing livestock production sector in Nigeria, with a growth of about 29% in 2006 alone, and with prospects of continued growth.

This is because demand for fish is on the increase with population growth, while catches from fisheries are on the decline, even globally (FAO, 2007).In Nigeria smoked fish products are the commonest form of fish product for consumption. Out of the total of 194,000 metric tons of dry fish produced in Nigeria, about 61% of it was smoked. One of the greatest problems affecting the fishing industry all over the world is fish spoilage. In high ambient temperature of the tropics, fresh fish have the tendency to spoil within 12 to 20 h (Clucas and Ward, 1996). Attempt has been made to reduce fish spoilage to the minimum through improved preservation techniques. Preservation and processing methods explore ways by which spoilage are stopped or slowed down to give product a longer shelf life. Fisheries have been the main source of livelyhood for the population of fishing communities in Nigeria and a vital sector of the economy by employing more than 6 million fisher folks in Nigeria (Fish for All Summit, 2005; Fregene and Bolorunduro, 2009) in terms of fish production, processing and distribution. Entire family (men, women and children) in the fishing communities are engaged in the sector. The catch from these fisheries plays an important source of animal protein in peoples' diets.

Seasonal fluctuation in food availability and household responses to this insecurity has been observed to influence individual consumption patterns (Longhurst, 1986; Fregene and Bolorunduro, 2009). In fishing communities, during the off-fishing season, which is usually in the rainy season (July to September), fish catch is low. This is because of the increased level of turbid water and strong wind, which hinder the fishermen without outboard engine from going far out to sea. In an attempt to make a living, they have resulted to the exploitation of generally fragile environment, thereby leading to a cycle of low production, low income, and poverty and being food in-secured (National Institute of Oceanography and Marine Research, (NIOMR), 1989; Federal Office of Statistics, (FOS), 1999). As a result of inadequate purchasing power (income), fisher folks often experience a food in-secured period (Fregene, 2002; Fregene and Bolorunduro, 2009).

Small scale fishing communities in Nigeria, as elsewhere in the world are vulnerable to exploitation due to poverty and uncertainty of their income. The seasonality of fish catch coupled with inadequate processing capacity has resulted in high post-harvest losses, which diminish benefits accruing to small-scale operators. Realizing that fisher folks are not a homogenous group of people, there is the need for a comparative study of fishing communities.

Therefore, the main objective of this study was to carry out a survey on traditional fish smoking activities in Lagos State, Nigeria in order to examine the socio-economic characteristics, role of women and expenditure pattern as they relate to variation in poverty level in among fish processors in fishing communities in Lagos State.

METHODOLOGY

Survey

The study was carried out in Lagos State, which has 22.5% of Nigeria's coastline and occupies an area of 3,577 square kilometre mass with 786.94 square kilometre or 22% of it being lagoons and creeks, in Lagos, Ikorodu, Badagry and Epe local government areas (Udo and Mamman, 1993). Purposive sampling technique was used for surveying. Structured questionnaire was administered to 200 respondents by enumerators. Data were also collected through field observation and on the spot assessment to collect information on socio-demographic and environmental health data of selected 'smoked fish' processors. Descriptive statistics used include measures of distribution, central tendency and dispersion respectively.

Sampling sites

The sampling sites for this study include Agbalata, Ajido, Asakpo, Boguru, Fvanoveh, Gberefun/Yovoyan, Gbetrome, Ilaje, Kofegameh, Pako, Afuye, Bodin Yawa, Idale, Igbodun, Ilogun, Mejona, Oluwo, Okorisan, Orita, Orogoro from two Local Government Areas (Badagry and Epe) of Lagos State.

Area of study

Using a current geopolitical map of Nigeria (Figure 1), Lagos State lies to the south-western part of Nigeria and has boundaries with Ogun State both in the north and east. It is bordered on the west by the Republic of Benin and in the south, stretches for 180 km. along the coast of the Atlantic Ocean. It therefore has 22.5% of Nigeria's coastline and occupies an area of 3,577 km^2 land mass with about 786.94 km^2 (22%) of it being lagoons and creeks. The state is endowed with marine, brackish and fresh water ecological zones with varying fish species that provide productive fishing opportunity for fishermen (Udo and Mamman, 1993). Two local government areas (Badagry and Epe Local Government) were covered because they are highly densed fish processing centres.

RESULTS AND DISCUSSION

Table 1 shows the characteristics of processors in the 20 processing centers surveyed. Majority of the processors

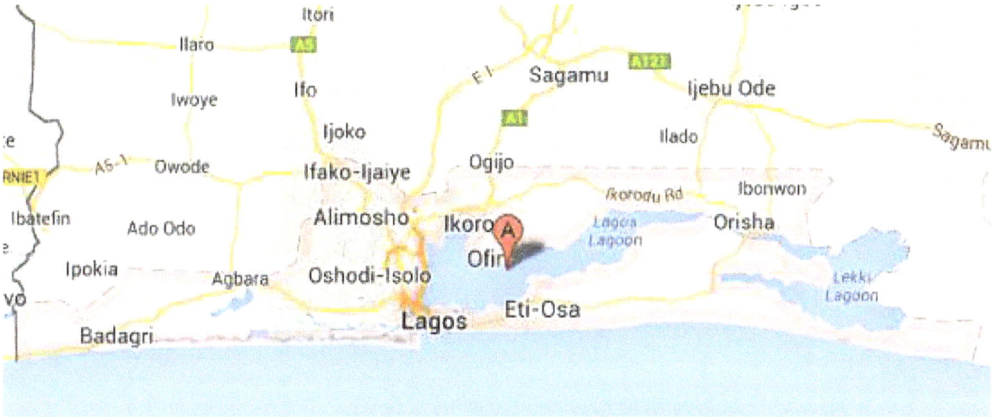

Figure 1. Map of Lagos State (http://nigerianfinder.com accessed Jan, 2015).

Table 1. Characteristics of the 20 study smoking/processing centres.

Parameters	Number observed (%)
Age of processors	
Old women	111(55.5)
Young women	89(44.5)
Educational level of processors	
No schooling	21(10.5)
Primary school	103(51.5)
Post secondary school	76(38)
Packaging	
Basket with dry plantain leaves	186(93)
Basket without polypropylene	14(7)
Method of sewage disposal	
Pit latrine	125(62.5)
Water-carriage system	10 (5)
None	65(12.5)
Source of processing water	
Spring	30(15)
Pipe-borne	20(10)
Stream	100(50)
Bore-hole	30(15)
Rain	20(10)
Normal cooking fuel	
Firewood	196 (98)
Coal/charcoal	1(0.5)
Others	3(1.5)
Source of income	
Husband	15(7.5)
Wife	175(87.5)
Others	10(5)

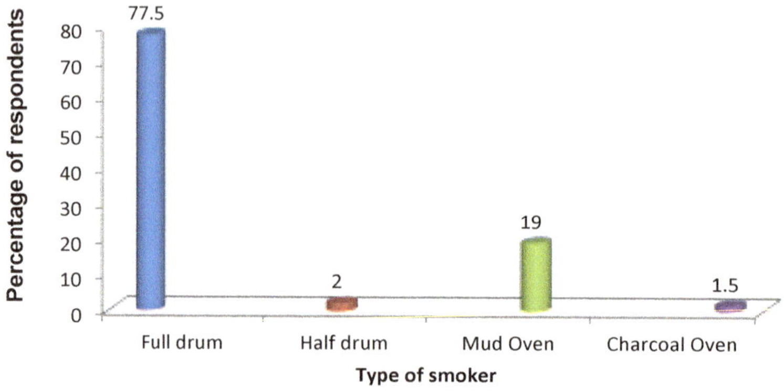

Figure 2. Distribution of type of smoker.

(55.5%) are old women, 51.5% had primary school education while 38% had post-primary school education. The relationship between household socio-economic characteristics and food processing has been amply demonstrated in the literature (Fregene and Bolorunduro, 2009). For example, using educational level of the processors and availability of household amenities as proxies for socio-economic status, it is apparent that most of the households were relatively poor. This has significant implications for fish processing in general and for fish hygiene behaviour in particular (Fregene and Bolorunduro, 2009). Education is also related to employment and income which influence access to household amenities and facilities, including those related to fish hygiene and environmental health. Result shows that 93% of the processors packaged the fish in Basket with dry plantain leaves while only 10% used Basket with polypropylene. Survey of the processing centres shows that 62.5% used pit latrine as method of waste disposal, 5% used water-carriage system and 12.5% had no means of waste disposal. Most (50%) of the process-sors used stream water for processing, 15% used spring and boreholes.

The processing sites in all the processing centres are mostly located in places where they remain a threat to food safety. In all the processing sites surveyed, there is no adequate drainage and waste disposal systems. The facilities provided were designed and constructed in a place closer to the processing sites which makes it having a high risk of contaminating the smoked fish. Most of the processing sites have no storage facilities for both raw materials and finished product. The processors were not applying a good quality control system into this critical aspect of the fish smoking/processing. The fresh fish are not normally inspected and sorted before processing so as to segregate fish which is evidently unfit for human consumption. Also, there supposed to be a way of protecting fresh fish from damages such as bruising which can easily initiate contamination by pests or microorganisms and enzymatic activity. Majority (98%) used firewood as normal cooking fuel while only 0.5% used charcoal. Majority (77.5%) of the processors used full drum as smoker, 2.0% used half drum (Figure 2), while 19.0% used mud oven and the rest (1.5%) used charcoal oven. Average capacity of a full drum as smoker used by processors was 71.42 kg, half drum smoker has capacity 66.67kg, while mud oven has capacity of 82.11 kg and the charcoal oven has capacity of 100.0 kg. Majority (87.5%) of processors contributed a minimum of N1, 600.00 and maximum of 480000.00 as start-up capital for the business and the mean contribution was N51, 324.86. The other processors (12.5%) contributed a minimum of N14, 000.00 and maximum of N600, 000.00 and their mean contribution was N55, 775.26.

The study shows that fish processors are operating in rural and urban areas and the activity appears to be increasing in popularity. Fish processors operate on a range of scales and cheap and readily available smoking kilns were used. Family labour plays a critical role, but micro-enterprises which employ casual labour are also common.

For some enterprises fish processing constitutes a full-time activity. They also operate on a part-time basis, using family labour. Majority (98.0%) of the processors practiced manual operations while (2.0%) practiced mechanical operation. Every processor used eviscerating, washing, filleting and de-scaling and (99.0%) of the proces-sors used non concrete floor while (1.0%) used mould floor. Majority (92.0%) of the processors were from urban communities (Figure 3) while (8.0%) of them were from rural communities and (17.0%) of the processors from rural communities processed Bonga shad type of fish and followed by (12.0%) of the processors processed Silver catfish. Very few processors (1.0%) used disinfect-tants for their processing facilities and environment (Figure 4). The facilities used for processing are not properly cleaned always and this is a threat to food safety. The cleaning system in the entire processing site is not adequate.

The materials provided for cleaning food are not adequately and suitably designed and these can easily arbor pathogenic organisms. Moreso, there is no facilities for adequate supply of potable water for cleaning. There is

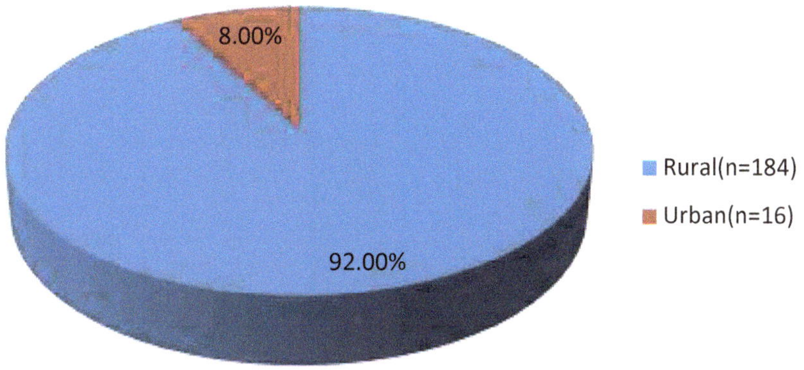

Figure 3. Distribution of processing centres/communities.

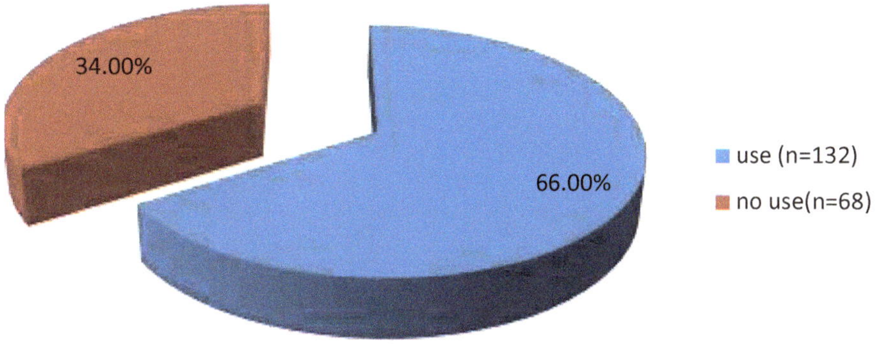

Figure 4. The use of chemicals by processors.

no adequate method of controlling the pests in all the processing sites. The containers used for processing are not always washed immediately after use, and this allows the pests free movement to operate and this can result in malicious or accidental contamination of food. Due to the traditional method of smoking fish, there are no facilities available for personal hygiene which can assist in ensuring that an appropriate degree of personnel hygiene is maintained and to avoid contaminating the fish. There was a high awareness concerning personal hygiene among the processors. People who came directly or indirectly into contact with the fish are not likely to contaminate the fish due to: maintaining of appropriate degree of personal cleanliness and behaviour and operating in an appropriate manner. Purchasing of smoked fish from market vendors poses a considerable health risk.

The reasons for this are apparent from observational data on hygiene practices in the market. Smoked fish are often displayed openly on the tray in very poor sanitary environments. The prevalence of flies at the markets and the apparent lack of facilities for food protection suggest a high potential for contamination. Smoked fish are also subjected to repeated contamination from the unwashed hands of vendors, and the materials used for wrapping, such as reusable polythene bags, waste paper and baskets, may also be a source of contamination. Figure 5 shows that 80, 61.5 and 28% of processors respectively from year 2010, 2011 and 2012 claimed that 30 to 40% of household income was spend on food, while 14, 31 and 60% of processors respectively from year 2010, 2011 and 2012 claimed that 41 to 50% of household income was spend on food and 5.5, 7.5 and 12% of processors respectively from year 2010, 2011 and 2012 claimed that >60% of household income was spend on food. The study shows that fish smoking makes an important contribution to household security in all processing centres. As described earlier, previous understanding amongst researchers was that fish processing was predominantly a rural activity; less integrated in market systems than other agricultural pro-ducts (Fregene and Bolorunduro, 2009). The baseline survey found quite a different situation. In processing centres such as Oluwo and Agbalata, and even more remote places such as Idale, fish processors have long-standing connection to urban markets and market their produce on a relatively large scale.

Conclusion

The baseline assessment revealed that using educational level of the processors and availability of household amenities as proxies for socio-economic status, it is

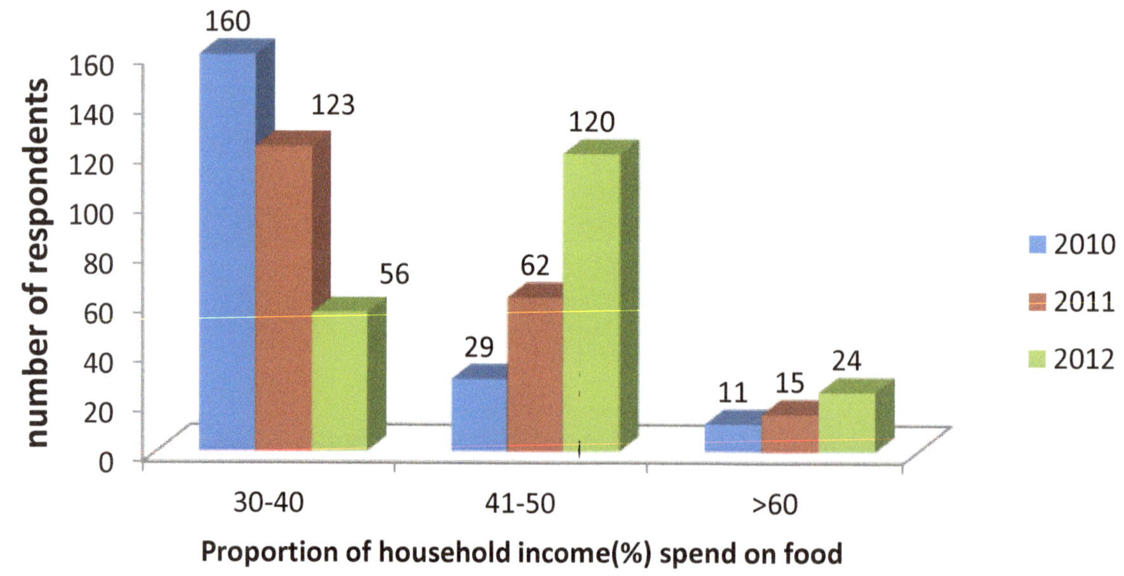

Figure 5. Proportion of household income spent on food.

apparent that most of the households were relatively poor. Majority of the processors (55.5%) are old women, 51.5% had primary school education while 38% had post-primary school education. Education is also related to employment and income which influence access to household amenities and facilities, including those related to fish hygiene and environmental health. The study shows that fish smoking makes an important contribution to household security in all processing centres. The study also found that fish processors are operating in rural and urban areas and the activity appears to be increasing in popularity. Fish processors operate on a range of scales. Family labour plays a critical role, but micro-enterprises which employ casual labour are also common. For some enterprises fish processing constitutes a full-time activity. They also operate on a part-time basis, using family labour. It was also found that the structure and condition of processing sites are below safety standard due to the following reasons: The floors are not constructed to allow adequate drainage and cleaning; the operations are made in open places without a constructed processing plant.

Conflict of interest

The authors have no conflict of interest.

REFERENCES

Abolagba OJ, Melle OO (2008). Chemical composition and keeping qualities of a Scaly Fish Tilapia (Oreochromis niloticus) Smoked with two Energy Sources. Afr. J. Gen. Agric. 4(2):113-117.

Clucas IJ, Ward AR (1996): Post Harvest Fisheries Development. A Guide to Handling, Preservation, Processing and Quality. Chatham Maritime, Kent ME4TB, United Kingdom. Pp. 665.

da Silva LVA (2002) Hazard Analysis Critical Control Point (HACCP), Microbial Safety, and Shelf Life of Smoked Blue Catfish (Ictalurus furcatus). A Master of Science in Food Science Thesis, the Graduate Faculty of the Louisiana State University and Agricultural and Mechanical College.

da Silva LVA, Prinyawiwatkul W, King JM, No HK, Bankston JD Jr, Ge B (2008) Effect of preservatives on microbial safety and quality of smoked blue catfish (Ictalurus furcatus) steaks during room-temperature storage. Food Microbiol. 25(8):958-963.

FAO (2007) Fish Processing in Africa. FAO, Fish .Rep.

FAO (2013) Post-harvest changes in fish Food and Agriculture Organization of the United Nations for a world without hunger, Fisheries and Aquaculture Department.

Federal Office of Statistics (FOS) (1999) Poverty and Agricultural Sector in Nigeria, Federal Office of Statistics, 33 pp.

Fregene, BT (2002) Poverty Assessment in Fishing Communities in Lagos State, Nigeria, Unpublished Ph. D. Thesis University of Ibadan, Ibadan 226p.

Fregene BT, Bolorunduro P I (2009). Role of Women in Food Security and Seasonal Variation of Expenditure Pattern in Coastal Fishing Communities in Lagos State. Journal of Agricultural Extension Vol. 13 (2) December 2009 pp 21-33

Fish for All Summit (2005). Presentation of Nigeria's Fisheries Resources Abuja, Nigeria, 22-25 August 2005.

FDA, Department of Health and Human Services, (2001). Pathogen Growth & Toxin Formation as a Result of inadequate Drying. In Fish & Fisheries Products Hazards & Controls Guidance: Third Ed. Chapter 14, p.191.

Goktepe I, Moody MW (1998). Effect of modified atmosphere package on the quality of smoked catfish. J. Muscle Foods 9:375-389.

Ikutegbe V, Sikoki F (2014). Microbiological and biochemical spoilage of smoke-dried fishes sold in West African open markets. Food Chem. 161:332–336.

Longhurst R (1986). Household Food strategies in Response to Seasonality and Famine, IDS Bulletin Vol. 17 No. 3 Institution of Development Studies, Sussex Map of various states and their local government areas. Nigerianmuse.com, accessed Dec. 3, 2014

Udo RK, Mamman AB (1993). Nigeria: Giant in the Tropics Vol. 2.Gabumo Publishing Co. Ltd. Lagos Nigeria.

Anti-oxidative biochemical properties of extracts from some Chinese and Thai rice varieties

Chanida Saenkod[2], Zhonghua Liu[1,2], Jianan Huang[1,2] and Yushun Gong[1,2]

[1]Key Laboratory of Tea Science of Ministry of Education and Hunan Province and Key Laboratory of Crop Germplasm Innovation and Utilization, Hunan Agricultural, University, Changsha, Hunan 410128, People's Republic of China.
[2]National Research Center of Engineering and Technology for Utilization of Botanical Functional Ingredients, Changsha, Hunan 410128, People's Republic of China.

Total phenolic compounds (TPC), total flavonoid contents (TFC) and antioxidant activity of extracts from 8 varieties of Chinese and Thai rice grains were evaluated. Extraction was done using water (at 25°C), hot water (at 50°C) and 70% ethanol. Highest TPC and TFC were seen in all rice varieties after extraction with 70% ethanol. The Heimi variety of Chinese rice showed the highest TPC and TFC (634.83 and 158.47 mg/kg, respectively) followed by Jing Nian variety, also from China, Dok Kam and Niaow Deang varieties both from Thailand, respectively. Highest antioxidant activity was observed in Heimi followed by Jing Nian, Dok Kam and Niaow Deang varieties. The color of Heimi and Jing Nian is black, and that for Dok Kam and Niaow Deang is red, so TPC and TFC were correlated with the color of rice. However, no correlation was observed with the Nang Dum variety from Thailand even though it has a black color. Its TPC and TFC were found to be statistically the same as that of the white-colored Ji Nong Da, T-You 597 and Xiang Wan Nuo 1 Hao varieties from China. Antioxidant activities from each solvent system extraction (water, hot water and 70% ethanol) were also found to be directly correlated with the quantity of TPC and TFC. The TPC had positive correlations with DPPH scavenging (R^2 = 0.9155, 0.9358 and 0.9625), reducing power (R^2 = 0.9799, 0.8521 and 0.9733) and inhibition of lipid peroxidation (R^2 = 0.7621, 0.7429 and 0.9610). TFC also had positive correlations with DPPH scavenging (R^2 = 0.9098, 0.9246 and 0.9477), reducing power (R^2 = 0.9916, 0.9745 and 0.9844) and inhibition of lipid peroxidation (R^2 = 0.7548, 0.7178 and 0.8283), respectively.

Key words: Rice grain, bioactive compounds, total phenolic, flavonoids, antioxidant component.

INTRODUCTION

Antioxidants for oxidative stress protection from natural sources such as grain crops, vegetables and fruits are becoming a profitable alternative as compared to synthetic antioxidants that have adverse effects. Free radicals have been claimed to induce oxidative stress in various cell components including proteins, lipids and DNA, ultimately leading to certain diseases such as cancer, cardiovascular diseases and ageing and inflammatory diseases.

Rice (*Oryza sativa* Linn.) is the most important cereal crop in the world, either directly as human foods or indirectly as animal feeds. In China, rice is the staple food for more than 60% of the Chinese people, accounting for 35% of all rice grain production (Cao et. al., 2010). Thailand, Rice is one of the main food and source of nutrition for most of Thai population. Rice is rich in many nutrient components including carbohydrates, proteins, certain fatty acids and micronutrients (vitamins and trace

Table 1 Description of the rice varieties used in the study.

Rice Variety	Abbreviation	Description	Color	Source
Heimi	BB	Brown Heimi	Black	China
Ji Nong Da 37	BJ	Brown Ji Nong Da	White	China
T-You 597	BT	Brown T-You 597	White	China
Xiang Wan Nuo 1 Hao	BW	Brown Xiang Wan Nuo 1 Hao	White	China
Jing Nian	BY	Brown Jing Nian	Black	China
Niaow Deang	BND	Brown Niaow Deang	Red	Thailand
Nang Dum	BN	Brown Nang Dum	Black	Thailand
Dok Kam	BD	Brown Dok Kam	Red	Thailand

minerals). It is also a good source of many bioactive non-nutrient compounds, known as antioxidants, including phenolic compounds.

Several studies indicate that phytochemicals are bioactive compounds that include phenylpropanoids, tannins, lignins, γ-oryzanol, tocotrienols, tocopherols, phenolics compounds and flavonoids. Most common groups of phenolic compounds are flavonoids which are water-soluble plant pigments with many colors (Hansakul et al., 2011). Phenolic compounds and flavonoid contents are potential antioxidative phytochemicals that can act as metal ion chelators, free radical scavengers and reducing agents thus offer human health benefits, which also can be found in pigmented rice (Srisawat et al., 2010; Lum and Chong, 2012). According to previous studies, antioxidant activity and phenolic content of black rice is greater than that of white rice (Higashi-Okai et al., 2008; Muntana and Prasong, 2010). In addition, pigmented rice extracts have been reported to effectively decrease inflammation and oxidative stress as well as atherosclerotic lesions (Ling et al., 2001; Xia et al., 2003).

The main objective of this study was to compare the total phenolic, total flavonoid and antioxidant components of extract from different varieties of Chinese and Thai rice grains using water (at 25°C), hot water (at 50°C) and 70% ethanol for extraction.

MATERIALS AND METHODS

Eight rice grain samples (*O. sativa* L.) were used in this study, as described in Table 1. The four varieties of de-hulled samples were obtained from different growth site in China, and were given the terms Brown Heimi (BB), Brown Ji Nong Da (BJ), Brown T-You 597 (BT) and Brown Xiang Wan Nuo 1 Hao (BW). One variety was purchased from a supermarket in China and was described as Brown Jing Nian (BY). Other three varieties were obtained from the southern of Thailand and were given the terms Brown Niaow Deang (BND), Brown Nang Dum (BN) and Brown Dok Kam (BD).

Extraction

One hundred grams (100 g) of rice grains from each variety were soaked in solvents, namely water at 25°C, hot water at 50°C and 70% ethanol at 25°C; ratio 1:5 (w/v) in a serial manner. Extraction was done for 40 min using an ultrasonic bath. A constant temperature was set throughout the extraction. The extracts were centrifuged at 5,000 rpm for 15 min and then the supernatants were collected. Each solvent extraction was repeated twice and each of the extract solution was combined and dried under vacuum rotary evaporator. All dried extracts were kept in the freezer (-20°C) until used for analysis.

Determination of total phenolic compounds (TPC)

The total phenolic content of extracts was determined using the Folin-Ciocalteau phenol reagent following the method of Iqbal et al. (2005). A 0.2 mL of each sample (concentration 0.2 mg/mL) was added to 0.8 mL of freshly diluted 10-fold Folin–Ciocalteau phenol reagent (Sigma, St. Louis, USA). After 10 min, 2 mL of 7.5% $NaCO_3$ was added to the mixture. The reaction was allowed to complete for 1 h in the dark at room temperature then measured at 765 nm on a spectrophotometer (UV-Vis 2450 Shimadzu, Japan). Gallic acid (Tianjin Kermel Chemical Reagent Development Center, Wangtai, China) was used as a standard and the results were calculated as gallic acid equivalents (mg/kg) of the sample. The procedure was conducted in triplicate and the results were averaged.

Determination of total flavonoid contents (TFC)

Total flavonoid contents of each sample were determined using colorimetric method with some modifications (Vichapong et al., 2010). Briefly, 0.2 mL of extract solution was mixed with 0.3 mL of 5% $NaNO_3$. After 5 min, 0.3 mL of 10% $AlCl_3$ was added then incubated for 6 min and 2 mL of 1 mol/L NaOH was subsequently added. The absorbance was measured immediately at 510 nm. The total flavonoid content was calculated from the calibration curve of catechin standard (Hefei Hiromi Biotechnology Co., Ltd., China). The measurements were calibrated to a standard curve of prepared catechin solution, and were expressed as 1 mg catechin equivalent per 100 g rice (mg/kg).

Scavenging effect on DPPH radical

Radical DPPH scavenging activity was determined using the method of Lai et al. (2009) with some modification. A 0.1 mL of the extraction solution (1 mg/mL in dH_2O) in test tube was mixed with 3.9 mL of 95% ethanol and 1 mL of DPPH solution (1 mM in 95% ethanol). The mixture was kept at room temperature for 30 min prior to measurement of the absorbance at 517 nm (A_{517nm}). All measurements were done in triplicate. The scavenging ratio was

Table 2. Total phenolic compounds (mg/kg) of different rice varieties using different solvents.

Rice varieties	Solvent		
	Water (at 25°C)	Hot water (at 50°C)	70% ethanol (at 25°C)
BB	433.99±0.00[a]	610.92±0.00[a]	634.83±0.00[a]
BJ	41.213±0.77[b]	45.53±0.54[b]	65.78±1.97[bc]
BT	41.87±0.79[b]	45.97±0.00[b]	54.86±0.68[b]
BW	42.33±0.46[b]	43.70±0.46[b]	72.62±3.42[c]
BY	286.21±0.46[c]	434.45±4.18[c]	462.45±0.60[d]
BND	101.99±1.72[d]	137.51±2.05[d]	270.49±3.99[e]
BN	45.97±1.72[b]	49.16±0.82[b]	91.90±8.06[f]
BD	135.69±0.23[e]	170.60±1.65[e]	382.07±0.68[g]

Values are the mean of triplicate measurements ± standard error (Means±SE); (n = 3). Mean followed by different superscript letters in a column are significantly (*p*≤0.05) different from each other.

calculated using the formula:

$$\text{DPPH scavenging (\%)} = 1 - \frac{A_{517nm,\ sample}}{A_{517nm,\ control}} \times 100$$

Determination of reducing power

According to the method of Lai et al. (2009), 2.5 mL of the extract solution (1 mg/mL in dH_2O) was mixed with phosphate buffer (2.5 mL, 0.2 M, pH 6.6). Then, each sample was mixed with 2.5 mL of 1% $K_3[Fe(CN)_6]$ and incubated at 50°C for 20 min. About 2.5 mL of 10% trichloroacetic acid were added and the mixture was centrifuged for 10 min. The upper layer of the solution 2.5 mL was mixed with distilled water 2.5 mL and 0.5 mL of 1% $FeCl_3.6H_2O$ and absorbance was measured at 700 nm, indicative of the reducing power, which was recorded on the final mixture. The procedure was done in triplicate.

Inhibition of lipid peroxidation

The procedure followed the method of Badmus et al. (2011). Briefly, 0.5 ml of egg yolk homogenate (10% v/v) was added to each sample extract of 0.1 mL (10 µL/ml). The volume of the sample was made to 1 ml by adding distilled water and then 0.05 ml of $FeSO_4.7H_2O$ was added. After 30 min, 1.5 ml of 0.8% thiobarbituric acid in 1.1% SDS was added. The mixed solution was then incubated at 95°C for 1 h. After cooling, 5 mL of butanol was added and the mixture was centrifuged at 3,000 rpm for 10 min. The absorbance of the organic upper layer was measured at 532 nm and the percentage of lipid peroxidation inhibition was calculated with the formula:

$$\text{Inhibition on lipid peroxidation (\%)} = 1 - \frac{A_{532nm,sample}}{A_{532nm,control}} \times 100$$

Statistical analysis

All experiments were conducted in triplicates and the results were expressed as mean ±SE. Significance of mean differences between chemical compositions and bioactive compounds within different rice varieties were statistically compared using an analysis of variance (ANOVA) at the 5% probability level ($p \leq 0.05$) with individual pairwise comparisons made using Tukey's HSD test through an SPSS program version 17.

RESULTS AND DISCUSSION

Total phenolic compounds (TPC) and total flavonoid contents (TFC)

Several studies have reported the use of different solvent systems to extract rice such as hexane, methanol, ethanol-water (70:30 v/v) and water at room temperature (Srisawat, et al., 2010). In this study, different solvents were used namely, water (at 25°C), hot water (50°C) and 70% ethanol (at 25°C) to extract Chinese and Thai rice varieties. The total phenolic compound in different rice varieties using different solvent extraction was investigated following a modified Folin-Ciocalteau phenol reagent method. The results were expressed as gallic acid equivalents (Table 2). In extract composition, samples extracted using 70% ethanol (at 25°C), hot water (50°C) and water (at 25°C) had the TPC content range from 54.85 to 634.83, 43.69 to 610.92 and 41.21 to 433.95 mg/kg, respectively. Statistics of TPC values of all crude extracts from different rice varieties with the solvent applied suggests a significant difference at 95% confidence level. But there was no significance between BJ, BT, BW and BN from water and hot water extracted. However, the levels of TPC were slightly different from those studied by Lum and Chong (2012) where the quantity of TPC in Malaysia rice ranged from 22.59 to 329.53 mg/kg. In addition, BB rice, a black rice variety from China, contained the highest TPC from all the solvents extraction. These data indicate that black rice has higher total phenolic compound than red and white rice, which is similar to those reported by Muntana and Prasong (2010) that high phenolic compounds can be found in pigmented rice such as red and black rice.

The same effect and trend can be observed in the total flavonoid contents (TFC) in all rice varieties (Table 3) where samples extracted with 70% ethanol (at 25°C)

Table 3. The total flavonoid compounds (mg/kg) of different rice varieties using different solvents.

Rice varieties	Solvent		
	Water (at 25°C)	Hot water (at 50°C)	70%ethanol (at 25°C)
BB	79.47±1.05^{a}	143.15±0.32^{a}	158.47±3.46^{a}
BJ	15.95±0.41^{bc}	16.63±0.02^{be}	17.06±0.03^{b}
BT	13.28±0.49^{bf}	18.51±1.78^{b}	21.92±2.25^{b}
BW	18.74±1.03^{c}	19.38±1.00^{b}	20.01±1.34^{b}
BY	74.37±0.57^{d}	113.30±0.48^{c}	123.68±0.42^{c}
BND	26.17±0.73^{e}	35.91±0.16^{d}	66.71±0.16^{d}
BN	11.65±0.99^{f}	13.56±0.21^{e}	16.98±0.78^{b}
BD	34.15±0.58^{g}	42.45±0.28^{f}	89.38±0.72^{e}

Values are the mean of triplicate measurements ± standard error (Means±SE); (n = 3). Mean followed by different superscript letters in a column are significantly ($p \leq 0.05$) different from each other.

shows that the TFC range from 16.98 to 158.47 mg/kg, 13.56 to 143.15 mg/kg for hot water (at 50 C) and 11.65 to 79.47 mg/kg for water (at 25°C) were significantly different ($p \leq 0.05$) between each other, except BT, BW from hot water extracted, and BJ, BT, BW and BN from 70% ethanol extracted were not significance. The result showed that TFC is lower than total flavonoid content of the ethanolic extract from Thai rice (Hansakul et al., 2011). Even though all the black rice varieties had the highest TFC than red and white rice varieties, some red rice varieties accessions still had lower contents than white varieties. This association was also observed by Shen et al. (2009), who reported that the flavonoids content of black rice was higher than that of red and white rice. However, some flavonoid contents in red rice were lower than that of white rice. It is possible that other flavonoid compounds such as flavonols and flavanones, rather than anthocyanins, were higher in white rice than those in red rice.

Among all the rice extracts, the 70% ethanol crude extracts contained more TPC and TFC than water (at 25°C) and hot water (at 50°C) crude extracts in the same rice variety. Ethanol can effectively increase permeability of cell wall thus facilitate efficient extraction of polar substances which when added to water can cause further solubility of polar substances in the cell (Veronique, 2006; Anwar and Przybylski, 2012).

Antioxidant components

The scavenging effect on DPPH radical of rice extracts were evaluated using the DPPH method (Figure 1A). Absorbance at 517 nm of DPPH radical in this method was recorded at 0 to 10 min intervals from initiation of the reaction. Scavenging capacity was similar at the beginning of the reaction and was changed with the increase of the reaction time until it stabilized at 10 min. The DPPH radical scavenging capacity varied from 6.28 to 55.86% for water (at 2°C) crude extracts, 7.99 to 68.58% for hot water (at 50°C) crude extracts and 12.81 to 70.82% for 70% ethanol (at 25°C) crude extracts with significant difference ($p \leq 0.05$). The highest scavenging capacity was found in 70% ethanol extract of BB variety (70.82%) as compared to other rice varieties and solvent system. Higher percentage of DPPH scavenging is correlated with higher antioxidant capacity (Sultana et al., 2009).

The reducing power of different extracts from all varieties of rice at 1 mg/mL concentration is depicted in Figure 1B. In this assay, the absorbance values of rice extract using water (at 25°C), hot water (at 50°C), and 70% ethanol (at 25°C) at 700 nm were, respectively, 0.13 to 2.29, 0.10 to 1.66 and 0.11 to 0. 49, which were significantly different ($p \leq 0.05$) between each other. But there was no significance in BJ, BT, BW and BN from water and 70% ethanol extracted. The highest value of absorption in rice extracts was observed in 70% ethanol extract (at 25°C) of BB variety at 2.29. Reducing power in BB variety was higher than other varieties which were found to be the same with the TPC, TFC and scavenging effect on DPPH radical. In addition, the reducing power of rice bran extracts was related to the total phenolic compound contents.

Inhibitory capacity of all rice varieties extracts was determined using linoleic acid test system that can inhibit hydrogen peroxide production in the peroxidation of linoleic acid. The inhibitory activity of rice extracts against lipid peroxidation was shown to decrease absorbance at 532 nm. This is due to the diminution of the lipid oxidation products of linoleic acid, specially the conjugated dienes. In this study, Figure 1C shows that the inhibition of lipid peroxidation of crude extract using 70% ethanol (at 25°C) was the most promising among the extracts examined accounting for 67.86% in BB variety. This can be attributed to the superior antioxidant activities of the sample extracted using 70% ethanol (at 25°C) that exhibited comparatively greater antioxidant compositions than samples extracted with water (at 25°C) and hot water (at 50°C). Statistics of lipid peroxidation inhibition values of all crude extracts from different rice varieties with the

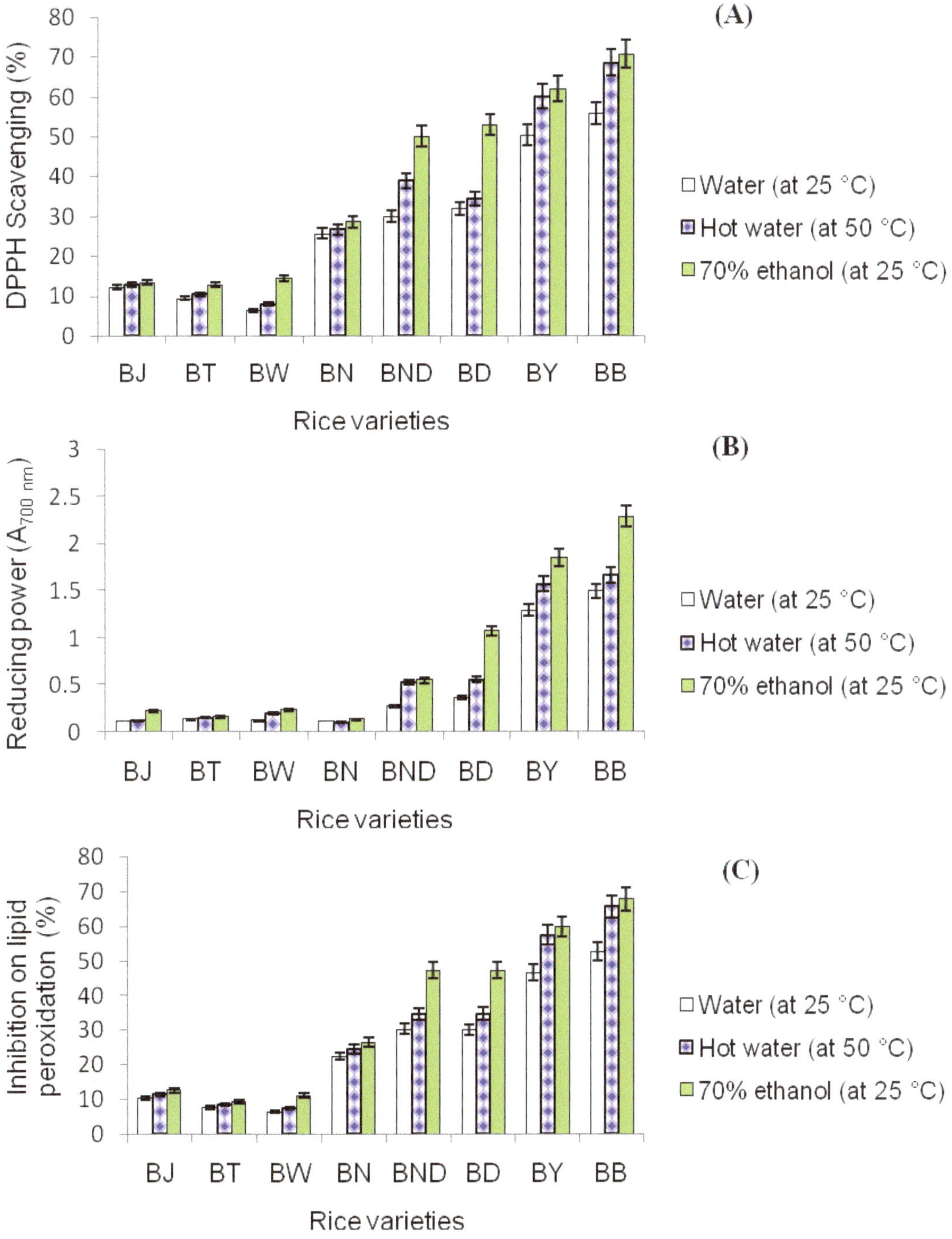

Figure 1. Antioxidant components of rice gRain extracts; (A) scavenging capacity on DPPH radical, (B) reducing power and (C) inhibition on lipid peroxidation. Values are the mean of triplicate measurements ± standard error (Means±SE) in each solvent extract; Statistical significance was determined by Tukey's HSD test ($p \leq 0.05$).

solvent applied suggests a significant difference at 95% confidence level. However, there was no significance between BB and BY extracted with water.

Total phenolic compounds, total flavonoids component and antioxidant activity (DPPH scavenging, reducing power and inhibition of lipid peroxidation) were used to determine the antioxidant activity of the extracts from different rice varieties. An integration of different methods is necessary because their actions and natural antioxidants from plant materials are complex. Furthermore, the correlation coefficient (R^2) between TPC or TFC and antioxidant activities (DPPH scavenging, reducing power and inhibition of lipid peroxidation) in each rice variety extract was determined. The antioxidative activities (DPPH scavenging activity, reducing power, and inhibit on lipid peroxidation) as a function of total phenolic compounds and total flavoniod contents has a positive correlation coefficient (R^2) (data not show). According to Butsat and Siriamornpun (2010), the antioxidant activity of different Thai Rice varieties was positively correlated with total phenolic contents.

Dietary antioxidants protect against reactive oxygen species in the human body. An increased intake of antioxidants may therefore have a number of health effects, such as reducing the incidence of cancer and cardiovascular diseases (Diplock et al., 1998; Halliwell and Gutteridge, 1999). Due to the detection of many new bioactive compounds in food with possible antioxidant activity, and the increased interest in the relationship between antioxidants and disease risks and mechanisms, there is an urgent need to establish the antioxidant capacity in different foods, especially the rice crop which constitute the main food for populations in different countries.

Conclusions

In all (Chinese and Thai) varieties of rice used in this study, ethanol was shown to be the best solvent for extracting phenolic compounds, while the least effective was the use of water (25°C). Same observation was also found after extracting flavonoid compounds. The variety of rice that showed the highest phenolic and flavonoid compounds after extraction was Heimi (BB) from China, followed by Jing Nian (BY), also from China, Dok Kam (BD) and Niaow Deang (BND) both from Thailand, respectively. The pigment of BB and BY is black while red for BD and BND, which according to Shen et al. (2009), TPC and TFC has a correlation with the color of rice. However, the correlation was not observed with the BN variety from Thailand even though it has a black color, its TPC and TFC was found to be almost insignificantly different from the BJ, BT and BW from China that has white color. Antioxidant activity was also found to be correlated with the quantity of TPC and TFC where BB has the highest, followed by BY, BD and BND, respectively.

The findings of this study suggest that nutritional values of these rice varieties extract contains bioactive compounds, may have potential to protect against diseases. Therefore, these rice should be developed in food applications as health products, food industry and pharmaceutical to add value to rice.

ACKNOWLEDGEMENT

The authors would like to thank Asst. Prof. Raumjit Nokkoul of King Mongkut's Institute of Technology Ladkrabang, Chumphon Campus the southern of Thailand for her help in providing Thai rice samples.

REFERENCES

Anwar F, Przybylski R (2012). Effect of solvents extraction on total phenolics and antioxidant activity of extracts from flaxseed (*Linum usitatissimum* L.) Acta Sci. Pol. Technol . Aliment 11(3):293-301.

Badmus JA, Adedosu TO, Fatoki JO, Adegbite VA, Adaramoye OA, Odunola OA (2011). Lipid peroxidation inhibition and antiradical activities of some leaf fractions of *Mangifera indica*. Acta Polo. Pharm. 68(1):23-29.

Butsat S, Siriamornpun S (2010). Phenolic acids and antioxidant activities in husk of different Thai rice varieties. Food Sci. Technol. Int .16:329-36.

Cao LY, Zhan XD, Chen SG, Feng Y, Wu WM, Shen XH, Cheng SH (2010). Breeding Methodology and Practice of Super Rice in China. Rice Science 17:87-93.

Diplock AT, Charleux JL, Crozier-Willi G, Kok FJ, Rice-Evans C, Roberfroid M, Stahl W, VinaRibes J (1998). Functional food science and defence against reactive oxidative species. Brit. J. Nutr. 80:77-112.

Halliwell B, Gutteridge JMC (1999). Free radicals in biology and medicine. Oxford, UK: Oxford University Press.

Hansakul P, Srisawat U, Itharat A, Lerdvuthisopon N (2011). Phenolic and Flavonoid Contents of Thai Rice Extracts and Their Correlation with Antioxidant Activities using Chemical and Cell Assays. J. Med. Assoc. Thai. 94(7):122-130.

Higashi-Okai K, Ishida E, Nakamura Y, Fujiwara S, Okai Y (2008). Potent antioxidant and radical-scavenging activities of traditional Japanese cereal grains. J UOEH 30:375-89.

Iqbal S, Bhanger MI, Anwar F (2005). Antioxidant properties and components of some commercially available varieties of rice bran in Pak. Food Chem. 93(2):265-272.

Lai P, Li KY, Lu S, Chen HH (2009). Phytochemicals and antioxidant properties of solvent extracts from Japonica rice bran. Food Chem. 117:538-544.

Lum MS, Chong PL (2012). Potential antioxidant properties of pigmented rice from sabah, Malaysia. IJANS 1(2):29-38.

Muntana N, Prasong S (2010). Study on total phenolic contents and their antioxidant activities of Thai white, red, and black rice bran extracts. Pak. J. Biol. Sci. 13:170-174.

Shen Y, Jin L, Xiao P, Lu Y, Bao JS (2009). Total phenolics, flavonoids, antioxidant capacity in rice grain and their relations to grain color, size and weight, J. Cereal Sci. 49(1):106-11.

Srisawat U, Panunto W, Kaendee N, Tanuchit S, Itharat A, Lerdvuthisopon N, Hansakul P (2010). Determination of phenolic compounds, flavonoids, andantioxidant activities in water extracts of Thai red and white rice cultivars. J. Med. Assoc. Thai. 93(7):83-91.

Sultana B, Anwar F, Ashraf M (2009). Effect of Extraction Solvent Technique on the Antioxidant Activity of Selected Medicinal Plant Extracts. Molecules 14:2167-2180.

Veronique S (2006). Initial and Bulk Extraction, in Natural Products Isolation, ed. by Satyajit DS, Zahid L and Alexander IG. New Jersey, U.S.A. pp. 27- 46.

Vichapong J, Sookserm M, Srijesdaruk V, Swatsitang P, Srijaranai S (2010). High performance liquid chromatographic analysis of phenolic compounds and their antioxidant activities in rice varieties. LWT-Food Sci. Technol. 43(9):1325-1330.

Xia M, Ling WH, Ma J, Kitts DD, Zawistowski J (2003). Supplementation of diets with the black rice pigment fraction attenuates atherosclerotic plaque formation in apolipoproteine deficient mice. J. Nutr. 133:744-51.

Ling WH, Cheng QX, Ma J, Wang T (2001). Red and black rice decrease atherosclerotic plaque formation and increase antioxidant status in rabbits. J. Nutr. 131:1421-1426.

4

Antimicrobial activity of lactic acid bacteria isolated from fermented milk products

Adejumo T. O.

Department of Microbiology, Adekunle Ajasin University, P.M.B. 01, Akungba-Akoko, Ondo State, Nigeria.

The bacteriocins produced by lactic acid bacteria (LAB) which acted as antimicrobial substances or preservative component on fermented food product (yoghurt, kunu and fufu) was investigated against selected indicator food spoilage causing bacteria and antibiotics. The agar-well diffusion assay was employed to investigate the antagonistic activity against indicator organisms (*Staphylococcus aureus*, *Shigella* spp., *Salmonella typhi* and *Escherichia coli*). *Lactobacillus plantarum, Lactobacillus bulgaricus, Lactobacillus delbrueckii* sub *bulgaricus, Lactobacillus amylophilus, Lactococcus lactis* and *Leuconostoc mesenteroides* were isolated from yoghurt, fufu and kunu. *L. lactis* showed the largest inhibition zone against *S. typhi* and *S. aureus* as 19 and 14 mm, respectively, while *L. amylophilus* showed antimicrobial activity against all the indicator organisms. Isolates from fufu generally showed more inhibition on *S. aureus*, *S. typhi* and *E. coli,* while *L. plantarum* showed no inhibitory zone against the indicator organisms. All the isolates were susceptible to ciprofloxacin, but resistant to ampiclox, zinnacef and amoxacillin, while variable susceptibility/resistance to pefloxacin, gentamycin rocephin, streptomycin, septrin and erythromycin was observed. LAB showed inhibitory properties against the food spoilage and pathogenic organisms tested, and have been reported to produce various antimicrobial compounds including organic acid, hydrogen peroxide, di-acetyl and bacteriocin. This study is important in risk assessment related to fermented food products, as well as the potential use of these LAB isolates in pharmaceutical industries.

Key words: Lactic acid bacteria, fermented foods, yoghurt, 'kunu', 'fufu.

INTRODUCTION

Lactic acid bacteria (LAB) are widely distributed in nature. They are typically involved in a large number of the spontaneous food fermentation, and they have been extensively studied for their use in the production of indigenous fermented foods (Holzapfel et al., 1995). Members of LAB produce bacteriocins and bacteriocins-like substances which may inhibit growth of spoilage and pathogenic microorganisms (Klaenhammer, 1988). Lactic acid fermentation of cereal-based foods and mammary glands derivatives (milk) is a traditional technology in Africa and has long been used in the processing of different foods (Mensah, 1997; Oyewole, 1997). Furthermore, LAB contributes to the enhancement of the organoleptic attributes of foods, as well as to their preservation and microbial safety (Caplice and Fitzgerald, 1999;Calderon et al., 2001).

The preservative action of starter culture in food and beverage systems is due to a range of antimicrobial metabolites produced during the fermentation process, which include many organic acids such as lactic, acetic and propionic acids which provide an acidic environment unfavorable for the growth of many pathogenic and spoilage microorganisms (Caplice and Fitzgerald, 1999; Rattanachaikunsopon and Phumkhachorn, 2010). These acids compounds generally exert their antimicrobial effect by interfering with the maintenance of cell membrane potential, inhibiting active transport, reducing intracellular pH and inhibiting a variety of metabolic functions (Ross et al., 2002). Others are low-molecular-mass (LMM) compounds such as hydrogen peroxide, carbon dioxide, diacetyl (2,3-butanedione), uncharacterized compounds, and high-molecular-mass compounds like bacteriocins (Jay, 1982; Piard and Desmazeaud, 1992). The antimicrobial-producing LAB may be used as protective cultures to improve the microbial safety of foods and also play an important role in the preservation of fermented foods, which is usually achieved by inhibition of contaminating spoilage bacteria such as *Pseudomonas* and foodborne pathogens such as *Staphylococcus aureus*, *Salmonella* spp. and *Listeria monocytogenes* (Buckenhuskes, 1993; Brinkten et al., 1994; Olasupo et al., 1995). LAB have the ability to trap mycotoxins, some might be used to reduce the availability and toxicity of toxins in the gastrointestinal tract of humans and animals, and might therefore be used as a probiotic agent (Dalié et al., 2010).

In Nigeria, yoghurt, fufu and Kunu-zaki are part of most common lactic acid fermented products popular with consumers. Consumption of these fermented foods has many advantages including enhanced nutritional value, digestibility, therapeutic benefits and safety against pathogens (Oranusi et al., 2003). Yoghurt is a smooth, fermented milk product that evolved empirically some centuries ago through the growth of thermophilic LAB such as *Streptococcus thermophilus and Lactobacillus bulgaricus* which ferment the lactose to produce lactic acid. It has a characteristic acidic taste possessing 0.95-1.5% and pH ranging from 3.7-4.2 with viable and abundant fermenting microorganisms (Adolfsson et al., 2004). Fufu is a fermented white paste made from cassava, it is ranked next to gari as an indigenous food of most Nigerians in the South. Fufu is made by steeping whole or cut peeled cassava roots in water to ferment for maximum of three days, during the steeping, fermentation decrease the pH, softens the roots and help to reduces the potentially toxic cyanogenic compound (Agbor-Egbe and Lape Mbome, 2006, Egwim et al., 2013).

Kunu is a product obtained from cereal grains such as sorghum (*Sorghum vulgare*), maize (*Zea mays*) and millet (*Pennisetum typhoideum*). It is a Nigerian drink made of germinated grain, mostly millet. These grains can be used singly or in combination. Sorghum and millet are the most common combinations used in the ratio of 1:2 (w/w). It is a non-alcoholic cereal beverage commonly consumed by the people of the northern Nigeria, though it is becoming popular in other parts of the country (Gaffer et al., 2002).

In Nigeria, cassava is processed into garri and fufu, pellets for compounding animal feeds, kpokpo garri and also into instant aromatized (fermented) flour (Oyewole, 1997). Fufu (variants of the name include foofoo, foufou, fufuo) is a staple food of the Asante, the Akyem, Bono and Fante peoples of the Akan ethnic group of Ghana and is eaten in Guinea (Guinean cuisine). It is made by boiling starchy food crops like cassava, yams or plantains and then pounding them into a dough-like consistency. In this context, the aim of this work was to identify the bacterial isolates from fermented sorghum (kunu), cassava (fufu) and milk product (yoghurt), and to determine the antimicrobial properties of the isolates against foodborne pathogens.

MATERIALS AND METHODS

Collection of samples

Three samples of yoghurt, fufu and kunu-zaki were bought from Akungba market in Akungba Akoko. The samples were carefully labeled. They were brought to the laboratory and processed within 6 h of collection.

Foodborne pathogens used

The food and human pathogenic organisms used include *Staphylococcus aureus*, *Shigella* spp., *Escherichia coli* and *Salmonella typhi*. They were obtained from Adekunle Ajasin University Microbiology Laboratory.

Analysis of samples

The pour plate method was used to isolate microorganisms producing antimicrobial activity from samples of fufu, kunu and yoghurt. Each sample was serially diluted using sterile distilled water as diluents (Speck, 1976). Nine milliliters of distilled water were put into test tubes. One milliliter of each of yoghurt, fufu and kunu-zaki sample were added into the first test tube and thoroughly mixed. Using a different sterile pipette, 1 ml from the first test tube was pipetted into the second test tube already containing 9 ml of distilled water. This procedure continued until the last dilution. Using the pour plate assay, 1 ml of the last dilution of each sample was pipetted into Petril dishes, which were subsequently filled with MRS agar. The plates were mixed and after solidification, they were incubated anaerobically at 37°C for 48 h.

After incubation, the representative colonies on the plates were sub-cultured on fresh MRS agar to obtain pure cultures of the isolates. The pure cultures were then transferred into MRS agar slants for biochemical identification.

Microbiological analysis

Isolation of bacteria

After incubation time, the different culture plates were examined for microbial growth; subcultures were made to get discrete colonies, which were then stored at 40°C for further biochemical investigations in order to identify microorganisms. Bacterial isolates were characterized to generic level and were possible to the species level on the basis of their cultural features (shape, colour, edge, elevation) and morphological features such as motility, gram staining, cell arrangement and shape, and biochemical features.

Preparation of cell-free supernatant

The cell-free supernatants were prepared based on methods by Schillinger and Lucke (1989). The culture extract of the producer strain were obtained from 72 h culture grown on MRS broth. The cultures were then centrifuged at 4.3008 kg force for 15 min. The supernatant was used immediately.

Antibiotic susceptibility tests

A modified method of Bauer-Kirby disk diffusion assay was used for the study. The antibiotics discs were tested on sterile Muller Hilton Agar, which was put into sterile Petri dishes and waited for solidification. A suspension of the isolated organisms was spread over the agar plates, and incubated at 37°C for 1 h. A forceps was used to transfer each antibiotic disc on the plate and incubated for 24 h at 37°C. Ten antibiotics including Pefloxacin, Gentamycin, Ampiclox, Zinnacef, Amoxacillin, Rocephin, Ciprofloxacin, Streptomycin, Septrin and Erythromycin were tested. Each disc was firmly pressed to ensure complete contact with the agar surface. The discs were evenly distributed so that they were not too close to one another. In an inverted position, the plates were incubated at 37°C for 24 h. The results were expressed as inhibition zones around the disc. These were measured, each antibiotic disc was grouped as susceptible or resistance by comparing the measured diameter with the standard given in the manufacturer's instruction.

In vitro inhibition test with indicator organisms

The antimicrobial activity of the isolated LAB (cell free filtrate) against *E. coli*, *Shigella* sp., *Salmonela* sp. and *S. aureus* was performed using the well diffusion assay. Elimination of inhibitory substances like hydrogen peroxide, organic acids and bacteriophages to ensure that the inhibition was caused by only bacteriocin was performed. This was carried out by using MRS containing 0.2% glucose and preparation of cell free supernatant at pH 6.5. The pathogenic indicator bacteria were incubated in Muller Hinton broth at appropriate temperature for 24 h.

Petri dishes containing 20 ml of Muller Hinton agar were prepared previously, another broth was prepared and incubated for 4 h and inoculated with 0.1 ml of 4 h broth culture of pathogenic bacteria. Once solidified, the dishes were stored for 2 h in a refrigerator. Three wells were made and filled with 2 ml of cell-free filtrate. Incubation of the Petri dishes was done at 37°C for 24 h. The antimicrobial activity was determined by measuring the zone inhibited around the wells (Zhennai, 2000).

Identification of bacterial isolates

Identification of isolates was based on cultural, morphological and biochemical characteristics following standard methods.

Gram staining

The method described by Carpenter (1977) and Thomas (1973) were used. Smears of the isolates were prepared and heat fixed on clean grease free slides. The smears were stained for one minute with crystal violet. This was washed out with a gentle running tap water. The slides were flooded with dilute Gram's iodine solution. This was washed off with water and the smears were decolorized with 95% alcohol until the blue colour no more dripped out (about 30 s). The smears were then counter stained with saffranin solution for about 10 s. Finally, the slides were washed with tap water; air dried and observed under oil immersion objectives.

Biochemical tests

Catalase test

This test was performed to demonstrate which of the isolates could produce the enzyme catalase that release oxygen from hydrogen peroxide. This test is usually used as an aid to differentiate *Staphylococci* from *Streptococci* and to differentiate other catalase positive organism from catalase negative. The method employed here was that described by Speck (1976). The catalase production was indicated by the prompt effervescence of oxygen due to the fact that the enzyme aids in the conversion of hydrogen peroxide into water and oxygen bubbles (in the form of effervescence). Effervescence of gas as a white froth indicates a catalase positive reaction while the absence of the effervescence showed negative reaction (AOAC, 2000).

Motility test

To demonstrate the ability of isolates to be motile, this indicating the possession of movement structures (flagellum), was investigated by using semi-solid medium. This medium was prepared and the test performed. The isolates were used to sub-inoculate the medium to a depth of about two-third of the medium and incubated overnight at 37°C. Motility of an organism was indicated by growth beyond the inoculation line, while growth confined to the line of inoculation was regarded as being non-motile.

Sugar fermentation

Each of the isolates was tested for their ability to ferment a given sugar with the production of acid and gas or acid only. Since most bacteria especially Gram-negative bacteria utilize different sugars as source of carbon and energy with the production of both acid and gas, or acid only the test is used as an aid in their differentiation. The growth medium used was peptone water and the method used was that described by Kirk et al. (1975). Peptone water was prepared in a conical flask and the indicators bromocresol purple was added. The mixture was dispensed into test tubes containing Durhams tubes. The tubes with their content were sterilized by autoclaving at 121°C for 15 min. One percent solution of the sugar was prepared and sterilized separately at 115°C for 10 min. This was then aseptically dispensed in 5 ml aliquot volume into the tubes containing the peptone water and indicator. The tubes were inoculated with young culture of the isolates and incubated at 37°C. Acid and gas production or acid only were observed after about 24 h incubation. Acid production was indicated by the change of the medium from light green to yellow color, while gas production was indicated by the presence of gas in the Durham's tubes.

Coagulase test

Slide and tube method was used (Carpenter 1977). In slide test, a

Table 1. Lactic acid bacteria isolated from different food samples.

Fermented food	LAB
Yoghurt	*Lactobacillus plantarum* *Lactobacillus bulgaricus* *Lactococcus lactis*
Fufu	*Lactobacillus delbrueckii* sub *bulgaricus* *Lactobacillus amylophilus* *Lactococcus lactis*
Kunu-Zaki	*Lactobacillus plantarum* *Lactococcus lactis* *Leuconostoc mesenteroides*

loop full of the isolate was mixed with human plasma and allowed to stand for some minutes. Particles indicating agglutination was used as indication of coagulase reaction. In the tube method, plasma was added unto a culture of the isolate in peptone water in bijou bottles. The bottles were incubated at 37°C for 24 h. A clumping/agglutination of the plasma were used to indicate presence of coagulase.

Citrate utilization

In actual practice, Koser's citrate medium containing 'citric acid' serves as the exclusive carbon source. It detects the ability of an organism to use citrate as the sole source of carbon and nitrogen. Evidently, the ability as well as the efficacy for the 'citrate utilization' (the prevailing substrate) is adequately indicated by the production of measurable turbidity in the medium. In carrying out this test, the isolate was stabbed into test tube containing the citrate medium from a 24 h culture. Simmon Citrate Agar was prepared and dispensed into each clean test tube, sterilized, and allowed to solidify in a slanted position. An inoculum from the broth culture was picked with an inoculating loop and incubated unto the surface of the slanted Citrate Agar prepared in each test tube. It was then incubated at 37°C for 5 days and examined. Those that changed color from green (original color of prepared medium) to blue or yellow were considered positive, while those that retained the green color were negative.

Starch hydrolysis test

A medium containing starch was used. After inoculation and overnight incubation, iodine reagent was added to detect the presence of starch. Iodine reagent complexes with starch to form a blue-black color in the culture medium. An inoculation from a pure culture is streaked on a sterile plate of starch agar. The inoculated plates were incubated at 35-37°C for 24 h. Iodine reagent was then added to flood the growth. Clear halos surrounding colonies is indicative of their ability to digest the starch in the medium due to the presence of alpha-amylase.

Production of hydrogen sulphide (H_2s)

There are several sulphur containing amino acids like cystein, methionine that may decompose certain organisms to yield hydrogen sulphide gas among the products of microbial degradation. Lead acetate was duly incorporated into the culture media which eventually turned into either black or brown due to the formation of lead sulphide (AOAC, 2000).

Spore stain

The malachite green staining method was used. The staining was carried out as described by Carpenter (1977). Smears of the pure isolates were made on grease-free glass slide and heat fixed. The slides were flooded with 5% v/v malachite green solution. The slides were flamed in such a way that the stain steamed but did not boil. The slides were then allowed to stand for 5 min. The stain was then washed out in running tap water. The smears were counter-stained with saffranin for 30 s. It was stained with saffranin, blotted dried and examined under the oil immersion objective. The spores stained green while vegetative cells stained red.

RESULTS

The results show the LAB isolated from different fermented dairy products: yoghurt and the fermented foods (fufu and kunu). Six LAB were isolated from the samples, *Lactobacillus plantarum, L. bulgaricus* and *Lactobacillus lactis* were obtained from yoghurt, *Lactobacillus delbrueckii* sub *bulgaricus, Lactobacillus amylophilus* and *L. lactis* were obtained from fufu, while *L. plantarum, L. lactis* and *Leuconostoc mesenteroides* were isolated from kunu (Table 1). *L. plantarum* was obtained from yoghurt and kunu, while yoghurt, fufu and kunu showed presence of *L. lactis.*

The morphological and biochemical characteristics of the bacterial isolates are shown in Table 2. The antibiotic resistance patterns of the isolates are presented in Table 3. All the isolates were susceptible to ciprofloxacin, but resistant to ampiclox, zinnacef and amoxacillin, while variable susceptibility/resistance to pefloxacin, gentamycin rocephin, streptomycin, septrin and erythromycin was observed. Table 4 shows inhibitory properties of the metabolites of the selected LAB against the target foodborne pathogenic organism (*S. aureus, Shigella* spp., *Salmonella* typhi and *E. coli*). The metabolites possess antagonistic activities against all the organisms used in this work.

L. plantarum isolates obtained from yoghurt and kunu was observed to be inactive against any of the indicator organisms. *L. lactis* showed the largest inhibition zone against *S. typhi* and *S. aureus* as 19 and 14 mm respectively, while *L. amylophilus* showed antimicrobial activity against all the indicator organisms. Isolates from fufu generally showed more inhibition on *S. aureus*, *S. typhi* and *E. coli.*

DISCUSSION

In this study, six LAB were isolated from traditionally fermented and diary food product: yoghurt, fufu and kunu obtain from Akungba. The results of the present

Table 2. Morphological and biochemical characteristics of the bacterial isolates.

Shape	Pigmentation	Surface	Margin	Elevation	Gram stain	Cell morphology	Motility	Catalase	Starch	Manitol	Fructose	Glucose	Sucrose	Lactose	Identified organisms
Circular	Creamy	Rough	Entire	Convex	+	Rod	-	-	-	+	+	+	+	+	*Lactobacillus plantarum*
Circular	Whitish	Shiny	Entire	Convex	+	Long rod	-	-	-	+	+	+	-	-	*Lactobacillus bulgaricus*
Circular	Creamy	Rough	Entire	Flat	+	Cocci	-	-	-	+	+	+	-	-	*Lactococcus lactis*
Circular	Whitish	Shiny	Entire	Flat	+	Rod	-	-	-	+	+	+	-	+	*Lactobacillus delbrueckii sub bulgaricus*
Circular	Creamy	Smooth	Entire	Flat	+	Short rod	-	-	-	-	+	+	-	-	*Lactobacillus amylophilus*
Circular	Creamy	Smooth	Entire	Convex	+	Cocci	-	-	-	-	+	+	-	-	*Lactococcus lactis*
Circular	Creamy	Rough	Entire	Flat	+	Rod	-	-	-	+	+	+	-	-	*Lactobacillus plantarum*
Circular	Creamy	Smooth	Entire	Flat	+	Cocci	-	-	-	-	+	+	+	-	*Lactococcus lactis*
Circular	Whitish	Shiny	Entire	Flat	+	Cocci	-	-	-	+	+	+	+	+	*Leuconostoc mesenteroides*

-: Negative, +: positive.

Table 3. Antibiotic susceptibility discs test

Lactic acid bacteria	Inhibition zone (mm)									
	PEF	GEN	APX	ZNF	AMX	RCP	CPX	STP	SEP	ERT
Lactobacillus bulgaricus	S	R	R	R	R	R	S	R	R	R
Lactobacillus plantarum	R	S	R	R	R	S	S	R	S	R
Lactococcus lactis	S	R	R	R	R	S	S	S	R	R
Lactobacillus delbrueckii sub bulgaricus	R	R	R	R	R	R	S	R	R	R
Lactobacillus amylophilus	S	R	R	R	R	R	S	R	S	R
Lactococcus lactis	R	R	R	R	R	S	S	S	S	R
Lactobacillus plantarum	S	S	R	R	R	S	S	R	R	R
Lactococcus lactis	S	R	R	R	R	R	S	S	R	S
Leuconostoc mesenteroides	R	R	R	R	R	R	S	R	S	R

R- Resistance (no value), S- Susceptible (value ranges from 4-32 mm); PEF- Pefloxacin, GEN- Gentamycin, APX- Ampiclox, ZNF- Zinnacef, AMX- Amoxacillin, RCP- Rocephin, CPX- Ciprofloxacin, STP- Streptomycin, SEP- Septrin, ERT- Erythromycin.

study is in accordance with the report of Odunfa and Adeyele (1985), who revealed that members of LAB could be detected in a variety of habitats including fermented foods. Alli et al. (2010) reported the presence of *L. acidophilus, L. bulgaricus, Lactobacillus cremoris, L. fermentum,*

Table 4. Inhibition of indicator organisms by LAB.

Sample	Bacteria isolated	Test organisms (mm)			
		Staphylococcus aureus	***Shigella spp***	***Salmonella typhi***	***Escherichia coli***
	Lactobacillus bulgaricus	NI	NI	9	10
Yoghurt	*Lactobacillus plantarum*	NI	NI	NI	NI
	Lactococcus lactis	8	6	7	NI
	Lactobacillus delbrueckii sub bulgaricus	8	NI	12	10
Fufu	*Lactobacillus amylophilus*	8	4	8	7
	Lactococcus lactis	14	NI	19	11
	Lactobacillus plantarum	NI	NI	NI	NI
Kunu-Zaki	*Lactococcus lactis*	9	7	NI	6
	Leuconostoc mesenteroides	8	NI	8	NI

Value shows the mean number of triplicate well zone (mm); NI- No zone of inhibition.

L. lactis, Lactococcus mesenteroides, Micrococcus acidophilus, Rhodosporium spp., *Saccharomyces bayanus, Saccharomyces cerevisiae, Streptococcus lactis* and *Streptococcus thermophilus* in yoghurt samples, out of which only *L. bulgaricus, L. plantarum* and *L. lactis* were obtained in this study.

The identification carried out for representative LAB strains from the fermented food products demonstrated the dominance of *L. amylophilus, L. bulgaricus, L. delbrueckii, L. lactis* and *L. mesenteriode*. These identified *Lactobacillus* species were in accordance with those earlier identified from similar fermented food products (Halm et al., 1993; Wakil et al., 2004).

Lactobacilli grow best in highly nutritive substrates. They use the nutrients in the substrate for their own metabolism and cell growth and multiply in food (from one million per millilitre to one billion per millilitre). They are present in the fermented food not only as viable cells and non-colony forming units, but also with the primary and secondary metabolites produced during the fermentation process (Robinson, 1991). The MRS medium used was selective for the isolation of *Lactobacillus* species since they are extremely fastidious. Lindquist (1998) reported that a medium that would support their growth must contain a fermentable carbohydrate and many growth factors. Gilliland and Speck (1977) had earlier reported that *Lactobacilli* showed stronger antibacterial properties against Gram-positive bacteria (*S. aureus* and *Clostridium perfringens*) than Gram-negative bacteria (*E. coli* and *Salmonella typhimurium*). *L. lactis* from yoghurt, fufu and kunu inhibited *S. aureus* in the order of 8, 14 and 9 mm with respect to diameter of inhibition zone. *L. bulgaricus, L.* lactis and *L. amylophilus* also inhibited *E. coli* in order of 10, 7 and 6 mm. Among the lactobacilli, there has been great interest in *L. plantarum*, due to the potential applications of the microorganisms as a starter bacterium for a variety of fermented foods. The bacteriocin produced from *L. plantarum* has been found to be inhibitory towards closely related LAB, particularly the mesophilic and thermophilic lactobacilli. In this study, *L. plantarum* produced no inhibition zone against the tested microorganisms.

Several studies have shown that pathogens such as enterotoxigenic *E. coli, Shigella flexneri, S. typhimurium* and *B. cereus* are adversely affected when present in traditional fermented foods (Kingamkono et al., 1995; Kunene et al., 2000; Obadina et al., 2006). Some of the antimicrobial properties exhibited by these fermented foods may be as a result of the low pH of the food as well as metabolites produced by microorganisms such as LAB involved in the fermentation. The pathogens used in this study were sensitive to the LAB metabolites. Brooks et al. (1998) reported rapid development of resistance by *Staphylococcus* sp. to antimicrobial agents. However, the *S. aureus* used in this study was sensitive to six LAB used against it.

Antimicrobial activity of LAB may be due to decrease in pH, depletion of nutrients and production of antimicrobial compounds (Olsen et al., 1995), including bacteriocins (Parente and Ricciardi, 1999), and various organic acids such as lactic acids, acetic acid. Microbial food safety is an increasing public health concern worldwide and many Gram negative bacteria such as *E. coli, Klebsiella* sp. together with Gram positive bacteria such as *S. aureus* have been implicated in food borne diseases (Mead et al., 1999). The LAB isolated from yoghurt, fufu and *kunu-zaki* in this study probably produced different antimicrobial compounds, the quantity of which might vary with time. Collins et al. (1983) also noted the inhibition of *Psuedomonas fragi* and *S. aureus* against other microorganisms by hydrogen peroxide by some LAB strains which contribute to their inhibitory activity. Liasi et al. (2009) emphasized that research on antimicrobial substances produced by LAB, had led to their potential use as natural preservatives, which may be used to combat the growth of pathogenic microorganisms in the food industry.

Conclusion

The promising results of bacteriocins produced by LAB from fermented food underline its important role in improving food quality and increasing safety, by using it as starter/protective cultures to combat the growth of food-borne pathogens and spoilage microorganisms.

Conflict of Interests

The author(s) did not declare any conflict of interests.

REFERENCES

Adolfsson O, Meydani SN, Russell RM (2004). Yoghurt and gut function. Am. J. Clin. Nutr. 80:245-256.

Agbor-Egbe T, Lape Mbome I (2006). The effects of processing techniques in reducing cyanogen levels during the production of some Cameroonian cassava foods. J. Food Compost Anal. 19:354-363.

Alli JA, Oluwadun A, Okonko IO, Fagade OE, Kolade AF, Ogunleye VO (2010). Microbial Assessment and Microbiological Quality of Some Commercially Prepared Yoghurt Retailed in Ibadan, Oyo State, Southwestern Nigeria. Br. J. Dairy Sci. 1 (2): 34-38,.

AOAC (Association of Official Analytical Chemist) (2000). Official Methods of Analysis, Association of Official Analytical Chemists. 17th Ed. Gaithersburg, USA.

Brinkten B, Minekns M, Vossen MV, Leer RJ, Huisin't Veld JHJ (1994). Antimicrobial activity of lactobacilli. J. Appl. Bacteriol. 77:140-148.

Brooks GF, Butel JS, Morse SA (1998). Medical Microbiology. 21st Edn. Norwalk. Conn. Appleton and Lange, USA. p. 740.

Buckenhuskes HJ (1993). Selection criteria for lactic acid bacteria to be used as starter cultures for various food commodities. FEMS Microbiol. Rev. 12:253-272.

Calderon M, Loiseau G, Guyot JP (2001). Nutritional improvements and simplified cultivation medium to study growth and energetic of a sourdough lactic acid bacterium *Lactobacillus fermentum Ogi* E1 during heterolactic fermentation of starch. J. Appl. Microbiol. 90:1-9.

Caplice E, Fitzgerald GF (1999). Food fermentations: Role of microorganisms in food production and preservation. Int. J. Food Microbiol. 50:131-149.

Carpenter PC (1977). Microbiology 4th edition W.B. Saunders Company Philadelphia. pp. 401-402.

Collins MD, Farrow JA, Phillips BA, Kandler O (1983). *Streptococcus garvieae* sp. nov. and *Streptococcus plantarum* sp. nov. J. Gen. Microbiol. 129(11):3427-3431.

Dalié DKD, Deschamps AM, Richard-Forget F (2010). Lactic acid bacteria - Potential for control of mould growth and mycotoxins: A review. Food Control 21:370-380.

Egwim E, Amanabo M, Yahaya A, Bello M (2013). Nigerian Indigenous fermented foods: Processes and Prospects. In: Mycotoxin and Food Safety in Developing Countries. Chapter 7 Pages 153-180.

Gilliland S, Speck M (1977). Antagonistic action of *Lactobacillus acidophilus* towards intestinal and foodborne pathogens in associative cultures. J. Food Prot. 40:820-823.

Halm M, Lillie A, Srensen AK, Jakobsen M (1993). Microbiological and aromatic characteristics of fermented maize dough for *kenkey* production in Ghana. Int. J. Food Microbiol. 19:135-143.

Holzapfel WH, Geisen R, Schillinger U (1995). Biological preservation of foods with reference to protective cultures, bacteriocins and food-grade enzymes Int. J. Food Microbiol. 24 (3):343-362.

Jay JM (1982). Antimicrobial properties of diacetyl. Appl. Environ. Microbiol. 44:525-532.

Kingamkono R, Sjogreen E, Svanberg U, Kaijser, B. (1995). Inhibition of different strains of entero pathogens in lactic fermenting cereal gruel. World J. Microbiol. Biotechnol. 11:299-303.

Kirk CJC, Peel RN, Kershew RJ (1975). Basic Medical Laboratory Technology. Pitman Medical Publishing Ltd, London. pp. 121 -122.

Klaenhammer TR (1988). Bacteriocins of lactic acid bacteria. Biochimie 70(3):337-349.

Kunene NF, Geornaras I, von Holy A, Hastings JW (2000). Characterization and determination of origin of lactic acid bacteria from a sorghum-based fermented food by analysis of soluble proteins and amplified fragment length polymorphism fingerprinting. Appl. Environ. Microbiol. 66:1084-1092.

Liasi SA, Azmi TI, Hassan MD, Shuhaimi M, Rosfarizan M, Ariff AB (2009) Antimicrobial activity and antibiotic sensitivity of three isolates of lactic acid bacteria from fermented fish product, Budu. Malays. J. Microbiol. 5(1):33-37.

Lindquist J (1998). Laboratory Manual for the Food Microbiology Laboratory (Online). Madison: University of Wisconsin. http://www.splasmmo.net/bact102/102xsal.html (accessed 29 June 2006.

Mead PS, Slutsker L, Dietz V, McCaige LF, Bresee JS, Shapiro C, Griffin PM, Tauxe RV (1999). Food related illness and death in the United States. Emerg. Infect. Dis. 5:607-625.

Mensah P (1997). Fermentation—the key to food safety assurance in Africa. Food Control 8:271-278.

Obadina AO, Oyewole OB, Sanni LO, Tomlins KI (2006). Bio-preservative activities of *Lactobacillus plantarum* strains in fermenting Casssava „fufu". Afr. J. Biotechnol. 5(8):620-623.

Odunfa SA, Adeyele S (1985). Microbiological changes during the traditional production of ogi baba, a West African fermented sorghum gruel. J. Cereals Sci. 3:173-180.

Olasupo NA, Olukoya DK, Odunfa SA (1995). Studies on bacteriocinogenic *Lactobacillus* isolates from selected Nigerian fermented foods. J. Bacteriol. Microbiol. 39:181-186.

Olsen A, Halm M, Jackobsen M (1995). The antimicrobial activity of lactic acid bacteria from fermented maize (*kenkey*) and their interactions during fermentation. J. Appl. Bacteriol. 79:506-512.

Oranusi SU, Umoh VJ, Kwaga JKP (2003). Hazards and critical control points of *Kunu-zaki*, a non-alcoholic beverage in Northern Nigeria. Food Microbiol. 20:127-132.

Oyewole OB (1997). Lactic fermented foods in Africa and their benefits. Food Control 8:289-297.

Parente E, Ricciardi A (1999). Production, recovery and purification of bacteriocins from lactic acid bacteria. Appl. Microbiol. Biotechnol. 52: 628-638.

Piard JC, Desmazeaud M (1992). Inhibiting factors produced by lactic acid bacteria: 2. Bacteriocins and other antibacterial substances. Lait 72:113-142.

Rattanachaikunsopon P, Phumkhachorn P (2010). Lactic acid bacteria: their antimicrobial compounds and their uses in food production. Ann. Biol. Res. 1 (4):218-228.

Robinson RK (1991). Therapeutic Properties of Fermented Milk. London, UK: Elsevier Applied Science. pp. 159-179.

Ross RP, Morgan S, Hill C (2002). Preservation and fermentation: past, present and future. Int. J. Food Microbiol. 79:3-16.

Schillinger U, Lucke F (1989). Antibacterial activity of *Lactobacillus sake* isolated from meat. Appl. Environ. Microbiol. 55:1901-1906.

Speck ML (1976). Compendium of Methods for the Microbiological Examination of Foods. American Public Health Examination Association Inc. pp. 563-567.

Thomas CGA (1973). *Medical Microbiology* 3rd *Edition,* Baucere Tindall, London. pp. 8-9.

Wakil SM, Bamgbose OO, Ilo EC (2004). Influence of fermentation time on the microbial profile, sensory attributes and shelf-life of "*kunu-tsamia*". Adv. Food Sci. 26 (2):52-55.

Zhennai Y (2000). Antimicrobial compounds and Extracellular polusaccharides produced by lactic acid bacteria: structures and properties. Academic Dissertation. Department of Food Technology, University of Helsinki, 61.

Physicochemical and nutritional characteristics assessement of two different traditional foods prepared with senescent plantain

Kra Kouassi Aboutou Séverin, Akoa Edwige, Mégnanou Rose-Monde, Yéboué Koffi, Akpa Eric Essoh and Niamké L. Sébastien

Laboratoire de Biotechnologies, Filière Biochimie-Microbiologie de l'Unité de Formation et de Recherche en Biosciences de l'Université de Cocody, 22 BP 582 Abidjan 22, Côte d'Ivoire.

This study investigated the physico-chemical and nutritional properties of two foods widely consumed by population. Thus, the products were purchased from the markets of 10 municipalities of Abidjan city, and several analyses such as moisture, ash, dry matter, fat, protein, vitamin C, total sugars, carbohydrate, starch, energy value, pH and titrable acidity, using common scientific methods of food characteristics determination were performed. The results showed that the product from the baked preparation has more nutrients in fat, protein, vitamin C, sugars, carbohydrates and starch than the boiled one. Consequently, the energy value of baked product was higher (69.52 ± 0.03 kcal/100 g of fresh material (FM)) than the boiled product (64.30 ± 0.03 kcal/100 g FM). The higher content in moisture (67.68 ± 0.55%) of boiled product was not favourable for this conservation. In addition, the mineral contents of boiled and baked products were not significantly different (P>0.05). In view to their mineral composition, the two products may be important sources of potassium (353.37 ± 3.25 and 358.34 ± 2.68 mg/100 g FM), phosphorus (36.560 ± 0.959 and 37.047 ± 0.595 mg/100 g FM) and magnesium (34.12 ± 1.88 and 36.93 ± 1.10 mg/100 g FM). Others minerals such as calcium (6.65 ± 0.34 and 6.30 ± 0.49 mg/100 g FM), iron (0.56 ± 0.10 and 0.69 ± 0.03 mg/100 g FM), sodium (salt) (2.793 ± 0.281 and 2.677 ± 0.177 mg/100 g FM), copper (0.077 ± 0.004 and 0.073 ± 0.003 mg/100 g FM), zinc (0.19 ± 0.05 and 0.23 ± 0.02 mg/100g FM) and manganese (0.31 ± 0.04 and 0.38 ± 0.04 mg/100 g FM) were not too affected or eliminated by the processes.

Key words: Plantain, baked and boiled food, traditional food consumption, food habit, physico-chemical and nutritional qualities.

INTRODUCTION

Plantain (*Musa paradisiaca*) usually treated as a vegetable and not a fruit (Anonymous, 2011), constitutes the fourth most important global food commodity after rice, wheat and maize (Frison and Sharrock, 1999; Bakry et al., 2002). This vegetable grows in tropical countries with world production of 33 million tonnes (Akubor et al., 2003; FAO, 2005) and also, contributes significantly to food and income security of people, particularly in developing countries. Many African countries use plantains as an inexpensive source of calories (Akubor et al., 2003) because it is an important starchy staple, invaluable source of carbohydrates (Swennen, 1990; IITA, 1998). Plantain provides a good source of nutrients for both human and animal consumption. In Côte d'Ivoire, an average of 1 510 778

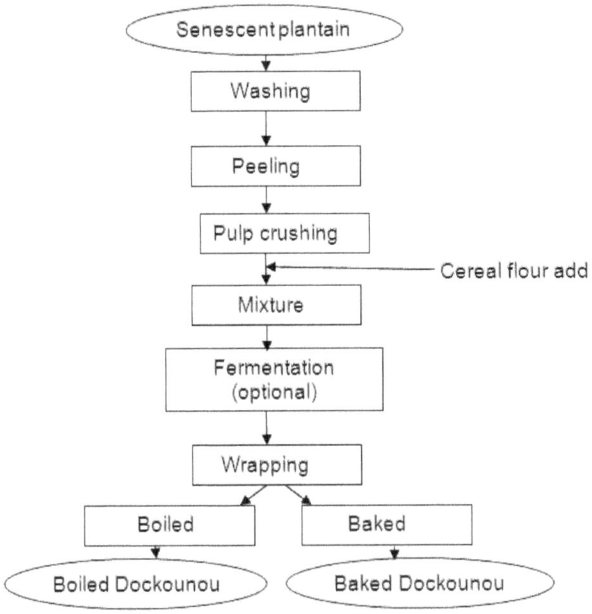

Figure 1. Manufacturing process of the "Dockounou".

tons of plantain are produced per year (Kuperminc, 1985; Séry, 1985). A major problem of plantain is that the fruits are highly perishable. Indeed, about 35 to 60% post harvest losses had been reported and attributed to lack of storage facilities and inappropriate technologies for food processing (Zakpaa et al., 2010; Atanda et al., 2011). The conservation of plantain requires processing into a more stable and convenient form (Ukhum and Ukpebor, 1991). So, several meals with plantain as chips, foufou, foutou, alloco, were prepared and consumed in Côte d'Ivoire (N'Guessan et al., 1993; Lassois et al., 2009). Very long time ago, to resolve the problems of post-harvests losses, farming women transformed overripe fruits of plantain into a meal called "Dockounou" with an empiric traditional process. Procedure of this manufacturing followed seven (7) steps according to these producers (Figure 1): washing of fruits, peeling, crushing, flour mixing, fermentation (optional), wrapping and cooking. The obtained product is package in banana *(Musa spp)* or *Thaumatococcus danieilii* or *Cola nitida* leaves. It is an inestimable food aid for farmers during the country works. Today, this preparation is much popularized and the products are sold in almost all the markets of big cities and consumed by all the Ivorian sectors of the society. Unfortunately, to date, there is no scientific data on this food which is integrated into urban food habits. Therefore, the aim of this research is to assess the physico-chemical and nutritional qualities of the "Dockounou" sold on the retail markets.

MATERIALS AND METHODS

The materials used for analyses were the boiled and the backed "Dockounou". They were purchased from the markets of 10 municipalities of the city of Abidjan and transported in the laboratory in iceboxes at 4 to 6°C. For each type of "Dockounou", the products of the collection are carefully crushed in a clear home-made mortar to obtain single homogeneous dough. Then a several samples of 5 g were prepared in sterile bags for the following analyses.

Physico-chemical and biochemical characteristics analyses

Moisture, dry matter and ash content were determined by using the AOAC (1980) methods while the pH was directly recorded with a pH-meter (Roucaire, Metr Ohm 632, Germany). The total titrable acidity was evaluated on purée of "Dockron" obtained from 40 g of paste. Raw proteins contents were measured by Kjeldahl method according to BIPEA (1976) using 6.25 as conversion factor. Lipids content was determined by extraction with hexane in a Soxtherm system during 6 h (BIPEA, 1976). Vitamin C content was obtained with the method of Tillmanns and Hirsch (1932). Total sugars contents were evaluated with the method of Dubois et al. (1956). As for carbohydrate and starch contents, they were calculated by using the expression described by Coulibaly (2008):

Carbohydrate content = 100 - (Moisture (%) + Ash (%) + Fat (%) + Protein (%))

Starch content = 0.9 (Carbohydrate (%) - total sugars).

Energy value was also calculated using the relation described by Atwater and Rosa (1899).

Mineral analyses

Minerals content (potassium, sodium, calcium, magnesium, iron, copper, zinc, manganese and phosphorus) of the samples was determined by using AOAC (1990) method. Five grams of samples were digested with a mixture of concentrated nitric acid, sulfuric acid and perchloric acid (10:0.5:2, v/v) and analyzed using an atomic absorption spectrophotometer (GBC 904AA; Germany). The total phosphorus was determined as orthophosphate by the colorimetric ascorbic acid method after acid digestion and neutralization using phenolphthalein as indicator reagent (APHA, 1995). The absorbance was read at 880 nm (Spectronic 21 D, Miltonroy, New York, USA). Blank analyses were performed before using distilled water.

Statistical analysis

The statistical analysis was performed using single factor analysis of variance (ANOVA) for all data and analyzed by the SPSS for Windows program, version 11.5. Comparisons of means were carried out by Duncan's multiple range test (DMRT) at the 5% significance level.

RESULTS

Physicochemical and nutritional characteristics

Table 1 shows the physicochemical characteristics of both products. All data showed, except for the ash content, were significantly different at 5% level for physic-chemical parameters contents of boiled and backed dockounou. The moisture content and acidity of boiled dockounou were higher (67.68 ± 0.55 > 60.40 ± 0.54%), (49.72 ± 1.60 > 40.87 ± 0.78 meq/100 g) than those of

Table 1. Physicochemical properties of boiled and baked dockounou (%).

Property	Boiling Dockounou	Baking Dockounou
pH (at 30°C)	5.28±0.03[b]	5.47 ±0.11[a]
Acidity (meq/100 g)	49.72±1.60[a]	40.87 ±0.78[b]
Moisture	67.68± 0.55[a]	60.40 ±0.54[b]
Dry matter	32.46±0.31[b]	39.60±0.54[a]
Ash	1.10±0.10[a]	0.99±0.01[a]

Each value is the mean of triplicate analyses. Different letters in the same line indicate statistical difference ($p \leq 0.05$).

baked dockounou. Conversely, the dry matter and pH value of boiled dockounou were lower (32.46 ± 0.31 < 39.60 ± 0.54%), (5.28 ± 0.03 < 5.47 ± 0.11) than those of baked dockounou. Vitamin C content of boiled dockounou (1.84 ± 0.34%) was higher than the baked dockounou one (1.12 ± 0.12%). Fat, protein, total sugars, carbohydrates and starch contents (0.280 ± 0.021, 1.50 ± 0.02, 15.26 ± 0.04, 37.37 ± 0.49 and 20.37 ± 0.41%) of baked dockounou were higher than the boiled dockounou values (0.027 ± 0.015, 1.21 ± 0.03, 14.73 ± 0.06, 29.47 ± 0.45 and 12.46 ± 0.48%) (Figure 2). Energy value of baked dockounou (69.52 ± 0.03 kcal) was higher than boiled product estimated to 64.30 ± 0.05 kcal.

Minerals composition

The results of the mineral characteristics for both products are depicted in Table 2. Data showed that there was no significant difference in the mineral contents of the boiled and the baked dockounou samples. Potassium is the most abundant mineral in both samples with estimated value of 353.37 ± 3.25 mg/100 g FM in the boiled product and 358.34 ± 2.68 mg/100 g FM in the baked product. Phosphorus and magnesium contents for boiled and baked dokcounou ranged from 36.560 ± 0.959 to 37.047 ± 0.595 mg/100 g FM and 34.12 ± 1.88 to 36.93 ± 1.10 mg/100 g, respectively. Copper values which range from 0.077 ± 0.00 to 0.073 ± 0.003 mg/100 g for boiled and baked products, respectively, were the lower mineral value content. The amount of calcium and sodium, respectively, ranged from 6.65 ± 0.34 and 2.793 ± 0.281 (boiled product) to 6.30 ± 0.49 and 2.677 ± 0.177 mg/100 g (baked product). The concentrations of iron were respectively 0.56 ± 0.10 and 0.69 ± 0.03 mg/100 g for boiled and baked dockounou. The estimated values for manganese and zinc ranged from 0.31 ± 0.04 and 0.19 ± 0.05 (boiled product) to 0.38 ± 0.04 and 0.23 ± 0.02 mg/100 g (baked product).

DISCUSSION

Concerning physicochemical properties of products obtained, their acidity value, moisture and ash contents were

Table 2. Mineral characteristics of boiled and baked "dockounou" (mg/100 g MF).

Mineral	Boiled Dockounou	Baked Dockounou
Potassium (K)	353.37±3.25[a]	358.34±2.68[a]
Calcium (Ca)	6.65±0.34[a]	6.30±0.49[a]
Sodium (Na)	2.793±0.281[a]	2.677±0.177[a]
Magnesium (Mg)	34.12±1.88[a]	36.93±1.10[a]
Iron (Fe)	0.56±0.10[a]	0.69±0.03[a]
Zinc (Zn)	0.19±0.05[a]	0.23±0.02[a]
Copper (Cu)	0.077±0.004[a]	0.073±0.003[a]
Manganese (Mn)	0.31±0.04[a]	0.38±0.04[a]
Phosphorus (P)	36.560±0.959[a]	37.047±0.595[a]

Each value is the mean of triplicate analyses. The same letter in the same line indicate no statistical difference ($p \geq 0.05$).

were more raised in the boiling process of dockounou, whereas pH value and dry matter content were weaker as reported by Ankrah (1974) and Adeyemi and Oladiji (2009).

According to Ahenkora et al. (1997), there is a gain in moisture and a loss of moisture with the boiling and baking processes, respectively. Indeed, boiling increased the moisture content of ripe plantains while baking decreased the moisture content. The increased water content of the boiled products is due to the cooking medium (water) (Ahenkora et al., 1997). Moisture content of food or processed product gives an indication of its shelf life and nutritive values. Low moisture content is a requirement for long storage life (Zakpaa et al., 2010; Adeyemi and Oladiji, 2009). Baked dockounou with less water content should have longer shelf life than boiled dockounou.

Ash content indicates presence of mineral elements in the sample and level of surface contamination. The values obtained for the both products were within the range for those published by Adeyemi and Oladiji (2009).

pH values give a measure of the acidity or alkalinity of the products, while titrable acidity of the cooked dockounou gives a measure of the amount of acid present as malic acid (Dadzie and Ochard, 1997). Values obtained were similar to those reported by Offem and Njoku (1993) and Dongo et al. (2011) for ripe plantain. Assessment of pH and titrable acidity of the cooking products with plantain are used primarily to estimate consumption quality and hidden attributes. Acids have an important contribution in both products as taste is mainly a balance between the sugar and acid contents, though ripe plantains are quite sweet (Offem and Njoku, 1993).

Baked dockounou is an important source of fat, protein, sugar, starch and carbohydrates than the boiled dockounou. Data obtained for these parameters were similar to those reported earlier (Ogazi, 1996; Jaffe et al., 1963; Marriott and Lancaster, 1983; N'Ganzoua et al., 2010). The nature of the processing medium influenced the levels

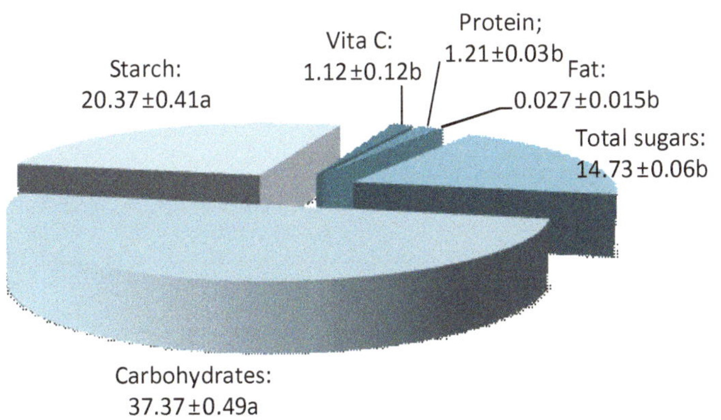

Figure 2a. Nutritional properties of baked dockounou.

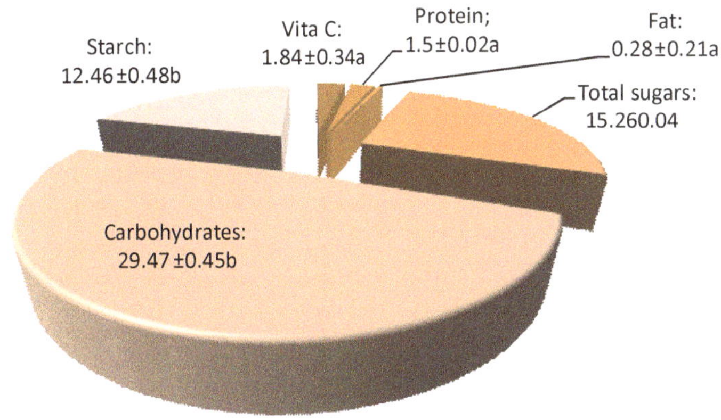

Figure 2b. Nutritional properties of boiled dockounou.

of fat and moisture in the products according to Ahenkora et al. (1997). Also, the energy delivered after consumption by the baked food is higher than the boiled product (69.52 ± 0.03 > 64.30 ± 0.05 kcal) and these data were in agreement with those proposed by Zakpaa and al. (2010) and Adeyemi and Oladiji (2009).

Boiling may result in a 20 to 30% loss of vitamin C (FAO, 1990). Swaminathan and Gangwar (1961) had estimated that the losses are due to leaching into the cooking water and the rest to destruction by heat. Vitamin C is the most thermolabile vitamin and is also easily leached into cooking water. In the case of baked food, it is known that the higher temperature leads to greater destruction of the vitamin (FAO, 1990). Generally, nutritional qualities of plantain fruits vary with the cooking processing method employed (Baiyeri et al., 2011). As concerning mineral characteristics, in general, the results obtained are consistent with the report by FAO (1990), Adeyemi and Oladiji (2009), Baiyeri et al. (2011), Ankrah (1974) and Welford-Abbey et al. (1988).

Minerals are usually lost through leaching into water (Lopez et al., 1980; Ahenkora et al., 1997) but the leaching loss on boiling could be minimized if the skin were retained, as reported by True et al. (1979). Wrapping played almost the same role as the skin and so, most of the minerals were retained in the boiled dockounou. This explain the fact that, there are no significant different between boiled and baked dockounou mineral values. According to Finglas and Faulks (1985), there are no losses in the case of copper and zinc.

Minerals can be broken down into two basic groups based on their requirement: macro and micro minerals. The macro minerals such as calcium, magnesium, sodium, potassium and phosphorus are needed in fairly substantial amounts for proper health, whereas, the micro minerals are needed in far smaller quantities and include substances such as zinc, iron, copper and manganese (FAO, 1990). Though minerals are required in trace quantities, these trace minerals function in a wide and varied form to regulate, balance and prevent the body from certain disease and improper functions.

Conclusion

Baked dockounou had more conserved nutritional values than the boiled one with regards to protein, starch, carbohydrates, vitamin C and fat during the cooking time. But nevertheless, the main part of these parameters was maintained during the boiling because of the wrapping. Cooking water increased the moisture content of boiled dockounou and then reduced its shelf life. The nutritional composition of plantain products depended on the processing method employed. Samples analyzed were considered as good source of K, P and Mg. Zn and Mn are other minerals of nutritional importance in bananas and this study has shown that their average values are adequate to support the nutritive value of both products. The results obtained in this study showed that in spite of the manufacturing process, both dockounou products could be a potential source of mineral elements supplement in the diet.

REFERENCES

Adeyemi OS, Oladiji AT (2009). Compositional changes in banana (Musa ssp.) fruits during ripening. Afr. J. Biotech. 8(5):858-859.

Ahenkora KM, Kye A, Marfo K, Banful B (1997). Nutritional composition of false horn *Apantu pa* plantain during ripening and processing. Afr. Crop Sci. J. 5(2):243-248.

Akubor PI, Adamolekun FO, Oba CO, Obari H, Abudu IO (2003). Chemical composition and functional properties of cowpea and plantain flour blends for cookie production. Plant Foods Hum. Nutr. 58(3):1-9.

Ankrah EK (1974). Chemical studies of some plant wastes from Ghana. J. Sci. Food Agri. 25(10):1229-1232.

Anonymous (2011). Differences Between Plantain and Banana http://www.differencebetween.net/object/comparisons-of-food . March 2012.

AOAC (1980). Association of Official Analytical Chemists: Official Methods of Analysis, Horwitz W (Ed.). 13th Edn.

AOAC (1990). Association of Official Analytical Chemists. 16th ed. Arlington, Va, USA: AOAC.

APHA (1995). American Public Health Association. Standard methods for examination of water and wastewater. 19th ed. Washington DC.

Atanda SA, Pessu PO, Agoda S, Isong IU, Ikotun I (2011). The concepts and problems of post-harvest food losses in perishable crops. Afr. J. Food Sci., 5(11):603-613.

Atwater W, Rosa E (1899), A new respiratory calorimeter and the conservation of energy in human body, II-physical, 9:214-251.

Baiyeri KP, Aba SC, Otitoju GT, Mbah OB (2011). The effects of ripening and cooking method on mineral and proximate composition of plantain (*Musa* sp. AAB cv. 'Agbagba') fruit pulp. Afr. J. Biotech. 10(36):6979-6984.

Bakry F, Didier C, Ganry J, Le-Bellec F, Lescot T, Pinon A, Rey J.Y, Teisson C, Vannicre M (2002). Fruits species. In: CIRAD (ed). Guide of agronomist, Paris. pp. 960-974.

BIPEA (1976). Bureau international d'études analytiques. Recueil de méthodes d'analyses des communautés européennes. BIPEA, Gennevillier, pp. 51-52.

Coulibaly N (2008). Caractérisation physicochimique, rhéologique et analyse sensorielle des fruits de quelques cultivars de bananiers (Musa AAB, AAAA, AAAB). Thèse de Doctorat, UFR des Sciences et Technologies des Aliments, Université d'Abobo-Adjamé, Côte d'Ivoire, p.180.

Dadzie BK, Orchard JE (1997). Routine Post-Harvest Screening of Banana/Plantain Hybrids: Criteria and Methods *Inibap Technical Guidelines*, p.75.

Dongo RK, Dick E, Fatogoma S, Camara B, Kone D (2011). Preserving treatments effect on the physicochemical properties of the plantain stored at an ambient temperature. Agric. Biol. J. North Am. 2: 761-766.

Douglas HM, Sylvia MB, Turner BS, William HS (1996). Physiological and Chemical Changes during Ripening of Costa Rican Bananas Harvested in Different Seasons. J. Am. Soc. Hort. Sci., 121 (6):1157-1161.

FAO (1990). Food and Agriculture Organization of the United Nations. Roots, tubers, plantains and bananas in human nutrition, Rome,

FAO (2005). Food and Agricultural Organization. Food and agricultural indicators. ESSA Oct., 2005. FAO Rome. Http://www.fao.org/es/ess/top/country.html. Visited in 22/11/2009.

Finglas PM, Faulks RM (1985). A new look at potatoes. Nutr. Food Sci. (92): 12.

Frison E, Sharrock S (1999). The economic, social and nutritional importance of banana in the world. In: Picq, C., Foure, E., Frison, E.A (Eds). Banana and Food Security International Symposium, Douala, Cameroon, 10-14 November.

IITA (1998). International Institute of Tropical Agriculture. Plantain and Banana Improvement Program-Annual Report for 1997. International Institute of Tropical Agriculture, Onne, Nigeria, p.19.

Jaffe WG, Chavez JF, Keigman BD (1963). Concerning the nutrient value of bananas and plantain. Achives Venez Nutr., 13: 9-23.

Kouamé RN, Brahima C, Emmanuel D (2010). Evaluation des changements physico-chimiques caractérisant le mûrissement au cours de l'entreposage de trois variétés de bananes *Musa spp.* (AAB, cv. Corne 1 ; AAA, cv. Poyo et AA, cv. Figue Sucrée). Sciences et Nature, 7(2): 155-163.

Kuperminc O (1985). Saisonnalités de commercialisation de la banane plantain en Côte d'Ivoire, Fruits, 43 :359-368.

Lassois L, Jean-Pierre B, Haïssam J (2009). La banane : de son origine à sa commercialisation. Univ. Liège - Gembloux Agro-Bio Tech. Plant Pathology Unit. Passage des Déportés, 2. B-5030 Gembloux (Belgium). Biotech. Agron. Soc. Environ. 13(4), 575-586.

Lopez A, Williams HL and Cooler FW (1980). Essential elements in fresh and in canned sweet potatoes. J. Food Sci., (45): 675-678,681.

Marriot J, Lancaster PA (1983). Bananas and plantain. Chan, H. T. (ed.) In: Handbook of tropical foods. Marcel Dekker, New York, pp.85-143.

N'guessan A, Yao N, Kehe M (1993). La culture du bananier plantain en Côte d'ivoire. Spécial bananes II: systèmes de production du bananier plantain. Fruits 48(2):133-143.

Offem JO, Njoku PC (1993). Mineral distribution in the fruits of the plantain plant (*Musa paradisiaca*) in relation to mode and degree of maturation. Food Chemistry 48(1): 63-68.

Ogazi PO (1996). Plantain: production, processing and utilization. Paman and Associates publishers, Okigwe, Nigeria, pp.1-29.

René KD, Emmanuel D, Sorho F, Brahima C, Daouda K (2011). Preserving treatments effect on the physicochemical properties of the plantain stored at an ambient temperature. Agri. Biolo. J. North Am. 2(5): 761-766.

Séry D (1985). Le plantain en Côte d'Ivoire: production, transport et stockage, IRFA, document de reunion annuelle. p. 5.

Swaminathan K, Gangwar BML (1961). Cooking losses of vitamin C in Indian potato varieties. Indian Potato J. (3):86-91.

Swennen R (1990). Plantain cultivation under West African condition. A Reference Manual, International Institute of Tropical Agriculture (IITA), Ibadan, pp. 1-24.

Tillmanns J, Hirsch P (1932). Neber das vitamin C, *Biol. Ztschr.*, pp. 250-312.

True RH, Hogan JM, Augustin J, Johnson SR, Teitzel C, Show RL (1979). Changes in the nutrient composition of potatoes during home preparation III. Minerals. Am. Potato J., (56): 339-350.

Ukhum ME, Ukpebor IE (1991). Production of instant plantain flour, sensory evaluation and physico-chemical changes during storage. Food Chemistry, 42:287-299.

Welford-Abbey LI, Victor OTO (1988). Effect of ripening on the chemical composition of plantain peels and pulps (*Musa paradisiaca*). J. Sci. Food Agri. 45(4): 333-336.

Zakpaa HD, Al-Hassan A, Adubofour J (2010). An investigation into the feasibility of production and characterization of starch from "apantu" plantain (giant horn) grown in Ghana. Afr. J. Food Sci. 4(9):571-577.

6

Quantification of lactic acid bacteria and bifidobacteria in goat milk based yoghurts with added water-soluble soy extract

Danielle Cristina Guimarães da Silva, Danilo Florisvaldo Brugnera and Luiz Ronaldo de Abreu

UFLA, Lavras, MG, Brasil.

This work was carried out with the objective of evaluating the microbiological aspects, during 29 days of storage, through the quantification of viable cells and probiotic bacteria enumeration, of strawberry flavored yoghurts produced with goat milk with water-soluble soy extract and *Bifidobacterium animalis* ssp. *lactis* probiotic culture during processing. Determination of the viable cell count during yoghurt storage showed that the number of lactic bacteria found was below the expected only in the treatment in which there was water-soluble soy extract addition, not presenting microbiological viability. The other yoghurt treatments were viable during storage. The enumeration of *Bifidobacterium animalis* ssp. *lactis* showed that the yoghurt treatments presented microbiological viability during storage. We concluded that the addition of water-soluble soy extract interfered negatively with the production of goat milk based yoghurts.

Key words: Yoghurt, water-soluble soy extract, lactic acid bacteria, *Bifidobacterium*.

INTRODUCTION

Yoghurt is a very popular fermented milk product produced by lactic acid fermentation of milk by addition of a starter culture containing *Streptococcus thermophilus* and *Lactobacilus delbrueckii* ssp. *bulgaricus*. It is a very versatile product that suits all palates and meal occasions (Isleten and Karagul-Yuceer, 2006). Besides the traditional culture, strains of other probiotic organisms, such as lactobacilli and bifidobacteria, have been used in fermented products as potential health promoters. The supplementation of fermented products with probiotic bacteria becomes beneficial by providing better use of the lactose, anticarcinogenic activity and intestinal infection control. Probiotics are referred to as "live microorganisms, which when administered in adequate amounts confer a health benefit on the host" (FAO/WHO, 2001; Allgeyer et al., 2010).

Strains of *L. acidophilus* and of *Bifidobacterium lactis* predominate in commercial probiotic products (Tabasco et al., 2007). The presence of multiple species in these products makes the differential enumeration between probiotic bacteria and starter cultures necessary. Numerous means have been proposed for selectivity and differential enumeration of lactobacilus and bifidobacteria in bacterial population mixtures.

In relation to other types of milk, goat milk presents advantages such as smaller size fat globules, low allergenic properties (Martín-Diana et al., 2003), a balance of essential amino acids, high levels of calcium, selenium, phosphate and rich in vitamins A and B. However, goat milk is deficient in folic acid and vitamin D. Furthermore, a certain therapeutic value in human nutrition has been attributed to goat milk (Alférez et al.,

2006). The use of goat milk as an excellent food source is undeniable. It has beneficial effects on health maintenance, physiological functions, in the nutrition of children and elderly people, and according to some authors, can be consumed without negative effects of suffering from cow milk allergy. This highlights the market potential of goat milk (Ribeiro and Ribeiro, 2010).

The elaboration of the yoghurts based on goat milk, can present significant alterations in its rheological properties, such as low consistency and a tendency towards whey separation (syneresis). Therefore, to obtain satisfactory results with fermented goat milk products, the addition of stabilizers is recommended. The enrichment of the dry matter content and/or the protein content are standard measures used to avoid syneresis and to improve the texture of the yoghurt. As such, the soy proteins, specifically, the water-soluble soy extract (WSSE), deserves prominence to improve the nutritional value of the product and to affect the gel structure formation of the yoghurt (Silva et al., 2012).

The nutritional deficiency of goat milk can be improved by the lactic fermentation process. According to Hugenholtz (2008), many lactic bacteria seem to produce some vitamins, where the fermented product is enriched as a result of the bacterial production. The fermented milk products are reported as containing high amounts of folates, as a result of the additional folate production via bacteria. Therefore, from the choice of viable starter cultures, the natural fortification of a milk product can be undertaken. Silanikove et al. (2010) reported that the quality and safety of goat milk are optimized to ensure consumer confidence owing to the growing interest in existing and new goat dairy products worldwide.

Based on the above, this research had the objective of evaluating the growth of total lactic acid bacteria and *B. animalis* ssp. *lactis* in goat milk based yoghurts with added water-soluble soy extract and probiotic culture during 29 days of refrigerated storage.

MATERIALS AND METHODS

Characterization of the raw material

The milk used in the experiment was through the milking of Saanen breed females. After milking, the milk was cooled immediately to 5°C, transferred to previously sanitized polypropylene milk cans and transported to the dairy plant where the processes were conducted.

The physiochemical analyses were carried out with milk samples in triplicate, to proof its quality. It consisted of the determination of the pH by direct potentiometry in a digital pHmeter (Instituto Adolfo Lutz, 1985), titratable acidity, density, fat and total soluble solids percentage (Brazil, 2006).

Preparation of the yoghurt

The methodology used in this study for development of the yoghurt was based on that described by Tamine and Robinson (1991) as shown in Figure 1. The yoghurts were prepared and identified with letters according to their processing specifics (addition of WSSE adjusted to the protein level of the milk at a concentration of 20%, resulting in the supplementation of 14.8 g/L of water-soluble soy extract and *B. animalis* ssp. *lactis* probiotic culture (2%), as expressed below:

1. Yoghurt A: without the WSSE addition, without addition of probiotic culture.
2. Yoghurt B: addition of WSSE (20%), without the probiotic addition.
3. Yoghurt C: without addition of WSSE with added probiotic culture.
4. Yoghurt D: with WSSE (20%) addition with added probiotic culture.

After analysis of the milk quality, WSSE was added in B and D treatments. Later, all the treatments were pasteurized at 80°C for 30 min and then cooled to 43°C. The different combinations of bacterial cultures: *S. thermophilus*, *L. delbrueckii* ssp. *bulgaricus* and *B. animalis* ssp. *lactis,* were inoculated. The final fermentation point was determined when the yoghurts reached pH 4.6 and were then removed from the oven and cooled to 15°C for the addition of the strawberry pulp and stored at 4°C. The microbiological analyses of the products were conducted on the 1st, 8th, 15th, 22nd and 29th day, post-production.

Viable lactic bacteria count during storage at 4°C

The lactic bacteria count was conducted by the pour plate method, with an overlay, adding 1 mL of diluted inoculum and adding a small amount of MRS agar in the Petri dishes. After solidification of the medium, an overlay was added, seeking the creation of a 15% CO_2 atmosphere, followed by incubation at 30°C for 5 days. After the required incubation time, the count was conducted in Petri dishes that presented between 25 and 250 colonies (Tebaldi et al., 2007).

Probiotic count during storage at 4°C

For the enumeration of *B. animalis* ssp. *lactis* the MRS medium was used supplemented with 1% raffinose, 0.05% lithium chloride and 0.05% cysteine, sterilized by filtration in a 0.45 mm membrane. There was no use of antibiotics and the inoculation technique used was the pour plate, with overlay. After the inoculation, the Petri dishes were incubated inverted in jars containing anaerobiose generator (Anaerobac) at 45°C for 72 h (Tabasco et al., 2007).

Statistical analysis

The experiment was conducted in a completely randomized design (CRD), in a 2 x 2 x 5 factorial (2 soy extract concentrations, 2 treatments with probiotic culture addition at 5 storage times). Three repetitions were carried out for each treatment. The effects of the different treatments were appraised by variance analysis (ANOVA), followed by the Scott-Knott Test, to 5% of significance, to express the differences in significant cases. The evaluation of the storage time of the yoghurts was analyzed through linear regression, after transformation of the data into logarithms (log). The variance analyses, test of averages and linear regression were carried out in the R software (R Development Core Team, 2009).

RESULTS AND DISCUSSION

Characterization of raw material

The results of the physiochemical analysis of the goat

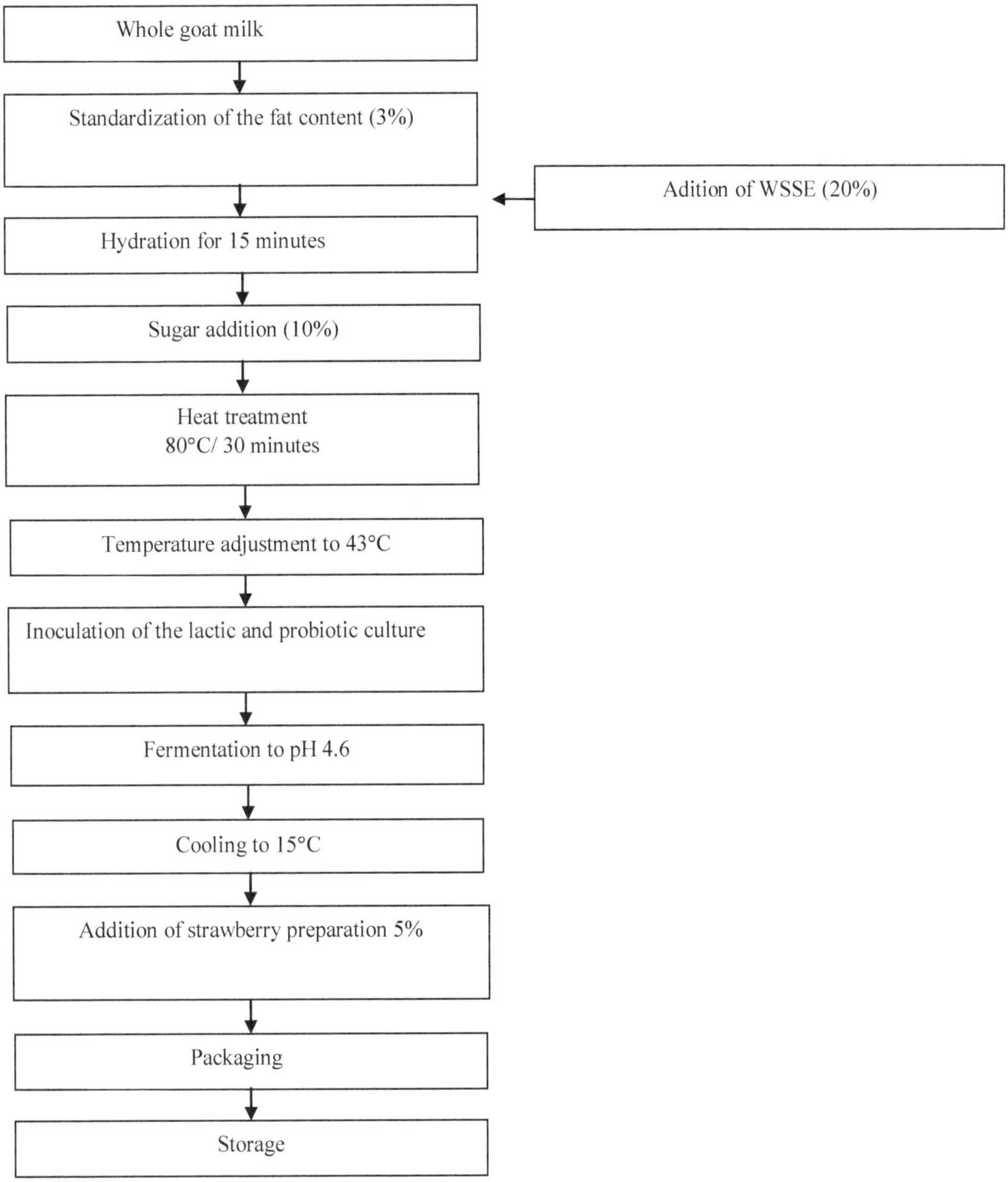

Figure 1. Flowchart of yoghurt preparation.

milk used in the yoghurt treatments are 6.68 from pH, 0.17 g lactic acid /100 mL of product from titratable acidity, 1.02 g L^{-1} from density, 3.1% from fat and 11.5% from total soluble solids. These results were similar to those found by Lora et al. (2006). The average value found for the goat milk acidity was 0.17 g lactic acid/100 mL of product. Such an average is within the standards required by the legislation (Brazil, 2000a).

The average fat value obtained in the raw milk by the Gerber method was above the minimum (3%) required by the legislation for whole milk, density of the goat milk was within the norms of the current legislation (Brazil, 2000a).

Viable lactic acid bacteria count

Table 1 presents the average values (CFU/mL) of the traditional lactic bacteria (viable cells of *S. thermophilus* and *L. delbrueckii* ssp. *bulgaricus*) counts of the strawberry flavor goat milk based yoghurts with added WSSE and of the probiotic culture during the storage period of 29 days at 4°C.

The viable lactic cell count remained 2.6 x 10^7 to 1.8 x 10^7 for the control yoghurt. The maintenance of the number of viable cells during the 29 day storage period under cooling at 4°C met the values established by the

Table 1. Average count of the number of viable lactic acid bacteria from goat milk yoghurt samples with added WSSE and probiotics during storage time (CFU/ml).

Yoghurt	(CFU/ml)				
	Day 1	Day 8	Day 15	Day 22	Day 29
A	2.6×10^{7} a	2.3×10^{7} a	1.8×10^{7} a	2.5×10^{7} a	1.8×10^{7} a
B	3.2×10^{6} b	2.2×10^{6} b	5.0×10^{6} b	4.1×10^{6} b	3.0×10^{6} b
C	9.0×10^{6} c	4.8×10^{6} c	1.15×10^{7} a	3.3×10^{7} a	3.1×10^{7} a
D	7.3×10^{6} c	5.0×10^{6} c	4.1×10^{6} b	8.3×10^{6} c	2.5×10^{6} b

Means followed by the same letter do not differ statistically among themselves, by Scott-Knott Test Average, to 5% probability. Treatment: A (without addition of WSSE and without addition of probiotic culture), B (with addition of WSSE and without addition of probiotic culture), C (without addition of WSSE and with addition of probiotic culture), D (with addition of WSSE and with added probiotic culture).

existing Brazilian legislation which, according to the Identity and Quality Standards (PIQ) of Fermented Milk, Resolution No. 5, November 13, 2000 (Brazil, 2000b), the total viable lactic bacteria count should be at least 10^{7} CFU/mL in the final product, during the entire validity period.

According to Silva (2007), studies have shown that the yoghurt bacteria (*S. thermophilus* and *L. delbrueckii* ssp. *bulgaricus*) survive well in the product during the shelf life. *L. bulgaricus* is the bacterium mainly responsible for the post-acidification of the yoghurts, a factor that can alter the microbial count; on the other hand, this microorganism contributed considerably to the production of aromatic compounds, especially acetaldehyde, characteristic of the yoghurt.

In yoghurt B, in which there was the WSSE addition, the viable lactic cell count remained between 2×10^{6} to 3.0×10^{6} CFU/mL. From these results we observed that the product did not meet the values established by the legislation (Brazil, 2000b), which establishes that the total viable lactic bacteria count should be at least 10^{7} CFU/mL in the final product, during the entire validity period. The excess lipids originating from the water-soluble soy extract can inhibit fermentation and microbial growth. This inhibition can occur due to the toxic effect of the unsaturated fatty acids on the bacterial cells. This toxic effect is associated with a change of the lipid composition and the physiochemical properties of the bacterial cell membranes (Kozloski, 2002).

The goat milk based yoghurt treatment codified by the letter C, with added *B. animalis* ssp. *lactis* probiotic culture, presented a sufficient count, being considered microbiologically viable according to the values established by the current law. Such law establishes that the total viable lactic bacteria count should be at least 10^{6} CFU/mL in the final product, during the entire validity period, when there is bifidobacteria inoculation in fermented milk. During the entire yoghurt storage period, the viable lactic cell count remained between 9.0×10^{6} and 3.1×10^{7}. In spite of the presence of multiple microbial species in a same food product leading to low microorganism counts, this was not observed in this study. According to Tamine and Robinson (1991), the pH value is involved with the metabolic activity of the bacteria, and may favor a particular group to the detriment of another.

The results of the determination of the viable lactic bacteria count during 29 days of storage show that the goat milk based yoghurt with added water-soluble soy extract and probiotic culture presented a count between 7.3×10^{6} and 2.5×10^{6}. Therefore, this product presented microbiological viability during the storage time studied, an appropriate count according to current law being observed, which establishes that the total count of viable lactic bacteria should be at least 10^{6} CFU/mL in the final product, during the entire validity period when there was inoculation beyond the traditional yoghurt bacteria, bifidobacteria.

In the evaluation of the yoghurts, it was observed that there was significant influence of the time and the treatments in the value of the viable cell count, according to the graph in Figure 2.

Studies have shown that the yoghurt bacteria (S. *thermophilus* and *L. bulgaricus)* survive well in the product during shelf life. According to Tamine and Robinson (1991), the pH resulting from the bacterial metabolic activity could favor a certain group, to the detriment of the other. In the case of yoghurt, the bacteria of the *Lactobacillus* genus grow and tolerate lower pH values than those belonging to the genus *Streptococcus*. The degree of fall of the viable cell numbers during the storage is quite variable, ranging from a slight decrease to a fall of 4 logarithmic cycles at the end of product storage (Vinderola and Reinheimer, 2000).

B. lactis culture count

The survival of the probiotic bacteria in fermented milk products depends on several factors, such as the lineage used, interaction among the species present, culture conditions, chemical composition of the medium (carbohydrate source), final acidity, milk solids content, nutrient availability, growth promoters and inhibitors, sugar con-

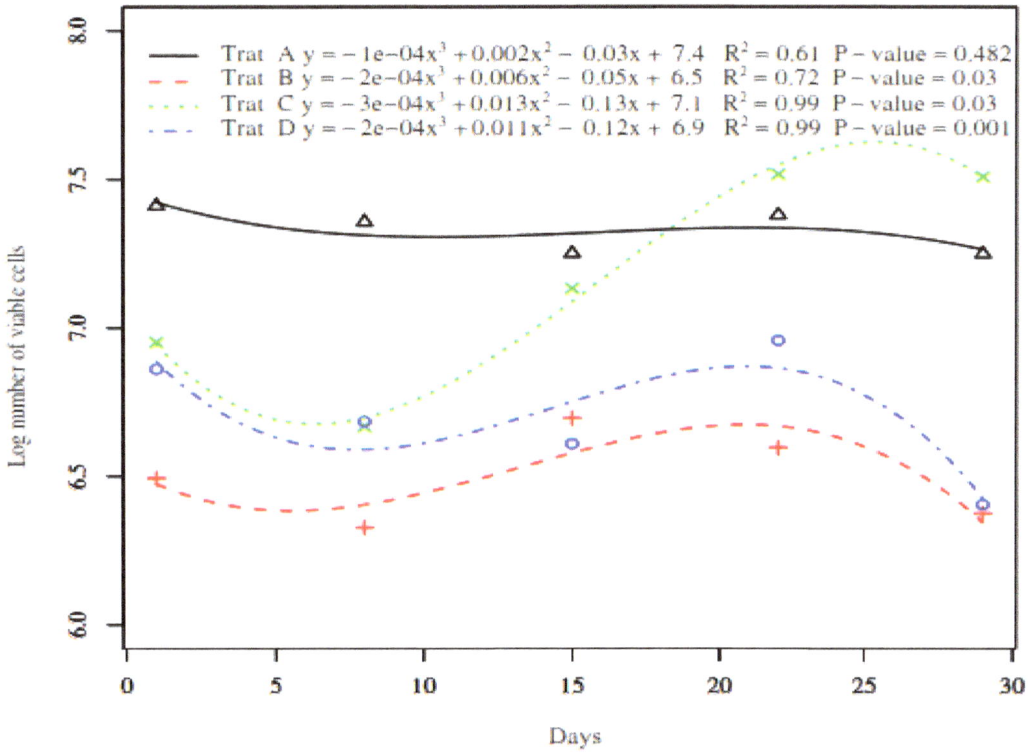

Figure 2. Regression model for the viable cell count (log CFU / mL) versus storage time at 4°C, in yoghurts. Treatment: A (without addition of WSSE and without addition of probiotic culture), B (with addition of WSSE and without addition of probiotic culture), C (without addition of WSSE and with addition of probiotic culture), D (with addition of WSSE and with added probiotic culture).

Table 2. Average viable cell count of *B. animalis* ssp. *lactis* in goat milk yoghurt samples with added WSSE and probiotic culture during the storage time (CFU/ml).

Yoghurt	CFU (ml)				
	Day 1	Day 8	Day 15	Day 22	Day 29
C	1.6×10^8 a	6.5×10^7 a	5.6×10^7 a	6.7×10^7 a	3.1×10^7 a
D	1.1×10^8 a	3.5×10^7 a	3.1×10^7 a	3.8×10^7 a	2.7×10^7 a

Means followed by same letter do not differ statistically among themselves, by Scott-Knott Test Average, to 5% probability. Treatment: C (without addition of WSSE and with addition of probiotic culture), D (with addition of WSSE and with added probiotic culture).

centration (osmotic pressure), dissolved oxygen (especially for the *Bifidobacterium* ssp.), amount inoculated, incubation temperature and storage temperature duration (Lourens-Hattingh and Viljoen, 2001; Silva, 2007).

The average enumeration of viable *B. animalis* ssp. *lactis* cell colonies in the yoghurt treatments can be seen in Table 2. The average viable cell count of the *B. lactis* probiotic microorganism remained between 1.6×10^8 and 3.1×10^7 CFU/mL for the yoghurt C, without WSSE addition, and the average count was 1.1×10^8 to 2.7×10^7 for the yoghurt D, with addition of water-soluble soy extract. From the ANOVA results, it can be observed that the average bifidobacteria count was not significantly influenced by the addition of soy extract. It is worth noting that the average viable cell count of the *B. animalis* ssp. *lactis* probiotic microorganism in the treatment without soy extract addition was slightly higher, however not significant, during the storage time of the yoghurts in comparison with the treatment with addition of this product. These results were contrary to the expected, because, according to Tamine et al. (1995), the soy contains prebiotics, such as rafinose and stachyose that are natural bifidobacteria growth promoters. It was

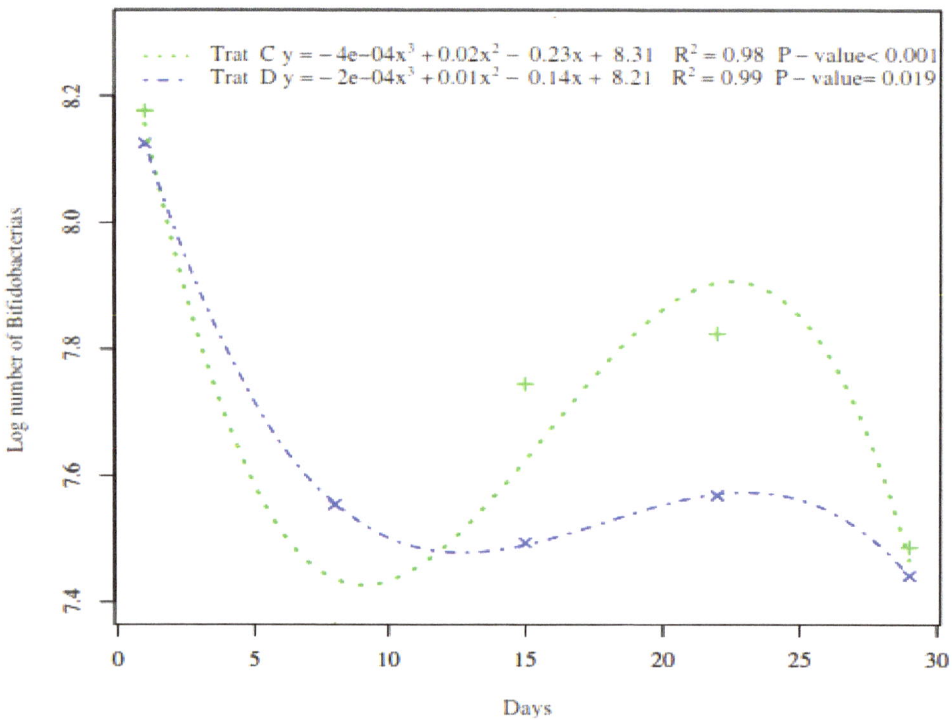

Figure 3. Regression model for bifidobacteria counts (log CFU/mL) in function of storage time at 4° C, in yoghurts. Treatment: C (without addition of WSSE and with addition of probiotic culture), D (with addition of WSSE and with added probiotic culture).

observed that the time influenced the bifidobacteria count in a significant way, according to the graph in Figure 3. The quality of a probiotic product is usually determined by the level, viability and amount of the probiotic cells in the food. This has been proposed as guarantee of the beneficial effects to human health (Schossler, 2009). Silva (2007) verified viability when evaluating species of *Bifidobacterium* and *L. acidophilus* probiotic bacteria in yoghurt development with the probiotic inulin.

According to Technical Regulation of Isolated Bioactive and Probiotic Substances with Allegation of Functional and/or Health Properties, Resolution RDC No.2, January 2002, probiotic is understood as live microorganisms capable of improving the intestinal microbial balance producing beneficial effects to the health of the individual (Brazil, 2002). Some benefits related to probiotic use are: stabilization of the intestinal function, reduction of infection by *Helicobacter pylori*, that is associated to gastritis and peptic ulcers and lactose intolerance symptoms relief. Other advantages in the use of these microorganisms are: the conservation of the milk through the production of lactic acid and possibly antimicrobial compounds; the production of flavor and aroma compounds and other metabolites that will provide a product with sensorial properties desirable to the consumer; the improvement of the nutritional value of the food, for instance, through the liberation of free amino acids or vitamin synthesis; the provision of therapeutic properties and special prophylaxis for cancer prevention and control of serum cholesterol levels. The potential benefits are resultant from the bacterial development and action during the fermented food production (Parvez et al., 2006).

Conclusions

This study showed that the viable cell count in the yoghurts was satisfactory without addition of water-soluble soy extract and also for those with added *B. animalis* ssp. *lactis* probiotic culture during the period appraised, therefore, the yoghurts were considered in accordance with the current law. However, the treatment with only added water-soluble soy extract was not shown viable according to the legislation; therefore it cannot be considered a yoghurt in that respect. The addition of the *B. animalis* ssp. *lactis* probiotic culture in the yoghurts with goat milk was shown viable during the storage time.

REFERENCES

Alférez MJM, López-Aliaga I, Nestares T, Díaz-Castro J, Barrionuevo M, Ros PB, Campos MS (2006). Dietary goat milk improves iron bioavailatability in rats with induced ferropenic anaemia in comparison with cow milk. Int. Dairy J. 16:813-821.

Allgeyer LC, Miller MJ, Lee SY (2010). Sensory and microbiological quality of yogurt drinks with prebiotics and probiotics. J. Dairy Sci. 93:4471-4479.

Brazil (2000a). Ministry of Agriculture, Livestock and Supply. Normative Instruction n. 37 of 08 November 2000. Technical Regulation identity and quality of goat milk. Official Gazette, Brasilia, 09 November.

Brazil (2000b). Ministry of Agriculture, Livestock and Supply. Standards of identity and quality of fermented milks. Official Gazette [of] the Federative Republic of Brazil. Brasilia, Resolution no. 5, DE13 November 2000b. 9 p. November 27. Section I.

Brazil (2002). National Agency for Sanitary Vigilance. RDC Resolution No. 2 of January 7, 2002. Creates the Technical Regulation of Bioactive Substances and Probiotics Isolates with Alleged Functional Properties and Health Official Gazette [of] the Federative Republic of Brazil, Brasilia, DF, 7, January.

Brazil (2006). Ministry of Agriculture, Livestock and Supply. Official analytical methods for physico-chemical control of milk and dairy products. Normative Instruction. 68 of December 12.

FAO/WHO (2001). Health and nutritional properties of probiotics in food including powder milk live lactic acid bacteria. Report of a Joint FAO/WHO Expert Consultation on Evaluation of Health and Nutritional Properties of Probiotic in Food Including Powder Milk with Live Lactic Acid Bacteria. WHO, Geneva, Switzerland.

Hugenholtz J (2008). The lactic bacterium as a cell factory for food ingredient production. Int. Dairy J. 18:466-475.

Instituto Adolfo Lutz (1985). Analytical standards of the Instituto Adolfo Lutz. 3.ed. São Paulo: IMESP, v. 1, 533p.

Isleten M, Karagul-Yuceer Y (2006). Effects of dried dairy ingredients on physical and sensory properties of nonfat yogurt. J. Dairy Sci. 89:2865-2872.

Kozloski GV (2002). Biochemistry of ruminants. Santa Maria: UFSM, 140 p.

Lora SCP, Prudêncio ES, Benedet HD (2006). Sensory evaluation of ice cream made with goat milk. Semina Ciênc. agrar. 27:221-230.

Lourens-Hattingh A, Viljoen BC (2001). Yogurt as probiotic carrier food. Int. Dairy J. 11: 1-17.

Martín-Diana AB, Janer C, Peláez C, Requena T (2003). Development of a fermented goat's milk containing probiotic bacteria. Int. Dairy J. 13:827-833.

Parvez, KA, Malik KA, Kang SA, Kim HY (2006). Probiotics and their fermented food products are beneficial for health. J. Appl. Microbiol. 100:1171-1185.

R Developmente Core Team (2009). R: a language and environment for statistical computing. Vienna: R Foundation for Statistical Computing.

Ribeiro AC, Ribeiro AC (2010). Specialty products made from goat milk. Small Rumin. Res. 89:225-233.

Schossler LS (2009). Feasibility study of probiotic microorganism (Bifidobacterium lactis) applied to cooked meat product. Dissertation (MSc in Food Science and Technology) - Federal University of Santa Maria, Santa Maria. p. 76.

Silanikove N, Leitner G, Merin U, Prosser CG (2010). Recent advances in exploiting goat's milk: Quality, safety and production aspects. Small Rumin Res. 89:110-124.

Silva DCG, Abreu LR, Assumpção, GP (2012). Addition of water-soluble soy extract and probiotic culture, viscosity, water retention capacity and syneresis characteristics of goat milk yogurt. Cienc. Rural 42:545-550.

Silva SV (2007). Development of probiotic yoghurt with prebiotic. Dissertation (MSc in Food Science and Technology) - Federal University of Santa Maria, Santa Maria, 2007. p. 106.

Tabasco R, Paarup T, Janer C, Pelaez C, Requena T (2007). Selective enumeration and identification of mixed cultures of Streptococcus thermophilus, Lactobacillus delbrueckii subsp. bulgaricus, L. acidophillus, L. paracasei and Bifidobacterium lactis in fermented milk. Int. Dairy J. 17:1107-1114.

Tamine AY, Marshall VMR, Robinson RK (1995). Microbiological and technological aspects of Milk fermented by Bifidobactéria. J. Dairy Res. 62:151-187.

Tamine AY, Robinson RK (1991). Yogurt: ciencia y tecnología. Zaragoza: Acribia, p. 368.

Tebaldi VMR, Resende JGOS, Ramalho GCA, Oliveira TLC, Abreu LRA, Piccoli RH (2007). Avaliação microbiológica de bebidas lácteas fermentadas adquiridas no comércio varejista do sul de Minas Gerais. Ciênc. Agrotec. 31:1085-1088.

Vinderola CG, Reinheimer JA (2000). Enumeration of Lactobacillus casei in the presence of L. acidophilus, bifidobacteria and lactic starter bactéria in fermented dairy products. Int. Dairy J. 10:271-275.

Effect of processing on the lectin and trypsin inhibitor content of *Plukenetia conophora* seeds as it affects growth performance and nutrients metabolism in rat

Kuku Adenike[1] , Alabi Mercy Olufunmilayo[1] and Ibironke Samson Ishola[2]

[1]Department of Biochemistry, Obafemi Awolowo University, Ile-Ife, Nigeria.
[2]Department of Food Science and Technology, Obafemi Awolowo University, Ile-Ife, Nigeria.

The study evaluated the effect of processing on lectin and trypsin inhibitor contents of *Plukenetia conophora* seeds and assessed such effect on growth and nutrient metabolism in rats. Mature conophora seeds were divided into three groups based on processing. The seeds were sun-dried and pulverized, and portions from each group were subjected to proximate composition analysis and measurement for the contents of trypsin inhibitor and lectin. The dried samples were incorporated into diets of experimental rats. Albino rats were randomly divided into four groups and fed with the control or experimental diets for a period of 21 days. During this period, feces and urine were collected, and body weights of the animals were measured daily. At the end of the experimental period, blood samples were collected from the rats for heamatological analysis, then the animals were sacrificed and some key organs were excised for histopathological analysis. The results showed that trypsin inhibitor and lectin contents of the raw seeds were higher than the under processed and processed seeds. All measured parameters including gain in body weight, feed utilization efficiency, nutrient digestibility, nitrogen balance, nitrogen retention and heamatological parameters were markedly different among the three groups of animals fed on diets incorporated with the conophora seeds, in comparison with the control animals. Histopathological analysis indicated that the lung and the spleen were adversely affected in animals treated with the under processed and the unprocessed seeds with the severity of toxicity more pronounced in the unprocessed group. These findings showed that *P. conophora* seeds have high nutritive value but ingestion of the unprocessed seeds impaired growth, affected nutrients utilization and caused damage to the lung and the spleen in experimental animals.

Key words: Conophor seeds, trypsin inhibitor, lectin, processing, metabolism, growth.

INTRODUCTION

Legume seeds have high acceptability and utilization as an important source of energy and protein for both animal and human consumption in the developing countries. *Plukenetia conophora* Müll.-Arg, formerly known as *Tetracarpidium conophorum* Hutch. and Dalz. is a woody perennial climber that belongs to the family Euphorbiaceae. It has a long history as food plant and is grown by peasant farmers across West African rain forest and in India (Petrova, 1980). Its common name is African walnut and it is cultivated principally for the nuts which are cooked and consumed as snacks. There have been reports on the high nutrient potentials of conophora nuts/seeds (Ogunsua and Adebona, 1983) and also on the impact of traditional processing on the nutrient and sensory qualities of the nut (Adesioye, 1991). However, a bitter taste is usually observed upon drinking water immediately after eating conophora nut. This has been attributed to the presence of alkaloids and some other antinutritional factors. Many legume seeds contain lectins which are toxic when being fed to humans or laboratory animals.

The existence of these antinutritional factors affects the nutritional value and digestibility, and has been a major limitation in the utilization of many of these unconventional protein-rich and high caloric seeds. Animal diets containing legume seeds with lectin constituents, for example, have been shown to impair the absorption of various nutrients such as amino acids, carbohydrates etc. and elicit a variety of biochemical and histological perturbations in laboratory animals (Pusztai et al., 1981). This has been attributed primarily to the binding of the lectins to the surface of the intestinal epithelial cells causing among other alterations, a non-specific interference with the final digestion and absorption of nutrients (Thompson, 1993; Thompson et al., 1986). There has been some information that lectins and other antinutritional factors in legume seeds may be inactivated by processing methods such as soaking, sprouting, cooking or fermenting although not completely (Pusztai et al., 1991). Many legume lectins are relatively resistant to both heating and digestion. Some of these lectins have a high thermal stability (70°C) and do not completely degrade with cooking while some are relatively resistant to stomach acid and proteolytic enzymes (Kilpatrick et al., 1985). Consequently, they are able to pass through the gut and about 1 to 5% are absorbed into the bloodstream in humans, a significant amount to cause an immune response (Pusztai et al., 1990). Therefore, lectins are a danger when consumed in their raw state (unprocessed), or by persons deficient in stomach acid or proteolytic enzymes.

Reports from studies over the years have also looked at trypsin inhibitors in some food items such as grains, nuts, seeds, fruits and vegetables (Daniel, 2005). Proponents of plant-based diets generally believe these diets provide plenty of protein, but failed to take into account the fact that protein ingested is not the same as protein digested when protease inhibitors are in the picture. Without high quality and proper bioavailability of protein, growth, repair, immunity, hormone formation and all metabolic processes suffer. These protease inhibitors found in foods are resistant to inactivation by most processing methods, only the old-fashioned fermentation techniques come close to deactivating them while with all other processing methods, they remain active. Even small quantities of these protease inhibitors in diet can adversely affect people whose digestive capacities are already compromised by low acid levels, pancreatic insufficiency, bowel diseases, gluten intolerance and other health challenges (Rackis and Gumbman, 1981).

Protease inhibitors keep the pancreas from producing enough proteases, but the body compensates by increasing the number of pancreatic cells (hyperplasia) and their size (hypertrophy). Over time, the pancreas rises to the challenge and then recovers; however, when the pancreas is stressed, pancreatitis and even cancer become distinct possibilities (Liener, 1995). There were also reports on the chemical and functional characteristics of conophora nuts and the effect of processing on some of the antinutritional factors (Enujiugha, 2003).

In traditional medicine, the seeds are considered as herbs and are used to tonify kidneys, strengthen the back and knees, moisten the intestines as well as to motivate bowel movement by the use of concotions made from the raw seeds (Ayoola et al., 2011). It is also possible that the conophora seed may be utilized for animal feeds because of its ready availability and cost. By incorporating the seeds (processed and unprocessed) into animal diets, it is of interest to assess the effect of ingestion of the seeds on the growth performance and nutrient metabolism in rat as well as to determine if the effects could be ameliorated by processing such as cooking.

MATERIALS AND METHODS

Seed collection and processing

Mature unshelled conophora seeds (Figure 1) were purchased from Better Life Market in Ile-Ife in the month of June. The seeds were authenticated at the IFE Herbarium, Department of Botany of Obafemi Awolowo University. The seeds were divided into three groups. Group A was the raw uncooked seed (unprocessed), group B was heat- treated at 60°C for 45 min (under processed) and group C was cooked for 2 h at 100°C (processed). Seeds from each group were carefully cracked, sliced separately to about 1 to 2 cm thickness using a kitchen knife with steel blade, sun-dried and milled into powder using Warring blender. The dried sample powders were stored in screw-capped bottles at 4 to 6°C until use.

Proximate composition analysis of seed samples

The proximate composition of the seed powder was determined according to Official Methods of Analysis (AOAC, 1995). The dry matter was determined by oven-drying at 105°C for 16 h; crude fat by petroleum ether extraction, crude fiber by digestion with H_2SO_4 and NaOH and the crude protein by Kjeldahl method. Nitrogen free extract (CHO) was calculated by taking the sum of values for crude protein, crude fat, ash, crude fiber and subtracting from 100. Analyses were done in triplicates.

Analysis of antinutrients

The content of trypsin inhibitor was determined by the procedure of Smith et al. (1980). One gram of the dried seed sample was extracted by shaking with 50 ml of 10 mM NaCl and left overnight at 4°C. The pH of the resulting slurry was adjusted from 9.4 to 9.6 with 1M HCl or 1M NaOH, and left overnight. The extract was diluted with water so that 1 ml produced trypsin inhibition of between 40 and 60%. Bovine trypsin (20 µg/ml) was used as standard. The assay was carried out as follows:

Figure 1. *Plukenetia conophora* seeds released from the pod.

Reagent blank: 2.0 ml distilled water, 2.0 ml of standard trypsin,
Sample blank: 2.0 ml diluted sample extract, 2.0 ml distilled water,
Sample(s): 1.0 ml diluted sample extract, 1.0 ml distilled water, 2.0 ml standard trypsin.

After mixing and pre-heating to 37°C for 10 min, 5.0 ml benzoyl-DL-arginine-P-nitroanilide (BAPNA) solution (previously warmed to 37°C) was added to each tube, mixed and incubated for 10 min at 37°C. 1 ml acetic acid (30% v/v) was then added to stop the reaction. The solution in each tube was filtered and the *p*-nitrophenol released was measured spectrophotometrically at 410 nm.
Trypsin inhibitor activity (TIA) was expressed as:

$$\frac{2632 \times D \times \Delta A}{S} \text{ mg trypsin inhibited g}^{-1} \text{ sample}$$

Where: D = dilution factor

S = sample (gram)

ΔA = change in absorbance = A_{sample} - A_{blank}

Where: D is dilution factor; S is sample (gram); ΔA is change in absorbance = A_{sample} - A_{blank}.

The lectin content in the saline extract of the seed was determined by hemagglutination assay according to the method of Kuku and Eretan (2004). The concentration of protein in the extract was measured by the method of Lowry et al. (1951). The hemagglutination unit of activity (HU) was taken as the reciprocal of the highest dilution (titre) of the extract showing visible agglutination of erythrocytes (rabbit). Hemagglutinating activity was expressed as HU/mg protein. All determinations were carried out in triplicates.

Bioassay

The dry seed sample powders (unprocessed, under processed and processed) were incorporated into the diets of the experimental animals. The diets were formulated using the American Institute of Nutrition (AIN) method as described by Reeves et al. (1993). The diet originally contained 50% maize starch, 10% potato starch, 15% corn oil, 5% mineral mixture and 5% vitamin mixture. The test and control diets were formulated by substitution of maize starch with the amount of a particular protein (casein) to give 10% protein requirement. Silicic acid was added to mimic animal food. The composition of the diets is shown in Table 1. Albino rats weighing 80 ± 5 g (six weeks old) bred from the same colony were obtained from the Animal House, Faculty of Pharmacy of the Obafemi Awolowo University. The rats were maintained on a standard stock pellet diet and then fed with the casein control diet for a period of one week and allowed to acclimatize under a controlled atmosphere (temperature, relative humidity and a fixed 12- h light/dark cycle). Only those rats which had a regular food intake and matched on the basis of body weight during the adaptation period were subsequently used in the experiment.

For this experiment, twenty rats were randomly divided into four groups of five animals each and were individually housed in Techniplast metabolism cages (Biotech, Clackmannanshire) fitted with a feeding tunnel to prevent food spillage and ensure minimal

Table 1. Composition of diets (g/kg).

Ingredient	Control diet	Diet incorporated with processed seed	Diet incorporated with under processed seed	Diet incorporated with unprocessed seed
Casein	125	-	-	-
Conophor seed	-	350	391	450
Corn starch	375	240	130	100
Glucose	150	150	150	150
Corn oil	150	60	130	175
Potato starch	100	100	100	100
Vitamin/mineral premix	100	100	100	100
Silicic acid	0.4	0.4	0.4	0.4

contamination of the feces and urine samples with food. The rats were fed on different diets as follows: group 1 was fed on a test diet incorporated with raw (unprocessed) seed powder; group 2 was fed on diet containing under processed powder and group 3 on diet incorporated with the processed seed powder. A control group was fed on the reference caseinated diet formulation (200 g casein/kg). The diets were isoenergetic and were given in powdered form for a period of 21 days. Water was available *ad libitum* during this period, meanwhile, the care and use of laboratory animals followed the institutional guidelines of Obafemi Awolowo University, Ile-Ife.

Food consumption was recorded daily. Body weight was recorded daily for rats from each group before feeding in the morning. To determine the nutritional efficiency of diets, urine and feces were collected daily. All samples were stored at -4°C. At the end of the experimental period, blood samples were collected from the rats for hematological analysis. The animals were sacrificed by cervical dislocation. The gastrointestinal tract was thoroughly rinsed out with a large amount of distilled water to remove food and feces. Other organs including spleen, heart, kidney and liver were excised and fixed immediately in 10% formyl saline for histological analysis.

Growth performance assay

Body weight and feed consumption were recorded in rats before feeding in the morning. Body weight gain and feed conversion ratio (Fcr) were calculated.

Nutrient digestibility determination

The samples of experimental diets or feces were homogenized using a mortar and pestle and analyzed by standard methods of the AOAC (1995). The nutrient digestibility was measured by the method of Noreen and Salim (2008). The dry matter was determined by oven-drying at 105°C for 16 h; crude fat by petroleum ether extraction, crude fiber by digestion with H_2SO_4 and NaOH and the crude protein by Kjeldahl method. Nitrogen free extract (CHO) was calculated by taking the sum of values for crude protein, crude fat, ash, crude fiber and subtracting from 100. Analyses were done in triplicates.

Nitrogen balance and nitrogen retention assay

Intake nitrogen, fecal nitrogen and urinary nitrogen were analyzed as total nitrogen by Kjedahl method. Nitrogen balance and nitrogen retention were calculated according to the following formulas:

Nitrogen balance = intake nitrogen - urinary nitrogen - fecal nitrogen
Nitrogen retention = (intake nitrogen - urinary nitrogen - fecal nitrogen)/intake nitrogen

Haematological and histopathological analysis

The packed cell volume (PCV), hemoglobin concentration and white blood cell (WBC) count of the blood samples collected from both control and experimental animals were determined using standard hematological methods (Coles, 1986). The organs collected from the animals including spleen, heart, lungs, kidney, liver and small intestine were fixed in 10% formyl saline for 24 h, dehydrated in ascending concentration of ethanol (50%, 70%, 90%, then twice in 100%) for interval of 1 h to enable the tissue to be embedded in paraffin (Drury and Wallington, 1980). The tissues were sectioned to 6- micron thin films using a rotary microtome and stained with Hematoxylin and Eosin, which were then examined with a zeiss EM light microscope.

Statistical analysis

Data were expressed as mean ± SEM using Graph Pad Prism Graphical –Statistical package version 5 (30 days demo version). Statistical analysis was performed using ANOVA, followed by significant difference test for comparisons between individual groups. The non-parametric Dunnett Comparison Test was applied to discriminate differences in variables with 5% level of significance ($p<0.05$).

RESULTS

Proximate composition of conophora seeds

The proximate chemical composition (% dry weight) of freshly harvested conophor nuts is presented in Table 2. The crude protein content of the processed seed (28.62%) was higher than that of the unprocessed seed (22.24%). The values for the crude fat, crude fibre and ash content of the processed seeds were higher than those of the unprocessed seeds. However, the carbohydrate (CHO) level and the ash content of the unprocessed seeds were higher than those of the processed seeds.

Table 2. Proximate composition (% dry weight) of conophor seeds.

Seed	Protein	Moisture	Fat	Ash	Crude fiber	CHO	Dry matter
Un processed	22.24± 1.75	15.85± 0.87	38.90±1.92	2.55± 1.45	2.18±0.17	20.29± 0.82	84.15±0.13
Under processed	25.56± 1.00	21.70± 0.50	30.02± 1.75	2.20 ±0.17	1.90± 0.42	14.00±0.25	70.68±0.17
Processed	28.62 ± 0.05	30.04 ± 0.75	27.46± 0.62	2.81± 0.32	1.59± 0.01	11.57 ±1.25	69.96± 0.90

Values are the means of triplicate determinations expressed on a dry weight basis.

Table 3. Effect of processing on trypsin inhibitor and lectin content of conophor seeds.

Conophor seed	Trypsin inhibitor (mg/g)	Heamaglutinating activity (HU/mg)
Unprocessed	7.32 ± 2.20^{c}	2048
Under processed	4.72 ± 0.10^{b}	64
Processed	2.61 ± 0.07^{a}	2

Values are the means of triplicate determinations. Means bearing different superscripts are significantly different ($p < 0.05$).

Content of trypsin inhibitor and lectin in conophora seeds

Table 3 shows significant differences in the content of trypsin inhibitor and lectin among the processed and unprocessed seeds. The content of trypsin inhibitor in unprocessed seed (7.32 mg/g) was highest compared with that of under processed and processed seed. Processing brought about a reduction in the trypsin inhibitor content of conophora seeds. The lectin content expressed in HU/mg protein of unprocessed seeds (2048) was higher than that of under processed (64), and was extremely low for processed seeds (2).

The effects of conophora seeds on growth performance

The body weight and feed conversion ratios (Fcr) were significantly reduced ($p < 0.05$) in the animals fed with diets incorporated with under processed and unprocessed seeds as shown in Figures 2A and B. There was no body weight gain in the group fed on diet with unprocessed conophora seed rather there was weight loss, while that of the group fed on diet with processed seed was higher than those of the under processed group. Feedstuff by gain weight ratio in processed group was the highest and was significantly different from other treatment groups ($p < 0.05$).

The effect of conophora seeds on nutrient digestibility

The nutrient digestibility was lower in the experimental diets compared to the control group (Figure 3A). The digestibility of dry matter, protein and carbohydrate were lower in unprocessed group and higher in processed group with significant differences between the values ($p < 0.05$). There was, however, a significant increase in fat digestibility, which was highest in unprocessed group and lowest in the processed group ($p < 0.05$).

The effect of conophora seeds on nitrogen metabolism

The nitrogen balance and retention were significantly lower in treatment groups compared with the control group as shown in Figure 3B and C. The striking difference in these parameters in the three treatment groups could be attributed to the different levels of antinutritional factors. Nitrogen balance and nitrogen retention among treatment groups were highest in the processed group and the lowest in the unprocessed group, and were significantly different from the control and the under processed groups ($p < 0.05$).

The effect of conophora seeds on blood parameters

Table 4 shows the results of the packed cell volume (PCV), haemoglobin concentration (Hb) and white blood cell count (WBC) in rats in both control and treatment groups. There was a significant difference between the control and treatment groups ($p < 0.05$). The values were highest in processed group and lowest in group fed diet incorporated with unprocessed seed.

Histophatological analysis

Histopathological analysis revealed that only the lung, spleen and the small intestine of the animals in the treatment groups were affected while the liver, the heart and the kidney were essentially normal.

Figures 4, 5 and 6 show the photomicrographs of sections of the lung, the spleen and the small intestine of the rats in the control and treatment groups. In rats fed on

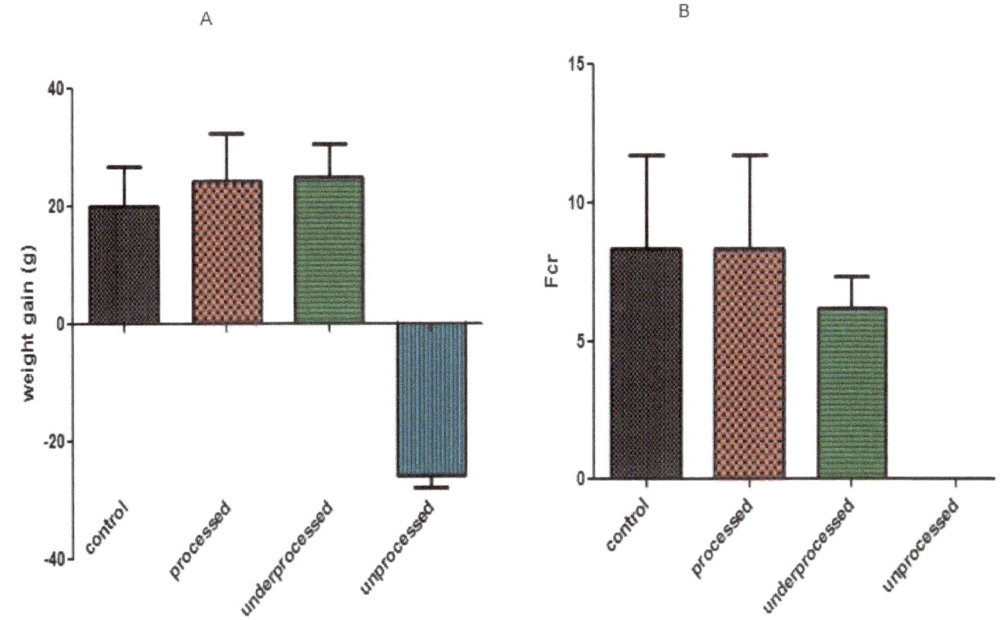

Figure 2. (A) Weight gain and (B) feed conversion ratio (Fcr) in rats fed diet incorporated with *P. conophora* seeds.

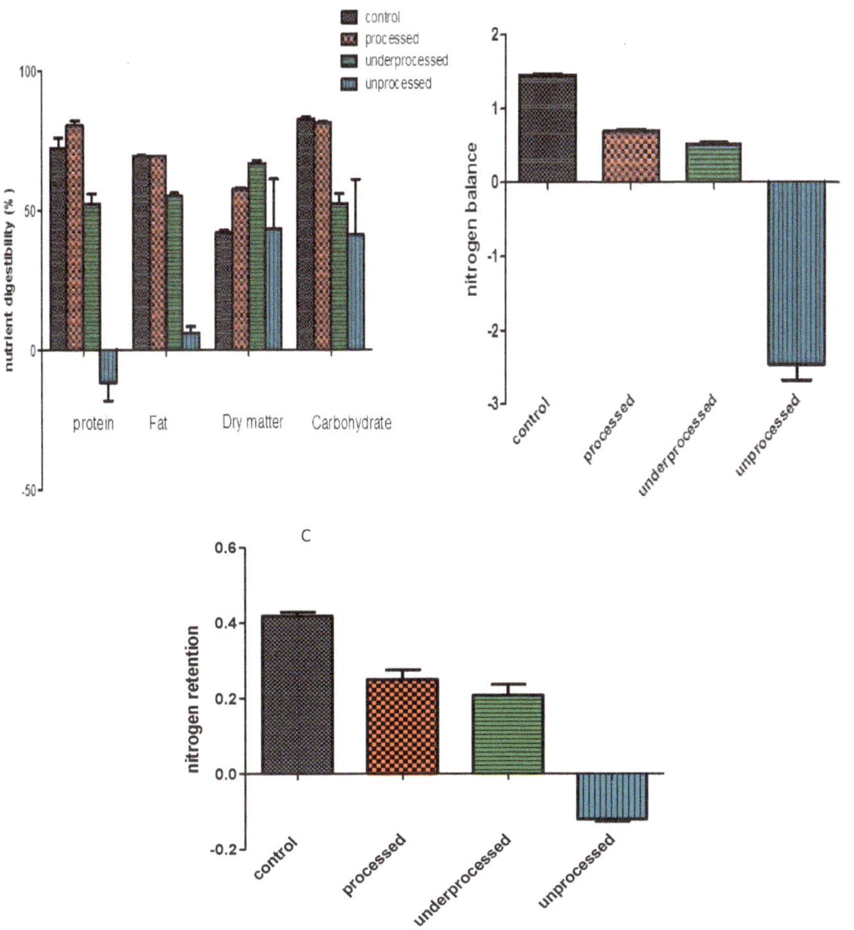

Figure 3. Nutrient digestibility (A), Nitrogen balance (B) and Nitrogen retention (C) in rats fed diet incorporated with *P. conophora* seeds.

Table 4. Hemoglobin, packed cell volume and white blood cell count of control and test animals.

Group	Hb (g/dL)	PCV (%)	WBC (10^9/L)
Control	12.76 ± 0.53^a	38.40 ± 1.60^a	10280 ± 1860^a
Unprocessed (raw)	4.3 ± 1.76^b	11.00 ± 4.59^c	660 ± 293^d
Underprocessed	11.08 ± 0.85^a	33.20 ± 2.56^b	4690 ± 517^c
Processed	11.52 ± 1.29^a	34.60 ± 3.89^a	7520 ± 2032^b

Values are expressed as Mean ± SD (n = 5). Means bearing different superscripts on the same column are significantly different (p< 0.05).

diet incorporated with unprocessed seeds, there was a reduction in the respiratory portion of the lung section due to the thickening or inflammation of the inter alveolar septum as compared with the lung of animals fed on normal diet in which respiratory bronchioles were well defined with numerous alveoli and alveolar sacs appearing on the pulmonary parenchyma giving the lung a spongy appearance. The spleen of rats fed on diet with the unprocessed seed showed highly reactive splenic follicles marked by reduction in the number of lymphoid aggregation of white pulp and several lymphocytes are scattered in the red pulp while in the control animals, the white pulp are discrete white nodules embedded in a highly vascularized red matrix (red pulp).

The transverse section of the jejunum of rats fed with unprocessed seeds showed degenerative changes in the mucosa and the muscular walls with most of the intestinal villi eroded together with the crypts of lieberkuhn as against the architecture of a normal jejunum observed for the control animals.

DISCUSSION

The proximate composition of food crop is a major index of the nutritional potential of that crop. The proximate composition of *Plukenetia conophora* seeds obtained in this study differed from those reported by other workers. Enujiugha and Ayodele-Oni (2003) reported *P. conophora* seeds to have 29.09% protein, 48.90% oil, 12.58% carbohydrate and 6.34% fibre. Odoemelam (2003) reported 26.3% protein and 46.5% fat. The FAO (2000) reported that the raw seeds contained 22.7% crude protein, 56.0% fat (ether extract), 3.70% fibre and 9.10% carbohydrate. The results showed that conophor nut is an oil seed with high oil content and adequate protein to satisfy the calorie and protein needs of the consuming populations. The results of this study showed 22.24% protein, 38.90% fat, 20.29% carbohydrate and 2.18% crude fibre. These values were lower than those of earlier studies. These differences may be attributed to distinct genetic varieties, or accompanied with climatic, environmental and geographical factors (Wang et al., 2000). The study also showed that processing by cooking had a marked effect on the content of antinutritional factors like trypsin inhibitor and lectin. The content of antinutritional factors in unprocessed (raw) seed was significantly higher (p< 0.05) than in the under processed and was lowest in the processed seed. Many studies have confirmed that daily gain in body weight and feed utilization efficiency are lower in animals fed with diets containing trypsin inhibitor and lectin (Liener, 1996).

In the present study, such two characteristic parameters of rats fed on diet incorporated with raw unprocessed seed were significantly lower than those fed on the under processed seed, while those animals fed on the processed seed had the highest values. One of the possible reasons for this result is that trypsin inhibitor leads to the loss of endogenous nitrogen, and the other is that lectin could combine with small intestine epithelium and induce constitutional and functional changes in the small intestine (Pusztai et al., 1991). Lienier (1996) reported that 50% of growth inhibition in rats fed with raw soya bean was attributed to lectin, 40% to trypsin inhibitor and 10% to other antinutritional factors. Schulze et al. (1993) discovered that by adding 2.4 g/kg units of Trypsin Inhibitor (TI) to a control diet, reduction in the growth of animals by 13% was observed and it was about 3-fold lower (32%) with the addition of 7.2 g/kg TI. This indicated that the effect of trypsin inhibitor on animal growth was related to its level in the diet. Grant et al. (1995) also reported that the effect of lectin on growth performance of animals changed with its dosage. Li (2003) observed no obvious change of growth performance in rats fed on diet containing lectin in the range of 0 to 1.2 mg/g within 20 days, but the growth performance was lower to a decrease by 23% compared with control rats from the effect of diet having lectin above 2.0 mg/g. In the present study, the results on body weight gain together with the feed utilizetion efficiency of animals interpreted that the higher the antinutritional factor content was, the lower the body weight gain and feed utilization efficiency were. It further demonstrated that heat treatment of *P.* conophora seed had such a remarkable effect on the trypsin inhibitor and lectin that it led to significant differences in body weight gain and feed conversion ratio in the treatment groups.

Digestion and absorption of nutrients could be measured by two parameters, including nutrient digestibility and deposit. In the present study, feeding on raw unprocessed seed diet caused a significant decrease in dry matter, protein and carbohydrate digestibility. Similar results

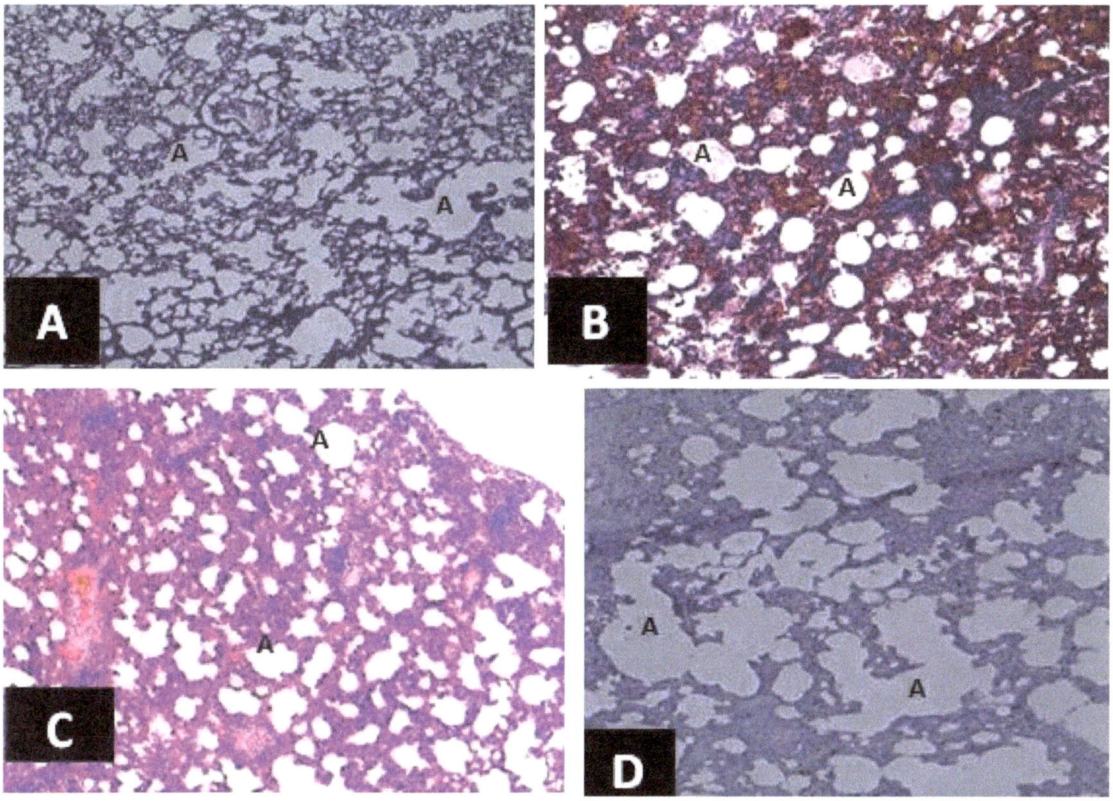

Figure 4. Photomicrographs showing transverse sections of the lung of the control and experimental animals; A (Control): The transverse section of the lung of a rat fed with the control diet. The respiratory bronchioles are well defined. Numerous alveoli (A) and alveolar sacs appear on the pulmonary parenchyma giving the lung a spongy appearance (x 100); B (Unprocessed): The transverse section of the lung of rat fed diet incorporated with the unprocessed conophora seeds. There is a reduction in the respiratory portion of the lung section due to the thickening or inflammation of the inter alveolar septum. (x 100); C (Under processed): The transverse section of the lung of rat fed diet incorporated with the under processed conophora seeds. The lung shows morphological similarities to the lung of the rat fed with the unprocessed seed (x 100); D (Processed): The section through the lung of rat fed with diet incorporated with the processed conophora seeds. The lung shows morphological similarities to the lung of control rat (x 100).

were previously reported (Qin et al., 1996). Nutrient digestibility was decreased in animals such as pig, chicken, rat, mouse and so on, fed with trypsin inhibitor, and the decrease of protein digestibility was more pronounced (Chunmei et al., 2010; Gilani et al., 2012). Li et al. (1996) reported that protein digestibility was decreased by 20 to 40% in animals fed with diets containing high levels of trypsin inhibitor (raw soybean) compared with those fed with diets containing such a factor at low levels (heated soybean or soybean meal). This study found that protein digestibility was significantly decreased in the animals fed diet containing the under processed and unprocessed (raw) seed compared with the control group. This could be explained by the findings of Qin (2003) that lectin could combine with a specific receptor (polyose) of the epithelial cell surface in the small intestine wall, destroying the brush border mucosa structure of the small intestine, interfering with the function of many enzymes in the brush border mucosa, so as to decrease protein utilization efficiency. Previous study had also indicated the effect of lectin on nitrogen metabolism (Li et al., 1998).

The effect was found to be obvious, and lectin mainly increased the effluence of endogenous nitrogen in the way that it lowered nitrogen balance and nitrogen retention. The increase of lectin content in diet resulted in a continuous increase in nitrogen loss, and a continuous decrease in the nitrogen balance and nitrogen retention. The level of trypsin inhibitor also affects these two parameters. Hagenleister and Barth (1993) reported that with an increase of trypsin inhibitor content in the diet, endogenous nitrogen in chymus was obviously increased whereas nitrogen balance and nitrogen retention were obviously decreased. These results indicated that trypsin inhibitor and lectin could hamper nitrogen deposit, thereby affecting digestion and absorption of the nutrient. In the present study, similar results were obtained. Trypsin inhibitor and lectin led to a significant decrease of

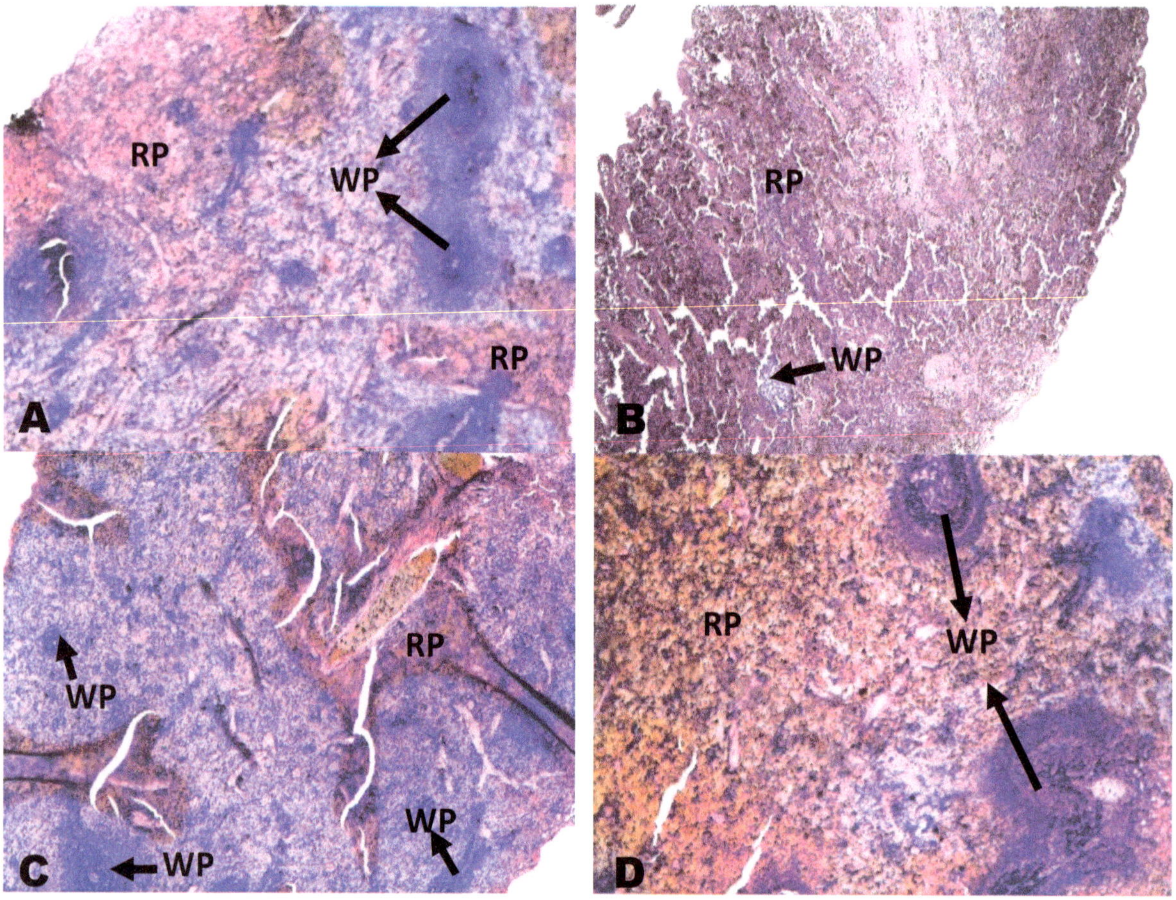

Figure 5. Photomicrographs showing sections of spleen of the control and experimental animals; A, A section of the spleen of the control rat. Note the white pulp (WP) which are discrete white nodules embedded in a highly vascularized red matrix called red pulp (RP) (x100); B, A section of the spleen of the rat treated with unprocessed conophora seed. The section showed highly reactive splenic follicles marked by reduction in the number of lymphoid aggregation of white pulp and several lymphocytes are scattered in the red pulp (x100); C, A section of the spleen of the rat treated with under processed conophora seed. There are few number of small sized white pulp (WP) embedded in the vascularized red pulp (RP). A few lymphocytes are also seen scattered in the red matrix (x 100); D, A section of the spleen of the rat treated with processed conophora seed. Note the white pulp (WP) consisting of lymphoid aggregations embedded in the vascularized red pulp (RP) (x 100).

nitrogen balance and nitrogen retention in animals in the treatment groups compared to those in the control group. The extent of the decrease in the raw unprocessed group was the highest, while that of the processed group was the lowest. These data showed that the higher the level of trypsin inhibitor and lectin content in the diet, the lower the nutrient digestibility, nitrogen balance and nitrogen retention.

A progressive decrease in blood parameters was observed for the different treated groups. Rats fed on diets incorporated with unprocessed seeds had the lowest values in the PCV, Hb concentration and WBC count. Hematocrit in terms of PCV measures the percentage of red blood cells of total body blood and a decrease in PCV is consequently an indicator of anaemia (Kjeldsberg, 2000). The PCV levels of rats treated with processed seed diet were within the normal range as those of normal control diet. This suggested that the rats had adequate Fe and other nutrients needed to maintain homeostasis of hemoglobin. However, there was significant decrease in the PCV of animals fed with diet comprising of under processed and unprocessed seeds. The total white blood cell count and the differential count helps to determine the state of the immune system of an organism. Disruption in the normal values predisposes one to pathogenic invasion. The reduction in WBC count in the rats fed with unprocessed seeds may be due to the damage to the spleen, thereby affecting the immune system (Walker et al., 1990). The abnormal spleen morphology of the animals corroborated the consequence of markedly low WBC count. The histopathological analysis showed inflammation of the lungs of rats fed with under processed and processed conophora seeds. The alveolar wall thickening may be caused by emphysema, edema,

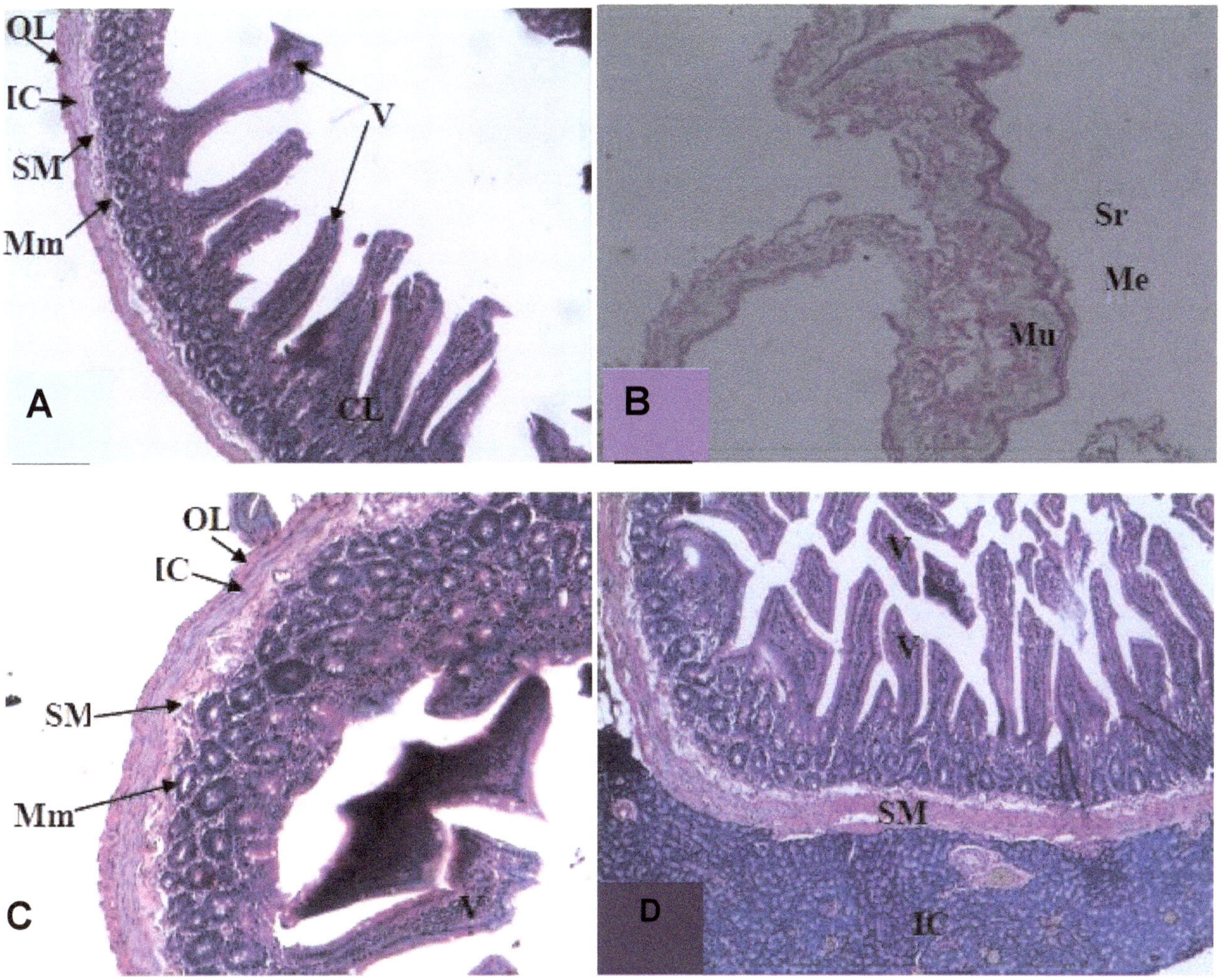

Figure 6. Photomicrographs showing sections of small intestine of the control and experimental animals; A (Control): The transverse section of the small intestine (jejunum) of normal control rat showing the general architecture of a normal jejunum. The intestinal villi (V) are lined by a simple columnar epithelium which is continuous with that of the Crypts of Lieberkuhn (CL). The muscularis mucosa (Mm) lies immediately beneath the base of the crypts. Also note the inner circular (IC) and outer longitudinal (OL) layers of the muscularis externa. The peritoneal aspect of the muscularis is invested with serosa (Sr) (x 100); B (Unprocessed): The transverse section of the jejunum of rats fed with unprocessed conophora seed. The wall of the jejunum is very thin as degenerative changes appear in its sub mucosa (SM) and the muscular wall. Most of the intestinal villi are eroded together with the crypts of lieberkuhn (x 40); C (Under processed): The transverse section of the jejunum of rats fed with under processed conophora seed. The intervilli distance is increased in this section indicating erosion/degeneration of some villi. The sub mucosa, crypts and muscular layer appear normal (x100); D (Processed): The transverse section of the jejunum of rats fed with processed conophora seed. The section appears normal (x 100).

interstitial infiltration with neutrophils and macrophages, as well as air-space cellularity (Vadiviel et al., 2010). This could be as a result of malnutrition due to the effect of lectin in hindering nutrient uptake or the trypsin inhibitor in reducing protein utilization.

The apparent depopulation of splenic follicles in the treated rats may also be related to mobilization of the lymphocytes into the blood stream in response to the presence of lectin in the diet fed to rats in the under processed and unprocessed groups (Mebius and Kraal, 2005). Hart et al. (1988) showed that epithelial cell microvilli particularly are affected by lectin exposure, which initiates disruption and shedding of these membrane rich surface projections. In this study, the small intestine morphology of rats fed with unprocessed and underprocessed seeds showed degenerative changes in the mucosa and muscular wall with most of the microvilli eroded which could be attributed to the lectin in the diet. These villi and microvilli provide the small intestine a large surface area for absorption of nutrients. Reduction in the number of the villi could result in improper digestion and malabsorption of nutrients.

No visible pathological lesions were observed in the kidney, liver and heart of the experimental animals.

Conclusion

This study concluded that *P. conophora* seeds have high nutritive values, but their utilization could be affected by the two antinutritional factors, including lectins and trypsin inhibitor present in the seeds. The seeds when incorporated into animal diets impaired growth and affected nutrient metabolism. In addition, ingestion of the seeds caused damage to the lung, the small intestine and the spleen. The latter effect impaired the cellular immune system of the animals. All of these effects could, however, be reduced or removed if the seeds are well cooked by heating.

REFERENCES

Adesioye HO (1991). The effect of processing and storage on the chemical and sensory quality of conophor nut. Nig. Food J. 9: 33-38.

AOAC (1995). Official Method of Analysis, 16th Edition. Association of Official Analytical Chemists, Washington.

Ayoola PB, Onawumi OO, Faboya OOP (2011). Chemical evaluation and nutritive values of *Tetracarpidium conophorum* (Nigerian walnut) seeds. J. Pharm. Biomed. Sci. 11(15): 1-5.

Chunmei G, Pan H, Sun Z, Qin G (2010). Effect of soybean variety on antinutritional factors content, and growth parameters and nutrients metabolism in rat. Int. J. Mol. Sci. 11(3): 1048-1056.

Coles EH (1986). Veterinary Clinical Pathology, 4th Edition. W.B.Saunders Co. Philadelphia. pp. 262 – 265.

Daniel KT (2005). The whole soy story. New Trends, Indiana. pp. 195-212.

Drury RAB, Wallington EA (1980). Carleton's histological technique. 5th Edition.Oxford; Oxford University Press.

Enujiugha VN, Ayodele-Oni O (2003). Evaluation of nutrients and some anti-nutrients in lesser known under-utilized oilseeds. Int. J. Food Sci. Technol. 38: 525-528.

Enujiugha VN (2003). Chemical and functional characteristics of conophor nut. Pak. J. Nutr. 2: 335-338.

Food and Agricultural Organization (FAO) (2000). Production Yearbook. Food and Agricultural Organization of the United Nations, Rome. pp. 1-18.

Gilani GS, Xiao WC, Cockell KA (2012). Impact of antinutritional factors in food proteins on the digestibility of protein and the bioavailability of amino acids and on [protein quality. Br. J. Nutr. 108(2): 315-332.

Grant G, More LJ, McKenzie NH, Dorward PM, Buchan WC, Telek L, Pusztai A (1995). Nutritional and haemagglutination properties of several tropical seeds. J. Agric. Sci. 124: 437-445.

Hagenleister HC, Barth A (1993). The influence of soybean trypsin inhibitor(s) on absorption of exogenous and loss of endogenous protein. In : Recent Advances of Research in Antinutritional Factors in Legume Seeds; van der Poel AFB, Huisman J, Saini, HS, editors.; Pudoc: Wageningen, The Netherlands. pp. 179–182.

Hart CA, Batt RM, Saunders JR, Getty B (1988). Lectin-induced damage to the enterocyte brush border. An electron-microscopic study in rabbits. Scand. J. Gastroenterol. 23: 1153–1159.

Kilpatrick PC, Pusztai A, Grant G, Graham C, Ewen SWB (1985). Tomato lectin resists digestion in the mammalian alimentary canal and binds to intestinal villi without deleterious effects. FEBS Lett. 185(2): 299-305.

Kjeldsberg CR (2000). Practical diagnosis of hematologic disorders 3rd Edition. Chicago ASCP Press.

Kuku A, Eretan OB (2004). Purification and Partial Characterisation of a Lectin from the fresh leaves of *Kalanchoe crenata* (And.) Haw. J. Biochem. Mol. Biol. 37(2): 229 – 233.

Li SF, Yang LJ, Huo GC, Feng M Sh, Ma JX (1996). Study on the tolerance of broilers to dietary trypsin inhibitor [in Chinese]. Chin. Poult. 3: 8–10.

Li SY, Willem CS, William RC (1998). Response of nutrient digestibility to feeding diets with low and high levels of soybean trypsin inhibitors in growing pigs. J. Sci. Food Agric. 76: 357–363.

Li ZT (2003). Determination, purification and antinutritional effects of soybean agglutinin in rats.China Agricultural University, Beijing, China, pp. 35–41.

Liener IE (1996). Effects of processing on antinutritional factors in legumes: the soybean case. *Archivos Latinoamericanos de Nutricion* 44: 48S-54S.

Liener IE (1995). Possible adverse effects of soybean anticarcinogens. J. Nutr. 125: 744S-750S.

Lowry O H, Rosenbrough NJ, Farr AL, Rondall RJ (1951). Protein measurement with the Folin phenol reagent. J. Biol. Chem. 193: 265-273.

Mebius RE, Kraal G (2005). Structure and function of the spleen. Natl. Rev. Immunol. 5:606–616.

Noreen U, Salim M (2008). Determination of nutrient digestibility and amino acid availability of various feed ingredients for *Labeo rohita*. Int. J. Agric. Biol. 57: 313-324.

Odoemelam SA (2003). Chemical composition and functional properties of conophor nut flour. Int. J. Food Sci. Technol. 36:729-734.

Ogunsua AO, Adebona MB (1983).Chemical composition of *Tetracarpidium conophorum*. Food Chem. 10: 173-7.

Petrova UV (1980). Studies of chemical composition and correlations between chemical characteristic in some walnut varieties, Naoka Publishing Co. Ltd. London.

Pusztai A, Ewen SWB, Grant G, Peumans WP, van Damme EJM, Rubio L, Bardocz, S (1990). Relationship between survival and binding of plant lectins during small intestinal passage and their effectiveness as growth factors. Digestion 46: 308–316.

Pusztai A, Clarke EM, Grant G, King T (1981).The toxicity of *Phaseolus vulgaris* lectins: nitrogen balance and immunochemical studies. J. Sci. Food Agric. 32: 1037-1046.

Pusztai A, Watt WB, Stewart JC (1991). A comprehensive scheme for the isolation of trypsin inhibitors and agglutinins from soybean seeds. J. Sci. Food Agric. 39: 862-866.

Qin G X (2003). The anti-nutritional factors in the feed and their eliminating methods. Pig Poult. 23: 10–13.

Qin G, ter Elst ER, Bosch M W, van der Poel AFB (1996). Thermal processing of whole soya beans: Studies on the inactivation of antinutritional factors and effects on ileal digestibility in piglets. Anim. Feed Sci. Technol. 57: 313–324.

Rackis JJ, Gumbmann MR (1981). Protease inhibitors physiological properties and nutritional significance. In Robert L. Ory, ed. Antinutrients and Natural Toxicants in Foods (Westport, CT, Food and Nutrition Press). pp. 203-238.

Reeves PG, Nielsen FH, Fahey GC Jr (1993). AIN-93 purified diets for laboratory rodents: final report of the American Institute of Nutrition ad hoc writing committee on the formulation of the AIN-76A rodent diet. J. Nutr. 123: 1939-1951.

Schulze H, Verstegen MWA, HuismanJ, Van Leeuwen P, van den Berg JWO (1993). Nutritional effects of isolated soya trypsin inhibitor on pigs: In Recent Advances of Research in Antinutritional Factors in Legume Seeds: van der Poel, A.F.B., Husman, J., Saini, H.S., Eds., Pudoc: Wageningen, The Netherlands. pp. 195-199.

Smith C, Megen W, Twaalfhaven L, Hitchock C (1980). The determination of trypsin inhibitor levels in food stuffs. J. Sci. Food Agric. 34: 341-350.

Thompson LU (1993). Potential health benefits and problems associated with antinutrients in foods. Food Res. Int. 26: 131- 149.

Thompson LU, Tenebaum AV, Hui H (1986). Effect of lectins and the mixing of proteins on rate of protein digestibility. J. Food Sci. 51: 150-152.

Vadiviel A, Mani K, Kandan K, Garesan D, Sakanran M (2010). Antioxidant and antihyperlipidemic effect of *S.nigrum* fruit on the experimental model age. Chronic ethanol toxicity. Pharmacogn. Mag. 6:42-50.

Walker K, Hall WD, Hurst JW (editors) (1990). Clinical Methods: The History, Physical, and Laboratory Examinations. 3rd Edition Butterworths, Boston.

Wang SP, Qin GX, Gong Q, Yang, LY (2000). Effect of antinutritional factors in full fat soybean on the performance of broilers (in Chinese). J. Jilin Agric. Univ. 22: 81-86.

A survey on existing practices adopted in *Dambu* production and utilization in some northern states of Nigeria

Agu, H. O.[1] , Badau, M. H.[2], Abubakar, M. U.[2] and Jideani, V. A.[3]

[1]Department of Food Science and Technology, Federal Polytechnic Bauchi, P.M.B. 0231 Bauchi, Nigeria.
[2]Department of Food Science and Technology, University of Maiduguri, P.M.B. 1069 Maiduguri, Nigeria.
[3]Department of Food Science and Technology, Cape Peninsula University of Technology, Cape Town 8000, South Africa.

A survey was conducted in relation to the production process of *dambu*. *Dambu* is a staple food for the Fulanis and Hausas. A majority of the *dambu* dealers (42.7%) were less than 25 years old. A proportion of 78.7% of processors of *dambu* were females. The processors/consumers (37.4%) indicated that millet was the major raw material for *dambu* production. A significant difference (p = 0.01) existed among respondents on the variety of millet chosen for *dambu* production. The study reveals that 'gero' is the variety in common use. Spices are indispensable as an ingredient in *dambu* production with ginger being the single most important spice (p = 0.07). The traditional pounding method for processing millet into flour is still very much used. The processors (58.9%) and retailers (41.0%) generally agreed that the method of marketing *dambu* is by hawking in transparent low density polyethylene package. Sun drying as indicated by 35.3% of the respondents was the most common method of storing left-over *dambu*. Total percentage of 27.4, 39.2 and 33.4% were deduced for the poor, the middle and the rich class consumption of *dambu*, respectively. With increasing influence of advertising upon customers, small food processing enterprises making *dambu* will have to improve the packaging and preservation of their products as to survive the competition. The implications of these are highlighted and a possible solution of optimizing the *dambu* production process is recommended.

Key words: Ginger, cloves, *dambu*, millet, packaging, shelf-life.

INTRODUCTION

In tropical Africa, cereal grains are milled and used to produce different types of food which are known by various names in different parts of the African continent as thin porridge (ogi) (Apena et al., 2006), thick porridge *fura* (Jideani et al., 1995; Umoh, 2003; Filli et al., 2007), baked or fried fermented bread such as *mawè* (Hounhouigan et al., 1993), *masa* (Ayo et al., 2008), snacks (Vidyavati et al., 2004), alcoholic beverage (burukutu) (Igyor et al., 2006) and non-alcoholic beverages such as *kunun zaki* (Gaffa and Jideani, 2001; Ayo and Obeya, 2004; Gaffa et al., 2004). In West Africa, particularly in the Northern part of Nigeria, one of such cereal-based products is called *dambu* - a steamed non-fermented granulated dumpling generally made from pearl millet (Nkama et al., 1999; Agu et al., 2007, 2008). It is known as *dambu* among the Hausas while among the Fulanis, it is known as "nyamri" (Haruna, 2003). *Dambu* is similar to *fura* – a semi- solid dumpling cereal meal, but differs only in the production process (Jideani and Wedzicha, 1994).

Dambu is produced mainly from moistened pearl millet flour, blended with spices and steamed for 20 min. using two fold systems, which involves pouring the mixed flour with spices into a sieve and placed on an open pot that contains little water. Care is taken for the sieve not to touch the water. As the water in the pot boils, the steam cooks the *dambu* on the sieve. It has coarse particles resembling moistened couscous. It is sprinkled into fermented skimmed milk (*nono*), or fermented whole milk (*Kindirimo*) and sugar may be added to taste. It is a popular mid-day meal called "*dambu* da nono" in Nigeria (Nkama et al., 1999).

Dambu contains energy of 257 kcal, moisture of 37%, ash 1.1%, crude protein 10.7%, fat 3.4%, crude fibre 2.0%, and carbohydrate 45.9% per 100 g. It also contains essential and non- essential amino acids with major and minor mineral elements (Agu et al., 2007). *Dambu* has been produced with different grain types and pearl millet was rated high based on aroma, texture and overall acceptability (Agu et al., 2007) and based on this, using different pearl millet cultivars will be productive.

Dambu is produced at home both for family and commercial consumption. Most *dambu* producers use the traditional method involving wooden mortar to dehull or mill the grains. Pearl millet products like *fura* and *dambu* are sold from calabash containers to consumers without appropriate packaging (Jideani et al., 2001a; Agu et al., 2008). Good packaging not only serves as food container but protects and carries the necessary information about the food product. It is particularly important in countries with tropical and humid climates, where food deterioration is most rapid (UNIFEM, 1993). *Dambu* is hawked by the local producers. The processors and retailers of *dambu* are only concerned with having a calabash or enamel porcelain container for their food. The practice of opening and closing of the container during sales subjects the food to microbial contamination. *Dambu* has a limited shelf-life of one day at ambient temperature. Usually, a day after the production, *dambu* shows visible mould growth on the surface. The short shelf-life has always been a major deterrent to large scale production. Thus, improving the processing, packaging and storage life of *dambu* are of interest before food manufacturers can think of large scale production. This study was conducted to establish the materials and the ingredients as well as the production process for *dambu* from the processors and consumers. This will provide baseline information for food manufacturers and a basis for improving this process.

MATERIALS AND METHODS

Survey area

States in northern Nigeria chosen for the study included Bauchi, Plateau, Kaduna, Kano, Gombe, Jigawa, Borno, Adamawa and Katsina. These are the major *dambu* producing and consuming states in Nigeria (Jideani et al., 2001).

Methods

The study data was collected using a structured questionnaire. The questionnaires were designed and pretested before being distributed to the potential respondents. The questionnaires were divided into sections in order to collect information on:

1. Socio demographic particulars,
2. Raw materials for *Dambu* production,
3. Steps involved in primary processing,
4. Steps involved in secondary processing mode of display,
5. Equipment used and its cost details,
6. Cost of inputs, packaging, mode od display, shelf life storage methods product utilization,
7. By product utilization and,
8. Personnel needed

Data collection and analysis

The processors, retailers and consumers of *dambu* in Bauchi, Plateau, Kaduna, Kano, Gombe, Jigawa, Borno, Adamawa and Katsina states were distributed a total of 1000 questionnaires. The questionnaires were completed by the respondents in their homes, markets and offices. There was agreement on when the completed questionnaires could be picked up. Where the respondents could not speak English, interviews were conducted in the local language with the assistance of an interpreter. Questionnaires (827) representing 82.7% were retrieved. The remaining 17.3% were either reluctant to return their questionnaire or could not be located.

Data from the questionnaires were analyzed by tally and the number of the tallies for each question was entered into a MINITAB (version 11, 1996) and tabulated as percentages. The Kruskal-Wallis H test (a non-parametric one-way ANOVA) was used to compare the data.

RESULTS AND DISCUSSION

Demography of respondents

Demographic characteristics in relation to the status of

Table 1. Demographic Characteristics as related to the processor retailer/consumer status.

Question	Processor	Retailer	Processor/Retailer	Consumer	Processor/Consumer	Total	P-value
	n =216	n = 83	n = 211	n = 143	n = 174	n = 827	
Sex							
Male	46 (21.3)	22(26.5)	67 (31.8)	77 (53.8)	78 (44.8)	290 (35.1)	
Female	170 (78.7)	61 (73.5)	144 (68.2)	66 (46.2)	96 (55.2)	537 (64.9)	0.07
Age							
≤25	106 (49.1)	37 (44.6)	79 (37.0)	58 (40.6)	73 (41.9)	353 (42.7)	
26-35	70 (32.4)	27 (32.5)	43 (20.4)	49 (34.3)	40 (23.0)	229 (27.7)	
36-45	30 (13.9)	9 (10.8)	56 (26.5)	36 (25.2)	15 (8.6)	146 (17.7)	
>46	10 (4.6)	10 (12.0)	33 (15.6)	-	46 (26.4)	99 (12.0)	0.01
Education							
None	55 (25.5)	12 (14.5)	74 (35.1)	37 (25.9)	39 (22.4)	217 (26.2)	
Primary	59 (27.3)	28 (33.7)	64 (30.3)	26 (18.27)	34 (19.5)	211 (25.5)	
Sec/Tech/Com	36 (16.7)	7 (8.4)	40 (19.0)	33 (23.1)	12 (6.9)	128 (15.5)	
Vocational	42 (19.4)	20 (24.1)	28 (13.3)	21 (41.7)	44 (25.3)	155 (18.7)	
University/Poly	24 (11.1)	16 (19.3)	5 (2.4)	26 (18.2)	45 (25.9)	116 (14.0)	0.23
Ethnicity							
Kanuri	31 (14.4)	21 (25.3)	5 (2.4)	56 (39.2)	28 (16.1)	141 (17.0)	
Hausa	84 (38.9)	40 (48.2)	93 (44.1)	32 (22.4)	46 (26.4)	295 (35.7)	
Fulani	101 (46.8)	22 (26.5)	113 (53.6)	55 (38.5)	100 (57.5)	391 (47.3)	0.06

Figures in parentheses are percentages; Kruskal - Wallis one-way ANOVA test.

the respondents are presented in Table 1. A total of 827 respondents were used for the study. Processors /retailers were those who produced *dambu* and either hawked it or sold it on a wholesale basis. Retailers- those who do not process *dambu* but bought from processors to sell were also surveyed. Processors/consumers were those who produced *dambu* for consumption purposes. Consumers were those not involved in processing but purchased *dambu* for consumption (Jideani et al., 2001). The respondents were comparable in their demographic characteristics ($p < 0.05$) for any variable.

The respondents (35.1%) of those directly involved with processing and retailing of *dambu* did not have a formal education. This may contribute to improper hygienic practices during production and handling of *dambu* which could lead to product with potential health hazards to the consumers. A significantly ($p < 0.05$) greater proportion (53.6%) of the processor/retailers and consumers of *dambu* were Fulanis, followed by the Hausas (44.1%). *Dambu* is therefore a staple food for both Hausas and Fulanis. Majority of the *dambu* dealers were less than 25 years old. A proportion of 78.7% of processors of *dambu* were females. The processing and marketing of *dambu* is known to be a woman's trade.

The study reveals that no male considered himself to be a processor. Those men who know the processing method for *dambu* either watched their mothers make it when they are younger or watched their wives. It was however surprising to note that there were some men who were retailers (26.5%) and 44.5% were processors /consumers. More men (53.8%) consumed *dambu* compared to women (46.2%). This could be because men are usually the bread winners in Africa and in these cultures have the right to spend their money anyhow and whenever they want and for anything of their choice. The women who in this culture depend on what is made available to them by their husbands, are concerned primarily with the feeding of the family and therefore, would have limited money to purchase *dambu* for consumption except when the female has made it part of the family menu (Jideani et al., 2001).

Raw materials and ingredients for *dambu* production

The major raw materials and ingredients used for *dambu* production are shown in Table 2. The processors /consumers (37.4%) indicated that millet was the major raw material for *dambu* production. The processors /consumers (36.8%) indicated that there were three

Table 2. Major raw materials and ingredients used in *dambu* production.

Question		Status: Processor	Retailer	Processor/ Retailer	Consumer	Processor/ Consumer	Total	P-value
		n =216	n = 83	n =211	n =143	n = 174	n = 827	
(a)	**Major grain**							
	Millet	59 (27.3)	33 (39.8)	70 (33.2)	77 (53.8)	64 (36.8)	303 (36.6)	0.01
	Sorghum	72 (33.3)	21 (25.3)	64 (30.3)	45 (31.5)	39 (22.4)	241 (29.1)	
	Maize	48 (22.2)	25 (30.1)	59 (28.0)	15 (10.5)	65 (37.4)	212 (25.6)	
	"Acha"	37 (17.1)	4 (4.9)	18 (8.5)	6 (4.2)	6 (3.4)	71 (8.6)	
(b)	**How many varieties?**							
	1	48 (22.2)	49 (59.0)	32 (15.2)	71 (49.7)	29 (16.7)	229 (27.7)	0.48
	2	58 (26.9)	19 (22.9)	23 (10.9)	52 (36.4)	13 (7.5)	165 (19.9)	
	3	53 (24.5)	6 (7.2)	56 (26.5)	16 (11.2)	64 (36.8)	195 (23.6)	
	4	42 (19.4)	_	87 (41.2)	4 (2.8)	16 (9.2)	149 (18.0)	
	All	15 (6.9)	9 (10.8)	13 (6.2)	_	52 (29.9)	89 (10.8)	
(c)	**Variety commonly used**							
	"Gero"	89 (41.2)	21 (25.3)	66 (90.4)	53 (37.1)	54 (31.0)	283 (34.2)	0.01
	"Dauro"	58 (26.9)	15 (18.1)	73 (34.6)	32 (22.4)	45 (25.9)	223 (27.0)	
	"Maiwa"	17 (7.9)	29 (34.9)	49 (23.2)	21 (14.7)	19 (10.9)	135 (16.3)	
	Pearl Millet	50 (24.3)	18 (21.7)	23 (10.9)	13 (9.1)	38 (21.8)	142 (17.2)	
	No Response	2 (0.9)	_	-	24 (16.8)	18 (10.3)	44 (5.3)	
(d)	**Can a substitute be used?**							
	Yes	143 (66.0)	49 (59.0)	178 (84.4)	55 (38.5)	75 (43.1)	500 (60.5)	0.27
	No	73 (33.8)	34 (41.0)	33 (15.6)	88 (61.5)	99 (56.9)	327 (39.5)	
(e)	**If yes what is the substitute?**							
	Sorghum	63(44.1)	12(24.5)	66(37.1)	12(21.8)	34(45.3)	187 (37.4)	0.8
	"Acha"	-	-	26(14.6)	8(14.5)	29(38.7)	63 (12.6)	
	Millet	55(38.5)	20(40.8)	54(30.3)	14(25.5)	10(13.3)	153 (30.6)	
	Maize	25(17.5)	17(34.7)	32(17.9)	17(30.9)	-	91 (18.2)	
	No response	-	-	-	4(7.3)	2(2.7)	6 (1.2)	
	Total	143	49	178	55	75	500	
(f)	**Reason for Substitute.**							
	Unavailability of millet	173 (80.0)	34 (41.0)	99 (46.9)	51 (35.7)	120 (69.0)	477 (57.7)	0.04
	Choice	31 (14.4)	27 (32.5)	38 (18.0)	22 (15.4)	54 (31.0)	172 (20.8)	
	Economy	12(5.6)	22 (26.5)	74 (35.1)	70 (49.0)	_	178 (21.5)	
(g)	**Effect of substitute on quality**							
	Colour	59 (27.3)	17 (20.5)	32 (15.2)	46 (32.2)	32 (18.4)	186 (22.5)	0.17
	Texture	71 (32.9)	10 (12.0)	83 (39.3)	21 (14.7)	32 (18.4)	217 (26.2)	
	Taste	40 (18.5)	24 (28.9)	68 (32.2)	27 (18.9)	53 (30.5)	212 (25.6)	
	Colour, Taste, Texture	36 (16.7)	8 (9.6)	17 (8.1)	41 (28.7)	29 (16.7)	131 (15.8)	
	Texture, Texture	10 (4.6)	24 (28.9)	11 (5.2)	8 (5.6)	28 (16.1)	81 (9.8)	

Table 2. Contd.

(h)	**Effect of substitute on demand**							
	Increase	125 (57.9)	31 (37.3)	33 (15.6)	57 (39.9)	52 (29.9)	298 (36.0)	
	Decrease	65 (30.1)	29 (34.9)	74 (35.1)	52 (36.4)	65 (37.4)	285 (34.5)	0.85
	Remain the same	26 (12.0)	23 (27.7)	104 (49.3)	34 (23.8)	57 (32.8)	244 (29.5)	
(i)	**Cereal used in combination**							
	Yes	140 (64.8)	60 (72.3)	114 (54.0)	42 (29.4)	147 (84.5)	503 (60.8)	0.22
	No	76 (35.2)	23 (27.7)	97 (46.0)	101 (70.6)	27 (15.5)	324 (39.2)	
(j)	**Preferred combination**							
	Millet & Maize	81 (57.9)	15 (25.0)	22 (19.3)	33 (78.6)	66 (44.9)	217 (43.1)	
	Millet & Sorghum	46 (32.9)	22 (36.7)	75 (65.8)	9 (21.4)	81 (55.1)	233 (46.3)	0.08
	Maize & Acha	13 (9.3)	23 (38.3)	17 (14.9)	_	_	53 (10.5)	
	Total	140	60	114	42	147	503	
(k)	**Purpose of combination**							
	To increase test	69 (31.9)	15 (18.1)	17 (8.1)	23 (16.1)	35 (20.1)	159 (19.2)	
	To make product brighter	82 (38.0)	12 (14.5)	46 (21.8)	59 (41.2)	49 (28.2)	248 (30.0)	0.66
	Absence of required grain	13(6.0)	34(41.0)	84 (40.0)	34 (23.8)	62 (35.6)	227 (33.5)	
	No response	52(24.1)	22(26.5)	64 (30.3)	27 (18.9)	28 (16.1)	193 (23.3)	
(l)	**Effect of combination on quality**							
	Colour	52 (24.1)	18 (21.7)	52 (24.6)	21 (14.7)	38 (21.8)	181 (21.9)	
	Texture	45 (20.8)	23 (27.7)	40 (19.0)	33 (23.1)	39 (22.4)	180 (21.8)	0.81
	Taste	69 (31.9)	36 (43.4)	27 (12.8)	27 (18.9)	42 (24.1)	201 (24.3)	
	Others	50 (23.2)	6 (7.2)	92 (43.6)	62 (43.4)	55 (31.6)	265 (32.0)	
(m)	**Effect of combination on demand**							
	Increase	68 (31.5)	34 (41.0)	48 (22.7)	53 (37.1)	21 (12.1)	224 (27.1)	
	Decrease	108 (50.0)	28 (33.7)	96 (45.5)	39 (27.3)	100 (57.5)	371 (44.9)	0.18
	Remain the same	40 (18.5)	21 (25.3)	67 (31.8)	51 (35.7)	53 (30.5)	232 (28.0)	
(n)	**Other ingredients used**							
	Potatoes	21 (9.7)	19 (22.9)	28 (13.3)	18 (12.6)	92 (52.8)	178 (21.5)	
	Ginger	63 (29.2)	18 (21.7)	66 (31.3)	24 (16.8)	20 (11.5)	191 (23.1)	0.71
	Pepper	25 (11.6)	24 (28.9)	54 (25.6)	47 (32.9)	33 (18.9)	183 (22.1)	
	Cloves	107 (49.5)	22 (26.5)	63 (29.9)	54 (37.8)	29 (16.7)	275 (33.2)	
(o)	**Proportion added**							
	Cloves/ginger/pepper							
	1:01:01	73 (33.8)	80 (96.4)	192 (91.0)	74 (51.7)	82 (47.1)	501 (60.6)	
	2:01:02	45 (20.8)	3 (3.6)	7 (3.3)	18 (12.6)	43 (24.7)	116 (14.0)	0.002
	4:01:01	93 (43.1)	-	12 (5.7)	51 (35.7)	49 (28.2)	205 (24.8)	
	No response	5 (2.3)	-	-		-	5 (0.6)	
(p)	**Commonest spice used**							
	Ginger	46 (21.3)	57 (68.6)	53 (25.1)	43 (30.1)	84 (48.3)	283 (34.2)	

Table 2. Contd.

	Cloves	78 (36.1)	8 (9.6)	44 (20.9)	63 (44.1)	63 (36.2)	256 (31.0)	
	Pepper	84 (38.9)	13 (15.7)	62 (29.4)	26 (18.2)	21 (12.1)	206 (24.9)	0.07
	Ginger/Pepper	8 (3.7)	5 (6.02)	52 (24.6)	11 (7.7)	6 (3.4)	82 (9.9)	
(q)	**Is it added for any purpose?**							
	Yes	99 (45.8)	35 (42.2)	82 (38.9)	94 (65.7)	92 (52.9)	402 (48.6)	0.68
	No	117 (54.2)	48 (57.8)	129 (90.2)	49 (34.3)	82 (47.1)	425 (51.4)	

Figures in parentheses are percentages; Kruskal - Wallis one-way ANOVA test.

varieties of millet, while (7.5%) indicated only two varieties. The number of varieties of millet varies from one state to another. For instance, in Bauchi State, there are three varieties of pearl millet in the market. These are 'gero', 'dauro', or 'maiwa' and 'gayamba'. In the other states, 'gero' and 'dauro' are very common. A significant difference ($p = 0.01$) existed among respondents on the choice of variety of pearl millet for *dambu* production. Majority of the respondents indicated that 'gero' is the variety in common use.

Although millet is the major cereal for *dambu* production, some other substitute grain can also be used alone or in combination with millet (Table 2). The choice of a substitute grain was not significant ($p = 0.27$). Among other substitute grain, sorghum was a significant grain of choice. The present study results are in agreement with results of Agu et al. (2007, 2008) which stated that different cereal grains could be used for *dambu* production. The significant reason ($p = 0.04$) given for the choice of a substitute grain was unavailability of millet. Other reasons included (i) the processors choice (ii) not sufficient money to purchase millet.

Other ingredients added to cereal flour for *dambu* production were sweet potatoes, as sweeteners; ginger, pepper and cloves as flavouring, potatoes were added as a sweetener. There was no significant difference ($p = 0.71$) in the choice of these ingredients. The quantity of these spices added was generally small. However, some of the respondents who estimated the ratio of the mixture gave 1:1:1, 2:1:1 and 4:1:1 for cloves, ginger and pepper respectively. There was a significant ($p = 0.05$) difference in the proportion of these spices. According to Agu et al. (2007, 2008) spices like ginger, cloves and pepper are used for *dambu* production in various proportions.

Most of the respondent indicated that the spices were added mainly for flavour purposes and other reasons given where for its medicinal and sedative properties. According to Zaika (1988) and Norman (1990), spices are one of the various strongly flavoured or aromatic substance of vegetable origin obtained from tropical plants commonly used as condiments and possess significant antimicrobial activity. These spices include cloves, ginger, pepper and nutmeg (Norman, 1990). The levels of the spices added in *dambu* are obviously not sufficient to produce a presser-vative effect. However, the properties of these spices if used in an overall approach (Giese, 1994), could be effec-tive in extending the shelf-life of *dambu*.

The production process for making *Dambu*

The production process for making *dambu* can be divided into primary and secondary processing (Figure 1). Primary processing are those to which the cereal grain is subjected but the product is still not directly consumable (Table 3). The objective is to perform a clean separation of the pericarp (bran) from the rest of the grain. In spite of the increasing number of dehulling machines, the producers of *dambu* prefer the traditional hand pounding of dehulling the grain and producing the flour. The respondents (25.8%) believed that decorticating (dehulling) of the grain was achieved by pounding the moistened grain with the butt end of a heavy wooden pole (pestle) in a mortar to knock off the outer seed coat. The beaten mixture was washed to separate the bran. The respondents (40.6%) indicated that the step of pounding was very important and needed improvement. The respondents (24.9%) suggested the use of a dehuller as alternative to local but wooden mortar and pestle approved by 45.1% of the respondents.

The endosperm obtained was allowed to drain in a sieve for 15 - 20 min, spread under the sun for 2 h as to further reduce the moisture content to 10 – 15% before grinding to flour by second pounding. Some of the respondents (51.4%) own grinding machines (metal but foreign) while 48.6% own metal but local. The respondents (processors) charge between 20 - 50 naira for grinding. The processors (55.1%) pay around N50 for grinding. The grinding of the flour was coarse then sieved using 600 µm.

Secondary processing are those processes to which the end product of primary processing (flour) undergoes to obtain *dambu* and it is shown in Table 3. This involved processes such as mixing, steaming, cooling and packaging. After mixing flour with the spices, it then poured into a sieve and placed on an open pot that contained little water. Care should be taken that the sieve

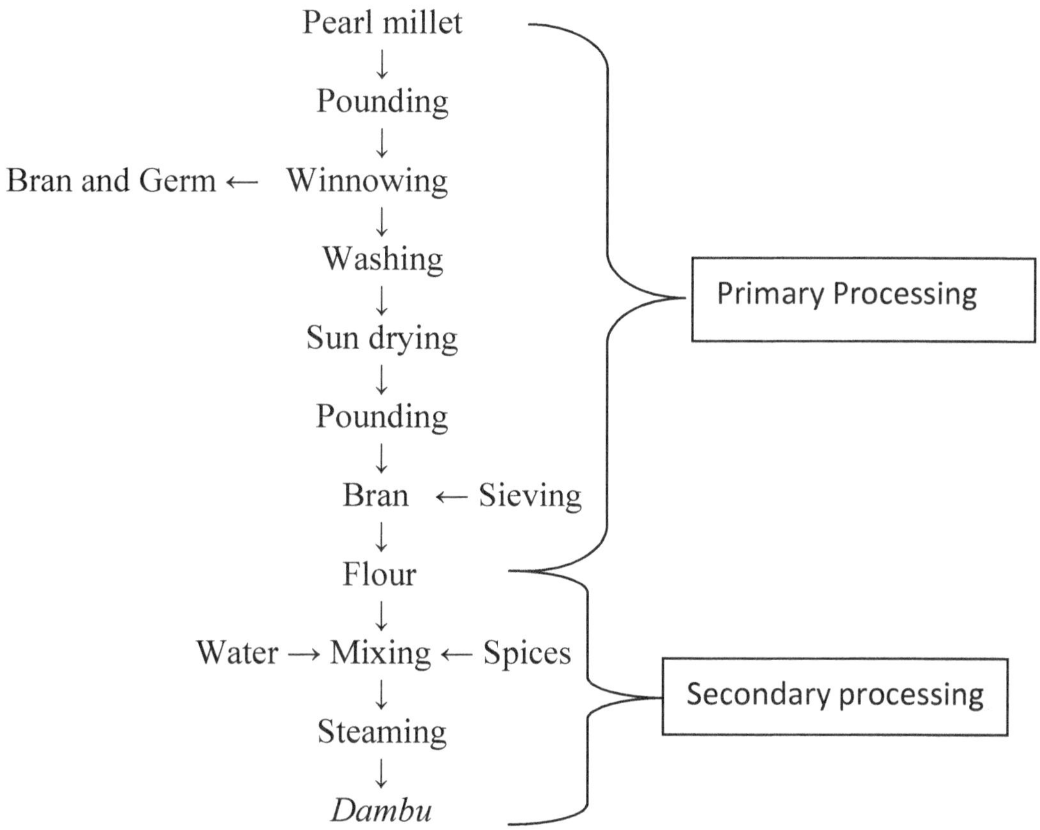

Figure 1. The Traditional Production of *dambu.*

does not touch the water. As the water in the pot boiled, the steam cooked the *dambu* on the sieve. The pot was covered and allowed to steam until the aroma was perceived. The respondents (28.3%) indicated that they mix between 30 – 60 S, while 32.6% indicated they steam between 20 – 45 min. The *dambu* after steaming was cooled between 1-5 min as indicated by 35.8% of the respondents.

The traditional pounding is a tedious task, which limits the use of the cereal. Enough pearl millet for a family meal (about 2.5 kg) takes two women about 1.5 h; converting the product into flour with a mortar and pestle requires an additional 2 h; sometimes more (NRC, 1996). Secondly, the moistening of the grain to facilitate removal of the bran and pounding into flour can result in slightly fermented flour leading to a modified flavour (Perten, 1983). Many of the respondents indicated that mechanical process needs to be improved in such a way to produce flour similar to that produced by the traditional method. The major criticism of the mechanical process is that dry grain must always be used, producing flour that is too dry and when used to produce *dambu,* does not result in the same taste as the traditionally processed flour (Jideani et al., 2001).

However, the production process for *dambu* can be optimized. Such optimization if achieved, would reduce the drudgery of women in *dambu* eating areas, improve health and family welfare and convert millet into a much more convenient grain and lead to food manufacturers getting involved in production of *dambu*. Moreover, with improvement in living standards in Nigeria and Africa at large, there has been an increasing demand for better quality food. Quality requirement for traditional *dambu* should consequently change. At present, the burden of the toil in millet processing is causing a silent rebellion against millet (Jideani et al., 2001) and the resulting *dambu*.

Packaging and distribution of *Dambu*

Packaging and distribution of *dambu* is shown in Table 4. The processors (58.9%) and retailers (41.0%) generally agreed that the method of marketing *dambu* is by hawking in a plastic (transparent low density polyethylene bag) while the processors (8.8%) and retailers (4.8%) indicated the use of calabash during hawking. At the vending point (market garage/park) the *dambu* are wrapped

Table 3. Primary and secondary processing of *dambu*.

Question		Status						
		Processor	Retailer	Processor/Retailer	Consumer	Processor/ Consumer	Total	P-value
		n = 216	n = 83	n = 211	n = 143	n = 174	n = 827	
(a)i	**Steps you follow in dehulling**							
	Winnowing, conditioning, pounding	64 (29.6)	0	91 (43.1)	16 (11.2)	42 (24.1)	213 (25.8)	
	Conditioning, pounding, drying	32 (14.8)	4 (4.8)	76 (36.0)	33 (23.2)	66 (37.9)	211 (25.5)	
	washing, winnowing, drying	88 (4.1)	56 (67.5)	28 (13.3)	15 (10.5)	19 (10.9)	206 (24.9)	
	No Response	32 (14.8)	23 (27.7)	16 (7.6)	79 (55.2)	47 (27.0)	197 (23.8)	1
(ii)	**Which step do you think improvement is Necessary?**							
	Winnowing	18 (8.3)	6 (7.2)	35 (16.6)	17 (11.9)	63 (36.2)	139 (16.8)	
	Pounding	83 (38.4)	59 (71.1)	76 (36.0)	44 (30.8)	74 (42.5)	336 (40.6)	
	Conditioning	32 (14.8)	_ (0)	25 (11.8)	36 (25.2)	_ (0)	93 (11.2)	
	Drying	8 (3.7)	_ (0)	19 (9.0)	21 (14.7)	28 (16.1)	76 (9.2)	
	No response	75 (34.1)	18 (21.7)	56 (26.5)	25 (17.5)	9 (5.2)	183 (22.1)	0
(b)i	**Steps in grinding to achieve desired quality**							
	Pounding & Sieving	93 (43.1)	45 (54.2)	65 (30.8)	43 (30.1)	69 (39.7)	315 (38.1)	
	Dehulling & Drying	78 (36.1)	8 (9.6)	31 (14.7)	72 (50.3)	28 (16.1)	217 (26.2)	
	Conditioning	42 (19.4)	30 (36.1)	87 (41.2)	13 (9.1)	47 (27.0)	219 (26.5)	
	Washing	3 (1.4)	_	28 (13.3)	15 (10.5)	30 (17.2)	76 (9.2)	0.04
(ii)	**Steps in which improvement is necessary to make grinding profitable**							
	Pounding & Sieving	36 (16.7)	43 (51.8)	65 (30.8)	88 (61.5)	83 (47.7)	315 (38.1)	
	Dehulling	79 (36.6)	32 (38.6)	83 (39.3)	35 (24.5)	46 (26.4)	275 (33.2)	
	Conditioning	56 (25.9)	8 (9.6)	33 (15.6)	16 (11.1)	27 (15.5)	140 (16.9)	
	Washing	45 (20.8)	_ (0)	30 (14.2)	4 (2.8)	18 (10.3)	97 (11.7)	0.01
(c)	**Steps in preparation of *dambu* with duration**							
i.	**Mixing**							
	30 - 60 S	53 (24.5)	15 (18.1)	72 (34.1)	31 (21.7)	63 (36.2)	234 (28.3)	
	1- 5 min	31 (14.4)	31 (37.3)	15 (7.1)	58 (40.6)	32 (18.4)	167 (20.2)	
	6 - 10 min	12 (5.6)	9 (10.8)	48 (22.7)	16 (11.2)	45 (25.9)	130 (15.7)	
	11 - 15 min	82 (38.0)	18 (21.7)	31 (14.7)	25 (17.5)	18 (10.3)	174 (21.0)	
	16 - 20 min	38 (17.6)	10 (12.0)	45 (21.3)	13 (9.1)	16 (9.2	122 (14.7)	0.22
ii.	**Steaming**							
	20 - 45 min	78 (36.1)	33 (39.8)	61 (28.9)	54 (37.7)	44 (25.3)	270 (32.6)	
	45 - 50 min	65 (30.1)	28 (33.7)	39 (.18.5)	68 (47.6)	33 (19.0)	233 (28.2)	
	50 - 60 min	59 (27.3)	12 (14.5)	83 (39.3)	13 (9.1)	15 (8.6)	182 (22.0)	
	1 - 12 h	14 (6.5)	10 (12.0)	28 (13.3)	8 (5.6)	82 (47.1)	142 (17.2)	0.44
iii.	**Cooling**							
	1 – 5 min	89 (41.2)	16 (19.3)	78 (37.0)	65 (45.4)	48 (27.6)	296 (35.8)	
	6 – 10 min	45 (20.8)	43 (51.8)	94 (44.5)	32 (22.4)	19 (10.9)	233 (28.2)	
	11 - 15 min	62 (28.7)	13 (15.7)	12 (5.7)	14 (9.8)	35 (20.1)	136 (16.4)	
	16 – 20 min	20 (9.25)	11 (13.3)	27 (12.8)	32 (22.4)	72 (41.4)	162 (19.6)	0.23
iv.	**Packaging**							
	10 – 20 min	4 1(19.0)	13 (15.7)	54 (25.6)	32 (22.4)	18 (10.3)	158 (19.1)	
	20 – 30 min	34 (15.7)	33 (39.8)	72 (34.1)	48 (33.6)	34 (19.5)	221 (26.7)	
	30 – 40 min	82 (38.0)	16 (19.3)	38 (18.0)	19 (13.3)	56 (32.2)	211 (25.5)	
	40 – 50 min	59 (27.3)	21 (25.3)	47 (22.3)	44 (30.8)	66 (37.9)	237 (28.6)	0.64

Figures in parentheses are percentages Kruskal - Wallis one way ANOVA test.

Table 4. Mode of marketing and storage stability of *dambu*.

Question		Status: Processor	Retailer	Processor/ Retailer	Consumer	Processor/ Consumer	Total	P-value
		n =216	n = 83	n =211	n =143	n = 174	n = 827	
(a)	**What packaging material is used for the product?**							
	Low density polyethylene	112 (51.9)	58 (69.9)	121 (57.3)	31 (21.7)	73 (42.0)	395 (47.8)	
	Paper	64 (29.6)	21 (25.3)	72 (34.1)	53 (37.1)	62 (35.6)	272 (32.9)	
	None	21 (9.7)	-	-	15 (10.5)	25 (14.4)	61 (7.4)	
	Calabash	19 (8.8)	4 (4.8)	18 (8.5)	44 (30.8)	14 (8.1)	99 (12.0)	0.01
(b)	**How do you market your product?**							
	Hawking	127 (58.8)	34 (41.0)	84 (39.8)	71 (49.7)	85 (48.9)	401 (48.5)	
	Whole sale	38 (17.6)	24 (28.9)	79 (37.4)	53 (37.1)	63 (36.2)	257 (31.1)	
	Consumption	46 (21.3)	25 (30.1)	48 (22.8)	19 (13.3)	26 (14.9)	164 (19.8)	
	No response	5 (31.3)	-	-	-		5 (0.6)	0
(c)	**Where do you sell your product?**							
	Market	102 (47.2)	136 (43.4)	116 (55.0)	44 (30.8)	93 (53.5)	391 (47.3)	
	Home	68 (31.5)	32 (38.6)	17 (8.1)	39 (27.3)	28 (16.1)	184 (22.2)	
	Garage/park	46 (21.3)	15 (18.1)	78 (36.8)	60 (42.0)	53 (30.5)	252 (30.5)	0.09
(d)	**How long it takes to sell the product?**							
	1 – 5 h	72 (33.3)	45 (54.2)	93 (44.1)	33 (23.1)	56 (32.2)	299 (36.1)	
	5 – 25 h	38 (17.6)	6 (8.4)	54 (25.6)	25 (17.5)	82 (47.1)	230 (27.8)	
	1 – 2 days	49 (22.7)	7 (8.4)	-	46 (32.2)	23 (13.2)	125 (15.1)	
	1 – 2 weeks	53 (24.5)	-	63 (29.9)	27 (18.9)	9 (5.2)	152 (18.4)	
	No response	4 (1.9)	-	1 (0.5)	12 (8.4)	4 (2.3)	21 (2.5)	0.01
(e)	**How is left over *dambu* preserved?**							
	Refrigeration	42 (19.4)	72 (86.8)	55 (26.1)	32 (22.4)	49 (28.2)	250 (30.2)	
	Sun-drying	97 (44.9)	5 (6.0)	82 (38.9)	45 (31.5)	63 (32.2)	292 (35.3)	
	Room Temperature	63 (29.2)	-	74 (35.1)	66 (46.2)	20 (11.5)	223 (27.0)	
	Others	-	-	-	-	42 (24.1)	42 (5.1)	
	No response	14 (6.5)	6 (7.2)	-	-	-	20 (2.4)	0
(f)	**How long can *dambu* be stored?**							
i.	**Refrigeration**							
	1 – 2 days	17 (16.2)	6 (35.3)	19 (23.2)	29 (40.0)	52 (57.4)	123 (34.4)	
	3 – 4 days	35 (33.3)	-	32 (39.0)	18 (28.6)	18 (19.8)	103 (28.8)	
	5 – 7 days	53 (50.5)	11 (64.7)	31 (37.8)	16 (25.4)	21 (23.7)	132 (36.9)	0.84
(ii)	**Sun-drying**							
	12 h	33 (45.2)	16 (36.4)	13 (24.5)	12 (31.6)	19 (17.0)	93 (35.8)	
	1 day	19 (26.0)	4 (9.1)	31 (58.5)	18 (47.4)	21 (39.6)	93 (35.8)	
	2 days and above	21 (28.8)	24 (54.6)	9 (17.0)	8 (21.1)	13 (24.5)	75 (28.8)	0.75
(iii.)	**Room Temperature**							
	1 day	15 (39.5)	3 (13.6)	34 (44.7)	16 (38.1)	12 (40.0)	80 (38.5)	
	2 days	18 (47.4)	19 (86.4)	17 (22.4)	5 (11.9)	3 (10.0)	62 (29.8)	

Figures in parentheses are percentages,Kruskal - Wallis one-way ANOVA test.

dambu migrating to towns and continents some distance from where *dambu* is made, a wider range of *dambu* packaging is required. Such packaging must be able to withstand transportation hazards, provide longer shelf-life and communicate necessary information about the product.

It takes about 1 - 5 h for 54.2% of the retailers to sell *dambu* worth N200.00. Some of the retailers (8.43%) would sell the same amount of *dambu* for 1 - 2 days. The rate of sale of *dambu* actually varies with the weather; peak of demand for *dambu* is usually the hot season (March to June) when the temperature range is 24.7 - 42.0°C.

Storage stability of *dambu*

Left over *dambu* can either be preserved by refrigeration, sun drying or at room temperature (Table 4). Sun drying as indicated by 35.3% of the respondents was the most common method of storing left over *dambu*. Choice of sun drying is not surprising since a majority of the processors/retailers do not have access to refrigerators. Storage of *dambu* under the sun was achieved by leaving the *dambu* in a desired container uncovered under the sun. Storage of *dambu* uncovered exposes the product to insects, flies and dust. The presence of dirt and dust on any food will increase the rate of spoilage because they carry microorganisms (UNIFEM, 1993). The habit of consuming the *dambu* without reheating can constitute a health hazard. The shelf-life of *dambu* stored under this condition was indicated by 35.6% of the respondents as 12 h, 35.6% of the respondents as 1 day (24 h) and 2 days and above by 28.7% of the respondents. It is known that shelf-life increases with decease in temperatures.

Storage at room temperature was achieved by leaving the *dambu* in a desired container (plate or calabash) uncovered. Storage of *dambu* without coverage predisposes the product to insect, flies and dust which will increase the rate of spoilage. This constitutes a health hazard. The variation in shelf-life depends on the ambient temperature. The best way to preserve *dambu* according to 36.9% of the respondents is by sun-drying. The respondents (33.5%) indicated it is by storing at room temperature and 28.8% at refrigeration temperature. Unfortunately, not many processors and consumers of *dambu* have access to refrigerators.

The two major signs of spoilage are off-odour and mould as indicated by 32.6 and 31.6% of the respondents respectively. Many of the respondents (32.9%) indicated that *dambu* stored under the sun was not safe for consumption after 1 day while majority indicated that it was after 4 - 7 days. Some of the respondents still believed that it can be consumed after this day.

Dambu Utilization Pattern

The common liquid in which the *dambu* are crumbled are 'nono' (fermented skim milk) and 'Kindrimo' (fermented whole milk) as indicated by 35.7 and 37.7% of the respondents respectively. Water and milk seem to be the alternative media for eating *dambu* (Tables 5 and 6). *Dambu* can be eaten alone as indicated by 57.9% of the respondents' that is, in the absence of 'nono', 'kindrimo', water or milk. *Dambu* can be eaten as a main meal when accompanied with 'nono' or 'kindrimo'. Total percentage of 27.4, 39.2 and 33.4% was deduced for the poor, the middle and the rich class consumption of *dambu*, respectively. Majority of the respondents indicated that in a paper, low density polyethylene bag or mashed into fermented skim or whole milk for immediate consumption. *Dambu* is distributed with minimum packaging (Jideani et al., 2001). Processors and retailers of *dambu* are primarily concerned with reducing wastage and having a container for their food. Their choice of suitable packaging is to provide protection during a short shelf life and for local distribution. With population growth and consumers of infants at the age of 0 - 3 years consume *dambu*. The poor consumed less *dambu* as compared to other classes. A possible reason may be that they lack the money to meet up with the increasing cost of *dambu* and the corresponding 'nono' or 'kindrimo'. The respondents (55.5%) indicated that 30 - 60% of each class consumed *dambu*. The middle class ate more of *dambu* than all the classes. The rich does not eat as much as expected since money was not their problem. The problem of the rich and elites (particularly non - Fulani or Hausa) seemed to be the unsatisfactory method with which *dambu* was handled. The study reveals that there is a market potential for *dambu* if the processing method, packaging and storage are improved.

Economics of *dambu* processing

The economics of *dambu* processing is shown in Table 6. The cost of equipment for *dambu* production may be regarded as capital expenditure in the business of *dambu* production. Total cost was less than ₦2,000 depending on the size of the equipment purchased. Moreover, the equipment are not used for *dambu* production only. They find use in other household chores. This made them quite profitable to the processor.

The price of the grain varied, being low during the harvesting period. The cost of fuel (firewood), which was most common in use ranged from ₦35 - ₦40. The cost of outsourcing dehulling and grinding was about N10. Processing of 2.5 kg of millet grain will produce *dambu* worth ₦400 - ₦600 with by-product (bran) either sold for feedstuff or fed to animals as well as the water obtained from washing. *Dambu* production therefore was quite economical. Apart from the profit obtained from the marketing of the product, the by-products are of economic importance.

Labour cost in *dambu* production was fairly cheap. Majority employs assistants during *dambu* production.

Table 5. *Dambu* Product Utilization

Question		Processor	Retailer	Processor/ Retailer	Consumer	Processor/Consumer	Total	P-value
		n = 216	n = 83	n = 211	n = 143	n = 174	n = 827	
(a)	**What is *dambu* eaten with?**							
	Soup	-	-	-	14 (9.79)	-	14 (1.7)	
	Water	-	-	-	4 (2.30)	-	4 (0.5)	
	"Madara"	86 (39.81)	16 (19.28)	6 (28.4)	35 (24.48)	59 (33.91)	202 (24.4)	
	"Nono"	73 (33.80)	45 (54.22)	77 (37.40)	17 (11.89)	81 (46.55)	295 (35.7)	
	"Kindirimo"	57 (26.39)	22 (26.51)	126(59.72)	73 (51.05)	34 (19.54)	312 (37.7)	0
(b)	**Can *dambu* be eaten alone?**							
	Yes	139 (64.35)	41 (49.40)	156(73.93)	90 (62.94)	53 (30.4)	479 (57.9)	
	No	77 (35.65)	42 (58.60)	55 (20.67)	53 (37.06)	121 (69.64)	348 (42.1)	0.36
(c)	**Is *dambu* eaten as a meal?**							
	Yes	138 (63.8)	31 (37.35)	95 (45.04)	41 (28.67)	80 (45.98)	385 (46.5)	
	No	78 (36.1)	52 (62.65)	116(54.98)	102 (71.33)	94 (54.04)	442 (53.4)	0.62
(d)	**If no, is it a snack?**							
	Yes	54 (69.23)	16 (30.77)	93 (80.17)	25 (24.51)	84 (84.3)	272 (61.5)	
	No	24 (30.77)	36 (69.23)	23 (19.83)	77 (75.49)	10 (10.63)	170 (38.5)	
	Total	78	52	116	102	94	442	0.3
(e)	**What classes of people eat *dambu*?**							
	Poor	74 (32.8)	19 (22.89)	44 (20.85)	32 (22.38)	61 (35.06)	227 (27.4)	
	Middle class	72 (33.53)	48 (57.83)	93 (44.08)	63 (44.06)	48 (27.59)	324 (39.2)	
	Rich	73 (33.80)	16 (19.28)	74 (35.07	48 (33.57)	65 (37.36)	276 (33.4)	
(f)	**Can you estimate the % of each?**							
	30 - 60%	118 (54.63)	60 (72.29)	163(77.15)	38 (26.57)	80 (45.98)	459 (55.5)	
	60 - 90%	68 (31.48)	15 (18.07)	42 (19.91)	71 (99.65)	66 (37.93)	262 (31.7)	
	90 - 100%	28 (12.96)	8 (9.69)	6 (2.81)	34 (15.87)	28 (16.99)	104 (12.6)	
	No response	2 (0.93)	_	_	_	_	2 (0.2)	
(g)	**Do infants eat *dambu*?**							
	Yes	115 (53.24)	45 (54.22)	142 (67.3)	78 (54.55)	92 (52.87)	472 (57.1)	
	No	101 (49.76)	38 (45.78)	69 (32.7)	65 (95.45)	82 (47.13)	355 (42.9)	0.26
(h)	**If yes at what age is *dambu* introduced?**							
	0 - 3 months	46 (40)	21 (46.67)	83 (58.45)	18 (23.08)	76 (82.61)	244 (51.7)	
	4 - 6 months	61 (53.04)	13 (28.89)	39 (27.46)	44 (56.04)	12 (13.04)	169 (35.8))	
	9 months - above	8 (6.96)	11 (24.44)	20 (14.08)	16 (20.51)	4 (4.35)	59 (12.5)	
	Total	115	45	142	78	92	472	
(i)	**If no what is the reason?**							
	Not ideal for them	34 (33.7)	7 (18.4)	23 (33.3)	9 (13.8)	3 (3.7)	76 (21.4)	
	Infant system cannot digest it	29 (28.7)	18 (47.7)	32 (36.4)	21 (32.3)	32 (39)	132 (37.2)	
	Constipation	23 (22.8)	13 (34.2)	14 (20.3)	35 (53.8)	47(57.3)	132 (37.2)	
	No response	15 (14.9)	_	_	_	_	15 (4.2)	
	Total	101	38	69	65	82	355	0.01
(j)	**Does *dambu* have any special attribute?**							
	Yes	120 (55.6)	53 (63.9)	93 (44.1)	71 (49.7)	89 (51.1)	426 (51.5)	
	No response	96 (32.4)	30 (36)	118 (55.9)	72 (50.3)	85 (51.1)	401 (48.5)	0.79

Figures in parentheses are percentages Kruskal - Wallis one-way ANOVA test.

Table 6. Economics of *dambu* processing.

		Status						
Question		Processor	Retailer	Processor/ Retailer	Consumer	Processor/Con sumer	Total	P-value
		n = 216		n = 211	n = 143	n = 174	n = 827	
(a)	**If yes describe the attribute**							
	Medicinal	85 (70.8)	44 (83)	69 (74.2)	19 (26.8)	71 (79.8)	288 (67.6)	
	sedative	35 (29.2)	9 (16.9)	24 (25.8)	52 (73)	18 (20)	138 (32.4)	
	Total	120	53	93	71	89	426	0.06
(b)	**What kind of by-product is derived?**							
	Bran	103 (47.7)	48 (57.8)	145 (68.7)	61(42.7)	122 (70.0)	479 (57.9)	
	Chaf	113 (52.3)	35 (42.2)	66 (31.3)	82 (57.3)	52 (29.9)	348 (42.1)	0.28
(c)	**How is by product used?**							
	Animal feed	160 (74.1)	55 (66.3)	133 (63)	69 (48)	38 (21.8)	455 (55.0)	
	Fire wood	56 (25.9)	28 (33.7)	78 (36.9)	74 (51.7)	136 (78)	372 (45.0)	0.59
(d)	**Any monetary value to a produce**							
	Yes	81 (37.5)	14 (16.9)	182 (86.3)	5 (3.5)	89 (51.1)	371 (44.9)	
	No	135 (62.5)	69 (83)	29 (13.7)	138 (96.5)	85 (48.9)	456 (55.1)	0.67
(e)	**If yes indicate the value**							
	Per Mudu 2 - 5	32 (39.5)	12 (85.7)	93 (51.1)	5 (100)	68 (76.4)	210 (56.6)	
	Per bag 50 kg	45 (55.6)	2 (14.3)	82 (45.1)	-	21 (23.6)	150 (40.4)	
	No response	4 (4.9)	-	7 (3.8)	-	-	11 (3.0)	
	Total	81	14	182	5	89	371	0.61
(f)	**Is anyone assisting you in processing?**							
	Yes	146 (676)	72 (86.7)	182 (86.3)	78 (54.5)	96 (55.2)	574 (69.4)	
	No	70 (32.4)	11 (13.3)	29 (12.7)	65 (45.5)	78 (44.8)	253 (30.6)	0.03
(g)	**If yes, describe**							
	Relative	40 (27.4)	38 (52.8)	84 (46.2)	63 (80.8)	43 (44.8)	268 (46.7)	
	Employees	21 (14.4)	3 (4.2)	14 (7.7)	1 (1.3)	21(21.9)	60 (10.5)	
	Children	63 (43.2)	31 (43.1)	84 (46.2)	14 (17.9)	32(33.3)	224 (39.0)	
	Other	22 (15.1)	-	-	-	-	22 (3.8)	
	Total	146	72	182	78	96	574	0
(h)	**Which of these in Q. (f) do you pay?**							
	Relative	20 (16.5)	14 (42.4)	43 (58.9)	21(77.8)	12 (25)	110 (36.4)	
	Employees	21 (17.4)	12 (39.4)	14 (19.2)	1 (3.7)	21 (43.8)	60 (19.9)	
	Children	51 (12.4)	16 (48.5)	16 (21.9)	5 (18.5)	15 (31.3)	67 (22.2)	
	No response	65 (53.7)	-	-	-	-	65 (21.5)	
	Total	121	33	73	27	48	302	0.8

Figures in parentheses are percentages Kruskal - Wallis one-way ANOVA test.

Out of this number, majority employs their relatives' and friends. The cost of employing these workers is usually minimal.

When *dambu* quality is improved, consumers can have confidence in the products and the producers can obtain better prices. Moreover, with the increasing influence of advertising upon customers, small food processing enterprises like *dambu* will have to improve the packaging and preservation of their products if they are to survive against competition (Jideani et al., 2001).

Conflict of interests

The authors did not declare any conflict of interest.

ACKNOWLEDGEMENTS

The authors are grateful to Miss Victoria Nnaji, Mr. Daniel Yakubu, Aliyu Ibrahim and Abdulraheem Saratu and others for helping in the distribution of the questionnaires and interviewing the respondents.

REFERENCES

Agu HO, Jideani IA, Yusuf IZ (2007). Nutrient and Sensory properties of *dambu* produced from different cereal grains. Nutr. Food Sci. 37(4):272-281.

Agu HO, Anosike AN, Jideani IA (2008). Physico-chemical and Microbial qualities of *dambu* produced from different cereal grains. Pak. J. Nutr. 7(1):21-26.

Apena A, Chinweike-Umeh SN, Abaelu AM (2006). Evaluation of improved *ogi* (Apik) with potential for the control of diarrhoea. Nig. Food J. 24(1):80-85.

Ayo JA, Obeya MC (2004). Effect of different cultivars of millet on the quality of *kunun-zaki*: A millet based non-alchoholic beverage. J. Food Sci. Technol. 41(4): 394-397.

Ayo JA, Agu H, Famoriyo OF (2008). Effect of different cereal grains on the quality of *masa*. Pak. J. Nutr. 7(4):530-533.

Filli KB, Nkama I, Abubakar UM, Jideani VA (2007). Optimization and amino acid profile of extruded millet fortified with Bambara nut for the manufacture of *fura*. 31st annual conference proceedings of Nigerian Inst. Food Sci. Technol. 67-69.

Gaffa T, Jideani IA (2001). Sensory evaluation of different levels of *Cadaba farinose* (*Dangara*) in *kunun-zaki* and determination of its position among other saccharifying agents. J. Food Sci. Technol. 38: 405-406.

Gaffa T, Jideani IA, Nkama I (2004). Chemical and physical preservation of *kunun-zaki* – A cereal based non-alchoholic beverage in Nigeria. J. Food Sci. Technol. 41:66-70.

Giese J (1994). Antimicrobials: assuring food safety. J. Food Technol.48:102-109.

Haruna US (2003). Personal communication. Food Science and Technology Department, Federal Polytechnic Bauchi, Bauchi State, Nigeria.

Hounhouigan DJ, Robert Nout MJ, Nago CM, Houben JH, Romberts FM (1993). Composition and microbiological and physical attributes of *mawè*, a fermented maize dough from Benin. Int. J. Food Sci. Technol. 28:513-517.

Igyor MA, Ariahu CC, Omale Y (2006). Chemical and microbilogical characterization of *burukutu* - a traditional Nigerian beer. Nig. Food J. 24(1):86-91.

Jideani VA, Wedzicha BL (1994). Reaction of sorbic acid in millet and sorghum doughs: reaction with thiols. J. Food Additives Contam. 11:539-548.

Jideani VA, Olakunmi OA, Jideani IA (1995). Shelf-life and predominating microflora of a non-fermented pearl millet dough (*fura*). Nig. Food J. 13:83-93.

Jideani VA, Nkama I, Agbo EB, Jideani IA (2001). Survey of *fura* production in some Northern states of Nigeria. Plant Foods Hum. Nutr. 56:23-26.

Nkama I, Angarawai IA, Badau MH (1999). Survey report on Food Technology of Millet. Lake Chad Research Institute Maiduguri, Nigeria, p. 53.

Norman J (1990). The Complete Book of Spices. Dorling Kindersley Ltd London, pp. 6-40.

NRC (1996). Lost Crops of Africa: *Grains*. National Research Council. National Academy Press Washington, D.C. 1: 77-125.

Perten H (1983). Practical experience in processing and use of millet and sorghum in Senegal and Sudan. Cereal Foods World 28:680-683.

Umoh VJ (2003). Safety and quality assessment of traditionally fermented soybean enriched millet based beverage, *fura*. Nig. Food J. 21:62-69.

UNIFEM (1993). Packaging. United Nation's Development Fund for Women. Food Cycle Technology source *Book* No.7 G.M.G., Italy, p. 9.

Zaika (1988). Spices and herbs: their antimicrobial activity and its determination. J. Food Safety 9:97-118.

Nutrients content and lipid characterization of seed pastes of four selected peanut (*Arachis hypogaea*) varieties from Ghana

Guy Eshun[1] , Emmanuel Adu Amankwah[2] and John Barimah[2]

[1]Department of Home Economics Education, Faculty of Science Education, University of Education, Winneba, Ghana.
[2]Department of Food Science and Technology, Faculty of Biosciences, Kwame Nkrumah University of Science and Technology, Kumasi, Ghana.

The proximate nutrients, calorific value, mineral nutrients and lipid characteristics of the seed pastes of four (4) peanut (*Arachis hypogaea*) varieties namely: *Sinkarzie*, *F-mix*, *JL 24* and *Manipintar* were investigated. Crude protein values ranged between 23.62% (±0.34) in *Sinkarzie* to 28.88% (±0.24) in *JL 24*. Crude fat ranged between 38.11% (±0.47) in *JL 24* to 48.79% (±0.20) in *Sinkarzie*. Carbohydrate content ranged between 11.54% (±0.18) for *Sinkarzie* to 19.65% (±0.55) in *JL 24*. The energy content in kCal/100 g sample ranged between 537.06 in *JL 24* to 581.54 in *Manipintar*. Calcium contents (mg/100 g) ranged between 23.14 (±0.49) in *Manipintar* to 128.25 (±2.90) in *F-mix*. At room temperature (25°C), all the oils from the four groundnut varieties were pale yellow coloured liquids with nutty odour. Saponification value (mg KOH/g oil) ranged from 144.70 (±0.63) in *JL 24* to 208.97 (±0.58) in *Sinkarzie*. Iodine value (mg/100 g) ranged from 79.10 (±0.15) in *Sinkarzie* to 86.93 (±0.26) in *F-mix*. Percentage free fatty acids (% FFA) ranged from 8.09% (±0.20) in *Manipintar* to 14.21% (±0.10) in *F-mix*. The results indicate that while the groundnut seeds could be used in fighting malnutrition, the oils obtained from the seeds could also be used for domestic and industrial purposes.

Key words: Ghana, peanut, proximate analysis, lipid characterization.

INTRODUCTION

Groundnut is a major annual oilseed crop and a good source of protein (Asibuo et al., 2008). The oil is mainly used as cooking oil and for the production of soap, margarine and cosmetics (Ong et al., 1995 as cited by Anyasor et al., 2009). Literature has reported many health benefits associated with the consumption of peanuts including cancer inhibition (Awad et al., 2000 as cited by Adjou et al., 2013).

To improve the nutritional quality of cereal-based traditional diets in Africa, the use of peanut flour as a protein supplement has often been suggested (Singh and Singh, 1991). Peanut can be roasted or fried to serve as a snack and the paste is used as margarine or butter (Atasie et al., 2009). The paste is also used as a thickener in soups and stews. It can also be consumed in the form of peanut cake (*kulikuli*) which is made from peanut after oil extraction (Adjou et al., 2012). Peanut protein is increasingly becoming important as food and feed sources, especially in developing countries where protein from animal sources are not within the means of majority of people. The seed has several uses as whole seed or processed to make peanut butter, oil, soups, stews and other products. The cake has several uses in feed and infant food formulations. Groundnut provides considerable amounts of mineral elements to supplement the dietary requirements of humans and other farm animals (Asibuo et al., 2008).

The chemical composition of groundnut seeds has been evaluated in some countries. However, studies on the chemical composition of groundnut cultivars grown in Ghana have not been conducted (Asibuo et al., 2008). This study was therefore carried out to find the chemical composition of the seeds and the physicochemical properties of the oils of four (4) peanut (*A. hypogaea*) varieties namely: *Sinkarzie*, *F-mix*, *JL 24* and *Manipintar* to enable peanut users make an informed decision for a specific use for the peanut variety they have.

MATERIALS AND METHODS

Source of raw materials

The seeds of peanut used in this study were obtained from the Savannah Agricultural Research Institute, Tamale in the Northern region of Ghana. Four varieties of peanut (*Sinkarzie*, *F-mix*, *JL 24* and *Manipintar*) which were in their shells and stored at room temperature at the Savannah Agricultural Research Institute were collected for the study.

Sample analysis

Proximate nutrients determination

The peanut shells were removed and the seeds were milled with a laboratory miller (Tecator 1093 Sample mill, Sweden) and proximate nutrient analyses were carried out on the milled peanut sample.

Moisture content

Moisture content was determined by the method of the Association of Official Analytical Chemists (AOAC, 1990) by drying the sample in a thermostatically controlled oven to constant weight. Five grams (5.0 g) of the sample was accurately weighed into a previously cleaned, dried and weighed glass crucible. The crucible with its content was put into the oven at 103°C for 12 h. The sample was then cooled in a desiccator and weighed. The loss in weight expressed as a percentage of the initial weight of the sample gives the moisture content of the sample.

Ash content

Ash was determined by the method of Association of Official Analytical Chemists (AOAC, 1990). A 2.0 g sample was weighed into a previously dried and weighed porcelain crucible. The crucible with its content was placed in a furnace preheated to 600°C for 2 h. The sample was allowed to cool in the furnace to 250°C. The crucible and the ash were then transferred into an oven at 100°C for 30 min cooling. After this period, the crucible with its content was cooled in a desiccator. The crucible with its content was weighed. The weight of the ash was expressed as a percentage of the initial weight of the sample.

Crude protein content

Crude protein was determined by the method of the Association of Official Analytical Chemists (AOAC, 1990). Two grams (2.0 g) of the sample was weighed into a digestion flask and 0.5 g of selenium catalyst was added. Twenty five (25) ml of concentrated H_2SO_4 was added and the flask was shaken to mix the contents. The flask was then placed on a digestion burner for 8 h and heated until the solution turned green and clear. The sample solution was then transferred into a 100 ml volumetric flask and made up to the mark with distilled water.

Twenty five millilitres (25 ml) of 2% boric acid was pipetted into a 250 ml conical flask and two drops of mixed indicator (20 ml of bromocresol green and 4 ml of methyl red) solution was added; and into the decomposition chamber of the distillation apparatus was added 15 ml of 40% NaOH solution. Ten millilitres (10 ml) of the digested sample solution was then introduced into a Kjedahl flask. The condenser tip of the distillation apparatus was then dipped into the boric acid contained in the conical flask. The ammonia in the sample solution was then distilled into the boric acid until it changed completely to bluish green. The distillate was then titrated with 0.1 N HCl solution until it became colourless. The percent total nitrogen and crude protein were calculated using a conversion factor of 6.25.

Crude fat content

Crude fat was determined based on the Sohxlet extraction method of AOAC (1990). A 250 ml quickfit round bottom flask was washed and dried in an oven at 103°C for 25 min and allowed to cool to room temperature before it was weighed. Two grams (2.0 g) of the sample was weighed into a muslin thimble. This was inserted into the extraction column with the condenser connected. Two hundred millilitres (200 ml) of the extracting solvent (petroleum ether, boiling point 40 to 60°C) was poured into the round bottom flask and fitted into the extraction unit. The flask was then heated with the aid of electrothermal heater at 60°C for 2 h.

Losses of solvent due to heating were checked with the aid of the condenser so that it cooled and refluxed the evaporated solvent. After extraction, the thimble was removed and the solvent salvaged by distillation. The flask containing the fat and residual solvent was placed on a water bath to evaporate the solvent followed by a further drying in an oven at 103°C for 30 min to completely evaporate the solvent. It was then cooled in a desiccator and weighed. The fat obtained was expressed as a percentage of the initial weight of the sample.

Crude fibre content

Crude fibre was determined by the method of the Association of Official Analytical Chemists (AOAC, 1990). The defatted sample (from crude fat determination) was transferred into a 750 ml Erlenmeyer flask and 0.5 g of asbestos was added. Two hundred millilitres (200 ml) of boiling 1.25 % H_2SO_4 was added and the flask was immediately set on a hot plate and condenser connected to it. The content was brought to boil within 1 min and the sample was digested for 30 min.

At the end of the 30 min, the flask was removed and the content was filtered through a linen cloth in a funnel and subsequently washed with boiling water until the washings were no longer acidic. The sample was washed back into the flask with 200 ml boiling 1.25 % NaOH solution. The condenser was again connected to the flask and the content of the flask was boiled for 30 min. It was then filtered through the linen cloth and thoroughly washed with boiling water until the washings were no longer alkaline. The residue was transferred to a clean crucible with a spatula and the remaining particles washed off with 15 ml ethanol into the crucible.

The crucible with its content was then dried in an oven overnight and cooled in a desiccator and weighed. The crucible with its content was then ignited in a furnace at 600°C for 30 min, cooled and reweighed. The loss in weight gave the crude fibre content and was expressed as a percentage of the initial weight of the sample.

Carbohydrate content

Total percentage carbohydrate was determined by the different method as reported by Onyeike et al. (1995). This method involves adding the total values of crude protein, crude fat, crude fibre, moisture and ash constituents of the sample and subtracting it from 100. That is:

Percentage carbohydrate = 100 - (% moisture + % ash + % crude protein + % crude fat + % crude fibre)

Minerals determination

Calcium, iron and magnesium

Calcium, iron and magnesium were determined by Atomic Absorption Spectrophotometry (Agte et al., 1995). One gram (1 g) of the sample was dry ashed in a muffle furnace at 550°C for 5 h until a white residue of constant weight was obtained. The minerals were extracted from the ash by adding 20.0 ml of 2.5% HCl, heated to reduce the volume to 7.0 ml, and this was transferred quantitatively to a 50 ml volumetric flask. It was diluted to the mark (50 ml) with distilled water, stored in clean polyethylene bottles and magnesium content determined using atomic absorption spectrophotometer.

Phosphorus

The phosphorus determination was done based on the method described by Jackson et al. (1974). Two grams of the sample was ashed in muffle furnace preheated to 600°C for 2 h. The ash was dissolved in 5 ml of 5 M H_2SO_4. Four milliliters (4 ml) of 2% ascorbic acid and 4 ml of 4% ammonium molybdate were added to the resulting solution and shaken for uniform mixing. The absorbance of each sample was determined with a UV spectrophotometer.

Characterization of peanut oil

Extraction of peanut oil

The peanut seeds were milled with a laboratory miller (Tecator 1093 Sample mill, Sweden) and dried to constant weight in a thermostatically controlled oven at 105°C. The dried paste was transferred into a thimble and oil extraction carried out using petroleum ether with Soxhlet apparatus. The extracting solvent (petroleum ether) was evaporated leaving the concentrated oil sample for analysis.

Physical parameters of peanut oil

Colour and state of the oils

At room temperature (25°C), all the oils from the four peanut varieties were liquids with nutty odour. They were pale yellow coloured liquids. They were bright and clear with a light consistency.

Specific gravity

Specific gravity of the oil was determined according to AOCS method Cc 10a-25 (1993). The specific gravity bottle was placed in a water bath maintained at 25°C and filled with distilled water. It was removed, wiped dry (outside the bottle) and weighed. The bottle was emptied, dried and again placed in water bath at 25°C. The bottle was refilled with the oil and made to stay in the water bath for 30 min. It was then removed, cleaned and wiped (outside the bottle) completely dry and weighed. The specific gravity at 25°C was calculated as follows:

$$\text{Specific gravity at 25/25°C} = \frac{weight\ of\ bottle\ and\ oil\ at\ 25°C - weight\ of\ bottle\ at\ 25°C}{weight\ of\ bottle\ at\ 25°C}$$

Refractive index

Refractive index of the oil was determined according to the AOCS method Cc 7-25 (1993). Several drops of the oil were placed on the lower prism of an Abbe refractometer (Model: Bellingham + Stanley Limited, 60/70 ABBE Refractometer, England) which was also adjusted to the same temperature as that of the sample. The prisms were closed and tightened firmly with the screw head, ensuring that the sample came to the same temperature of the instrument. The instrument was adjusted until the most distinct reading possible was obtained and the refractive index read.

Chemical parameters of groundnut oil

Percentage free fatty acid

Percentage free fatty acid was determined using the recommended method of the American Oil Chemists' Society (AOCS, 1993). One point four grams (1.4 g) of oil was weighed into a flask containing 15 ml of hot neutralized alcohol and 0.4 ml of phenolphthalein indicator was then added. The content was titrated with 0.5 N NaOH. Percentage free fatty acid value was calculated (as oleic acid) using the formula:

$$FFA\,(as\ oleic) = \frac{V \times N \times 28.2}{W}$$

Where, V = volume (ml) of NaOH solution, N = normality of NaOH solution, W = weight of oil sample.

Iodine value

Iodine value was determined according to AOCS method (1993). An amount of 0.2 g of the oil was accurately weighed into a 500 ml flask. Fifteen millilitres (15 ml) of carbon tetrachloride was added to the sample and swirled to ensure that the sample completely dissolved in it. Twenty five millilitres (25 ml) of Wij's solution was then pipetted into the flask containing the sample. The flask was stoppered and swirled to ensure complete mixing.

The sample was then placed in the dark for 30 min at room temperature. The flask was removed from storage and 20 ml of 10% KI solution added, followed by 150 ml of distilled water. The mixture was titrated with 0.1 N $Na_2S_2O_3$ solution, adding gradually and with constant and vigorous shaking until the yellow colour had almost disappeared. One and half millilitres (1.5 ml) of starch indicator solution was added and the titration was continued until the blue colour disappeared. A blank determination was conducted simultaneously.

The iodine value was calculated using the formula:

$$Iodine\ value = \frac{(B - S) \times N \times 12.69}{Weight\ of\ oil}$$

Table 1. Proximate and energy composition of raw peanut (fresh weight basis).

Constituent	*Sinkarzie*	*F-mix*	*JL 24*	*Manipintar*
Moisture (%)	7.54^{bc} (±0.39)	8.99^{a} (±0.61)	7.99^{b} (±0.15)	7.13^{c} (±0.05)
Ash (%)	2.96^{a} (±0.09)	2.45^{c} (±0.02)	2.68^{b} (±0.05)	2.77^{b} (±0.10)
Crude protein (%)	23.62^{d} (±0.34)	25.51^{c} (±0.14)	28.88^{a} (±0.24)	26.71^{b} (±0.26)
Crude fat (%)	48.79^{a} (±0.20)	47.56^{b} (±0.36)	38.11^{d} (±0.47)	46.71^{c} (±0.32)
Crude fibre (%)	5.55^{a} (±0.06)	3.25^{b} (±0.06)	2.69^{d} (±0.06)	3.10^{c} (±0.04)
Carbohydrate (%)	11.54^{d} (±0.18)	12.24^{c} (±0.29)	19.65^{a} (±0.55)	13.58^{b} (±0.06)
Energy (kCal/100 g)	579.72^{a} (±1.84)	579.07^{a} (±3.97)	537.06^{b} (±2.62)	581.54^{a} (±1.71)

Values are means ± standard deviations of triplicate determinations. Values in the same row having the same superscript letters are not significantly different ($p > 0.05$).

Where, B = blank titre value, S = sample titre value and N = normality of $Na_2S_2O_3$.

Saponification value

Saponification value was determined according to AOCS (1993). Two grams (2.0 g) of the oil was weighed into a flask. Twenty five millilitres (25 ml) of alcoholic KOH was pipetted and allowed to drain for about 1 min into the mixture. A blank determination was prepared and determined simultaneously with the sample. A condenser was connected to the flask and the mixed sample was allowed to boil gently and steadily for 45 min for complete saponification. The flask and condenser were then cooled but not sufficient to form a gel. The condenser was disconnected and 1 ml of phenolphthalein indicator was added to the content of the flask. The solution was titrated with 0.5 N HCl until the pink colour just disappeared. The saponification value was calculated using the formula:

$$Saponification\ value = \frac{(B - S) \times 56.1 \times N}{weight\ of\ oil\ sample}$$

Where, B = Blank titre value, S = sample titre value and N = normality of HCl.

Statistical analysis

All determinations were done in triplicates and the data analyzed by analysis of variance (ANOVA) at 5% level of significance using the programme ASSISTAT Version 7.5 beta (2011). Means were separated using the Duncan's multiple range statistical test.

RESULTS AND DISCUSSION

Proximate nutrients composition and calorific values of peanut

Table 1 presents results of proximate analyses and calorific (energy) values of raw peanuts. The moisture contents of the peanut varieties ranged from 7.13% (±0.05) in *Manipintar* to 8.99% (±0.61) in *F-mix*. Significant differences ($p < 0.05$) were observed among the four peanuts varieties for moisture content. Ash was highest in *Sinkarzie* with a value of 2.96% (±0.09) and lowest in *F-mix* with a value of 2.45% (±0.02). *JL 24* and *Manipintar* did not differ significantly ($p > 0.05$) in ash contents. However, the ash contents of *JL 24* and *Manipintar* differed significantly ($p < 0.05$) from the other peanut varieties. The ash content obtained fell within the range of ash content for nuts (0.8 to 3.5%) (Aurand et al., 1987 as cited by Nagre et al. (2012). Crude protein values for the four peanut varieties ranged from 23.62% (±0.34) in *Sinkarzie* to 28.88% (±0.24) in *JL 24*. The values obtained fell within the range of crude protein values (18.92 to 30.53%) for groundnut given by Asibuo et al. (2008).

The values for crude protein obtained are in close agreement with the value of 26.40% obtained by Onyeike and Oguike (2003). The figures are however lower than the 33.88% reported by Adjou et al. (2013). The crude protein content of the peanut varieties differed significantly ($p < 0.05$) from each other. *Sinkarzie* had the highest crude fat content of 48.79% (±0.20) and *JL 24* had the lowest crude fat content of 38.11% (±0.47). All four peanut varieties were significantly different ($p < 0.05$) in their crude fat contents. The crude fat values obtained lie within the range of crude fat values between 33.6 to 54.96% reported by Asibuo et al. (2008).

The fat content is important in diets as it promotes fat soluble vitamin absorption. It is a high energy nutrient and does not add to the bulk of the diet (Atasie et al., 2009). The high crude fat values obtained give indication that the peanut varieties could be used in improving palatability of foods in which they are incorporated. The high crude fat values also signify that these peanut varieties are viable sources of oil and will be suitable for commercial production of oil.

Crude fibre values for the peanut varieties ranged from 2.69% (±0.06) for *JL 24* to 5.55% (±0.06) for *Sinkarzie*. Crude fibre is a measure of the quantity of indigestible cellulose, pentosans, lignin and other components of this type present in food (Aurand et al., 1987). These fibres protect the body against colon cancer, diabetis and cardiovascular illnesses (Ponka et al., 2005). It provides bulk to food to relieve constipation (Appiah et al., 2011). Diet low in crude fibre is undesirable as it could cause constipation and such diets have been associated with diseases of the colon like piles, appendicitis and cancer

Table 2. Iron, calcium, magnesium and phosphorus contents of raw peanuts.

Constituent (mg/100 g)	*Sinkarzie*	*F-mix*	*JL 24*	*Manipintar*
Fe	2.71[d] (±0.03)	4.45[c] (±0.05)	6.72[b] (±0.25)	7.69[a] (±1.54)
Ca	47.96[c] (±1.34)	128.25[a] (±2.90)	85.46[b] (±2.49)	23.14[d] (±0.49)
Mg	8.40[b] (±0.24)	8.46[a] (±0.43)	8.42[b] (±0.34)	7.57[c] (±0.54)
P	161.96[c] (±3.67)	433.09[a] (±4.26)	288.59[b] (±3.08)	78.14[d] (±2.13)

Values are means ± standard deviations of triplicate determinations. Values in the same row having the same superscript letters are not significantly different (p > 0.05).

(Atasie et al., 2009). All four peanut varieties differed significantly ($p < 0.05$) from each other in their crude fibre contents. The values of crude fibre obtained indicate that the peanut varieties have the ability to give bulk to foods in which they are incorporated to maintain gut distension for normal peristaltic movement of the gut. Carbohydrate values for the four peanut varieties ranged from 11.54% (±0.18) in *Sinkarzie* to 19.65% (±0.55) in *JL 24*. Carbohydrates are the main source of energy for all animals, including human beings. In most parts of the world, 80% of the kilocalories consumed by humans are in the form of carbohydrates (Mehas and Rodgers, 1997). The values of carbohydrate obtained suggest that peanuts could be used to manage protein energy malnutrition since it also contains good amounts of protein and fat. Values of carbohydrate obtained for the four peanut varieties differed significantly ($p < 0.05$) from each other.

Calorific value was highest in *Manipintar* with a value of 581.54 kCal/100 g (±1.71) and lowest in *JL 24* with a value of 537.06 kCal/100 g (±2.62). *Sinkarzie*, *F-mix* and *Manipintar* did not differ significantly ($p > 0.05$) in their calorific values. These however differed significantly ($p < 0.05$) from the calorific value of *JL 24*. The high calorific values obtained indicate that peanuts could constitute a major source of energy for many of the world's poor and least privileged people.

Mineral composition of peanuts

The mineral (iron, calcium, magnesium and phosphorus) content of the peanuts are shown in Table 2. Mineral analysis showed a significant difference ($p < 0.05$) in iron contents among the four peanut varieties. Iron is an important element that is necessary in the haemoglobin of the red blood cells and myoglobin in the muscle (Thomas, 2002). According to Mehas and Rodgers (1997), iron is a vital part of red blood cells that carry and release oxygen. The values for iron ranged from 2.71 mg/100 g (±0.03) in *Sinkarzie* to 7.69 mg/100 g (±1.54) in *Manipintar*. Calcium contents ranged from 23.14 mg/100 g (±0.49) in *Manipintar* to 128.25 mg/100 g (±2.90) in *F-mix*. The calcium contents of the four peanut varieties differed significantly ($p < 0.05$) from each other. Calcium is one of the main components of teeth and bones. It also plays a role in blood clotting (Mehas and Rodgers, 1997). Magnesium contents ranged from 7.57 mg/100 g (±0.54) in *Manipintar* to 8.46 mg/100 g (±0.43) in *F-mix*. These values differed significantly ($p < 0.05$) from each other except for *Sinkarzie* and *JL 24* where there was no significant difference ($p > 0.05$) in magnesium contents. The magnesium contents of *Sinkarzie* and *Manipintar* significantly differed ($p < 0.05$) from the other peanut varieties. Magnesium is involved in making proteins and releasing energy, and helps hold calcium in the enamel of the teeth (Mehas and Rodgers, 1997). Magnesium helps in keeping the muscles relaxed and the formation of strong bones and teeth. It also helps to control blood pressure and is a nerve transmitter (Kolawole et al., 2007).

Phosphorus contents ranged from 78.14 mg/100 g (±2.13) in *Manipintar* to 433.09 mg/100 g (±4.26) in *F-mix*. All the peanut varieties differed significantly ($p < 0.05$) in phosphorus content from each other. Phosphorus is closely linked with calcium. The two minerals combine to form calcium phosphate, which give bones their strength and rigid structure (Mehas and Rodgers, 1997).

Physical parameters of peanut oil

The studied physical properties for peanut oil are presented in Table 3. At room temperature (25°C), all the oils from the four peanut varieties were liquids with nutty odour. They were pale yellow coloured liquids. They were bright and clear with a light consistency. The specific gravity of the oils at 25°C ranged between 0.89 (±0.00) for *Sinkarzie* and *F-mix* to 0.90 (±0.00) for *JL 24* and *Manipintar*, indicating that the oils are less dense than water. The specific gravity values obtained are well in line with the observation of Singh and Singh (1991) who determined the density of peanut edible oil to be 0.885 g/ml. No significant difference ($p > 0.05$) was observed for the specific gravity values for the four peanut varieties. Refractive index at 25°C ranged between 1.459 (±0.00) for *Manipintar* to 1.461(±0.00) for *F-mix*. No significant difference ($p > 0.05$) was observed for the refractive indices for *Sinkarzie* and *JL 24*. The refractive

Table 3. Physical properties of peanut oil from four different peanut varieties.

Variety	Specific gravity (at 25°C)	Refractive index (at 25°C)	State at 25°C	Colour	Odour
Sinkarzie	0.89^{a} (±0.00)	1.460^{ab} (±0.00)	Liquid	Pale yellow	Nutty
F-mix	0.89^{a} (±0.00)	1.461^{a} (±0.00)			
JL 24	0.90^{a} (±0.00)	1.460^{ab} (±0.00)			
Manipintar	0.90^{a} (±0.00)	1.459^{b} (±0.00)			

Values are means ± standard deviations of triplicate determinations. Values in the same column having the same superscript letters are not significantly different (p > 0.05).

Table 4. Chemical indices of peanut oil.

Property	*Sinkarzie*	*F-mix*	*JL 24*	*Manipintar*
Saponification value (mgKOH/g oil)	208.97^{a} (±0.58)	189.34^{c} (±0.52)	144.70^{d} (±0.63)	194.25^{b} (±0.57)
% Free fatty acids (% FFA)	9.70^{c} (±0.30)	14.21^{a} (±0.11)	12.80^{b} (±0.11)	8.09^{d} (±0.21)
Iodine value (mg/100 g)	79.10^{c} (±0.15)	86.93^{a} (±0.26)	80.37^{b} (±0.15)	86.61^{a} (±0.26)

Values are means ± standard deviations of triplicate determinations. Values in the same row having the same superscript letters are not significantly different (p > 0.05).

indices for *Sinkarzie* and *JL 24* however differed significantly ($p < 0.05$) from the refractive indices for the other peanut varieties. The refractive indices obtained for the peanut varieties are close to the value of 1.449 reported by Atasie et al. (2009) for peanut oil. Refractive index is used as a measure of purity and a means of identification of fats and oils, since each substance has a characteristic refractive index (Nielsen, 1994). Studies have shown that refractive index increases by a value of 0.001 as rancid odour is noticeable (Rosell, 1984).

Chemical parameters of peanut oil

The chemical parameters of the studied oils are shown in Table 4. The results indicate that the iodine values ranged from 79.10 mg/100 g (±0.15) in *Sinkarzie* to 86.93 mg/100 g (±0.26) in *F-mix*. The iodine values obtained in this study are consistent with that obtained by other workers like Onyeike and Oguike (2003) who obtained iodine value of 87.6 mg/100 g for groundnut oil from raw groundnut seeds. Asibuo et al. (2008) indicated that the iodine values of groundnut oils studied from twenty (20) peanut varieties ranged from 85.77 to 98.43 mg/100 g. Since the iodine values of the peanut oils from this study fell below 100, they could be classified as non-drying oils and this non-drying attribute qualifies them to be used in the paint industry (Akubugwo and Ugbogu, 2007). The iodine values obtained in this study indicate that the oils contain appreciable level of unsaturated bonds. Studies have revealed that the higher the degree of unsaturation (high iodine value), the greater the tendency of the fat to go through oxidative rancidity (Joseph, 1977). The oils should therefore be stored in such a way as to protect them from oxidative deterioration. The saponification values ranged from 144.70 mg KOH/g oil (±0.63) in *JL 24* to 208.97 mg KOH/g oil (±0.58) in *Sinkarzie*. Statistical analysis revealed a significant difference ($p < 0.05$) in saponification value for all the four peanut varieties. These saponification values indicate that the oils contain a large number of fatty acids of low molecular weight and hence useful in the soap industries and in the manufacture of lather shave creams (Onyeike and Oguike, 2003). The saponification value is only of interest if the oil is going to be used for industrial purposes as it has no nutritional significance (Asiedu, 1989).

Percentage free fatty acid (% FFA) ranged from 9.70% (±0.20) for *Sinkarzie* to 14.21% (±0.10) for *F-mix*. The statistical analysis revealed significant difference ($p < 0.05$) in percentage free fatty acid values for all the four peanut varieties. The free fatty acid value is used as a measure of oil deterioration. According to Bassir (1971), it is desirable that the free fatty acid content for a dietary lipid should lie within the limits of 0.0 to 3.0%. Free fatty acids (FFAs) are not present to any significant level in healthy plant cells but FFA levels of up to 15% (more in very bad cases) is usually found in commercial crude vegetable oils whether pressed or solvent extracted. These FFAs are present as a consequence of cell damage in vegetable tissue during harvesting, storage, transportation and must be removed during refining (Harmond, 1993 cited by Afolabi, 2008). This explains the very high % FFA values for the solvent extracted peanut oils which were studied. The oils were not refined after solvent extraction before carrying out the analysis of the oil. The physicochemical properties of the studied peanut oils indicate that the oils could be used for both domestic and industrial purposes.

Conclusion

The study has established the nutritive value of the peanut seeds which could be considered as good sources of protein and oil. They can therefore be used to fight against malnutrition, especially protein energy malnutrition. The study also identified peanut varieties with various quality attributes which could be used by manufacturers of peanut products to select the varieties with desirable quality attributes for their products. For soapmaking, it will be beneficial to use the oil from *Sinkarzie* because of its high saponification number. When products which require high protein contents are to be developed, *JL 24* will be useful because of its high protein content. The presence of calcium, magnesium and phosphorus is good indication that the peanut varieties studied are rich in the minerals for bone formation. The studied chemical properties of the peanut oils indicated that the oils have potential for development for use as domestic and industrial oils. The oils are non-drying and this qualifies them to be used in the paint industry. The large saponification values obtained indicate that the oils will be useful in the soap industries and in the manufacture of lather shave creams.

ACKNOWLEDGEMENT

Authors would like to thank the Savannah Agricultural Research Institute, Tamale, Ghana for the supply of the peanuts.

REFERENCES

Adjou ES, Dahouenon-Ahoussi E, Soumanou MM (2013). Investigations on the mycoflora and processing effects on the nutritional quality of peanut (*Arachis hypogaea* L. var. TS 32-1). J. Microbiol. Biotechnol. Food Sci. 2 (3):1025-1039.

Adjou ES, Yehouenou B, Sossou CM, Soumanou MM, de Souza CA (2012). Occurrence of mycotoxins and associated mycoflora in peanut cake product (*kulikuli*) marketed in Benin. Afr. J. Biotechnol. 11(78):14354 - 14360

Afolabi IS (2008). Chemical qualities of oils from some fresh and market vegetable crops within Kwara State of Nigeria. Niger. Soc. Exp. Biol. Biokemistri 20 (2):71-75.

Agte VV, Gokhale MK, Paknikar KM, Chiplonkar S (1995). Assessment of pearl millet versus rice-based diets for bio-availability of four trace metals. Plant Foods Hum. Nutr. 48:149-158.

Akubugwo IE, Ugbogu AE (2007). Physicochemical studies on oils from five Selected Nigerian plant seeds. Pak. J. Nutr. 6 (6):75-78.

Anyasor GN, Ogunwenmo KO, Oyelana OA, Ajayi D, Dangana J (2009). Chemical Analyses of Groundnut (Arachis hypogaea) Oil. Pak. J. Nutr. 8(3):269 - 272.

AOAC (1990). Official methods of analysis. 15th edition. Association of Official Analytical Chemists, AOAC Pub., Virginia, U.S.A. pp.69-84, 951-979.

AOCS (1993). Official methods and recommended practices of the American Oil Chemists' Society. AOCS Press, Washington, DC.

Appiah F, Asibuo JY, Kumah P (2011). Physicochemical and functional properties of bean flours of three cowpea (*Vigna unguiculata* L. Walp) varieties in Ghana. Afr. J. Food Sci. 5(2):100-104.

Asibuo JY, Arkromah R, Safo-Kantanka O, Adu-Dapaah HK (2008). Evaluation of nutritional quality of groundnut (*Arachis hypogaea* L.) from Ghana. Afr. J. Food Agric. Nutr. Dev. 8 (2):134-141.

Asiedu JJ (1989). Processing Tropical Crops. A Technological Approach. Macmillan Publishers, London. pp. 170-172.

Atasie VN, Akinhanmi TF, Ojiodu CC (2009). Proximate analysis and physico-chemical properties of groundnut (*Arachis hypogaea* L.). Pak. J. Nutr. 8 (2):194-197.

Aurand LW, Wood AE, Wells MR (1987). Food Composition and Analysis. Van Nostrand Reinhold, New York. pp. 20-23.

Bassir O (1971). Handbook of Practical Biochemistry 2nd edn. Ibadan University Press, Ibadan Nigeria. pp. 28-35.

Jackson E, Farinton DS, Henderson K (1974). The analysis of Agricultural Materials, 3rd edition, Ministry of Agriculture (Fish and Food). pp. 1-37.

Joseph AF (1977). Measuring Flavour Deterioration of Fats, Oils, and Foods. General Food Cooperation Technical Center, New York. pp. 1-7.

Kolawole OM, Kayode RMO, Akinduyo B (2007). Proximate and microbial analysis of burukutu and pito produced in Ilorin, Nigeria. Afr. J. Biotechnol. 6(5):587 - 590.

Mehas KY, Rodgers SL (1997). Food Science. The Biochemistry of Food and Nutrition. Third edition. Glencoe/McGraw Hill, Peoria. Illinois, USA. pp. 35, 169, 171.

Nagre RD, Ellis WO, Oduro I (2012). Yield and quality of Pycnanthus kombo kernel butter as affected by temperature-moisture interactions. Afr. J. Food Sci. 6 (8):224 - 231.

Nielsen SS (1994). Introduction to the Chemical analysis of Foods. Chapman and Hall, New York. pp. 93 - 107.

Onyeike EN, Oguike JU (2003). Influence of heat processing methods on the nutrient composition and lipid characterization of groundnut (*Arachis hypogaea*) seed pastes. Niger. Soc. Exp. Biol. Biokemistri 15 (1):34-43.

Onyeike EN, Olungwe T, Uwakwe AA (1995). Effect of heat treatment and defatting on the proximate composition of some Nigerian local soup thickeners. Food Chem. 53:173-175.

Ponka R, Fokou E, Fotso M, Achu MB, Tchouanguep FM (2005). Methods of preparation and the energy, protein and mineral values of three Cameroonian dishes: "corn chaff", "nnam owondo/ebobolo" and "nnam ngon/ebobolo". Afr. J. Agric. Nutr. Dev. 5(1):3.

Rosell B (1984). Vegetable oils and fats. Academic Press, NY, USA. pp. 263 - 265.

Singh U, Singh B (1991). Functional properties of Sorghum-Peanut Composite Flour. American Association of Cereal Chemists, Inc. Cereal Chem. 68 (5):460-463.

Thomas D (2002). A textbook of Biochemistry with clinical correlations. Fifth edition, John Wiley and Sons Inc. publication. pp. 1159 - 1160.

10

The physicochemical and sensory evaluation of commercial sour milk (*amasi*) products

Jeremia N. Moyane[1,2] and Afam I. O. Jideani[1]

[1]Department of Food Science and Technology, School of Agriculture, University of Venda, Private Bag X5050, Thohoyandou 0950, Limpopo Province, South Africa.
[2]Agricultural Research Council, Animal Production Institute, Gastrointestinal Tract Microbiology and Biotechnology Unit, Private Bag X2, Irene, 0062 South Africa.

The physicochemical and sensory properties of five commercially available sour milk (*amasi*) products- AoA, AoB, AoC, AoD and AoE were analyzed in 3 batches. Samples were collected from retail shops in Thohoyandou. The products were examined for *Escherichia coli*, lactic acid bacteria (LAB), total plate count (TPC), pH, titratable acidity (TA), colour, viscosity and sensory properties. Microbial analysis results for LAB ranged between 1.25×10^5 and 1.97×10^6cfu with sample AoA and AoB being the highest by 1.96×10^6 and 1.97×10^6 cfu, respectively. *E. coli* count ranged from 1.22×10^4 to 1.78×10^5 cfu with sample AoE and AoC being the lowest and highest ($p < 0.05$), respectively. TPC had the least number of counts with the values between 4.2×10^3 and 9.1×10^4 cfu. The pH values of the products ranged between 4.22 and 4.34 and TA ranged between 0.80 and 0.84. Colour measurements gave L* values of the products ranging from 33.77 to 40.19, while a* and b* values were 3.08 to 6.43 and 13.17 to 18.77, respectively. Differences existed in viscosities of the products that were not significantly different ($p < 0.05$). Sensory score for the overall acceptability indicated that consumers did like the sour milk, as the values ranged from 3.8 to 4.4. Although there were significant differences ($p < 0.05$) in terms of sweetness, smoothness and astringency, they did not affect the consumer acceptability of the products. Presence of *E. coli* in the sour milk products can be of health concern to consumers.

Key words: Sour milk, *amasi*, physicochemical, microbial.

INTRODUCTION

Traditionally, sour milk known as *amasi* among the *Zulus* in South Africa is a fermented product produced in clay pots and gourds which are used repeatedly through spontaneous fermentation of raw milk that occurs naturally at ambient temperature of ± 5°C (Bryant, 1949; Keller and Jordan 1990; Gadaga et al., 2001). Fermentation is one of the oldest methods for food preservation, which contributes to their flavor, appearance and texture (Mazahreh et al., 2008; Quasem et al., 2009). Hence fermented products are generally more attractive to the consumer than non-fermented ones.

However, due to the ever escalating human population size, and the advent of commercial starter cultures, commercial *amasi* is now produced from pasteurized milk, under controlled processing conditions. Commercially, *amasi* products are pasteurized before distribution and consumption, and the shelf life of the refrigerated beverages is 21 days (Beukes et al., 2001; Dlamini and Buys

Table 1. Definition of attributes constituting the sour milk (*amasi*) products.

Attribute	Definition
Sourness	Fresh sour smell, for example, fresh sour cream
Sweetness	Sweet sensation of sugar
Texture (thickness)	Resistance to flow or motion
Smoothness	Absence of detectable solid particles
Astringency	Dry feeling in the mouth, similar to mouth-feel after drinking dry red wine
Colour	Whiteness
Overall acceptability	How good is all combined attributes

2008; Burger, 2010). Examples of 2008; Burger, 2010). Examples of *amasi* products in the study location include Limpopo dairy maas, Clover Inkomazi, Save Mor*amasi*/ Spar maas, Moordriftmaas and Hygienik (*amaziakgomo*).

Sour milk is one of the products in which microorganisms are used during its production; therefore in cases where hygienic practices are not properly applied, microbial contamination may occur during processing, distribution of product and retail outlets. The presence of *Escherichia coli* and coliform groups of microorganisms in sour milk could pose a potential health problem. Monitoring the quality of dairy product is important, considering the safety of the consumers. Since milk and milk products are good vehicles for both beneficial and harmful microorganisms, some of which may have health and subsequent financial consequences to the consumer, it is important to independently monitor quality. This study, therefore, examined those physical, chemical, microbial and sensory parameters that determine quality of sour milk product.

MATERIALS AND METHODS

Five commercial sour milk (*amasi*) products were bought from shops in Thohoyandou, Limpopo Province, South Africa. These products were coded: AoA, AoB, AoC, AoD and AoE. The samples were refrigerated at 4°C at the Department of Food Science and Technology, University of Venda (Univen). The sensory evaluation, and the chemical, physical and microbial analyses were carried out in the laboratory. Titration for lactic acid concentration, pH determination and microbial count, colour and viscosity analysis were done.

Physical analysis

The viscosity was determined using DV-E Viscometer (Brookfield Engineering Laboratories model No. RVDVE230). Spindle no. DV-E 4 was used at a speed of 10 rpm and the reading on the viscometer was taken after 3 min for each sample (Kip et al., 2006). The colour was determined as outlined in HunterLab (2008), using a Hunter Lab, ColourFlex model No. 45/0 (Hunter Associates Laboratory, Inc 11491 Sunset Hills Road, Reston, VA 20190-5280).

Physicochemical analysis

The pH of the samples was determined using pH meter, (METTLER TOLEDO, Education Line model No. EL20). The pH probe was immersed inside the 50 ml beaker of sample. Determinations were done in triplicates.The instrument was calibrated using standard buffer solutions at pH 4 and 7. The lactic acid concentration was determined by weighing 10 g of *amasi* sample in a conical flask; 30 mL of distilled water were added to the sample and mixed well by shaking. Phenolphthalein indicator (5 drops) was added and 0.1 M NaOH from the filled burette was used to titrate the solution until it retained a very slight pink tinge after mixing. The volume of NaOH used was recorded. Each sample was titrated three times and the percentage of lactic acid was calculated using the equation (NZIFST, 2010):

$$\text{Acid (\%)} = \frac{\text{(ml of NaOH used) (concNaOH) (0.090 milli equivalent weight of lactic acid) (100)}}{\text{Weight of sample}}$$

Enumeration of lactic acid bacteria (LAB) and *E. coil*

Samples (1 mL) of each sour milk were serially diluted (10^{-1}, 10^{-2}, 10^{-3} and 10^{-4}) in 9 mL of buffered peptone water (BPW) and appropriate dilution was surface plated onto Sorbitol MacConkey agar (SMAC) (Oxoid) for *E. coil* and de Man Rogosa and Sharp (MRS) agar was prepared from the ingredients for Lactobacillus. MRS plates were incubated anaerobically at 37°C for 48 h for the enumeration of mesophilic lactobacilli and leuconostocs and at 42°C for the enumeration of thermophilic lactobacilli and streptococci, while SMAC plates were incubated aerobically for 24 h at 37°C (Dlaminiand Buys, 2008). Total viable counts were performed using plate count agar (Merck) incubated at 25°C for 24 h (Lefoka and Viljoen, 2009).

Sensory analysis of sour milk products

A panel of 10 judges familiar with the *amasi* from the Department of Food Science and Technology at Univen was constituted and used to evaluate the five products. The panelists were asked to score for sourness, sweetness, texture (thickness), smoothness, astringency, colour and overall acceptability of the products (Table 1) on a 5-point hedonic scale ranging from 5 to 1, where 1 is dislike very much, 3 is neither like or dislike and 5 like very much (Bille et al., 2002).

Statistical analysis

Products were analyzed in triplicates. Analysis of variance (ANOVA) was applied to the entire data set to determine the significance of the differences ($p < 0.05$) between the sour milk products using a Microsoft Excel for windows 7.

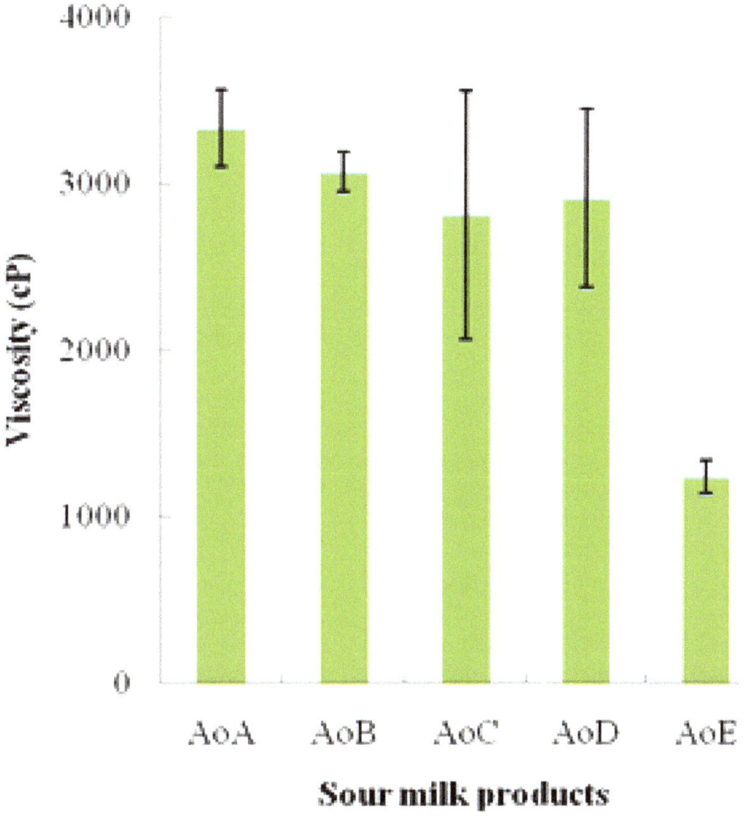

Figure 1. The viscosity of commercial sour milk (*amasi*) products.

RESULTS AND DISCUSSION

Physical analysis

The results presented are from the following physical analysis: viscosity and colour. The results for viscosity are shown in Figure 1. It seems that sample AoA had a higher viscosity ($p < 0.05$) as compared to sample AoE which had a low viscosity. Significant differences at $p < 0.05$ existed for the viscosity between the samples (*amasi*). Sample AoA showed the highest viscosity of 3330 cP, while sample AoE showed the lowest viscosity of 1240 cP (Figure 1). There were differences ($p < 0.05$) between sample AoA and sample AoB and AoE, whereas there was no difference in samples AoC and AoD. Sample AoB differed from sample AoC, AoD and AoE. Sample AoB, AoC and AoD had values of 3070, 2810 and 2915 cP, respectively. There was no difference in viscosity between sample AoC and AoE but there was difference in sample AoD. More work has been done to improve the *amasi* using either enrichment of the milk with 2.5% (w/v) skim milk or by dry matter ultra-filtration (Narvhus et al., 1998).

Ozerand and Atamer (1994) reported that acid development is required for fine curd formation. The study by Mustafa et al. (2001) on yoghurt recorded a viscosity value of 100 to 2825 cP. The LAB that produce exopolysaccharides are often used to increase the viscosity of stirred fermented milks, such as yoghurt and to decrease the susceptibility to syneresis (Ruas-Madiedo et al., 2002). Sample AoE had lowest viscosity and not surprising it also recorded lowest on consumer acceptance on texture and smoothness sensory attributes.

Large curd in sample such as AoA was observed as compared to sample AoE that suggests that viscosity could have been influenced by the size of the particles and whey proteins from denaturation caused by pH (Masson et al., 2010). This can be caused by unavoidable variations in factors such as temperature, pH and mechanical vibration during production, resulting in grainy texture and varying viscosities (Anonymous, 2010).

Eissa et al. (2010) stated that prolonged metabolic activity of microflora in yogurt causes changes in the micro-structure of the media and hence affecting viscosity. This may suggest that high values on the standard deviation were caused by the length of time each set of collected samples were stored and the number of microflora of the sour milk during analysis. The viscosity is affected by the state and concentration of fats, protein, temperature, pH and milk age (Park, 2007).

Results for colour analysis of sour milk samples are shown in Table 2. There was no significant difference ($p < 0.05$) between the L* values of the sour milk samples, with all the values ranging from 34 to 40 (Table 2). The L*

Table 2. The pH and colour parameters of commercial sour milk (*amasi*) products.

Sample	pH	Colour		
		L*	a*	b*
AoA	4.22 ± 0.18^{a}	33.77 ± 4.54^{a}	6.26 ± 3.29^{a}	16.70 ± 1.19^{a}
AoB	4.34 ± 0.30^{a}	35.29 ± 6.03^{a}	6.43 ± 0.08^{ab}	$18.77 \pm 0,71^{b}$
AoC	4.23 ± 0.12^{a}	38.67 ± 0.58^{a}	3.08 ± 0.49^{ac}	13.71 ± 0.30^{ac}
AoD	4.30 ± 0.14^{a}	38.04 ± 1.84^{a}	5.10 ± 0.26^{a}	16.02 ± 0.42^{ad}
AoE	4.34 ± 0.12^{a}	40.19 ± 1.13^{a}	3.78 ± 0.31^{ac}	16.58 ± 0.52^{a}

Mean values with their standard deviations. L* = Lightness (ranging from 0 = black to 100 = white); a* = redness and b* = yellowness. Values in the same column with different letters differ significantly from each other ($p < 0.05$).

indicate the black and white colour which ranges from L = 0 (black) to L = 100 (white) colour, and white is the predominant colour in sour milk. Again, there was no significant difference between the a* values except the significant difference between sample AoB and sample AoC and AoE ($p < 0.05$). The a* values represent the redness colour intensity and were below 6 in all the samples. The b* values indicate the yellowness and were at high level. The yellow is one of the properties of sour milk that appears to be yellowish in colour (Burger, 2010).

The study showed that, L* values ranged from 34 to 40 among the three batches, confirming that the sour milk has relatively dark colour when compared with standard white colour (L* = 100). Phillips et al. (1995) reported that L* value increased with the increase in fat content in milk. Whiteness in fluid milk is as a result of the presence of colloidal particles, such as milk fat globules and casein micelles, capable of scattering light in the visible spectrum (Fox and McSweeney, 1998).

Moreover the a* and b* values of sour milk were observed to be on the positive side, indicating that the product was slightly reddish yellow and the values ranged from 3.08 to 6.26 and 13.71 to 18.77, respectively.

Physicochemical analysis of *amasi* products

The pH of the samples shown in Table 2 ranged from 4.22 to 4.34 with an average of 4.29. Mean values in the same column with different letters differ significantly from each other ($p < 0.05$). Generally there was no significant difference between the samples in terms of pH values obtained (Table 2). The pH values of all the samples were between the range of 4.22 and 4.34; this is almost in the range of 3.6 and 4.2 discussed by Burger (2010). Dlamini and Buys (2008) also mentioned the pH of commercial *amasi* to be pH 4.4.

The use of pure cultures enables different acid flavors to be developed leading to a range of fermented (cultured) products (Kurmann et al., 1992). The pH values obtained from the samples make *amasi* drinkable since the ideal values are between 4.0 and 4.5 because bitter and acid sensations are not present at this interval (da Cruz et al., 2009). The low pH in the fermented milk offers a selective environment for yeast growth, but is unfavorable for most bacteria (Fleet, 1990). However, spoilage becomes evident when the yeast population reaches 10^5 to 10^6 cells/g (Fleet, 1990).

The TA data are represented in Figure 2. The TA values of the sour milk samples fall within the range 0.80 and 0.84%. These values were lower as compared to the TA of *amasi* discussed in the study by Dlamini and Buys (2008) which had the TA value of 1.4% with the pH of 4.0 at the initial stage. The organic acids (predominantly lactic acid) cause the pH of the sour milk to drop, killing the pathogenic bacteria and inhibiting the growth of many of the common spoilage micro-organisms (Gadaga et al., 2004).

Lactic acid bacteria (LAB) and *E. coil* in *amasi* products

The number of lactic acid bacteria (LAB) shown in Table 3 was high in all the sour milk samples (Table 3), the number of total plate count was less, while in all the sour milk samples, the number of *E. coli* ranged from 1.22×10^4 to 1.78×10^5. The survival of *E. coli* O157:H7 cells for up to several weeks in fermented dairy products, specifically cheese, sour cream, yoghurt, kefir and buttermilk, illustrates the potential health risks associated with post-processing contamination of even low levels of this organism in various dairy foods (Dineen et al., 1998). In the study by Dlamini and Buys (2008), *E. coli* O157:H7 was detectable in commercial *amasi* after 3 days at 7°C but not in traditional *amasi* processed at ambient temperature over the same period.

The number of LAB ranged between 1.25×10^5 and 1.97×10^6 cfu/mL with sample AoA and AoB being the highest by 1.96×10^6 and 1.97×10^6 cfu/mL, respectively. Beukes et al. (2001) reported that the lactic acid bacteria predominated the microbial population and numbers between 4.7×10^5 and 2.03×10^9 cfu/mL were recorded with mean of 108 cfu/mL on MRS, M17 and Rogosa agar in South African traditional fermented milks. In a study by Savadogo et al. (2004) on Fulani traditional fermented milk in Burkina Faso, they obtained almost similar results where the mean count of thermophilic bacteria (42°C) on MRS agar, 8.04×10^5 cfu/mL was less than the mean mesophilic count (35°C), 7.80×10^7 cfu/mL indicating the predominance

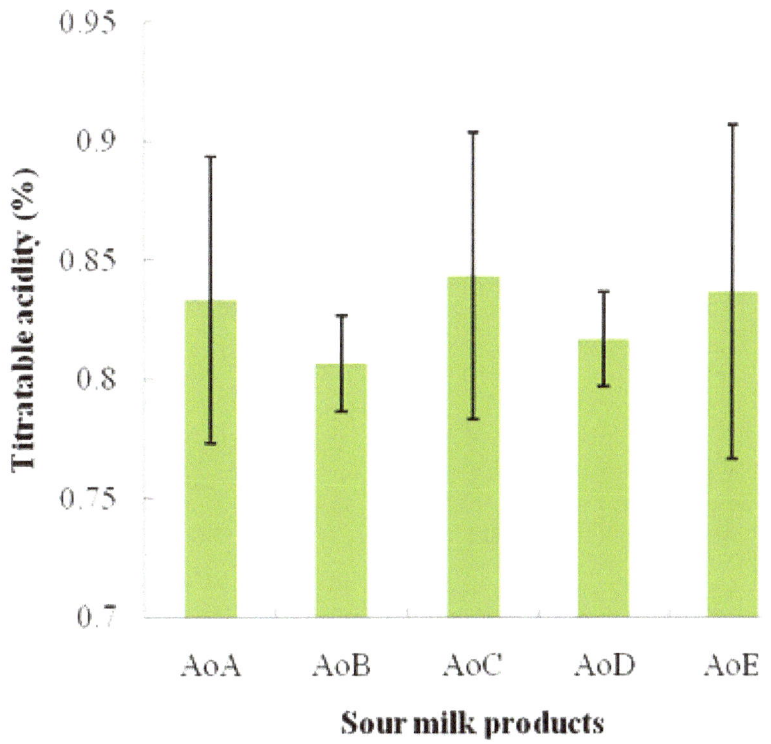

Figure 2. Titratable acidity (TA) of commercial sour milk (*amasi*) products.

Table 3. Microbial count of commercial sour milk (*amasi*) products.

Samples	LAB (cf/mL)	*E. coli* (cf/mL)	Total plate count (cf/mL)
AoA	$1.03 \times 10^6 \pm 0.98$	$7.80 \times 10^4 \pm 0.09$	$4.20 \times 10^3 \pm 2.40$
AoB	$1.97 \times 10^6 \pm 1.19$	$7.20 \times 10^4 \pm 1.10$	$9.09 \times 10^4 \pm 0.80$
AoC	$1.80 \times 10^6 \pm 0.74$	$1.78 \times 10^5 \pm 1.02$	$8.20 \times 10^3 \pm 0.90$
AoD	$4.81 \times 10^5 \pm 3.94$	$2.09 \times 10^4 \pm 0.41$	$2.66 \times 10^4 \pm 0.44$
AoE	$1.27 \times 10^6 \pm 0.09$	$1.22 \times 10^4 \pm 1.28$	$7.55 \times 10^4 \pm 1.75$

LAB = lactic acid bacteria.

of lactic acid bacteria. The mean count of coliforms was 0.98×10^4 cfu/mL for 25 samples. Dlamini and Buys (2008) also found that LAB counts in commercial *amasi* inoculated with AA and NA *E. coli* O157:H7 increased from 7.0 to 8.1 $\log_{10}$ cfu/mL. Differences in experimental conditions such as different fermentation temperatures and storage could have caused this difference.

High viable counts of LAB are necessary to get the desired acid production, which affects the product shelf-life (Helland et al., 2004). LAB also plays a major role in the development of flavour and aroma through the production of flavouring compounds such as diacetyl, acetoin and acetyaldehyde (Oberman and Libudzisz, 1998).

Consumer acceptability of *amasi* products

The mean scores for sensory properties of the sour milk samples are shown in Table 4. Five samples were evaluated for sourness, sweetness, texture, astringency, colour and overall acceptability. There were no significant differences observed between the samples in terms of texture, colour and overall acceptability (Table 4). However, there were some significant differences in sweetness, smoothness and astringency properties of the samples. Sample AoA was the most liked by the consumers than sample AoE. Sensory properties of foods offer quality control criteria.

The sensory score indicate that consumers did like the sour milk, as the overall acceptability ranges from 3.8 to 4.4 where 5 is like very much, and 1 being dislike very much (Table 4). The annual consumption of *amasi* is considerable in South Africa and an estimated 104 million liters were consumed in 2003 by black miners alone (McMaster et al., 2005). In this study, the most liked

Table 4. Sensory properties of commercial sour milk (*amasi*) products.

Sample	Sensory attributes						
	Sourness	Sweetness	Texture	Smoothness	Astringency	Colour	Overall acceptability
AoA	4.5± 0.97[a]	3.8± 1.23[ac]	4.0± 0.94[a]	4.3± 1.06[a]	4.3± 1.06[a]	4.4± 0.97[a]	4.4± 0.70[a]
AoB	3.7 ± 0.95[a]	3.1± 0.57[a]	3.8± 1.03[a]	3.7± 1.34[ab]	3.6± 1.51[ab]	4.2± 1.03[a]	4.0± 0.67[a]
AoC	3.7 ± 1.25[a]	2.7± 0.67[b]	3.8± 0.79[a]	3.6± 1.07[b]	4.0± 1.05[b]	4.3± 0.67[a]	3.8± 0.92[a]
AoD	3.7 ± 1.34[a]	4.0± 1.25[c]	3.9± 0.74[a]	4.3± 0.48[a]	3.8± 1.03[a]	4.5± 0.71[a]	4.3± 0.67[a]
AoE	3.3 ± 1.25[b]	3.3± 0.95[ad]	3.6± 1.07[a]	3.2± 0.79[bc]	3.1± 1.29[c]	4.0± 0.82[a]	3.8± 1.14[a]

Sensory characteristics were graded on the scale 1 = dislike very much; 5 = like very much. Mean values in the same column with different letters differ significantly from each other ($p< 0.05$).

sensory attribute was the colour with the scores between 4.0 and 4.5. The analysis of the aroma of dairy products is complex due the heterogonous nature of milk.

Sample astringency increased with storage time (Irigoyen et al., 2005). In contrast, Katsiari et al. (2002) found that storage did not significantly affect the sensory attributes of yoghurt samples. This can be true because the sour milk products examined in this study had different expiring dates. Moreover, their astringency was not affected by the storage period as the panelists did not judge the products based on the influence of storage period. However, significant differences existed between the examined sour milk products.

Conclusion

The outcome of this study showed the availability of *E. coli* in the sour milk products which can be of health concern to consumers. Lactic acid bacteria dominated the count in all the products in MRS media and the total plate count showed low bacteria count. However, the significant differences which existed on other parameters tasted did not affect the consumer acceptability as it was predicted.

REFERENCES

Anonymous (2010). LabPlusInternational. Technology to achieve smooth fermented milkproducts [wwwDodument] http://www.labintonline.com/index.php?id=1031andtx_ttproducts_pi1[product]=1593 Accessed 10-02-2011

Beukes EM, Bester BH, Mostert JF (2001). The microbiology of South African traditional fermented milks. Int. J. Food Microbiol. 63:189-197.

Bille PG, Ozuuko ATR, Ngwira T (2002). Sensory properties of traditionally-fermented buttermilk (Omashikwa) processed in Namibia. J. Food Tech. Afr. 7(2):52-54.

Bryant AT (1949). The Zulu People as they were before the White Man Came. 1st edition.Shuter and Shooter, Pietermaritzburg, South Africa, pp. 270-271.

Burger L (2010). Amasi's sweet success – August, [www document] URL http://www.dairyconnect.co.za/dairyMail/issues/aug2010/amazis_sweet_success.djhtml Accessed 07-10-2010.

da Cruz AG, Walter EHM, Cadena RS, Faria JAF, Bolini HMA, Frattini Fileti AM (2009). Monitoring the authenticity of low-fat yogurts by an artificial neural network. J. Dairy Sci. 92(10):4797-4804.

Dineen SS, Takeuchi K, Soudah JE, Boor KJ (1998). Persistance of *Escherichia coli* O157:H7 in dairy fermentation systems. J. Food Prot. 61:1602-1608.

Dlamini BC, Buys EM (2008). Adaptation of *Escherichia coli* 0157:H7 to acid in traditional and commercial goat milk amasi. J. Food Microbiol. 26(1):58-64.

Eissa EA, Mohamed Ahmed IA, Yagoub AEA, Babiker EE (2010). Physicochemical, microbiological and sensory characteristics of yoghurt produced from goat milk. Livestock Research for Rural Development.Volume 22, Article #137. [WWW Document]http://www.lrrd.org/lrrd22/8/eiss22137.htmAccessed 23-02-2011

Fleet GH (1990). Yeasts in dairy products. J. Appl. Bacteriol. 68:199-211.

Fox PF, McSweeney PLH (1998). Physical properties of milk. In: Dairy chemistry and biochemistry. Fox PF, McSweeney PLH (eds.), London: Blackie Academic and Professional. pp. 437-462.

Gadaga TH, Mutukumira AN, Narvhus JA (2001). Growth characteristics of Candida kefyr and two strains of *Lactococcus lactis* subsp. *lactis* isolated from Zimbabwean naturally fermented milk. Int. J. Food Microbiol. 70(1–2):11-19.

GadagaTH, Nyanga LK, Mutukumira AN (2004). The occurrence, growth and control of pathogens in African fermented foods. Afr. J. Food Agric. Nutr. Dev. 4(1):20-23.

Helland MH, Wicklund T, Narvhus JA (2004). Growth and metabolism of selectes strains of probiotic bacteria, in maise porridge with added malted barley.Intern. J. Food Microbiol. 91:305-313.

Irigoyen A, Arana I, Castiella M, Torre P, Ibáñez FC (2005). Microbiological, physicochemical, and sensory characteristics of kefir during storage. Food Chem. 90:613-620.

Katsiari MC, Voutsinas LP, Kondyli E (2002). Manufacture of yogurt from stored frozen sheep's milk. Food Chem. 77:413-420.

Keller JJ, Jordan I (1990). Fermented milks for the South African market.S.Afric. J. Dairy Sci. 22:47-49.

Kip P, Meyer D, Jellema RH (2006). Inulins improve sensoric and textural properties of low-fat yoghurts. Int. Dairy Sci. 16:1098-1103.

Kurmann JA, Rasic JL, Kroger M (1992). Encyclopedia of fermented fresh milk products. An international inventory of fermented milk, cream, buttermilk, whey, and related products. Van Nostrand Reinhold. ISBN 0-442- 00869-4 New York. USA.

Lefoka M, Viljoen BC (2009). The survival of microbial pathogens in dairy products. Master's thesis in Faculty of Natural and Agricultural Sciences, Department of Microbial, Biochemical and Food Biotechnology, University of the Free State, Bloemfontein, South Africa.

Masson LMP, Calado VMA, Deliza R, Rosenthal A (2010). Rheological behavior of fermented dairy beverages obtained from the ultra-high pressure homogenization (UHPH), International Conference on Food Innovation.Food Innova.Dijon, France.

Mazahreh AS, Al-Shawabkeh AF, Quasem JM (2008). Evaluation of the chemical and sensory attributes of solar and freeze-dried jameed produced from cow and sheep milk with the addition of carrageenan mix to the Jameed paste. Am. J. Agric. Biosci. 3(3):627-632.

McMaster LD, Kokott SA, Reid SJ, Abratt VR (2005). Use of traditional African fermented beverages as delivery vehicles for

Bifidobacterium lactis DSM 10140. Int. J. Food Microbiol. 102:231-237.

Mustafa G, Ahmet E, Huseyin T, Selahattin S (2001). Some quality criteria of yogurt made from added with antibiotic at different levels. Pak. J. BioSci. 4(7):886-887.

Narvhus JA, Østeraas K, Mutukumira T, Abrahamsen RK (1998). Production of fermented milk using a malty compound producing strain *of Lactococcus lactis* subsp.*lactisbiovar.diacetylactis* isolated from Zimbabwean naturally fermented milk. Int. J. Food Microbiol. 41:73-80.

NZIFST (2010). New Zealand Institute of Food Science and Technology. Food Science Experiments [www document] URL http://www.nzifst.org.nz/myfiles/Expt_5.__Sour_Cream_Chemistry_guide.doc Accessed 10-05-2010

Oberman H, Libudzisz Z (1998). Fermented milks. In: Microbiology of fermented foods. 2nd Ed. Wood BJB (Ed.), Blackie Academic and Professional.Thompson Science, 2-6 Boundary Row, London, UK. pp. 319-321.

Ozer HS, Atamer M (1994). Gida.19: 155-160. (As cited by) Mustafa G, Ahmet E, Huseyin T, Selahattin S, (2001). Some quality criteria of yogurt made from added with antibiotic at different levels. Pak. J. Biosci. 4(7):886-887.

Park YW (2007). Rheological characteristics of goat and sheep milk. Small Rumin. Res. 68(1-2):73-87.

Phillips LG, McGrift ML, Barbano DM, Lawless HA (1995).The influence of fat on the sensory properties, viscosity and colour of low fat milk. J. Dairy Sci. 78:1258-1266.

Quasem JM, Mazahreh AM, Afaneh IA, Omari A (2009). Solubility of solar dried Jameed. Pak. J. Nutr. 8(2):134-138.

Ruas-Madiedo P, Tuinier R, Kanning M, Zoon P (2002). Role of exopolysaccharides produced by *Lactococcus lactis* subsp. *cremoris* on the viscosity of fermented milks. Int. Dairy J. 12:689-695.

Savadogo A, Ouattara CAT, Savadogo PW, Ouattara AS, Barro N, Traore AS (2004). Microorganisms Involved in Fulani traditional fermented milk in Burkina Faso. Pak. J. Nutr. 3(2):134-139.

11

Effects of stocking density on haematological functions of juvenile African catfish (*Clarias gariepinus*) fed varying crude protein levels

Ajani, E. K.[1] , Setufe, S. B.[1] and Oyebola, O. O.[2]

[1]Department of Wildlife and Fisheries Management, University of Ibadan, Oyo State, Nigeria.
[2]Olabisi Onabanjo University, Ogun State, Nigeria.

This study was carried out to evaluate the growth performance and some haematological functions of *Clarias gariepinus* fed different levels of crude protein under varying stocking densities. Fish of even sizes were randomly selected and stocked in plastic circular tanks (0.05 x 0.03 x 0.03 m) at three stocking densities of 10 fish/m^3 (control), 15 fish/m^3 and 20 fish/m^3. The fish of each stocking density were fed either a diet containing 40% Crude protein (CP) or 45% CP with a feeding rate of 3% body weight twice daily. The experiment was replicated thrice and it lasted for 12 weeks during which haematological parameters and plasma biochemistry were measured. From this study, haemoglobin (Hb) and Red Blood Cell (RBC) slightly increased in all treatments but the variation was not significant in relation to protein level in the diets. Plasma glucose increased significantly ($P<0.05$) in relation to the stocking densities. The plasma protein showed insignificant variation in relation to the stocking density but the variation was more pronounced ($P<0.05$) at the lowest stocking density as the protein level in the diets varied. It can be concluded that enhancing feed quality especially protein level in the fish diet may ensure faster growth, stress reduction and improve health status of the fish.

Key words: Crowding stress, haematological functions, *Clarias gariepinus*, diet quality.

INTRODUCTION

Environmental stress is an important factor responsible for limiting fish performance under aquaculture conditions (Ellis et al., 2002). When fish are subjected to adverse environmental conditions, some endocrine and physiological alterations occur, often resulting in change in ability of the fish to survive, grow and reproduce (Barton and Iwama, 1991). Overcrowding is a common chronic stressor in aquaculture that can induce a prolonged elevation of cortisol levels, which may cause damaging consequences, and suppressed growth (Rowland et al., 2006). This effect has been attributed to factors such as decreased food consumption. The high stocking density also imposes increased energy demands that require fish to cope with metabolic adjustments, such as changes of gluconeogenic and glucolytic activities. Under such conditions, food consumption is reduced; the extra expenditure energy has to be met by the reserve, resulting in reduced growth (Rowland et al., 2006).

Crowding is a common aquaculture practice used to manage water usage or increase fish stocking density (Baras and Lagardere, 1995). However, the use of high stocking density as a technique to maximize water usage and thus increase stock production has also been shown to have adverse effect on growth. In many cultured fish species, growth has indirect relationship to stocking density and this observation is mainly attributed to social interact-tions (Holm et al., 1990; Haylor, 1991; Ma et al., 2006). Social interactions as a result of competition for food and/or space can negatively affect fish growth. On the other hand, the fish price is influenced by the market requirement such as size and production, which depends on their growth. In modern day commercial aquaculture the stocking density is an important indicator that determines the economic viability of the production system (Ellis et al., 2002).

Therefore by establishing the relationship between dietary protein level and crowding stress, monitoring of fish stocks and prediction on their haematological needs will be possible and enhanced. This study was carried out to assess the effects of crowding stress on the growth performance and some haematological functions of juvenile African catfish; *Clarias gariepinus* (a commercially impor-tant fish species in Africa) fed varying crude protein levels.

MATERIALS AND METHODS

Experimental procedure

This study was carried out in the Department of Wildlife and Fisheries Management laboratory, Faculty of Agriculture and Forestry, University of Ibadan. Four hundred *Clarias gariepinus* juveniles were purchased from a reputable farm - Rayak Fish farm Ibadan, Nigeria. The fish were transported in a well -oxygenated bag, which was half filled with fresh water to the Department of Wildlife and Fisheries Management laboratory. The average weight of the fish was 17.8 g±2.50 and their average length was 14.4 cm±1.20. The fish were kept to acclimatize for 2 weeks under laboratory condition in an indoor circular plastic tank. The fish were randomly distributed in 18 experimental tanks (each of 0.05x0.03x0.03 m) and the tank was filled with water at 60% capacity. The positioning of the tanks allowed a natural photoperiod of 12 ho sunlight and 12 hof darkness throughout the experiment. Water was changed daily to prevent fouling resulting from food residues. The source of water was the University of Ibadan Borehole Water Station. The fish were stocked into the tanks using three different stocking densities of 10 fish/m^3, 15 fish/m^3 and 20 fish/m^3 for treatments I and II respectively. The fish with lowest stocking density and fed with 40% CP diet served as the control based on recommendation of Akinwole (2007). The fish were fed with 40% crude protein and 45% crude protein diets for 12 weeks twice daily between the hours of 8.00 hand 1700 hat the rate of 3% body weight. Feeds were given in pellet form and the particle size increased periodically as fish grew. Fish were weighed initially using a weighing scale (g) before the commencement of the experiment and on a bi-weekly basis during the experiment. A calibrated measuring ruler (cm) was used to take the length of the fish before the experiment and bi-weekly during the experiment for 12 weeks.

The experimental diets were analyzed for proximate composition by using the methods prescribed by Association of Official Analytical Chemists (AOAC, 1990).

Haematological and biochemical analyses

This was carried out at the haematological laboratory of Veterinary Medicine Department University of Ibadan. Fish were starved for 24 h prior to sampling. Fish were anaesthetized with buffered MS222 (50 mg/l) and blood was collected with a hypodermic syringe from the caudal vein. Blood collection lasted less than 3 min in order to avoid stress-induced situation during sampling. The extracted blood was divided in two sets of eppendorf tubes. One set contained heparin, used as anticoagulant, for haematology (haemoglobin, hematocrit and red blood cell counting). The second set, without anticoagulant, was left to clot at 4°C and subsequently centrifuged at 5000 rpm for 10 min at room temperature. The collected serum was stored at –20°C for further assays (Stoskopf, 1993).

Physiological measurements

Red blood cells (RBCs, cells/µl) were counted under the light microscope using a Neubauer hemocytometer after blood dilution with phosphate-buffered saline solution. White blood cells were counted by using diluting fluids (3% aqueous solution of acetic acid plus 1% aqueous solution of gentian violet up to 1%). A 1:20 dilution was made and charged into Neubauer hemocytometer chamber for counting. Glucose was determined colorimetrically using glucose kits according to Trinder (1969). Total protein content in blood plasma was determined colorimetrically according to Henry (1964).

Statistical analysis

The data obtained were subjected to two-way ANOVA test and the differences between means were at the 5% probability level using Duncan's new multiple range test (DMRT). The software SPSS, version 10 (SPSS, Richmond USA) was used as described by Dytham (1999).

RESULTS

Gross composition of experimental diets at 40% and 45% CP is presented in Table 1 while Table 2 shows the haema-tological parameters of the test fish at various stocking density fed varying crude protein diets. Result from this study showed that there is no variation in the RBCs level with increasing stocking density; the RBCs level however increased significantly (P< 0.05) with increase in diet protein level. Also the Hb and Hcrt followed the same pattern as observed for the RBC in relation to the stocking density but however did not show any appreciable reduction with increase in diet protein level.

The plasma protein was significantly influenced by the crude protein elevation in the diets while high stocking density gave no significant variation in the plasma protein level. Generally, the plasma glucose level ranged from 90.00 to 126.28 mg/dl with lowest value recorded at lowest stocking density in the fish fed 40% crude protein diet. The plasma glucose level significantly increased

Table 1. Gross composition and chemical analysis of experimental diets using 40% and 45% crude protein.

Ingredient	(40% CP)	(45% CP)
Fish Meal	8.38	9.95
Groundnut Cake	25.14	29.85
Soybean Meal	33.52	39.81
Wheal offal	10.49	6.30
Yellow Maize	20.97	12.59
Salt	0.5	0.5
Vitamin Premix	0.25	0.25
Vegetable Oil	0.5	0.5
Mineral Premix	0.25	0.25
Total (kilogram) 100	100	
Chemical analysis (%)		
Moisture	8.61±0.02	8.52±0.03
Crude Protein	41.28±1.01	45.43±0.57
Crude Fat	4.79±0.002	5.86±0.02
Crude Fiber	2.97±0.02	2.84±0.02
Ash	14.88±0.001	15.34±0.01
Dry Matter	91.36±0.02	91.46±0.01
Nitrogen Free Extract	27.44±0.02	27.98±0.01

Vitamin premix contains; Vit. A, 20000000 I.U; D_3, 2000000 I.U; E, 200000 mg; K_3, 8000 mg;B_2, 30000 mg; B_1, 30000; B_6, 12000mg; B_{12}, 50 mg; C, 500000 mg; Niacin, 150000; pantothenic acid, 50000 mg; folic acid, 4000 mg; Biotin H12, 800 mg; choline chloride, 600000 mg; cobalt, 2000mg; copper, 4000 mg; iodine, 5000 mg; inositol, 200000 mg; iron, 40000 mg; manganese, 30000 mg; selenium (organic), 200 mg; zinc, 40000 mg; lysine, 100000 mg; methionine, 100000 mg and Antioxidant, 100000 mg.

Table 2. Haematological parameters of African catfish (*C. gariepinus*) fed different protein levels at two stocking densities.

Parameter	Treatment					
	40%CP			45% CP		
	SD1 (10 fish/tank)	SD2 (15 fish/tank)	SD3 (20 fish/tank)	SD1 (10 fish/tank)	SD2 (15 fish/tank)	SD3 (20 fish/tank)
RBCs (10^6/mm^3)	$1.68^a \pm 1.05$	$1.82^a \pm 0.04$	$1.94\ ^a \pm 0.04$	$1.74^a \pm 0.13$	$1.88^a \pm 0.14$	$2.05^a \pm 0.05$
Hb (g/dl)	$7.02^a \pm 0.82$	$7.50^a \pm 0.54$	$7.62^a \pm 0.54$	$7.90^a \pm 0.42$	$8.22^a \pm 0.20$	$8.90^a \pm 0.45$
Glucose (mg/dl)	$90.00^c \pm 8.25$	$102.24^b \pm 6.25$	$112^b \pm 3.42$	$122.40^a \pm 10.00$	$125.20^a \pm 8.22$	$126.28^a \pm 10.25$
Protein (g/dl)	$3.84^d \pm 0.20$	$4.10^c \pm 0.05$	$4.25^c \pm 0.45$	$4.90^c \pm 0.32$	$5.10^b \pm 0.58$	$6.05^a \pm 0.35$

The same letter in the same row is not significantly different at P<0.05. Means with the same superscript along the same row are not significantly different (P>0.05).

(P<0.05) in relation to stocking densities at 40% CP. A significant increase (P< 0.05) was however observed in the plasma glucose level between the two protein diets with higher values recorded in the fish fed diet containing 45% CP. There was significant elevation (P< 0.05) in the plasma glucose level in relation to stocking density as it increases with the increasing stocking density.

DISCUSSION

The use of heamatological parameters in assessing the stress level of teolost fish has been widely adopted in recent times (Barton and Iwama, 1991; Ajani, 2008). The use of these immune system parameters in the determination of alterations in fishes as a result of stress relating issues and the renew interest in understanding the fish body defense mechanism comes from the need to develop a good healthy management tools toward the sustenance of the rapidly growing aquaculture industry (Mehdi et al., 2010). The RBCs, Hb, and Hcrt levels in the *C. gariepinus* increased as the stocking density increased. This may be linked to the issue of length of study, space and water volume used in this study, which resulted in low level of fish activity during the experiment. This observation agreed with the report of McFarlane et al. (2004) and Akungur et al. (2007) who confirmed that any chronic stress induced by high fish density may be limited, but if the fish density increased for a long time or the fish growth was faster, the stress effect may be higher. It was observed that under longer stress time, growth reduction would be recorded especially at higher fish density, as fish will spend more energy on swimming activity at the expense of growth. Mehdi et al. (2010) reported that length of time may be a factor affecting the physiological functions as a result of crowding-related social stress due to increase in biomass with time.

Cortisol and glucose are two of the most common stress indicators but due to their high variability, they must be complemented with other stress indicators in order to have a more complete profile about the stress status of any fish (Martinez-Porchas et al., 2009) and its levels are enhanced under adverse situations (Grutter and Pankhurst, 2000). The plasma glucose of the fish increases with increase in stocking density. This increase may be attributed to the mobilization of glucose by the fish in response to crowding stress; this is generally used as a means of providing extra energy resources so at to enable the fish to overcome the disturbance (Arends et al., 1999; Carragher and Rees, 1994). Alterations in glucose metabolism are a common response to stress (Barton and Iwama, 1991; Braley and Andersson, 1992). On the other hand, the direct effect of the higher metabolic energy demands during stress is driven by an increase in oxygen transport efficiency by elevation of the haematocrit and red blood cell numbers (Acerete et al., 2004). Farghaly et al. (1973) and Wedemeyer and Yasutake (1977) observed that plasma proteins contribute significantly to the maintenance of the blood volume and the water content of the fluid in tissue. The colloidal plasma protein cannot diffuse through capillary membranes to the surrounding relatively protein-free tissues. In this way, they exert an osmotic pressure thus allowing a minimum liquid volume to be maintained in the capillary blood vessels (Verdegem et al., 1997). Plasma protein in this study reflected the protein feeding levels. Schippers et al. (1994) in their works with freshwater-reared rainbow trout did not find significant differences in total plasma protein levels in fish fed different protein levels.

Therefore from the results of this study, it can be concluded that African catfish can easily adapt to high stocking density without any serious effect on the health status or big response of the fish if the correct management procedures are adopted.

Conflict of interests

The authors did not declare any conflict of interest.

REFERENCES

Acerete L, Balasch JC, Espinosa E, Josa A, Tort L (2004). Physiological.responses in Eurasian perch *(Perca fluvatilis L.)* subjected to stress by transport and handling. Aquaculture 237:167-178.

Ajani F (2008). Hormonal and heamatological responses of *Clarias gariepinus* (Burchell 1822) to ammonia toxicity. Afr. J. Biotechnol. 7 (19):3466-3471.

Akinwole A.O. and Faturoti E.O. (2007): Biological performance of African catfish cultured in recirculating system in Ibadan. Aquaculture Engineering 36 (1):18-23.

Akungur N, Akusungur M, Akbuhit B, Kutlu I (2007). Effects of stocking density on growth performance, survival and food conversion ratio of Turbot *(Psetta maxima)* in the net cages on the southeastern coast of the black sea. Turk. J. Fish. Aquat. Sci. 7:147-152

AOAC (1990). Official Methods of Analysis. 15*th* edition. K. Helrich (Ed.). Association of Official Analytical Chemists Inc., Arlington,, VA.

Arends RJ, Mancera JM, Munoz JL, Wendelaar Bonga SE (1999). The stress response of the gilthead sea bream (*Sparus aurata* L.) to air exposure and confinement. J. Endocrin. 163:149-157.

Baras E, Lagardere JP (1995). Fish telemetry in aquaculture: review and perspectives. Aquac. Int. 3:77-102.

Barton BA, Iwama GK (1991). Physiological changes in fish from stress in aquaculture with emphasis on the response and effects of corticosteroids. Annu. Rev. Fish Dis. 10:3-26.

Braley H, Andersson TA (1992). Changes in blood metabolite concentrations in response to repeated capture, anesthesia and blood sampling in the Golden perch, *Macquaria ambigua*. Comp. Biochem. Physiol. 103A:445-450.

Carragher JF, Rees CM (1994). Primary and secondary stress responses in golden perch, *Macquaria ambigua*. Comp. Biochem. Physiol. 107A:49-56

Dytham C (1999). Choosing and using statistics: A Biologist's guide. Blackwell Science Ltd., London, United Kingdom, p. 147.

Ellis T, Ellis B, North AP, Scott NR, Bromage M, Porter M, Gold D (2002). The relationship between stocking density and welfare in farmed Rainbow trout. J. Fish Biol. 61:493-531.

Farghaly AM, Ezzat AA, Shabana MB (1973). Effect of temperature and salinity changes on the blood characteristics of *Tilapia zillii* G. in Egyptian littoral lakes. Comp. Biochem. Physiol. 46A:183-193.

Grutter AS, Pankhurst NW (2000). The effects of capture, handling and ectoparasites load on plasma levels of cortisol, glucose and lactate in the coral ref. fish *Hemigymnus melapterus*. J. Fish Biol. 57: 391-401.

Haylor GS (1991). Controlled hatchery production of *Clarias gariepinus* (Burchell 1922): growth and survival of fry at high stocking density. Aquacult. Fish. Manag. 22:405-422.

Henry RJ (1964). Colorimetric determination of total protein. In: Clinical Chemistry. Harper and Row Publ., New York, p. 181.

Holm JC, Refstie T, Bo S (1990). The effect of fish density and feeding regimes on individual growth rate and mortain rainbow trout (*Oncorhynchus mykiss*). Aquaculture 89:3-4.

Ma A, Chen C, Lei J, Chen S, Zhuang Z, Wang Y (2006). Turbot *Scophthalmus maximus*: stocking density on growth, pigmentation and feed conversion. (Abstract) Chin. J. Oceanol. Limnol. 24(3):307-312.

Martinez-Porchas M, Martinez-Cordova LR, Ramos-Enriquez R (2009). Cortisol and glucose: Reliable indicators of stress? Pan Am. J. Aquat. Sci. 4:158-178.

McFarlane WJ, Cubitt KF, Williams H, Rowsell D, Moccia R, Gosine R, McKinley RS (2004). Can feeding status and stress level be assessed by analyzing patterns of muscle activity in free- swimming rainbow trout (*Oncorhynchus mykiss* Walbaum)? Aquaculture 239: 467-484.

Mehdi N, Masood A, Behzan NH, Ali A, Mosa ZD, Hamid FL (2010). Physiological stress responses in Kutum *Rutilus frisii kutum* subjected to captivity. Int. Aquat. Res. 2:55-60.

Rowland SJ, Mifsud C, Nixon M, Boyd P (2006). Effects of stocking density on the performance of the Australian freshwater silver perch (Bidyanus bidyanus) in cages. Aquaculture 253:301-308.

Schippers C, Ulloa Rojas JB, Booms GHR, Machiels MAM, Verdegem MCJ, Boon JH (1994). A dietary effect on some cellular and humoral parameters of rainbow trout (*Oncorhynchus mykiss*). Aquact. Fish. Manag. 25:659-659.

Stoskopf MK (1993). Clinical pathology. In Stoskopf M.K. (Ed.) Fish Medicine. Philadelphia, Pa: WB Saunders. pp. 113-131.

Trinder P (1969). Determination of glucose concentration in the blood. Ann. Clin. Biochem. 6:24.

Verdegem MCJ, Hilbrands AD, Boon JH (1997). Influence of salinity and dietary composition on blood parameter values of hybrid red tilapia, *Oreochromis niloticus* (Linnaeus) x *O. mossambicus* (Peters). Aquac. Res. 28:453-459.

Wedemeyer GA, Yasutake WT (1977). Clinical methods for the assessment of the effects of environmental stress on fish health. Technical Paper 89, U.S. Department of the Interior, Fish and Wildlife Service, Washington DC.

12

Evaluation of antioxidant potentials of *Monodora myristica* (Gaertn) dunel seeds

Akinwunmi Kemi Feyisayo and Oyedapo Oluboade Oluokun

Department of Biochemistry, Faculty of Science, Obafemi Awolowo University, Ile-Ife, Osun State, Nigeria.

The antioxidant activity of the flavonoid fraction of the seed extract of *Monodora myristica* (Gaertn) Dunel was evaluated. The *M. myristica* seed was extracted with aqueous ethanol (70%) to produce the crude extract. Phytochemical analysis of *M. myristica* ethanolic extract revealed the presence of secondary metabolites, such as alkaloids, flavonoids, saponins, cardiac glycosides and tanins. Fractionation of the crude extract by solvent-solvent extraction using n-hexane, dichloromethane, ethylacetate and butanol yielded the EtOAC fraction which tested positive for flavonoid as the flavonoid fraction (FF). The FF exhibited potent and appreciable anti-oxidant potentials with maximum DPPH-radical scavenging activity (41.20 ± 0.12 %), hydroxyl radical scavenging activity (46.34 ± 0.24%). It also exhibited a significant $p < 0.05$ reduction of Fe^{3+} to Fe^{2+} (64.64 ± 0.11%). It exhibited a dose dependent protective effect against free radical induced haemolysis of red blood cells with maximum protection of 37.00 ± 0.13 and 46.23 ± 0.41% inhibition of lipid peroxidation and free radical generation in liver homogenate. The result suggests that flavonoid fraction of *M. myristica* seed possessed significant antioxidant properties.

Key words: *Monodora myristica,* haemolysis, 1,1-dipheny-2-pycryl hydrazine (DPPH), flavonoid fraction.

INTRODUCTION

It is ironic that oxygen, an element indispensible for life can, under certain situations have several deleterious effects on the human body. Most of the potentially harmful effects of oxygen are due to the formation and activity of a number of chemical compounds known as reactive oxygen species. The radicals derived from oxygen represent the most important class of radical species in living systems (Grassi et al., 2009). Reactive oxygen species includes hydrogen peroxide (H_2O_2), superoxide radical anion (O_2^-), hydroxyl radical ($^{\cdot}OH$), nitric oxide ($NO^{\cdot}$), singlet oxygen ($^1O_2$) and hypochlorous acid (HOCl). These species are commonly called "free radicals" because they contained one or more unpaired electrons in their atomic or molecular orbital and thus have the tendency to donate oxygen to other substances (Bagchi and Puri, 1998; Halliwell, 2006). This unpaired electron(s) usually gives a considerable degree of reactivity to the free radical.

Reactive oxygen species are formed under normal physiological conditions or from external sources such as X-rays, ozone, cigarette smoking, air pollutants and industrial chemicals (Bagchi and Puri, 1998; Jae-young et al., 2005; De Groot and Rauen, 1998). These reactive oxygen species are considered to contribute decisively to a wide variety of degenerative processes and diseases such as cancer, atherosclerosis, neurodegenerative diseases (Parkinson's disease, Alzhemeir), hypertension, diabetes mellitus, dementia and reperfusion injury of the brain, liver as well as protein oxidation, DNA damage and lipid peroxidation (Valko et al., 2007; Finkel, 2005). However, controlled production of these radicals has important physiological roles and also serves useful purposes in human body. Several observations indicate that the oxygen radicals in living systems are probably necessary compounds in the maturation processes of cellular structures, white blood cells release free radicals to destroy invading pathogenic microbes as part of the body's defense mechanism against diseases (Halliwell, 1996; Morrissey and O'Brien, 1998; Miguel, 2009).

The biological systems have developed several effective antioxidant mechanisms to counteract the harmful effects of reactive oxygen species. Antioxidants are a group of substances which when present at low concentrations in relation to oxidisable substrates significantly inhibit or delay the oxidation and oxidative processes, while often being oxidized themselves (Kanner et al., 1999; Halliwell, 1990). The anti-oxidative principles include enzymes such as superoxide dismutase, catalase and GSH peroxidase (which degrades O_2^- and H_2O_2, respectively), but also not enzymatic compounds such as water soluble glutathione (GSH), polyphenols, proteins (Ferritin), trace elements (e.g. selenium, zinc), ascorbic acids (vitamin C) and the membrane-bound α-tocopherol (vitamin E) (Malterud et al., 1993, Abdou et al., 2010).

The plant *Monodora myristica* belongs to the custard apple family of flowering plants called Annonecea. It is widely distributed from Africa to Asia, Central and South America and Australia (Omobuwajo et al., 2003). It is native to West, Central and East Africa extending from Sierra Leone to Uganda, Kenya, Kongo and Angola (Keay, 1989). It grows well in the ever green forest zone of West Africa and most prevalent in the Sothern part of Nigeria (Adegoke and Akinsanya, 1970). Its local names include: Ehuru or Ehiri (Igbo), Ariwo (Yoruba), Jamaica nutmeg, African nutmeg, Calabash nutmeg, Airama. Ethnomedically, *M. myristica* is used in Ivory Coast to treat hemorrhoids, stomach ache and fibrie pain. The seeds are aromatic and are employed after grinding to a powder as condiments in food providing a flavour resembling that of nutmeg (Ekeanyanwu et al., 2010). The seeds are also used as an aromatic and stimulating addition to medicines and to snuff (Burkill, 1985; Ekeanyawu et al., 2010). When pulverized, the kernel is used to prepare pepper soup as stimulant to relieve constipation and control passive uterine hemorrhage in women immediately after child birth (Okafor, 1987; Udeala, 2000; Iwu, 2002).

Phytochemical studies of *M. myristica* seeds have shown that it is rich in alkaloids, glycosides, flavonoids, tannins, saponin and steroids (Adegoke et al., 1968; Uheghu et al., 2011; Ekeanyanwu et al., 2012). This study therefore sorts to evaluate the antioxidant potentials of *M. myristica* seeds.

MATERIALS AND METHODS

Plant materials collection and identification

Dried seeds of *M. myristica* were purchased from a local market in Ile-Ife, Nigeria. The seeds were identified and authenticated at IFE Herbarium, Obafemi Awolowo University, Ile-Ife, Nigeria with voucher specimen reference number 4734. The seeds were decocted to release the kernels which were later ground to fine powder with electronic blending machine.

Reagents and chemicals

All the reagents used were of analytical grade, obtained from the following Chemical Manufacturing Companies: British Drug House (BDH), Poole, U.K., Sigma Chemical Company, Louis, U.S.A., Fluka Chemical Company and Pharmacia Fine Chemicals, Uppsala, Sweden. Aluminium Oxide was from Associated Chemical Enterprises, South Africa.

Preparation of ethanolic extract of *M. myristica*

The ethanolic extract of seed of *M. myristica* was prepared according to the procedure earlier described by Oyedapo et al. (2004). Typically, 400 g of defatted seeds of *M. myristica* was suspended in aqueous ethanol (70% v/v, 1000 ml). The suspension was stirred occasionally and kept at room temperature for 24 h. The supernatant was collected by filtration through two layers of cheese cloth followed by filtration with filter paper Whatman number 2. The residue was washed with 70% (v/v) ethanol until the supernatant was clear. The filtrates were combined and concentrated under reduced pressure on rotatory evaporator at 40°C to give a brown residue termed ethanolic extract (EE).

Phytochemical screening of ethanolic extract of *M. myristica*

The ethanolic extract was subjected to phytochemical screening for the presence of secondary metabolites using standard procedure that was based on the methods described by Sofowora (2006), Evans (2002) and Oyedapo et al. (1999).

Estimation of flavonoid concentration

The concentration of flavonoid in the ethanolic extract of *M. myristica* was estimated spectrophotometrically according to the method of Sun et al. (1999). Typically, 0.5 ml of working solution of EE (5 mg/ml) was pipetted into test tubes in triplicates and diluted to 5.0 ml with distilled water after which 0.3 ml of 5% (w/v) $NaNO_2$, 0.3 ml of 10% $AlCl_3$ and 4.0 ml of 4% (w/v) NaOH were added to each of the test tubes. The reaction mixtures were incubated at room temperature for 15 min and the absorbance of the product was read at 500 nm against reagent blank. Rutin was used as standard flavonoid.

Estimation of total phenol concentration

The total phenol in the ethanolic extract of *M. myristica* was estimated using the Folin Ciocalteau's phenol reagent reaction as described by Yoshikana et al. (2002). Typically, 0.5 ml of 5 mg/ml of EE was pipetted into clean dried test tubes in triplicate and the volumes were adjusted to 1.0 ml with distilled water. Into each of the test tubes, 1.5 ml of Folin Ciocalteau's phenol reagent (1:10) was added. The reaction mixtures were incubated at room temperature for 5 min after which 1.5 ml of 10% (w/v) $NaHCO_3$ was added and the reaction mixture was further incubated for additional $1^1/_2$ hours. The absorbance was read at 725 nm against the reagent blank. Gallic acid was used as standard phenol.

Partitioning of ethanolic extract of *M. myristica*

Ethanolic extracts of *M. myristica* was partitioned by solvent-solvent extraction according to a procedure that was based on those reported by Aderogba et al. (2007) and Owoyele et al. (2008). Typically, EE (20 g) was taken up in hot distilled water (0.2 L). This was successively extracted with 1 L of n-hexane, dichloromethane, ethylacetate and n-butanol in a separating funnel (1 L) to afford four fractions namely n-Hex, DCM, EA and nB, respectively. The fractions were screened for flavonoids. Phytochemical screening, thin layer chromatography, staining with specific detection reagents together with chemical tests with 2,4-dinitrophenyl hydrazine revealed that ethyl acetate fraction contained the bulk of flavonoid of the seed of *M. myristica*. The fraction was evaporated to dryness and termed flavonoid fraction (FF).

Biochemical analysis

Assay of DPPH- radical scavenging activity

The DPPH-radical scavenging activity of flavonoids of *M. myristica* was assessed using a modified method of Blois (1985) as reported by Cakir et al. (2003). Typically, varying concentrations (0 to 350 µg/ml) of the extract of *M. myristica* was pipetted into clean dried test tubes in triplicates and the volumes adjusted to 1 ml with 10 mM acetate buffer, pH 4.5. This was followed by the addition of 2 ml of 0.2 mM DPPH solution in methanol. The reaction mixture was mixed thoroughly by inversion and then incubated in the dark for 30 min. The absorbance was read at 517 nm against the blank that contained 1 ml of 10 mM acetate buffer, pH 4.5 and 2 ml of 0.2 mM DPPH solution in methanol. The above procedure was followed for ascorbic acid (1 mg/ml) and rutin (1 mg/ml) standard. The percentage scavenging activity was evaluated using the expression:

$$\text{Percentage scavenging activity} = \frac{(\text{Absorbance of control} - \text{Absorbance of sample})}{\text{Absorbance of control}} \times 100$$

Assay of hydroxyl-radical scavenging activity

The ability of flavonoid fraction of *M. myristica* to scavenge hydroxyl radical was investigated by method of Halliwell et al. (1988) with slight modification. The varying concentrations of flavonoid fraction of *M. myristica* were pipetted into clean dried test tubes in triplicates and the volume in each tube was adjusted to 1 ml with distilled water.

The substrate (1 ml) consisting of 3 mM deoxyribose, 0.1 mM $FeCl_3$, 0.1 mM EDTA, 0.1 mM ascorbic acid and 2 mM H_2O_2 was added and followed by incubation at 37°C for 1 h. The reaction was terminated by the addition of 1.0 ml of TCA (1%) and TBA (1%) was later added. The mixture was incubated in boiling water at 100°C for 20 min. The tubes were removed, cooled and the absorbance was read at 532 nm against reagent blank. The percentage hydroxyl radical scavenging activity was calculated using the expression:

$$\text{Percentage inhibition} = \frac{A - B}{A} \times 100$$

Where, A = absorbance without the fraction and B = absorbance with fraction.

Inhibition of free radical induced oxidative heamolysis of red blood cells with $CuSO_4.5H_2O$ and H_2O_2

The ability of flavonoid fraction of *M. myristica* to inhibit free radical induced heamolysis of red blood cell was determined by a modified method of Amzal et al. (2008) as reported by Bode and Oyedapo (2011). The assay involved pipetting varying concentrations (0 to 350 µg / ml) of flavonoid fraction into clean dried test tubes in triplicate. The volumes in each test tube was made up to 1.5 ml with normal saline and was followed by the addition of 1 ml of 10 mM $CuSO_4.5H_2O$ and 1 ml of H_2O_2 (30% v/v) to each test tube. Freshly prepared 2% red blood cell (1 ml) was then added to the reaction mixture, mixed thoroughly and incubated at 37°C for 1 h. The tubes were cooled and centrifuged at 3000 rpm for 10 min and supernatants were carefully collected. The absorbance of released haemoglobin was read at 540 nm against the reagent blank. This procedure was repeated with rutin (1 mg/ml) as standard. The percentage inhibition was evaluated using the expression:

$$\text{Percentage inhibition} = \frac{A - B}{A} \times 100$$

Where, A is the absorbance without flavonoid fraction and B is the absorbance with flavonoid fraction

Assay of the reducing power

The ability of flavonoid fraction of *M. myristica* to reduce iron from the form (III) to the form (II) was assessed with the method of Hinneburg et al. (2006). Aliquots of the flavonoid fraction (0.5 ml) was mixed with 1.25 ml of phosphate buffer (0.2 M, pH 6.6) and 1.25 ml of a 1% (w/v) aqueous potassium hexacyanoferrate [$K_3Fe(CN)_6$] solution. After 30 min incubation at 50°C, 1.25 ml of 10% (w/v) trichloroacetic acid was added, mixed and shaken thoroughly. The suspension was centrifuged at 2000 rpm for 10 min. Then, 1.25 ml of the upper layer was mixed with 1.25 ml of distilled water and 0.25 ml of aqueous $FeCl_3$ (0.1% w/v). The absorbance was read at 700 nm against the reagent blank. The percentage reducing power was calculated by this expression:

Reducing power (RP) = $(A_m / A_b - 1) \times 100$

where A_m = Absorbance of reaction mixture; A_b = absorbance of blank mixture.

Assay of inhibition of lipid peroxidation

The inhibitory activity of flavonoid fraction of *M. myristica* on lipid peroxidation (LPO) was determined by thiobarbituric acid method. The mixture of $FeCl_2$-H_2O_2 (ratio 1:1) was used to induce peroxidation according to the method of Su et al. (2009) with slight modification. Typically, 0.2 ml of flavonoid fraction of *M. myristica* (1.25 mg/ml) was mixed with 1.0 ml of 1% (w/v) liver homogenate (1 g rat liver was homogenized to give 100 ml filtrate), then 50 µl of $FeCl_2$ (0.5 mM) and 50 µL of H_2O_2 (0.5 mM) were added. The reaction mixtures were incubated at 37°C for 1 h. This was followed by the addition of 1.0 ml of trichloroacetic acid (15%) and 1.0 ml of 2-thiobarbituric acid (0.67%) and the mixtures were heated in a boiling water bath for 15 min. The absorbance was read at 532 nm against the reagent blank. Quercetin and gallic acid were used as the positive controls. The percentage inhibition was calculated according to following equation:

Inhibition rate (%) = $\{1 - (A_1 - A_2)/A_0\} \times 100$

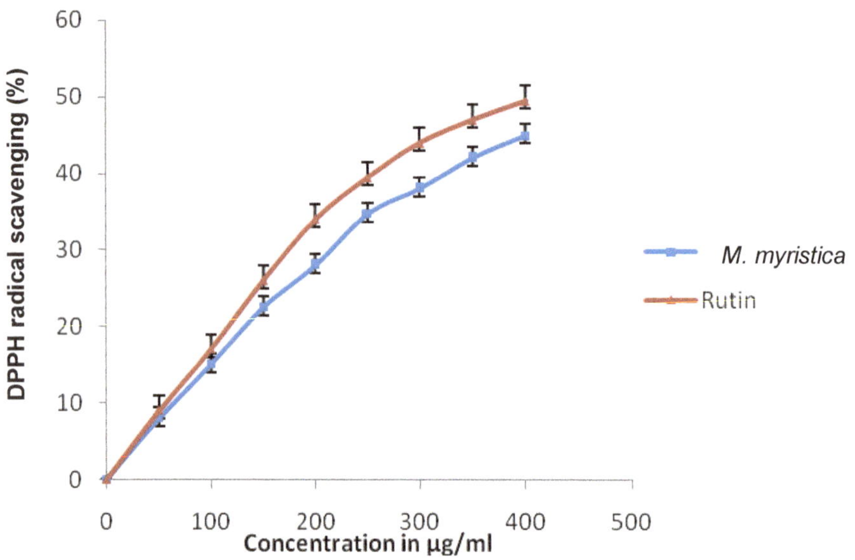

Figure 1. DPPH-radical scavenging activity of flavonoid fraction of *M. myristica* and rutin. Each value represents the mean ± SD of three parallel measurements.

Where, A_0 is the absorbance of the control (without extract), A_1 is the absorbance of the extract addition and A_2 is the absorbance without liver homogenate.

Statistical analysis

Data for the biochemical assays were expressed as mean ± SEM. Analysis was performed using GraphPad Instant 3 (version1.1, 2007).

RESULTS AND DISCUSSION

Numerous natural products are effective antioxidants and many medicinal plants with long history of use in folk medicine in different countries against a variety of diseases have turned out to be rich sources of antioxidant (Mantle et al., 2000; Pereira da Silva et al., 2000; Cho et al., 2003; Lee et al., 2002). Recently, great interest has been given to naturally occurring antioxidants which played prominent roles in inhibiting both free radicals and oxidative chain reactions within tissues and membranes (Nsimba et al., 2008). Antioxidant mechanisms are diverse and a variety of *in vitro* techniques has been developed. The proposed mechanisms of action of antioxidants include direct radical scavenging, inhibition of enzymes such as NO-synthase, xanthine oxidase, cyclooxygenase and lipooxygenase; iron chelation and direct inhibition of lipid peroxidation (Xanthopoulou et al., 2009). Therefore, it is better to employ different assays based on different mechanisms to evaluate the antioxidant capacity. In this study, DPPH radical scavenging, hydroxyl radical scavenging, the reducing power, inhibition of lipid peroxidation and inhibition of free radical induced haemolysis of red blood cell were employed to evaluate the antioxidant potential of flavonoid fraction of *M. myrisica*.

The scavenging of 1.1-diphenyl-2-pycrylhydrazine (DPPH) radical is a commonly used procedure which has been validated against several other assays for antioxidant activities (Malterud et al., 1993; Tilak et al., 2004). The procedure has been employed by various researchers to evaluate the free radical scavenging activity of molecules and plant extracts despite the fact that this chromogen does not solubilize easily, albeit it does generate a strongly coloured solution with methanol and ethanol as solvent (Han et al., 1999; Yan et al., 1998; Matsukawa et al., 1997). DPPH is a stable free radical because of delocalization of the spare electron over the molecule so that the molecules do not dimerise. DPPH radical are capable of accepting electrons or hydrogens to become diamagnetic molecule converting it to 1,1-diphenyl-2-pycrylhydrazine (Fargere et al., 1995). The reduction of DPPH radical was evident by the loss of deep violet colour to light yellow and concomitant decrease in absorbance at 517 nm by the antioxidant that donate hydrogen or electron (Kulisic et al., 2004). The degree of reduction in absorbance measurement was indicative of the radical scavenging power of the antioxidant. The flavonoid fraction (FF) of *M. myristica* was able to scavenge DPPH radical with an IC_{50} value of 412 µg/ml ($r = 0.9825$, $p < 0.0001$). Rutin used as positive control was able to scavenge DPPH with an IC_{50} values of 361 µg/ml ($r = 0.9808$, $p < 0.001$) (Figure 1). Evident from this study, a good linear relationship was observed between the decrease in absorbance of DPPH and different concentrations of flavonoids fraction of *M.*

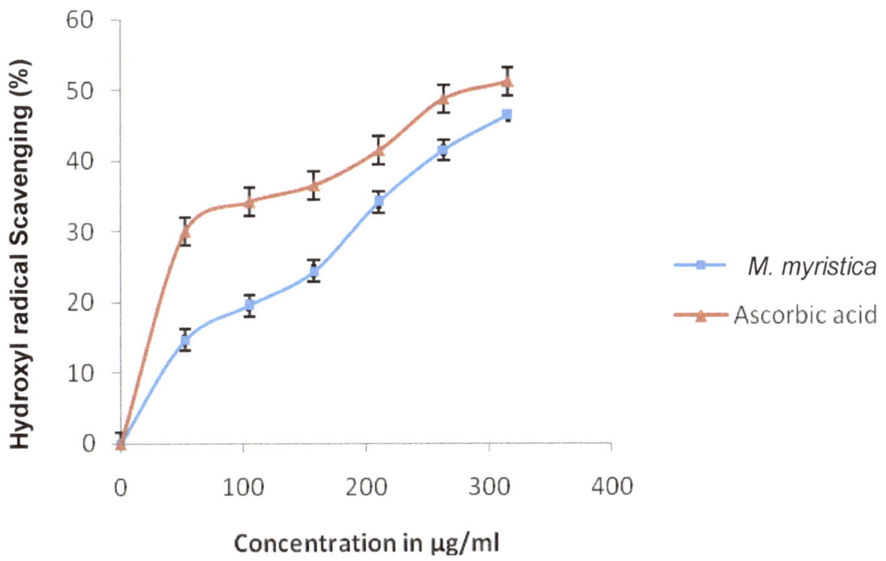

Figure 2. Hydroxyl radical scavenging activities of flavonoid fraction of *M. myristica* and ascorbic acid. Each value represents the mean ± SD of three parallel measurements.

myristica. The increased concentration of *M. myristica* showed increased DPPH radical scavenging activity. The DPPH-radical scavenging activity was concentration dependent and compared favorably with standard compound. Hydroxyl radical is a highly potent oxidant that reacts with almost all biological molecules found in living cells (Tanasait et al., 2010). The chemical activity of hydroxyl radical is the strongest among reactive oxygen species (Jae-Young et al., 2005).

Therefore, the removal of hydroxyl radical is probably one of the most effective defenses of a living body against various diseases. Fe^{2+} can react with hydrogen peroxide (H_2O_2) to produce the hydroxyl radical (OH) via the fenton reaction. In this study, the formation of hydroxyl radicals (OH) from Fenton reaction was quantified using 2-deoxyribose oxidative degradation. The ability of *M. myristica* to scavenge the generated hydroxyl radical was investigated by the inhibition of deoxyribose degradation.

The results reveal that the flavonoid fraction was able to inhibit deoxyribose degradation at maximum percentage inhibition of 46.34% at concentration of 315 μg/ml. The flavonoid concentration that caused 50% inhibition (IC_{50}) was estimated to be 329 μg/ml. Ascorbic acid used as standard antioxidant gave an IC_{50} value of 271 μg/ml (Figure 2). The flavonoids of *M. myristica* was able to scavenge hydroxyl radicals generated by Fenton reaction in a concentration dependent manner. The results of this study are in agreement with earlier reports of Abdou et al. (2010), Adefega and Oboh (2011) that extracts of *M. myristica* inhibited the decomposition of deoxyribose.

Ferric reducing antioxidant power is a novel antioxidant defense mechanism. It is used to assess the metal (iron exclusively) ions binding ability (Nsimba et al., 2008). The two mechanisms that are available and affect this property are electron transfer and hydrogen atom transfer (Dastmalchi et al., 2007).

The reducing power is generally associated with the presence of reductones which exerts antioxidant actions by breaking free radical chain by donating hydrogen atom (Prasad et al., 2010). FRAP assay measures the ability of antioxidant components to reduce Fe^{3+} / ferricyanide complex to ferrous form which has an intense bluish colour. The measurement of the formation of Perl's Prussian blue at 700 nm was used to monitor the Fe^{2+} concentration.

It was observed that flavonoid of *M. myristica* gave a maximum percentage reduction of Fe^{3+} to Fe^{2+} of 68.68% at 400 μg/ml. The ability of flavonoids of *M. myristica* to reduce Fe^{3+} to Fe^{2+} was found to increase as the concentration increases and compared favourably with 2,6-di-tert-butyl-4-methyl phenol BHT (Figure 3), a standard phenol.

The reaction of $CuSO_4.5H_2O$ with hydrogen peroxide generates hydroxyl radical. This generated hydroxyl radical is capable of oxidizing membrane lipids of the red blood cells resulting in haemolysis which is evident by the release of haemoglobin. Flavonoid of *M. myristica* was able to inhibit the action of free radical generated as a result of the reaction of $CuSO_4$ with H_2O_2 by causing a decrease in the amount of haemoglobin released. The result showed a concentration-dependent mode of inhibition which is comparable with the mode of inhibition of rutin, a standard flavonoid. The flavonoid fraction gave maximum percentage protection of 37% at concentration of 290 μg/ml while rutin gave a maximum percentage protection of 40% at a concentration of 300 μg/ml (Figure

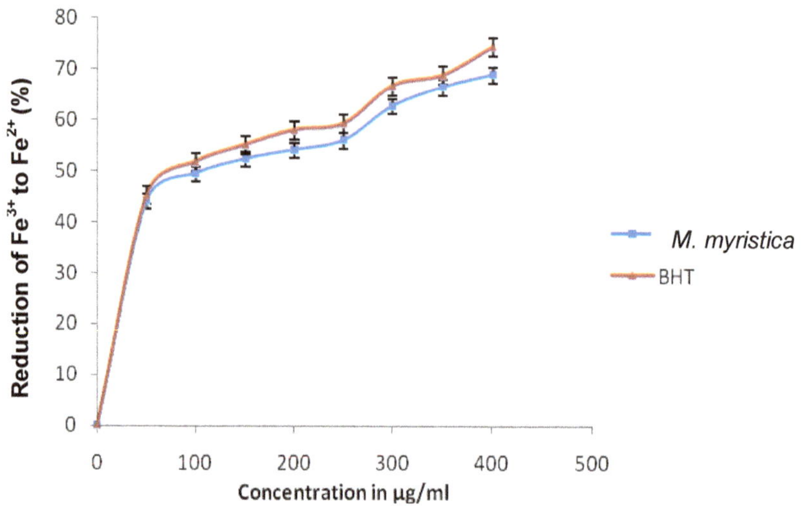

Figure 3. Ability of flavonoid of *M. myristica* and BHT to reduce Fe^{3+} to Fe^{2+}. Each value represented mean ± SD of n = 3.

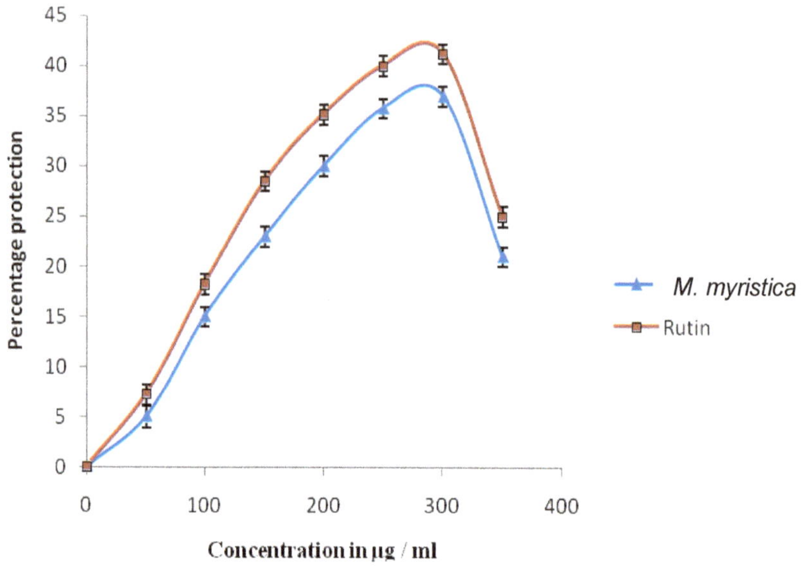

Figure 4. Ability of flavonoid fraction of *M. myristica* and rutin to inhibit free radical induced haemolysis of red blood cells. Each value represents the mean ± SD of three parallel measurements.

4). The result obtained in this study was in agreement with those of Liu et al. (2002), Amzal et al. (2008) and Bode and Oyedapo (2011) that certain plant metabolites protected erythrocytes from free radical induced haemolysis of red blood cells. However, information was limited on the ability of flavonoid of *M. myristica* to inhibit $CuSO_4$ induced haemolysis of red blood cells from the literature.

The role of peroxidative processes in disease is a subject of intense research interest. Lipid peroxidation of cell membranes is associated with various pathological events such as atherosclerosis, inflammation and liver injury (Roome et al., 2008). In this study, the percent inhibition of lipid peroxidation was quantified by measuring the reduction of TBARS in the presence of flavonoid fraction of *M. myristica*. The flavonoid fraction of *M. myristica* caused a decrease in the malondialdehyde produced. The production of MDA was used as a biomarker to measure the level of oxidative stress in organism (Cheeseman and Holley, 1993). MDA reacts

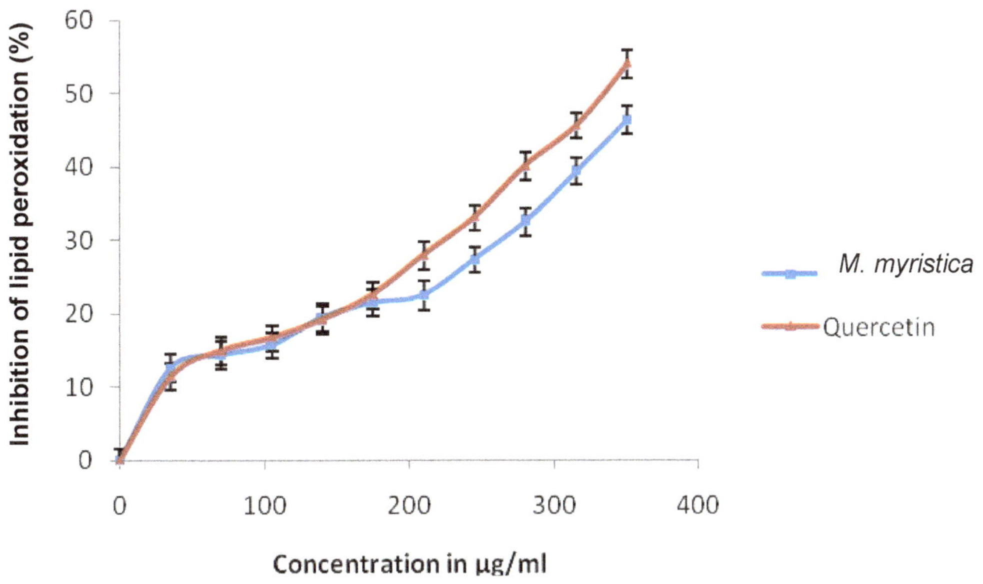

Figure 5. Percentage inhibition of lipid peroxidation in rat liver homogenate by flavonoid fraction of *M. myristica* and quercetin. Each value represented mean ± SD of n = 3.

with thiobabituric acid to produce a fluorescent product (TBARS). Flavonoid fraction of *M. myristica* was able to reduce TBARS produced thereby inhibiting lipid peroxidation with a maximum percentage inhibition of 46.23% at 350 µg/ml. This mode of inhibition compared favourably with quercetin, a standard flavonoid (Figure 5). This result was in agreement with that of Adefega and Oboh (2011) that extracts of *M. myristica* showed inhibitory effects on Fe^{2+}- induced lipid peroxidation in rat liver homogenates.

Conclusion

It was evident from this study that *M. myristica* seeds contained secondary metabolites with antioxidant activity which appeared to be significant in the reduction of free radicals in a dose dependent manner. Its phenolic and flavonoid contents could be a plausible explanation for the antioxidant activities exhibited by this spices. Therefore, its consumption could possibly boost the antioxidant defense system; thereby reduce the free radical status in humans.

REFERENCES

Abdou BA, Njintang YN, Scher J, Mbofung CM (2010). Phenolic compounds and radical scavenging potentials of twenty Cameroonian spices. Agric. Biol. J. North Am. 1(3):213-224.

Adegoke EA, Akinsanya A, Nagri SHZ (1968). Studies of Nigerian Medicinal plants: A- preliminary survey of plant Alkaloids. J. West Afr. Sci. Assoc. 13:13-17.

Adegoke E, Akinsanya A (1970). Studies of Nigerian Medicinal Plants. J. West Afr. Sci. Assoc. 13(2):15-65.

Aderogba MA, Mcgaw LJ, Ogundaini AO, Eloff JN (2007). Antioxidant activity and cytotoxicity study of the flavonol glycosides from *Bauhinia galpinii.* Nat. Prod. Res. 21(7):591-599.

Amzal H, Aloni K, Tok S, Errachidi A, Charof R, Cherrah Y, Benjouad A (2008). Protective effect of saponins from *Argania spinosa* against free radical induced oxidative haemolysis. Fitoterapia 79: 337-344.

Bagchi K, Puri S (1998). Free radicals and antioxidants in health and diseases. East. Med. Health J. 4(2):350-360.

Blois S (1985). Antioxidant determination by use of stable free radicals. Nature 29:1199-1201.

Bode SO, Oyedapo OO (2011). Biological activities ad phytoconstituents of the lower plant *Platycerium angolense*, Welwen Hook. J. Med. Plant Res. 5:1321-1329.

Burkill HM (1985). The Useful Plants of West Tropical Africa. Vol. 1, Royal Botanic Gardens,*Kew,* ISBN: 094764301X. pp. 119, 960.

Cakir A, Mari A, Yildinn A, Duru ME, Harmender M, Kazaz C (2003). Isolation and Characterization of antioxidant Phenolic compounds from the aerial part of *Hespericum hyssopifolium,* L. by activity guided fractionation. J. Ethnopharm. 87:73-83.

Cho EJ, Yokozawa T, Rhyu DY, Kim SC, Shibahara N, Park JC (2003). Study on the inhibitory effects of Korean medicinal plants and their main compounds on 1, 1-dipheny-2-picrylhydrazyl radical. Phytomedicine 10:544-551.

De Groot H, Rauen U (1998). Tissue injury by reactive oxygen species and the protective effects of flavonoid. Fundam. Clin. Pharmacol. 12: 249-255.

Ekeanyanwu CR, Ugu IG, Nwachuku UP (2010). Biochemical characteristics of the African nutmeg, *Monodora myristica.* Agric. J. 5 (5):303-308.

Ekeanyanwu RC, Etienajirhevwe OF (2012). Invitro antihelmintic potentials of Xylopia aethiopica and Monodora myristica from Nigeria. Afr. J. Biochem. Res.6(9):115-120.

Evans WC (2002). Textbook of Pharmacology. Elsevier Science Limited. pp. 229-248.

Fargere T, Mohammed A, Michel D, Bernard B (1995). Determination of peroxides and hydroperoxides with 2,2, diphenyl-1-picrylhydrazyl (DPPH).Application to ozonized ethylene vinlyl acetate copolymers (EVA). Eur. Polym. J. 31:489-497.

Finkel T (2005). Radical Medicine: treating ageing to cure disease. Nat. Rev. Mol. Cell Biol. 6:971-976.

Grassi D, Desideri G, Tiberti S, Ferri C (2009). Oxidative stress, endothelial dysfunction and prevention of cardiovascular diseases. AgroFOOD Industry li-tech. 20:76-79.

Halliwell B (1990).How to characterize a biological antioxidant. Free Rad. Res. Com. 9:1-32.

Halliwell B (1996). Antioxidants in human health and diseases. Ann. Rev. Nutr. 16:33-49.

Halliwell B (2006). Reactive species and antioxidants. Redox biology is a fundamental theme of aerobic life. Plant Physiol. 141: 312-322.

Halliwell B, Hoult JRS, Blake DR (1988). Oxidants, inflammation and anti inflammatory drugs. FASEB J. 2:501-508.

Han KH, Lee EJ, Sung MK (1999). Physical characteristics and antioxidative capacity of major seaweeds. J. Food Sci. Nutr. 4:180-183.

Hinneburg I, Dorman HJD, Hiltunen R (2006). Antioxidant activities of extracts from selected culinary herbs and spices. Food Chem. 97:122-129.

Iwu MM (2002). Evaluation of the antihepatotoxic activity of the biflavonoids of *Garciana kola* seeds. J. Ethnopham. 21:14-19.

Jae-Young J, Pyo-jam P, Se-kwon K (2005). Antioxidant activity of a peptide isolated from Alaska Pollack (*Theragra chalcogramma*) frame protein hydrolysate. J. Food Res. Int. 38:45-50.

Kanner PH, Guyton KE, Gorospo M (1999). Oxidative stress and the molecular biology of antioxidant defense (Scandolios, J. G. ed). Cold spring Harbor laboratory press, New York. pp. 247-272.

Keay RWJ (1989).Trees of Nigeria. Claridon Press Oxford. Vol II issue 3. pp. 5.

Kulisic T, Radonic A, Katalinic V, Milos M (2004). Use of different methods for testing antioxidative activity of oregano essential oil. Food Chem. 85(4):633-640.

Lee MJ, Maliakal P, Chen L, Meng X, Bondoc FY, Prabhu S, Lambert G, Mohr S, Yang CS (2002). Pharmacokinetics of tea and (+) - epigallocatechin- 3-gallate by humans: Formation of different metabolites and individual variability. Cancer Epidermiol. Biomarker Prev. 11:1025-1032.

Mantle D, Eddeb F, Pickering AT (2000). Comparison of relative antioxidant activities of British medicinal plants species invitro. J. Ethnophamacol.72: 47-51.

Matsukawa R, Dubinsky Z, Kishimoto E, Masaki K, Masuda Y, Tokeuchi T (1997). A comparison of screening methods for antioxidant activity in seaweeds. J. Appl. Physiol. 9:29-35.

Miguel L (2009). Distribution and Biological activities of flavonoid Lutiolin. Mini Rev. Med. Chem. 9: 31-59.

Morrissey PA, O'Brien NM (1998). Dietary antioxidants in health and disease. Int. Dairy J. 8:463-472.

Nsimba RY, Kikuzaki H, Konishi Y, (2008). Antioxidant activity of various extract fractions of *Chenopodium quinoa* and *Amaraathus* sp. seeds. Food Chem. 106:760-766.

Okafor JC (1987). Development of Forest tree crops for food supply in Nigeria. Forest Eco. Manag. 1:235-247.

Omobuwajo TO, Omobuwajo OR, Sanni LA (2003). Physical properties of calabash nutmeg (*Monodora myristica*) seeds. J. Food Eng. 57: 375-381.

Owoyele BV, Oguntoye SO, Dare K, Ogunbiyi BA, Arubuola EA Soladoye AO (2008). Analgesic, anti-inflammatory and antipyretic activities from flavonoid fractions of *Chromolaena odorata*. J. Med. Plant Res. 2(9):219-225.

Oyedapo OO, Akinpelu BA, Orefuwa SO (2004). Antiinflammatory effects of *Theobroma cacao*, L. root extract. J. Trop. Med. Plants (Malaysia). 5(2):161- 166.

Oyedapo OO, Sab FC, Olagunju JA (1999). Bioactivity of fresh leaves of *Lantana camara*. Biomed. Lett. 59:175- 183.

Pereira da Silva A, Rocha R, Silva CML, Mira L, Daurte MF, Florencio MH (2000). Antioxidant in medicinal plant extracts. A research study of the antioxidant capacity of *Crataegus,Hamamelis* and *Hydrastis*. Phytother. Res. 14:612-616.

Sofowora A (2006). Phytochemical screening. In: Medicinal Plants and Traditional Medicine in Africa. Second Edition, Spectrum Books Limited Ibadan, Nigeria. pp. 150-153.

Su XY, Wang Z, Liu J (2009). In vitro and in vivo antioxidant activity of Pinus koraiensis seed extract containing phenolic compounds. Food Chem. 117:681-686.

Sun PX, Yie LK, Zhang ZL, Hu M, Lu L (1999). Colometric determination of the total content of the flavonoids in epimedium capsules. J. Shenyang Pharm. Univ. 16:68-70.

Uhegbu FO, Iweala EJ, Kanu I (2011).Studies on the chemical and anti nutritional content of some Nigerian Spices. Inter. J. Nutri. Metab. 3(6):72-76.

Valko M, Leibfritz D, Moncol J, Cronium MT, Mazur M, Telser J (2007). Free radicals and antioxidants in normal physiological functions and human disease. Int. J. Biochem. Cell Biol. 39:44-84.

Xanthopoulou MN, Fragopoulou E, Kalathara K, Nomikos T, Karantonis HC, Antonopoulou S (2009). Antioxidant and anti-inflammatory activity of red and white wine extracts. Food Chem. 120: 665-672.

Yan X, Nagata T, Fan X (1998). Antioxidative activities in some common seaweeds. Plant Foods Hum. Nutr. 52:253-262.

Yoshikana N, Nimoniya K, Shimoda H, Nishida N, Matsada H (2002). Hepatoprotective and antioxidative properties of *Salacia reticulate*: protective effects of phenolic constituents on CCl_4- induced liver injury in mice. Biol. Pharm. Bull. 23:72-76.

Effect of ten amino acids on elimination of acrylamide in a model reaction system

Miao Yu, Shiyi Ou, Deng Liumengzi, Caihuan Huang and Guangwen Zhang

Department of Food Science and Engineering, Jinan University, Guangzhou 510632, China.

Ten (10) amino acids (asparagine, aspartic acid, cysteine, glycine, glutamic acid, glutamine, methionine, leucine, tyrosine and lysine) were used to test their effect on elimination of acrylamide, when co-heating of 10 μmol of acrylamide respectively with 0.1 mmol of amino acids at 160°C for 15 min, acrylamide was eliminated from 7.4% (asparagine) to 94.4% (cysteine) at natural pH, and 11.4% (glutamic acid) to 94.8% (cysteine) at pH 7.0. It was found that cysteine, glysine and lysine showed much higher reactivity with acrylamide than the other seven amino acids. When the molar ratios of amino acid to acrylamide were increased to 60:1 for cysteine and glycine, and 80:1 for lysine, acrylamide in the reaction system was almost eliminated. The elimination efficiencies of the three amino acids for acrylamide were enhanced by addition of glucose. Liquid chromatography/time-of-flight mass spectroscopy showed that cysteine, glycine, and lysine adducted with two molecules of acrylamide, while the other amino acids except for glutamic acid (no adduct compounds were found) just adducted with one mole of acrylamide. This finding may partly explain why cysteine, glycine and lysine showed high capacity to reduce acrylamide content in high-temperature processing food.

Key words: Acrylamide, adduct, amino acids.

INTRODUCTION

Acrylamide is a neurotoxic compound and probable carcinogen found at significant levels in carbohydrate-rich processed foods. It is largely derived from heat-induced reactions between the amino group of the free amino acid asparagine and the carbonyl group of reducing sugars, such as glucose, during baking and frying (Friedman, 2003; Koutsidis et al., 2009).

Addition of other amino acids has been proposed as a mitigation strategy to reduce the levels of acrylamide in crisps, flat breads and bread crust (Vinci et al., 2012). Glycine, lysine and cysteine have received particular attention as additives that could potentially reduce acrylamide formation by either competing for available Maillard reaction inter-mediates or reacting with acrylamide itself through Michael addition (Adams et al., 2010; Friedman and Levin, 2008; Ou et al., 2010). Zamora et al. (2010) speculated that each molecule of amino acid could form adducts with one or two molecules of acrylamide, but they did not identify the later one.

In addition to the intended addition of amino acids into food, detectable amounts of various free amino acids have been found in food raw materials like potato, wheat and rice (Elmore et al., 2005; Mustafa et al., 2007; Saikusa et al., 1994).

In order to investigate the efficiency of different amino acids to eliminate acrylamide, 10 amino acids were used and reacted with acrylamide in model reaction system; and the type of adducts formed were determined using liquid chromatography/time -of-flight mass spectroscopy (LC/TOF-MS).

MATERIALS AND METHODS

Chemicals

Asparagine, aspartic acid, cysteine, glycine, glutamic acid, glutamine, methionine, leucine, tyrosine, lysine and glucose were purchased from Aladdin Reagents Database Inc. (Shanghai, China) and were of 99% purity. The acrylamide standard (>99.8%) was purchased from Sigma-Aldrich Co. (St. Louis, MO, USA). HPLC-grade methanol was purchased from J.T. Baker (USA).

Effect of amino acids on elimination of acrylamide and determination of their adducts

Aqueous solutions (4 mL) containing 10 μmol acrylamide and 0.1 mmol of a single amino acid species (asparagine, aspartic acid, cysteine, glycine, glutamic acid, glutamine, methionine, leucine, tyrosine and lysine), were heated in sealed stainless test tubes at 160°C for 15 min by incubation in a polyethylene glycol (MW 600) bath (Ou et al., 2008). The pH values of the samples were kept unadjusted or adjusted to 7.0 using 0.2 mol/L phosphate buffer before heating. At the end of reaction, the test tubes were cooled in an ice bath. The reaction solutions were decanted, and 6 mL of deionised water was added to wash the test tubes. The reaction and wash solutions were combined and then centrifuged at 8000 ×g for 20 min. Acrylamide and its amino acid adducts in the supernatant were characterised and quantified by HPLC (Figure 1), ESI-MS, and time-of-flight mass spectrometry (TOF-MS).

Effect of cysteine, glycine and lysine on elimination of acrylamide

Mixtures (4 mL) containing 10 μmol acrylamide and different amounts (0.1, 0.2, 0.4, 0.6 and 0.8 mmol) of single amino acids (glycine, cysteine and lysine), with or without the addition of 0.1 mmol of glucose, were reacted in sealed stainless test tubes at 160°C for 15 min. The pH was either left unadjusted or adjusted to 7.0 using 0.2 mol/L phosphate buffer. At the end of the reaction, the test tubes were cooled in an ice bath. The reaction solutions were decanted, and 6 mL of deionised water was added to each test tube. Specific reaction and wash solutions were combined and then centrifuged at 8000 ×*g* for 20 min. The residual acrylamide was determined by HPLC.

Determination of acrylamide

Quantification of acrylamide was performed on a Shimadzu LC-20AT system (Kyoto, Japan) equipped with a diode array detector and LC-solution software. An Agilent Zorbax SB Aq C_{18} column (4.6 mm x 250 mm, 5 μm; Waldbronn, Germany) was selected for acrylamide analysis. The injection volume was 5 μL. Elution was carried out at a flow rate of 0.5 mL/min under isocratic conditions at 30°C using 5% methanol aqueous solution as the mobile phase. Acrylamide was detected at 205 nm.

Determination of the adducts between amino acid and acrylamide using TOF-MS

The adducts of amino acids with acrylamide were determined using an Agilent 6210 LC/TOF-MS with an Agilent Zorbax SB-C_{18} column (2.1mm × 30mm, 3.5 μm). The TOF-MS operation parameters were as follows: drying gas (N_2) at 9 L/min flow rate; drying gas temperature, 300°C; nebulising gas (N_2) pressure, 30 psi; capillary voltage, 3500 V; skimmer voltage, 65 V and fragmentor voltage, 300 V. Elution was performed at a flow rate of 0.4 mL/min under isocratic conditions and using 10% methanol aqueous solution as the mobile phase. TOF-MS scanning results were collected as centroids from 100 to 2000 m/z. Two reference mass compounds, a lock mass solution including purine ($C_5H_4N_4$ at m/z 121.050873) and hexakis (1H, 1H, 3H-tetrafluoropentoxy)-phosphazene ($C_{18}H_{18}O_6N_3P_3F_{24}$ at m/z 922.009798) were used to perform real-time lock mass correction. ESI positive or negative ions were used to determine the molecular weight of the components in the products and identify amino acid and acrylamide adducts.

RESULTS AND DISCUSSION

Effect of amino acids on acrylamide elimination

Except for asparagine, aspartic acid and glutamine, the other amino acids tested obviously decreased acrylamide levels at natural pH (allowing the pH to change as the reaction proceeded). In addition, cysteine showed the highest efficiency in reducing acrylamide, with acrylamide disappearance of 94.6%, followed by glycine (72.1%) and lysine (69.6%). The starting pH of aspartic acid - acrylamide, glutamic acid -acrylamide, glutamine - acrylamide and lysine-acrylamide system were 2.85, 3.10, 4.54 and 9.67, respectively, and that of the rest systems were between 6.01 and 6.90. A buffered model reaction system is designed to eliminate impact of pH differences between each amino acid on test results. When reaction pH was kept constant at 7.0, the amount of residual acrylamide fell slightly in contrast with that of unbuffered reaction system (Table 1). The results showed that pH difference did not influence greatly on amino acid-acrylamide reaction.

Among the amino acids tested, cysteine, glycine and lysine were most effective on eliminating acrylamide but they could not completely deplete free acrylamide when the molar ratio was kept at 10:1. In the later research, different molar ratios of amino acids to acrylamide, from 10: 1 to 80: 1, were tested to investigate whether or not increasing amino acid concentration could further eliminate acrylamide. Acrylamide was decreased gradually as the increase of amino acid addition amount (Table 2). When the molar ratio rose to 60: 1 for cysteine and glycine or 80:1 for lysine, the amount of residual acrylamide in the reaction system was reduced to below the detection limit.

During the Maillard reaction, cysteine, glycine and lysine concentrations would be reduced while reacting with reducing sugars. To investigate whether Maillard reaction would influence the elimination effect of acrylamide by the three amino acids, glucose was added into the reaction system. The results showed that addition of glucose did not decrease but instead increased the acrylamide-reducing efficiency of these amino acids, except for lysine (Table 3).

Formation of adducts between amino acids and acrylamide

Acrylamide contains an α,β-unsaturated group which can

Table 1. Effect of amino acids on elimination of acrylamide in a model system added with 10 μmol of acrylamide and 0.1 mmol of amino acid heating at 160°C for 15 min.

Amino acid	Blank	Asn[a]	Asp	Cys	Gly	Glu	Gln	Met	Leu	Tyr	Lys
Residual acrylamide at nature pH (μmol)	10.28 ±0.23[b]	9.26±0.45	9.21±0.16	0.56±0.12	2.87±0.34	7.48±0.26	9.23±0.36	7.92±0.35	8.47±0.37	8.48±0.15	3.13±0.21
Residual acrylamide at pH 7.0 (μmol)	10.11±0.21	8.78±0.17	9.15±0.23	0.52±0.05	2.48±0.11	8.86±0.13	8.74±0.26	7.24±0.06	8.12±0.13	7.65±0.12	2.83±0.06

[a]AA, amino acid; Asn, asparagine; Asp, aspartic acid; Cys, cysteine; Gly, glycine; Glu, glutamic acid; Gln, glutamine; Met, methionine; Leu, leucine; Tyr, tyrosine; Lys, Lysine; [b]Means ± SD (n=3).

Table 2. Effect of different amount of cysteine, glycine and lysine on elimination of 10 μmol of acrylamide in 4 ml of aqueous solution heating at pH 7.0 for 15 min.

Amino acid	Residual acrylamide (μmol) after reaction with different amount of amino acids				
	0.1 mmol	0.2 mmol	0.4 mmol	0.6 mmol	0.8 mmol
Cysteine	0.50±0.03[a]	0.42±0.02	0.21±0.03	ND[b]	ND
Glycine	2.53±0.02	1.78±0.02	0.63±0.01	ND	ND
Lysine	3.23±0.04	1.81±0.04	1.23±0.03	0.67±0.01	ND

[a]Means ± SD (n=3); [b]Not determined (below 0.05 μg/mL).

Table 3. Effect of different amount of cysteine, glycine and lysine on elimination of 10 μmol of acrylamide heating with 0.1 mmol glucose at pH 7.0 for 15 min.

Amino acid	Residual acrylamide (μmol) after reacting with different amount of amino acids				
	0.1 mmol	0.2 mmol	0.4 mmol	0.6 mmol	0.8 mmol
Cysteine	0.46±0.04	0.31±0.07	0.14±0.02	ND	ND
Glycine	1.58±0.07	1.07±0.12	0.31±0.03	ND	ND
Lysine	3.68±0.16	2.73±0.13	1.27±0.07	0.13±0.01	ND

[a]Means ± SD(n=3); [b]Not determined (below 0.05 μg/ml).

reacts with different kinds of nucleophiles including amino acids through Michael addition. According to Zamora et al. (2010), one molecule of some amino acids can react with one molecule of acrylamide through the Michael reaction to form compound 1, and they speculated that one molecule of acrylamide can adduct with two molecules of some amino acids to form compound 2. However, they did not detect the compound 2 during the reaction of acrylamide with butylamine, glycine and lysine, respectively.

In this research, compound 1 was detected in the reaction mixture of acrylamide with all tested amino acids except for glutamic acid by TOF-MS (Table 4), and the compound 2 was successfully detected in the reaction system of acrylamide with cysteine, glycine and lysine, respectively (Table 4). Moreover, the ions at m/z 205 and 203 which

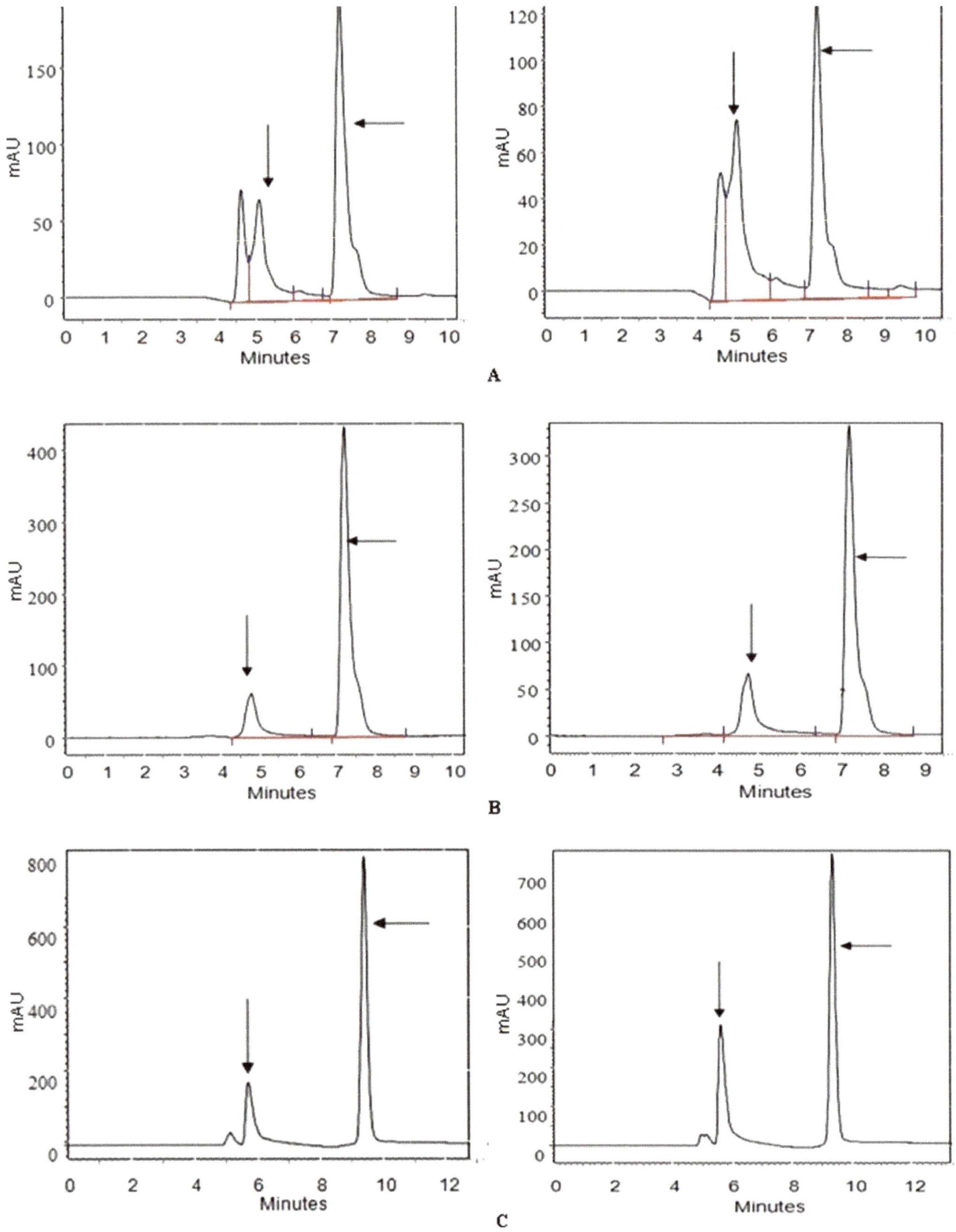

Figure 1. HPLC of the reaction products from 10 µmol acrylamide reacted with 0.1 mmol (left) and 0.2 mmol (right) cysteine (A), glycine (B), and lysine (C). The vertical arrow indicates the peak for the adduct formed from the reaction of acrylamide with amino acids. The horizontal arrow indicates the peak for acrylamide.

Table 4. Detected accurate mass and predicted chemical formulas of adduct 1 and adduct 2 by TOF-MS.

AA[a]	Ion	Calc m/z	m/z	Diff (ppm)	Formula
Asn	$(M+H)^+$	204.09788	204.09762	1.28	$C_7H_{13}N_3O_4$
Asp	$(M+H)^+$	205.0819	205.08143	2.29	$C_7H_{12}N_2O_5$
Cys	$(M+H)^+$	193.06414	193.06478	-3.35	$C_6H_{12}N_2O_3S$
Cys	$(M+H)^+$	264.10125	264.10160	-1.33	$C_9H_{17}N_3O_4S$
Gly	$(M-H)^-$	145.06187	145.06188	-0.13	$C_5H_{10}N_2O_3$
Gly	$(M-H)^-$	216.09898	216.0989	0.4	$C_8H_{15}N_3O_4$
Glu	-	-	-	-	-
Gln	$(M+H)^+$	218.1141	218.1124	-7.8	$C_8H_{15}N_3O_4$
Met	$(M+H)^+$	221.09544	221.09531	0.57	$C_8H_{16}N_2O_3S$
Leu	$(M+H)^+$	203.13902	203.13944	-2.09	$C_9H_{18}N_2O_3$
Tyr	$(M+H)^+$	253.11828	253.11844	-0.61	$C_{12}H_{16}N_2O_4$
Lys	$(M+Na)^+$	204.09788	204.09762	1.28	$C_9H_{19}N_3O_3$
Lys	$(M+H)^+$	289.18703	289.18707	-0.13	$C_{12}H_{24}N_4O_4$

[a]The full name was listed in the footnote of Table 1.

correspond respectively to the $[M+H]^+$ and $[M-H]^-$ of the adduct for aspartic with acrylamide were detected by ESI-MS in asparagine-acrylamide reaction system, suggesting that, during the heating process, asparagine may be hydrolyzed to aspartic acid which then reacts with acrylamide.

The TOF-MS results may partly explain why cysteine, glycine and lysine showed high capacity to reduce acrylamide in food system. As reported by Zamora et al. (2010), the compoud 1 is not stable and easy to decompose. Thus, the amino acids that can only form compound 1 with acrylamide would show low elimination efficiency for acrylamide because of decomposition of compound 1. However, cysteine, glycine and lysine can react with acrylamide to form compound 2, the later compound may be more stable than the compound 1, thus make these amino acids show higher elimination efficiency for acrylamide. According to Koutsidis et al. (2009), cysteine and glycine, which were able to form compound 2 tested in our research, could form relatively more amino acid-acrylamide adducts when compared with another 3 amino acids.

Moreover, analysis of the products of three amino acids reacting with acrylamide by HPLC at 205 nm revealed that, there were one or two peaks with shorter retention time than acrylamide in the chromatogram. The peak area increased as the peak area of acrylamide decreased (Figure 1), HPLC-MS results of the reaction products between acrylamide and cysteine, lysine and glycine respectively confirmed that they were adduct 1 and adduct 2. This finding may encourage us to detect the adducts in food system.

ACKNOWLEDGEMENTS

The present work was co-funded by the National Natural Science Foundation of China (No. 31071596) and the Minister of Science and Technology of China (No. 2012BAK01B03).

REFERENCES

Adams A, Hamdani S, Van Lancker F, Méjri S, De Kimpe N (2010). Stability of acrylamide in model systems and its reactivity with selected nucleophiles. Food Res. Int. 43:1517-1522.

Elmore JS, Koutsidis G, Dodson AT, Mottram DS, Wedzicha BL (2005). Measurement of acrylamide and its precursors in potato, wheat, and rye model systems. J. Agric. Food Chem. 53:1286-1293.

Friedman M (2003). Chemistry, biochemistry, and safety of acrylamide. A review. J. Agric. Food Chem. 51:4504-4526.

Friedman M, Levin CE (2008). Review of methods for the reduction of dietary content and toxicity of acrylamide. J. Agric. Food Chem. 56:6113-6140.

Koutsidis G, Simons SPJ, Thong YH, Haldoupis Y, Mojica-Lazaro J, Wedzicha BL, Mottram DS (2009). Investigations on the effect of amino acids on acrylamide, pyrazines, and Michael addition products in model systems. J. Agric. Food Chem. 57:9011-9015.

Mustafa A, Aman P, Andersson R, Kamal-Eldin A (2007). Analysis of free amino acids in cereal products. Food Chem. 105:317-324.

Ou SY, Lin QL, Zhang YP, Huang CH, Sun X, Fu L (2008). Reduction of acrylamide formation by selected agents in fried potato crisps on industrial scale. Innov. Food Sci. Emerg. 9:116-121.

Saikusa T, Horino T, Mori Y (1994). Distribution of free amino acids in the rice kernel and kernel fractions and the effect of water soaking on the distribution. J. Agric. Food Chem. 42:1122-1125.

Vinci RM, Mestdagh F, Meulenaer BD (2012). Acrylamide formation in fried potato products - Present and future, a critical review on mitigation strategies. Food Chem. 133:1138-1154.

Zamora R, Delgado RM, Hidalgo FJ (2010). Model reactions of acrylamide with selected amino compounds. J. Agric. Food Chem. 58:1708-1713.

Utilisation of common grain crops in Zimbabwe

Misheck Chandiposha, Chagonda Ignatius and Makuvaro Veronica

Department of Agronomy, Midlands State University, P. Bag 9055, Gweru, Zimbabwe.

Grain production in Zimbabwe is very important to food security of the country and therefore should be promoted. Potentially, knowledge on utilisation and value addition of grain crops can be viewed as strong strategies in promoting its production. Value addition results in various products from each grain crop and this may increase utilisation and profitability. In this review paper, an attempt is made to describe the utilisation and value addition of common grain crops to Zimbabwe which include maize, soybeans, wheat, sorghum, groundnuts, pearl millets, bambara nuts and cowpea.

Key words: Zimbabwe, grain crops, value addition, utilization.

INTRODUCTION

Grain crops like maize, soybeans, wheat, sorghum, and many others contribute immensely to food security of Zimbabwe. For example, the country needs about 1.8 million tons of maize annually while approximately 500,000 tons of wheat is required annually (Gono, 2008). Most of these grain crops have high content of carbohydrates, for example maize, sorghum and wheat, while other grain crops have high levels of protein and fats like soybeans, dry beans and ground nuts (Šramková et al., 2009). Carbohydrates and proteins are essential to human diet for energy and growth, respectively. However, despite the importance of grain crops, their production is facing a decline in Zimbabwe. The decline may be attributed to a number of factors that may include pests and diseases, poor management, climatic variability and poor pricing among others. In the short and long term, poor pricing of grain crops is significantly affecting choice of crop to grow by most farmers (Jayne, 1994). Since the sole buyer of most grain crops in Zimbabwe is Grain Marketing Board (GMB), the price that it offers to farmers is far below the cost of production. In addition, securing local market for most grain crops is a constraint in many developing countries including Zimbabwe. This discrepancy has led many farmers to shift from growing grain crops to cash crops like tobacco, cotton and others. The shift from grain production to cash crop production undermines food security and perpetuates poverty in developing countries. The gap between production and consumption levels is normally bridged by importation of grain by governments of developing countries especially maize and wheat. Importation of grain result in loss of much needed foreign currency that may be used in other critical sectors like health.

An alternative to stimulate an increase in production of grain crops and profitability among farmers is to promote value addition (Ja'afar-Furo et al., 2011). Value addition is simply defined as a process of increasing value and consumer appeal. According to Coltrain et al. (2000), value addition means improving the product economically by altering its current place, time and form characteristics. Value addition is divided into three ways; primary, secondary and high end processing. Primary processing involves proper cleaning, grading and packaging. While secondary processing encompasses basic processing, packaging and branding. Basic processing is normally carried after cleaning and grading, it involves grinding of grain into meal or other products. Packaging and branding gives the product uniqueness which is important in marketing. The final way of value addition is high end processing and this involves modern processing technology, which may be complex although it is high rewarding. A good example of high end processing can be drawn from extraction of ethanol from maize grain which may be used in production of bio – fuels. Value addition is much dependent on the grain crop since they differ in their chemical constituents.

The following section will look at value addition of grain crops that are commonly grown in Zimbabwe which include maize, soybeans, wheat, sorghum, groundnuts, pearl millets, bambara nuts and cowpea.

Maize

The first step in increasing value of maize grain is centred on proper drying of the crop to safe moisture levels (13 to 15% moisture content). This will reduce the spoilage due to insects and pathogens, thereby increasing shelf life of maize grain (Wilcke and Morey, 2009). The next step of adding value of grain involves cleaning, grading and packaging the product. Cleaning removes all extraneous material like weed seeds and soil thereby improving appearance and this may attract higher price. Grading of maize grain may be done according to size using sieves and this increases quality of the product. Processing should be done aiming for specific and unique product for purposes of branding. For example, maize can be grounded to produce livestock feeds, samp, roller and refined meal. Packaging of maize grain and its products is also significant in value addition. Research for different needs in packaging for the market is very important in value addition. There are various ways dry maize can be utilized. The grain may be popped in a hot plate and eaten as hot popped maize. Alternatively, the grain may be boiled and consumed as whole grain or the grain is partially grounded to produce mealie rice, sometimes called samp. The whole grain may be milled to produce mealie meal or flour which is used for various purposes. The meal is usually used to make cooked paste while the flour is used to prepare unleavened or leavened bread, biscuits and corn bread. The mealie meal can also be used to make traditional non-alcoholic and alcoholic beverages. Maize grain can be processed to produce livestock and poultry feeds. Similarly, maize plants especially after flowering stage can be used as fodder to dairy and draft cattle.

In the commercial industry, maize grain is used in the wet or dry milling to produce secondary and derived products from maize that include tortillas, maize flours, chips, snacks, breakfast cereal, maize oil, soft drinks, starch, ethanol for biofuels and wine like whisky (Mejia, 2003; John and Babynath, 2011).

Soybeans

Soybeans have various products that can be derived from its grain. When the grain is harvested in its fresh state, it can be used to produce soy foods like miso, tofu, tempeh, soy sauce traditional to countries like Taiwan, Japan, China, Indonesia, India, Thailand, Uganda and Nigeria (Thoenes, 2007). Fresh whole grain of soybeans can be processed to produce diary products substitutes that include soy milk, cheese, margarine, soy ice cream and soy cream cheese (Jaiswal, 2009). When the grain is dried, it can be primarily processed into soy flour, textured vegetable protein, soybean oil and soy meal. Soy flour is used to make soy bread and soy pasta. Textured vegetable protein is made up of different vegetable products and can be a substitute of meat (Jaiswal, 2009). Oil extracted from soybeans can be used for edible purposes and in industry as a source of key fat ingredient (Thoenes, 2007). When oil is extracted from soybeans, the by-product that is produced is soy meal which can be fed to livestock. However, this meal is believed to have anti-nutritional substances like trypsin inhibitors, saponins, phytoestrogens, glucinins, goitrogens, lectins, urease and other substances (CRC, 1983; Liener, 1994). The anti-nutritional factors can cause infertility to livestock like cattle and sheep (Adams, 1995). The anti-nutritional factors in soybeans can be reduced by roasting or heating the grain.

Wheat

Wheat is mainly added value in milling and bakery industries. Flour milling takes the major share of wheat processing and the main product is wheat flour. Most of wheat flour is sold to grain based food manufacturers including bakery industry. Some of the wheat flour can be begged and sold to ingredient distributors. Apart from milling wheat into flour, the crop product can be used to produce livestock feed, starch and ethanol (FAO, 2009). In the bakery industry, several products can be made depending on type of wheat flour (and wheat type). Although, blending of different types of wheat flour can be employed in milling industry to produce bakery products like pastries, sweet goods and cakes. The bakery products include bread (white, brown, whole wheat and other variety breads), rolls (hamburger, hotdog, muffins and bread type rolls), sweet yeast goods (doughnuts), soft cakes, pies, pizza crust, cookies and crackers (Lou and Wilson, 1998). In the flour milling industry, bran is an important by product with several uses (Hemery et al., 2007; Javed et al., 2012). Wheat bran can be used as an alternative to synthetic medium in fermentation processes (Pandey, 1992). Apart from that, wheat bran can be used as a supplement of nitrogen in medium for the production of enzymes like protease, amylase and glucoamylase (Vishwanatha et al., 2009). In addition, the compounds from wheat bran like lingocellulosic can be used to remove heavy metals like Cu II, Pb II and Cd II from industrial waste waters since the latter pose a health risk.

Another potential use of wheat bran is in bio ethanol production as an alternative to fossil fuel which is facing depletion (Shafiee and Topal, 2009; Manikandan and Viruthagiri, 2009). According to Gomez et al. (2003) bran of wheat can be used to supplement wheat flour and baked products with cheap vitamin and other nutrients. Apart from that, wheat bran can be used to make livestock feed formulations and its presence has been found to improve daily milk yields (Tahir et al., 2002).

Sorghum

Depending on sorghum variety, there are multiple functions the crop can be used for. Sorghum can be used as forage or silage crop to feed livestock and as a grain crop (Rao et al., 2010). In Zimbabwe, the main sorghum variety grown is Kafir (*Sorghum bicolor* Kaffrorum) and is mainly used as a grain crop. Sorghum grain can be used as a substitute to maize and wheat meals although it can also be used to supplement feed for poultry, horses and swine. However, sorghum with low tannin should be used to make livestock feed as their presence decrease feed efficiency by 5 to 30% (Rooney, undated). Sorghum grain can be added value by processing it into opaque beer and this is common in Zimbabwe and South Africa. Similar to other grain crops, sweet stalked and high energy sorghum can potentially be used in ethanol production and electricity (Celada, 2008). Sorghum can be processed into flour and may substitute wheat. The flour is used in the bakery industry to produce different products that include cakes, bread, biscuits flakes and other products. Value addition of sorghum also involves popping by exposing the grains to high temperature for a short time. Different methods of popping have been used and include hot air popping, gun puffing, popping in hot oil and microwave popping. Furthermore, popping sorghum is known to improve shelf life and nutritional quality of the grain.

Popped sorghum can further be used to make snack foods (Gundboudi, 2006). Sorghum has also been reported in starch production suitable for diabetic and obese people (Dicko et al., 2006). Sweet sorghum can also be used to make syrup (Mazumdar et al., 2012).

Groundnuts

Groundnuts can be used as freshly harvested seeds or dry kernels. Fresh harvested pods may be boiled or roasted with or without salt and served as a snack. Post-harvest technologies in dry groundnut are very fundamental to value addition of the crop. These technologies include adequate drying, cleaning, de-stoning and grading of ground nuts. The processing of dry groundnuts begins with shelling and can be done by hand or machinery; it also adds value to the crop. Nutritive groundnut haulm from shelling can be used to feed livestock while the hulls can be used to make compost manure. Shelled groundnuts can also be boiled or roasted with or without salt and served as a snack (Singh and Diwakar, 1993). The most popular value added product from groundnuts is peanut butter especially in Western countries. The production of peanut butter involves pre-cleaning, shelling, grading, roasting, blanching, grinding, cooling and packing. The butter is used as spread on bread and is used in the food industry to make candy, cookies and sandwiches (Kadam and Chavan, 1991). Another important valued added product of groundnuts is oil which is used in cooking or can be processed in margarine and soap making. The groundnut oil is also used in production of candles, cosmetics, suckericide, leather dressing, furniture and creams; although on a low scale.

Oil extraction results in a by-product called groundnut cake. Groundnut cake can be used in bakery industry to make biscuits, breads and cookies. The cake can be fermented to make the products easily digestible, tasty and nutritious. In addition, groundnuts can be processed into milk by soaking them in 1% sodium bicarbonate for 16 to 18 h. Groundnut milk when fermented can be used to make yoghurt, substitute up to 20% whole milk in ice cream production (Singh and Diwakar, 1993).

Pearl millet

Pearl millet tolerate stresses like drought, low fertility and high temperatures and is grown in areas where other cereal crops like maize cannot survive (Hanna and Cardona, 2001; Basavaraj et al., 2010). Therefore, value addition of pearl millet is very important to increase its production and utilisation. Similar to sorghum, pearl millet is grown for fodder or grain purposes. However, in world's poorest countries, pearl millet is mainly grown as a grain crop. The utilisation of pearl millet has remained very low possibly due to a number of anti- nutritional factors the grain crop has, for example, phytate and poly phenols that are found in the grain. Therefore, processing of pearl millet can add value and increase its utilisation. The first process is dehulling or decortication by removing outer layers of grain reduces fibre, ash and fat, increasing nutritional quality. Milling is the second process and it produces flour which can be used to make porridge. However, the shelf life of that flour is very short due to the presence of high fat content. This limitation can be reduced by moist heating and drying the grains followed by decortication. Blanching has also been reported to increase shelf of the flour without changing nutrient content of the flour (Chavan and Kachare, 1994). The process is achieved by submerging grain for shorter period and then drying. The flour from milling pearl millet is used to make bakery products like bread, biscuits and cakes. However, flour of pearl millet is not good for bakery due to absence of gluten and it makes dough of poor consistency (Badi et al., 1976).

Grain processing that involves dehulling and milling results in a by-product called bran that can be used to make edible oil and deoiled bran can be a source of dietary fibre. Dietary fibre can be used to make flakes for obese people. The third process of pearl millet is malting and is done by allowing limited germination of the grain under moist controlled environments. Malted pearl millet is normally used in the brewery industry to make beer. Other processes that reduces the anti-nutritional factors and increase shelf life include acid treatment (Hadimani and Malleshi, 1993) and dry heat treatment (Poonam, 2002). Another process of pearl millet that can improve digestibility and shelf life is parboiling. Parboiled grains can be cooked to produce rice like products. Other value added products of pearl millet are non-alcoholic beverages (mahewu), flakes, drinks and pops. Pearl millet grains can also be processed into livestock feeds while its stalks are used to roof traditional houses (NRAA, 2012).

Bambara nut

According to Hillocks et al. (2012), bambara nuts can be eaten while it is fresh by boiling the seeds. Dried seeds can be milled into flour which is used to make flat cakes, biscuits and porridge when mixed with other cereals. In Zimbabwe, the nuts are pounded and mixed with onions, tomatoes and oil to make a relish. In order to improve protein content of traditional weaning foods, the legume grain can be used to make bambara- fortified high protein fermented maize dough (Mbata et al., 2009). Wheat flour that is used in biscuit, bread and cake production lacks certain essential amino acids like lysine, tryptophan and threonine (Kent, 1975). Grain legumes that include bambara nuts can be used to make fortified wheat flours that are nutritive (Abu-salem and Abou-Arab, 2011). Similar to soybeans, bambara nuts can be used to produce milk by soaking overnight, dry frying the seeds, homogenising and removal of insoluble material (Brough et al., 2003). In order to increase value of the nuts, canning has been reported in Ghana and Nigeria (Hillocks et al., 2012).

Despite the high nutritional value of bambara nuts, anti-nutritional factors have been reported that include tannins, trypsin inhibitors and also poor dehulling properties. Fermentation reduced anti-nutritional factors while hot soaking improved dehulling properties (Barimalaa and Anoghalu, 1997).

Cowpea

Cowpea is predominantly grown in Africa as the crop can tolerate dry environments than other legumes like soybeans. Therefore, its utilization and value addition is important in Africa. This legume crop can be grown for fodder or grain purposes. When the crop is grown under stressful conditions, the foliage may be used to feed livestock as few pods will be produced. Fresh green cowpea pods can be used in the same way as snap beans while tender cowpea leaves can be served as a vegetable. Dry mature seeds can be boiled and served or canned or frozen to preserve. Dried cowpea grain can be milled into flour or meal which can be packaged in unbranded or branded packets (Gomez, 2004). The difference between flour and meal is that flour is more finely milled than meal. Cowpea meal can be used to produce ready to use cowpea dry meal, fritters which is used as a snack or breakfast (Nagai, 2008). A number of products can be made from cowpea flour that includes cakes, dumplings, rock buns, doughnuts, biscuits, pie, fritter, fried cowpea paste, thick cowpea porridge and many others (Randolph et al., 1981). Besides milling cowpea, innovative technologies like decortication, fermentation and extrusion has been incorporated in the past to produce weaning mixes and blending, new formulation and fortification (Nyankori, 2002), for example weanmix introduced in Ghana in 1987 (Nagai, 2008).

REFERENCES

Abu-salem FM, Abou-Arab AA (2011) Effect of supplementation of Bambara groundnut (*Vigna subterranean* L.) flour on the quality of biscuits. Afr. J. Food Sci. 5(7):376-383.

Adams NR (1995) Detection of the effects of phytoestrogens on sheep and cattle. J. Anim. Sci. 73(5):1509-1515.

Badi SM, Hoseney RC (1976) Use of sorghum and pearl millet flours in cookies. Cereal Chem. 53 (5): 733-738.

Barimalaa IS, Anoghalu SE (1997) Effect of processing on certain anti-nutrients of Bambara groundnuts (*Vigna subterranea*) cotyledons. J. Sci. Food Agric. 73: 186–188.

Basavaraj G, Parthasarathy RP, Bhagavatula S, Ahmed W (2010) Availability and utilization of pearl millet in India. J. SAT Agric. Res. 8.

Brough SH, Azam-Ali SN, Taylor AJ (2003) The potential of bambara groundnut in vegetable milk production and basic protein functionality systems. Food Chem. 47: 277–283.

Celada JRJ (2008) Estimating the potential returns to research and development from sorghum value added products in El Salvador and Nicaragua. Masters thesis. Kansas State University.Manhattan, Kansas

Chavan JK, Kachare DP (1994) Effect of seed treatment on lipolytic deterioration of pearl millet flour during storage. J. Food Sci. Technol. 31: 81–82.

Coltrain D, Barton D, Boland M (2000) Value Added: Opportunities and Strategies. Department of Agricultural Economics, Kansas State University

CRC (1983) CRC handbook of naturally occurring food toxicants. CRC Press, Boca Raton, FL, USA.

Dicko MH, Gruppen H ,Traoré AS, Voragen AGJ, van Berkel WJH (2006) Sorghum grain as human food in Africa: relevance of content of starch and amylase activities. Afr. J. Biotechnol. 5 (5): 384-395

FAO (2009) Agribusiness Handbook: Wheat flour. FAO. Rome, Italy

Gomez C (2004) Cowpea: Post-harvest operations. FAO. Rome, Italy

Gomez M, Ronda F, Blanco CA, Caballero PA, Apesteguia A (2003). Effect of dietary fibre on dough rheology and bread quality. Eur. Food Res. Technol. 216: 51-56.

Gono G (2008) Extraordinary Interventions by the Reserve Bank of Zimbabwe: Supplement to the second half Monetary Policy Review Statement. Reserve Bank of Zimbabwe, Harare, Zimbabwe.

Gundboudi ZA (2006) Nutritional and processing qualities of pop sorghum cultivars and value addition. Masters thesis. University of Agricultural Sciences, Dharwad

Hadimani NA, Malleshi NG (1993) Studies on milling, physico-chemical properties, nutrient composition and dietary fibre content of millet s. J. Food Sci. Technol. 30(1): 193-198.

Hanna WW, Cardona ST (2001). Pennisetums and sorghums in an integrated feeding system in the tropics. In: Tropical forage plants: Development and use (Rios AS and Pitman WD, eds.). Boca Raton, Florida, USA: CRC Press. pp. 193 -200.

Hemery Y, Rouau X, Lullien-Pellerin V, Barron C, Abecassis J (2007). Dry processes to develop wheat fractions and products with enhanced nutritional quality. J. Cereal Sci. 46: 327–347

Hillocks RJ, Bennett C, Mponda OM (2012) Bambara nut: A reviw of utilisation, market potential and crop improvement. Afr. Crop Sci. J. 20 (1): 1 – 16

Ja'afar-Furo MR, Bello K, Sulaiman A (2011) Assessment of the prospects of value addition among small-scale rural enterprises in Nigeria: Evidence from North-eastern Adamawa State. J. Dev. Agric. Econ. 3(3):144-149.

Jaiswal A (2009). Economics of Production and Value Addition to Soybean in Madhya Pradesh. MSc in Agriculture Thesis. University of Agricultural Sciences, Dharwad.

Javed MM, Zahoor S, Shafaat S, Mehmooda I, Gul A, Rasheed H, Bukhari SAI, Aftab MN, Ikram-ul-Haq (2012) Wheat bran as a brown gold: Nutritious value and its biotechnological applications. Afr. J. Microbiol. Res. 6(4) : 724-733.

Jayne TS (1994) Do high food marketing costs constrain cash crop production? Evidence from Zimbabwe. Economic development and cultural change 42:2.

John S, Babynath V (2011). Corn – The versatile crop. Science Tech Entreprenuer.

Kadam SS, Chavan JK (1991). Utilization of groundnut in India and scope for novel and alternative uses. In: Uses of tropical grain legumes: Proceedings of a Consultants Meeting, 2 7 - 3 0 Mar 1989, ICRISAT, India. pp. 277-285

Kent NL (1975). Technology of Cereals. Pergamon Press, Oxford. pp. 216.

Liener IE (1994) Implications of antinutritional components in soybean foods. Crit. Rev. Food Sci. Nutr. 34(1):31-67.

Lou J, Wilson WW (1998). Value-added wheat products: Analysis of markets and competition. Agricultural Economics report Number 386.North Dakota State University.

Manikandan K, Viruthagiri T (2009). Simultaneous saccharification and fermentation of wheat bran flour into ethanol using coculture of amylolytic *Aspergillus niger* and thermo tolerant *Kluyveromyces marxianus*. Front. Chem. Eng. China 3: 240–249.

Mazumdar DS, Poshadri A, Rao SP, Reddy RCH, Reddy, B.V.S., (2012) Innovative use of Sweet sorghum juice in the beverage industry. Int. Food Res. J. 19(4):1361-1366

Mbata T I, Ikenebomeh MJ, Ezeibe S (2009) Evaluation of mineral content and functional properties of fermented maize (generic and specific) flour blended with bambara groundnut. Afr. J. Food Sci. 3:107 – 112.

Mejia D (2003) Maize: Post-harvest Operation. FAO, AGST.

Nagai T (2008) Competitiveness of cowpea-based processed products: a case study in Ghana. Masters thesis. Michigan State University. USA.

NRAA (2012) Products, Diversification, Marketing and Price Discovery of Pearl Millet in India. Policy Paper No. 2 National Rainfed Area Authority, NASC Complex, DPS Marg, New Delhi-110012, India. pp. 48

Nyankori J (2002) Product life cycle model of cowpea based products in Ghana. Department of agricultural and applied economics. Clemson University. Clemson, USA.

Pandey A (1992) Recent process developments in solid-state fermentation. Process Biochem. 27:109–117.

Poonam (2002) Effect of acid and heat treatment on nutrient composition and shelf life of pearl millet *(Pennisetum glaucum)* flour. M.Sc thesis, CCS Haryana Agricultural University, Hisar, Haryana, India. pp.106.

Randolph M, Dake F, Owusu E, Manuh J (1981) Formulation for Utilization of Cowpea Flour. Food Research Institute, CSIR. Accra, Ghana.

Rao BD, Patil JV, Rajendraprasad MP, Reddy KN, Devi K, Sriharsha B, Kachui N (2010) Impact of Innovations in Value Chain on Sorghum Farmers. Agric. Econ. Res. Rev. 23: 419-426.

Rooney LW (undated) Overview: Sorghum and millet food research failures and successes. Soil and Crop Science Dept, Texas, USA.

Shafiee S, Topal E (2009). When will fossil fuel reserves be diminished? Energy. Policy. 37: 181-189.

Singh F, Diwakar B (1993) Nutritive value and uses of Pigeonpea and groundnut. Skill Development Series no. 14. ICRISAT, India

Tahir MI, Khalique A, Pasha TN, Bhatti JA (2002).Comparative evaluation of maize bran, wheat bran and rice bran on milk production of holstein friesian cattle. Int. J. Agric. Biol. 4: 559–560.

Thoenes P (2007). Soybean International Commodity Profile. Markets and Trade Division, Food and Agriculture Organization of the United Nations.

Vishwanatha KS, Appu RAG, Singh SA (2009). Acid protease production by solid-state fermentation using *Aspergillus oryzae* MTCC 5341: optimization of process parameters. J. Ind. Microbiol. Biotechnol. 85: 1849-1859.

Wilcke W, Morey RV (2009) Review: Natural-Air Corn Drying in the Upper Midwest. Regents of the University of Minnesota.

Physicochemical and bioactive properties of selected white yam (*Dioscorea rotundata*) varieties adapted to riverine areas of Nigeria

Alamu, E. Oladeji[1], Maziya-Dixon, Bussie[1], Okonkwo, C. Christian[2] and Asiedu Roberts[1]

[1]International Institute of Tropical Agriculture, Ibadan, Oyo State, Nigeria.
[2]International Fertilizer Development, Centre, Nigeria.

Yam (*Dioscorea spp.*) is a major food of cultural, economic and nutritional importance in Nigeria and throughout West Africa. In this sub-region, white yam (*Dioscorea rotundata*) is the most dominant and important species. This study is aimed at characterizing high yielding yam varieties of this species adapted to riverine areas and forest zones of Nigeria for physical and chemical characteristics. Eleven yam varieties collected from local farmers in the South-southern part of Nigeria were evaluated for their physicochemical characteristics, bioactive compounds (vitamin C, phytic acid and tannin), functional and pasting properties. Results indicated that there were significant varietal differences ($P<0.05$) among the parameters evaluated. The moisture contents of the investigated varieties ranged from 59.5 to 68.8%, ash content ranged from 1.39 to 2.93%, protein content from 1.96 to 4.90%, fat content from 0.356 to 3.39%, total free sugars from 1.05 to 7.02% and total starch from 33.9 to 75.7%. The bioactive content results show that vitamin C content ranged from 5.64 mg/100 g to 6.99 mg/100 g, phytate from 1.12 to 2.37% and tannin from 0.359 to 1.18 mg/g. The pasting properties results show that peak viscosity ranged from 215 to 470 RVU, trough viscosity from 198 to 385 RVU, breakdown viscosity from 8.71 to 84.5 RVU, final viscosity from 278 to 571 RVU, setback viscosity from 66.2 to 204 RVU; peak time ranged from 4.97 to 7.0 min and the pasting temperature from 61.7 to 62.6°C. This study shows that the physical and chemical characteristics of these high yield yam varieties were similar to those reported for most yam varieties in other parts of Nigeria and has a great potential as source of bioactive compounds and protein.

Key words: Yam, *Dioscorea rotundata,* riverine areas, varieties.

INTRODUCTION

Globally, yam is grown in many tropical regions but the main production centre is the savannah region of West Africa, where more than 90% of the crop is grown (Sanusi and Salimonu, 2006; IITA, 2009), with Nigeria being the main producer followed by Ghana and Cote d'Ivoire (FAO, 2008). Nigeria produces 60% of the world's yam, annually producing an average of 31 million metric tons (Bergh et al., 2012), although Food and Agriculture Organisation (FAO) estimates showed the annual production to be above 37 million metric tons (FAOSTAT, 2013).

Yam (*Dioscorea* spp.) are widely grown in West Africa (Coursey and Haynes, 1970; Waitt, 1963), they are an important source of carbohydrate for many people of the sub-Sahara region, especially in the yam zone of West Africa. In Nigeria, yam has had the second highest production level of any food crop after cassava (Bergh et al., 2012). According to Food and Agriculture Organisation (FAO) estimates, there has been steady rise of 12.6% in domestic consumption of yam between 1990 and 2009 (FAOSTAT, 2013). In addition to the food and market values of yams, they play a major role in socio-cultural life of smallholder households, such as the celebrated New Yam Festival and also the presentation of yams as gifts during marriage ceremonies (Opara, 2003; IITA, 2009).

Yams are grown in the coastal region in rain forests, wood savanna and southern savanna habitats. Common species of yam include *Dioscorea rotundata* (White yam), *Dioscorea cayenensis* (Yellow yam), *Dioscorea alata* (Water yam), *Dioscorea bulbifera* (Potato yam), *Dioscorea dumentorum* (Bitter yam), *Dioscorea esculenta* (Lesser yam), *Dioscorea opposita* (Chinese yam) and *Dioscorea trifida* (Cush-cush yam) (Opara, 1999). However, *D. rotundata* is the most important species particularly in the dominant yam production zone in West and Central Africa (IITA, 2009). Yam is composed mainly of starch (75-84% of the dry weight) with small amounts of proteins, lipids and most vitamins and is very rich in minerals (Lasztity et al., 1998), and low in sodium and saturated fat content (Bergh et al., 2012). Yam's richness in carbohydrate especially starch consequently has a multiplicity of end use. Yam, sweet in flavour, is consumed as boiled yam (as cooked vegetable) or fufu or fried in oil and then consumed. It is often pounded into a thick paste after boiling and is consumed with soup. It is also processed into flour for use in the preparation of the paste. Its use as an industrial starch has also been established as the quality of some of the species is able to provide as much starch as in cereals (Izekor and Olumese, 2010).

The average crude protein content of seven yam cultivars was 7.4%, which was higher than those reported for other tropical roots, and the protein from yam also showed a better amino acid balance for human nutrition (Baquar and Oke, 1976; Bradbury, 1988; Marcus et al., 1998). Modern researches have shown that yam extracts can reduce blood sugar (Hikino et al., 1986; Undie and Akubue, 1986) and blood lipid (Araghiniknam et al., 1996), inhibit microbe activity (Hu et al., 1996, 1999; Kelmanson et al., 2000) and show antioxidative activity (Farombi et al., 2000). Yams also have pharmaceutical usage as they contain a steroid sapogenin compound called diosgenin, which can be extracted and used as base for drugs such as cortisone and hormonal drugs (Opara, 2003). There is little information in the literature on the physical and chemical properties of the high yield yams from riverine areas of Nigeria. The main objective of this study was to characterize the high yielding yam varieties (*D. rotundata*) adapted to riverine areas and forest zones of Nigeria for physical and chemical characteristics.

MATERIALS AND METHODS

Yam tubers and sample preparation

Fresh eleven (11) white yam varieties (*D. rotundata)* designated as high yielding were collected from local farmers in the South-Southern part of Nigeria. The yam samples were thoroughly washed, peeled, oven dried at 60°C. All the dried samples were milled to 0.50 mm sieve size with Perten Laboratory Hammer Mill 3102 for further laboratory analysis.

Moisture content determination

Five grams of sample was weighed into known weight of Petri-dish. The weighed sample was put into the pre-set oven (Fisher Scientific Isotemp^{R} Oven model 655F) at 110°C for 3 h. The sample was removed and cooled in a dessicator to room temperature and the weight was noted after which it was returned into the oven at temperature of 110°C for 30 min until constant weight was obtained: AOAC (2004).

Ash content determination

Five grams of sample was weighed into a previously ignited and cooled silica dish. The dish was ignited first gentle and then at 600°C for 3 h in a muffle furnace (VULCAN^{TM} furnace model 3-1750). The dish and its content were cooled in a dessicator and the remaining residue due to the ash was reweighed.

Crude fat determination

Crude fat was determined by the method of AOAC (2004). This was determined using a Soxtec System I IT2 fat extractor. Crude fat was extracted from the sample with hexane, and the solvent evaporated off to get the fat. The difference between the initial and final weight of the extraction cup was recorded as the crude fat content.

Crude protein determination

Crude protein was determined by Kjeldahl method using Kjeltec^{TM} model 2300, as described in Foss Analytical manual, AB, (2003). The method involved digestion of the sample at 420°C for 1 h to liberate the organically bound nitrogen in the form of ammonium sulphate. The ammonia in the digest ammonium sulphate was then distilled off into a boric acid receiver solution, and then titrated with standard hydrochloric acid. A conversion factor of 6.25 was used to convert from total nitrogen to percentage crude protein (AOAC, 2004).

Pasting properties determination

Pasting properties (peak viscosity, trough viscosity, breakdown viscosity, setback viscosity, peak time and pasting temperature) were determined with a Rapid Visco Analyzer (RVA) (Newport Scientific RVA Super 3) equipped with software. 3.0 g of each sample was weighed in a vessel and 25 ml of distilled water was dispensed into a new test canister. Sample was then transferred into the water surface in the canister.

The paddle was placed into the canister and the blade vigorously jogged through the sample up and down ten times. The canister was then inserted into RVA machine. The measurement cycle (12 min) was initiated by pressing the canister inside the instrument. The temperature- time lag in the RVA was as follows: started at a temperature of 50°C for 1 min, heated from 50 to 95°C in 3 min and subsequently cooled to 50°C over a 4 min period. It was followed by a period of 1 min where the temperature was kept constant at 50°C.

Determination of swelling power and solubility

This was determined by the Leach et al., (1959) method. It involved weighing 1g of milled sample into 100ml conical flask, 15ml of distilled water was added and mixed gently at low speed for 5minutes. The slurry was heated in a thermostated water bath (Thelco model 83, USA) at 80°C for 40 min. During heating, the slurry was stirred gently to prevent dumping of the starch. The content was transferred into a pre weighed centrifuge tube and 7.5 ml distilled water was added.

The tubes containing the paste were centrifuged at 2,200 rpm for 20 min using Sorvall glc-1 table top centrifuge (model 06470, USA). The supernatant was decanted immediately after centrifuging into a preweighed can and dried at 100oC to constant weight. The weight of the sediment and weight of soluble were taken and recorded.

Determination of water binding capacity

This was determined using the method described by Medicalf and Gilles (1965). 2.5 g of each sample was weighed in a tared % 50 ml centrifuge tube and 37.5 ml of distilled water was added. The tube was capped and agitated on a wrist action shaker for 1 h. Centrifuged for 10 mins at 2.200g or approximately 7,500 rpm and decanted the water. The centrifuge tube with content was weighed and and the amount of water bound was calculated.

Determination of tannin (polyphenols)

Tannin content was determined by the Folin-Dennis colorimetric method described by Joslyn (1970) with modifications. 5 g sample was dispersed in 50 ml of distilled water and shaken. The mixture was allowed to stand for 30 min at 28°C before it was filtered through what man No 42 grade of filter paper. 2 ml of the extract was dispensed into a 50 ml volumetric flask. Similarly, 2 ml standard tannin solution (tannin acid) and 2 ml of distilled water were put in separate volumetric flasks to serve as standard and Folin reagent was added to each of the flask and the 2.5 ml of saturated Na_2CO_3 solution added. The content of each flask was made up to 50 ml with distilled water and allowed to incubate at 28°C for 90 min. Their respective absorbance was measured in a spectrophotometer at 260 nm using the reagent blank to calibrate the instrument at zero.

Determination of Phytic acid content

Phytate was determined by a combination of two methods. The extraction and precipitation of phytic acid was done according to the method of Wheeler and Ferrel (1971). Extraction was done by mechanical shaking of the mixture for 30min and then centrifugation for 15 mins at 3,500 rpm. A 10 ml aliquot of the supernatant was transferred into a 40 ml conical centrifuge tube. 4 ml Ferric chloride solution was added by blowing rapidly from the pipette. The tube and its contents were heated in a boiling water bath for 45 min. The suspension was centrifuged at 3500 rpm for 15mins and the supernatant carefully decanted.

The precipitate was washed twice by dispersing well in 25 ml of 3% TCA, heating in boiling water bath 5min and centrifuging. The volume was made up to 30 ml with distilled water and the mixture heated in boiling water bath for 30 min. The suspension was filtered hot, and precipitate washed with 60ml hot water. The filtrate was discarded and the precipitate from the paper was dissolved with 40ml hot 3.2 M HNO_3 into 100ml volumetric flask.

A 5 ml aliquot was transferred to another 100ml volumetric flask and diluted to 70 ml. 20 ml 1.5 M KSCN was added and volume made to 100ml. Absorbance was read (within 1 min) at 480nm. Iron in the precipitate was then measured according to the method of Makover (1970). A 4:6 Fe/P atomic ratio was used to calculate the phytic acid content.

Determination of apparent amylose content

The apparent amylose was determined by the method of Williams et al. (1985). Starch (500 mg) was defatted by standard AOAC (2004) method using hexane. The defatted starch (100 mg) was dispersed in ethanol (1 ml) and 1 M NaOH (9 ml). The volume was made up to 100 ml with distilled water and a 5 ml aliquot transferred to a volumetric flask containing water, 25 ml. 1 M acetic acid (0.5 ml) and 1 ml iodine solution (0.2% iodine in 2% potassium iodide) were added and the volume made up to 50 ml with water and absorbance recorded at 620 nm.

Starch and sugar contents

The starch and total sugars content were determined using a colorimetric method described earlier by Onitilo et al. (2007). This involves weighing 0.02 g of the sample into a centrifuge tube with 1 ml ethanol, 2 ml distilled water, and 10 ml hot ethanol. The mixture was vortexed and centrifuged at 2000 rpm for 10 min. The supernatant was decanted and used for determining sugar content while the sediment was hydrolyzed with perchloric acid and used to estimate starch content.

Phenol- sulfuric reagent was used for colour development and glucose standards were used for estimation of sugar. The absorbance was read with a spectrophotometer (Milton Roy Spectronic 601, USA) at 490 nm.

Stastical analysis

Data were subjected to analysis of variance (ANOVA) using SAS version 8e (SAS 2001) software at $P<0.05$.The least significance difference (LSD) test was used for mean comparison.

Table 1. Proximate composition (dry weight basis) of selected white yam varieties.

Variety	Moisture (fresh) (%)[a]	Dry matter (%)	Ash (%)	Protein (%)	Fat (%)	Sugar (%)
Ndoni Aggah	62.1±0.117[e]	37.9±0.117[b]	2.36±0.033[b]	2.42±0.134[e]	0.562±0.224[fgh]	1.72±0.211[f]
Oyonku	59.5±0.080[f]	40.5±0.080[a]	1.72±0.046[ef]	3.37±0.037[b]	2.53±0.157[b]	3.63±0.011[d]
Legbaka Gambari	62.7±0.456[d]	37.3±0.456[c]	1.68±0.006[f]	4.90±0.183[a]	0.454±0.023[gh]	1.05±0.121[g]
Ndoni Ekpe	62.0±0.263[e]	38.0±0.263[b]	1.95±0.019[d]	2.54±0.047[e]	0.356±0.011[h]	2.15±0.111[e]
Asoko	65.7±0.012[b]	34.3±0.012[e]	1.52±0.031[g]	2.36±0.235[e]	1.27±0.151[c]	6.28±0.193[b]
Legbaka Awana	63.7±0.059[c]	36.3±0.059[d]	2.13±0.081[c]	3.22±0.169[cc]	0.904±0.090[de]	7.02±0.012[a]
Ndoni Abi	62.1±0.242[e]	37.9±0.242[b]	2.93±0.106[a]	3.18±0.103[bc]	3.39±0.124[a]	2.34±0.219[e]
He - Abalo	63.0±0.111[d]	37.0±0.111[c]	2.18±0.063[c]	1.96±0.058[f]	0.673±0.034[efg]	1.61±0.056[f]
Alomako	68.8±0.042[a]	31.2±0.042[f]	2.08±0.037[c]	2.43±0.109[e]	0.992±0.007[d]	2.34±0.065[e]
Omin	68.7±0.168[a]	31.3±0.168[f]	1.83±0.047[e]	2.68±0.245d[e]	0.724+0.112[ef]	5.24±0.023[c]
Asukwu	62.2±0.068[e]	37.8±0.068[b]	1.39±0.008[h]	2.96±0.203[cd]	1.39±0.069[c]	3.60±0.159[d]
Means	63.7	36.3	1.98	2.91	1.20	3.36
SD	2.91	2.91	0.433	0.788	0.944	2.01
Min	59.5	31.2	1.39	1.96	0.356	1.05
Max	68.8	40.5	2.93	4.90	3.39	7.02
CV (%)	0.281	0.493	2.78	5.43	8.89	4.09
LSD(0.05)	0.399	0.399	0.122	0.352	0.238	0.307
Pr >F (entry)	***[b]	***	***	***	***	***

[a]Parameters mean value ±SD; [b]***, **, *- Significant at P<=0.001, P<=0.01 and P<=0.05, respectively; ns -not significant, P>0.05. Means with the same letter are not significantly different.

RESULTS AND DISCUSSIONS

The results of one way analysis of variance (ANOVA) showed that there were significant varietal differences (P<0.05) among the parameters evaluated. The proximate composition results show that the moisture content of the investigated varieties ranged from 59.5 to 68.8%, ash content ranged from 1.39 to 2.93%, protein content from 1.96 to 4.90%, fat content from 0.356 to 3.39%, free total sugars from 1.05 to 7.02% and total starch from 33.9 to 75.7% (Tables 1 and 2). The bioactive content results showed that vitamin C content ranged from 5.64 mg/100 g to 6.99 mg/100 g, phytate from 1.12 to 2.37%, and tannin from 0.359 mg/g to 1.8 mg/g (Table 3). The functional properties show that the swelling power ranged from 6.66 to 9.68, solubility ranged from 4.72 to 10.0% and water binding capacity ranged from 108 to 144% (Table 4). The pasting properties results show that peak viscosity ranged from 215 to 470 RVU, trough viscosity from 198 to 385 RVU, breakdown viscosity from 8.71 to 84.5 RVU, final viscosity from 278 to 571 RVU, setback viscosity from 66.2 to 204 RVU; peak time ranged from 4.97 to 7.0 min and the pasting temperature from 61.7 to 62.6°C (Table 5). The solubility index and water binding capacity ranged from 4.72 to 10.0% and observed among the water binding capacities of starches from the selected white yam varieties. The chemical composition of yam is characterized by a high moisture content and dry matter which is composed mainly of carbohydrate, vitamins as well as protein and minerals (Osunde, 2008; Ezeocha and Ojimelukwe, 2012). The total starch differed significantly among the different white yam varieties, ranging between 33.9 and 75.7%. *Alomako* variety had the lowest starch content. Decrease in amylose content was observed as amylopectin content increased, this is an indicator that one is a function of the other; however both properties are important in food preparation and development (Eke-Ejiofor and Owuno, 2012).

The amylopectin content of the different white yam varieties was between the range of 61.9 and 75.7%, *Asoko* variety had the highest amylopectin content while *Asukwu* had the lowest value. The amylose content was lower than the amylopectin content in the selected white yam varieties as it was also observed in *Ozibo*, another variety of *D. rotundata* (Okorie et al., 2011). Mali et al. (2004) observed that starch from yam had more amylose content than starches from corn and cassava. Moisture content in the selected varieties was high when compared with those reported in the literature and this might due to the environmental factor. Yam varieties with low moisture content are suitable for high yield flour production (Polycarp et al., 2012). *Oyonku* variety had the lowest moisture contentand therefore may be more suitable for flour production.

The protein values for the selected white yam varieties were significant and higher than the value reported for *D. rotundata* and *Colocasia esculenta,* majority of the selected white yam varieties also had lower fat contents as compared to that of cocoyam (Alinnor and Akalezi,

Table 2. Proximate composition (dry weight basis) of selected white yam varieties.

Variety	Starch (%)^{a}	DCHO (%)^{b}	TCHO (%)^{c}	Amylose (%)	Amylopectin (%)	TME (KJ)^{d}
Ndoni Aggah	58.2±0.442^{e}	60.0±0.653^{ef}	32.6±0.208^{ab}	25.3±0.121^{f}	74.7±0.121^{a}	605±7.16^{d}
Oyonku	58.1±2.43^{e}	61.8±2.42^{df}	32.9±0.159^{a}	27.3±0.174^{e}	72.7±0.174^{b}	701±3.82^{a}
Legbaka Gambari	62.6±0.815^{d}	63.6±0.936^{d}	30.3±0.610^{c}	28.3±0.108^{de}	71.7±0.108^{bc}	605±7.98^{d}
Ndoni Ekpe	57.4±0.669^{e}	59.5±0.781^{f}	33.2±0.186^{a}	33.4±0.109^{b}	66.6±0.109^{e}	609±4.31^{d}
Asoko	71.6±0.783^{b}	77.8±0.975^{b}	29.1±0.366^{d}	24.3±0.142^{f}	75.7±0.142^{a}	573±3.44^{f}
Legbaka Awana	75.7±0.333^{a}	82.7±0.345^{a}	30.1±0.062^{c}	27.7±0.078^{e}	72.3±0.078^{b}	590±0.482^{e}
Ndoni Abi	68.9±0.021^{c}	71.3±0.198^{c}	28.4±0.575^{e}	24.4±0.190^{f}	75.6±0.190^{a}	654±3.26^{b}
He - Abalo	49.8±0.657^{f}	51.4±0.714^{g}	32.2±0.266^{b}	30.0±0.200^{cd}	70.0±0.200^{cd}	595±2.21^{f}
Alomako	33.9±0.023^{i}	36.2±0.088^{i}	25.7±0.108^{f}	33.1±1.94^{b}	66.9±1.94^{e}	507±0.236^{g}
Omin	39.7±0.851^{h}	44.9±0.874^{h}	26.0±0.082^{f}	30.3±1.07^{c}	69.7±1.07^{d}	507±1.28^{g}
Asukwu	47.3±0.573^{g}	50.9±0.732^{g}	32.0±0.058^{b}	38.1±2.14^{a}	61.9±2.14^{f}	637±0.157^{c}
Means	56.7	60.0	30.2	29.3	70.7	598
SD	13.1	13.9	2.66	4.26	4.26	57.2
Min	33.9	36.2	25.7	24.3	61.9	507
Max	75.7	82.7	33.2	38.1	75.7	701
CV (%)	1.55	1.58	1.06	2.92	1.21	0.547
LSD(0.05)	1.96	2.11	0.717	1.90	1.90	7.29
Pr >F(Entry)	***^{e}	***	***	***	***	***

^{a}Parameters mean value ±SD; ^{b}DCHO = Digestible carbohydrate; ^{c}TCHO = Total carbohydrate^{d}; TME = total metabolisable energy, KJ = Kilojoules; ^{e}***, **, * -Significant at P<=0.001, P<=0.01 and P<=0.05, respectively. ns- not significant P>0.05. Means with the same letter are not significantly different.

Table 3. Bioactive contents (dry weight basis) of selected white yam varieties.

Variety	^{a}Vitamin C (mg/100 g)	Phytate (mg/g)	Tannin (mg/g)
Ndoni Aggah	6.35 ± 0.145^{d}	2.37 ± 0.04^{a}	0.64 ± 0.017^{ef}
Oyonku	6.04 ± 0.082^{e}	1.73 ± 0.044^{e}	0.68 ± 0.017^{d}
Legbaka Gambari	5.64 ± 0.071^{f}	1.12 ± 0.025^{h}	1.18 ± 0.019^{a}
Ndoni Ekpe	6.67 ± 0.076^{c}	2.29 ± 0.024^{ab}	0.74 ± 0.018^{c}
Asoko	6.99 ± 0.016^{a}	1.55 ± 0.021^{f}	0.89 ± 0.010^{b}
Legbaka Awana	6.83 ± 0.004^{abc}	1.40 ± 0.052^{g}	0.67 ± 0.000^{ed}
Ndoni Abi	6.73 ± 0.097^{bc}	1.75 ± 0.023^{e}	0.69 ± 0.001^{d}
He - Abalo	6.17 ± 0.068^{ed}	1.47 ± 0.053^{gf}	0.69 ± 0.008^{d}
Alomako	6.88 ± 0.064^{ab}	2.16 ± 0.047^{c}	0.36 ± 0.019^{g}
Omin	6.96 ± 0.071^{a}	1.91 ± 0.047^{d}	0.62 ± 0.027^{f}
Asukwu	6.12 ± 0.069^{e}	2.24 ± 0.025^{bc}	0.64 ± 0.026^{ef}
Means	6.49	1.82	0.708
SD	0.448	0.412	0.200
Min	5.64	1.12	0.359
Max	6.99	2.37	1.18
CV (%)	1.26	2.22	2.45
LSD(0.05)	0.182	0.090	0.039
Pr > F Entry)	***^{b}	***	***

^{a}Parameters mean value ±SD; ^{b}***, **, * -Significant at P<=0.001, P<=0.01 and P<=0.05 respectively; ns- Not significant at P>0.05. Means with the same letter are not significantly different.

2010). *Legbaka Gambari* variety had the highest protein content, while *Legbaka awana* variety had the highest sugar content. It was reported that yamsgenerally have considerably higher protein content than the 1.2-1.8% on

Table 4. Functional properties of flour from selected white yam varieties.

Variety	Swelling power[a]	Solubility (%)	Water binding capacity (WBC) (%)
Ndoni Aggah	8.69 ± 0.094[a]	7.85 ± 0.016[bh]	126 ± 0.812[abcd]
Oyonku	8.21 ± 0.158[cd]	6.67 ± 3.15[bc]	116 ± 0.132[cd]
Legbaka Gambari	7.07 ± 0.016[e]	4.72 ± 0.150[c]	133 ± 0.779[bcd]
Ndoni Ekpe	7.04 ± 0.128[e]	6.48 ± 0.117[bc]	124 ± 0.258[abc]
Asoko	7.94 ± 0.192[d]	7.94 ± 0.176[ba]	132 ± 1.01[abc]
Legbaka Awana	7.71 ± 0.203[d]	10.0 ± 1.68[a]	133 ± 0.341[a]
Ndoni Abi	8.80 ± 0.531[b]	7.88 ± 0.184[ba]	144 ± 0.053[abc]
He - Abalo	6.66 ± 0.557[e]	4.84 ± 2.01[c]	134 ± 0.208[abc]
Alomako	9.04 ± 0.029[b]	4.73 ± 0.011[c]	108 ± 27.6[d]
Omin	9.68 ± 0.037[a]	6.47 ± 0.689[bc]	143 ± 1.39[ab]
Asukwu	8.09 ± 0.258[d]	7.52 ± 0.327[abc]	130 ± 2.28[abc]
Means	8.09	6.83	129
SD	0.930	1.65	10.5
Min	6.66	4.72	108
Max	9.68	10.0	144
CV (%)	3.23	18.7	6.39
LSD(0.05)	0.582	2.85	18.4
Pr > F(Entry)	***[b]	*	*

[a]Parameters mean value ±SD; [b]***, **, * - Significant at P<=0.001, P<=0.01 and P<=0.05 respectively; ns- not significant at P>0.05; Means with the same letter are not significantly different.

dry weight basis as reported by Charles et al. (2005) for cassava. The protein contents observed in the selected varieties of white yam were however lower than reported value of 10.27% for *D. alata*. *He-Abalo* had the lowest protein value. The vitamin C contents of the varieties fall within the range of 4.00 - 18.00 mg for edible yam species (Osagie, 1992) and between 6.5 and 11 mg/100g as observed by Coursey and Aidoo (1966). *Asoko* variety had the highest vitamin C content. However, different cooking techniques may affect the retention of vitamin C in yams.

The ash contents observed in the selected white yam varieties were lower than those obtained from other *Dioscorea* spp. reported by Shanthakumari et al. (2008)

Peak viscosity indicates the water binding capacity of the starch or mixture occurs at the equilibrium point between swelling causing an increase in viscosity rupture and alignment causing its decrease (Adegunwa et al., 2011). It is also indicative of the strength of the pastes which are formed from gelatinization during processing in food applications (Eke-Ejiofor and Owuno, 2012). There was no significant difference between the pasting temperatures of the selected white yam varieties. Similar ranges were also observed by Adegunwa et al. (2011). Flours from all the selected *D. rotundata* varieties had peak viscosity higher than flour from *D. alata,* but only a few were higher than that of cassava flour (Babajide and Olowe, 2013). Swelling power, as described by Richard et al. (1991) is a factor of the ratio of amylose toamylopectin, *Omin* variety had the highest swelling power while *He-Abalo* variety had the lowest. *Asukwu* variety had the highest, peak, trough, breakdown and final viscosity values, while *Legbaka Gambari* variety had the highest setback viscosity value. The pasting temperature is the temperature at which the viscosity starts to rise (Liang and King, 2003), thus, the lower the pasting temperature, the faster the swelling. Flours from the selected *Dioscorea rotundata* varieties had higher pasting temperature than *D. dumetorum* flour and wheat flour (Eke-Ejiofor and Owuno, 2012).

The water binding capacity varied significantly among the different varieties studied (P<0.05). Tannins are water-soluble polyphenols that are present in many plant foods (Chung et al., 1998). These phenolic compounds usually interfere with iron absorption through a complex formation with iron in the gastrointestinal tract, decreasing the bioavailability of iron (Adegunwa et al., 2011), though tannins have anti-carcinogenic and anti-mutagenic potentials due to their antioxidative property (Chung et al., 1998). The tannin contents recorded in this study were however significant at P<0.001. The tannin content was similar to that observed by Adegunwa et al. (2011) and also falls within the range observed for other *Dioscorea* spp. (Shanthakumari et al., 2008). *Legbaka*

Table 5. Pasting properties of flour from selected white yam varieties.

Variety	Peak 1[a]	Trough 1	Breakdown	Final viscosity	Setback	Peak time	Pasting temperature
Ndoni Aggah	337 ± 9.37^{c}	300 ± 1.12^{c}	37.665 ± 8.25^{b}	402 ± 4.77^{d}	103 ± 3.66^{ef}	5.84 ± 0.049^{ef}	61.9 ± 0.141^{a}
Oyonku	231 ± 5.77^{h}	222 ± 6.42^{e}	8.71 ± 0.651^{f}	311 ± 5.19^{g}	89.0 ± 3.66^{fg}	$6.30 \pm 0 .042^{cb}$	61.9 ± 0.177^{a}
Legbaka Gambari	245 ± 2.71^{g}	213 ± 1.94^{e}	32.04 ± 0.764^{bc}	418 ± 3.30^{cd}	204 ± 1.24^{a}	$6.20 \pm 0 .000^{cbd}$	61.8 ± 0.247^{a}
Ndoni Ekpe	289 ± 4.72^{e}	266 ± 9.02^{d}	22.375 ± 4.31^{cde}	371 ± 3.71^{e}	104 ± 5.30^{ed}	5.53 ± 0.000^{f}	61.7 ± 0.354^{a}
Asoko	215 ± 3.95^{i}	198 ± 3.30^{f}	17.46 ± 0.651^{def}	278 ± 2.88^{h}	80.1 ± 0.420^{gh}	7.00 ± 0.000^{a}	62.3 ± 0.566^{a}
Legbaka Awana	275 ± 4.48^{ef}	264 ± 3.66^{d}	10.83 ± 8.13^{f}	330 ± 0.420^{f}	66.2 ± 4.07^{h}	6.44 ± 0.049^{b}	61.9 ± 0.071^{a}
Ndoni Abi	269 ± 4.13^{f}	253 ± 0.18^{d}	15.875 ± 3.95^{def}	370 ± 2.72^{e}	117 ± 2.54^{d}	5.90 ± 0.141^{ed}	61.9 ± 0.071^{a}
He - Abalo	311 ± 5.78^{d}	287 ± 2.83^{c}	23.75 ± 2.94^{cd}	431 ± 10.31^{c}	144 ± 13.1^{c}	6.90 ± 0.141^{a}	61.9 ± 0.035^{a}
Alomako	356 ± 5.78^{b}	344 ± 12.55^{c}	12.045 ± 4.07^{ef}	461 ± 19.50^{b}	117 ± 6.95^{d}	6.87 ± 0.191^{a}	62.6 ± 1.379^{a}
Omin	355 ± 12.90^{b}	341 ± 16.86^{b}	14.375 ± 3.95^{def}	423 ± 15.80^{c}	81.8 ± 1.06^{g}	6.03 ± 0.424^{ed}	61.9 ± 0.141^{a}
Asukwu	470 ± 4.77^{a}	385 ± 1.29^{a}	84.54 ± 6.07^{a}	571 ± 8.84^{a}	186 ± 10.1^{b}	4.97 ± 0.049^{g}	61.9 ± 0.035^{a}
Means	305	279	25.4	397	118	6.18	62.0
SD	72.9	59.3	21.6	80.2	44.0	0.621	0.256
Min	215	198	8.71	278	66.2	4.97	61.7
Max	470	385	84.5	571	204	7.00	62.6
CV (%)	2.13	2.27	18.4	2.12	5.50	2.63	0.748
LSD(0.05)	14.5	14.1	10.4	18.7	14.4	0.362	1.03
Pr > F(Entry)	***[b]	***	***	***	***	***	ns

[a]Parameters mean value ±SD; [b]***, **, * -Significant at P<=0.001, P<=0.01 and P<=0.05 respectively; ns -Not significant at P>0.05. Means with the same letter are not significantly different.

Gambari variety had the highest tannin content. Phytate has a strong ability to chelate multivalent metal ions,specially zinc, calcium and iron, making them biologically unavailable, it is also regarded as an anti-nutritional factor in the diet of humans because of their inability to utilize it (Bohn et al., 2008). The phytate values recorded in selected white yam varieties were also significant (P< 0.001). The presence of enzyme inhibitors in yams could affect digestion of starch and protein, limiting their utilization as food (Polycarp et al., 2012). This study observed low levels of tannin and phytate in the selected varieties of white yam.

Lower levels of tannin and phytate were also recorded in yam varieties studied by Polycarp et al. (2012) and are recommended as safe for food processing application.

Conclusion

The result of this study revealed that the physical and chemical characteristics of these yam high yielding varieties were similar to those reported for most yam varieties in other parts of Nigeria. These yam varieties have better functional and pasting properties, hence better poundability and mealiness. They also have a great potential as source of bioactive compounds and protein for the people in the zone.

Conflict of interests

The authors did not declare any conflict of interests.

REFERENCES

Adegunwa MO, Alamu EO, Omitogun LA (2011). Effect of processing on the nutritional contents of yam and cocoyam

tubers. J. Appl. Biosci. 46:3086-3092.
Alinnor IJ, Akalezi CO (2010). Proximate and Mineral Compositions of *Dioscorea rotundata* (white yam) and *Colocasia esculenta* (white cocoyam). Pak. J. Nutr. 9 (10): 998-1001.
Araghiniknam M, Chung S, Nelson-White T, Eskelson C, Watson RR (1996). Antioxidant activity of *Dioscorea Dioscorea* and dehydroepiandrosterone (DHEA) in older humans. Life Sci. 59:PL147-157.
Association of Official Analytical Chemists, AOAC (2004). Official Methods of Analysis of the Association of Official Analytical Chemists. Arlington. Virginia.
Babajide JM, Olowe S (2013). Chemical, functional and sensory properties of water yam - cassava flour and its paste. Int. Food Res. J. 20(2):903-909
Baquar SR, Oke OL (1976). Protein in Nigerian yams (Dioscorea spp.). Nutrient Reports International 14:237-248.
Bergh K, Orozco P, Gugerty MK, Anderson L (2012). Yam value chain: Nigeria. Evans School Policy and Analysis Research Brief: No. 207
Bohn L, Meyer AS, Rasmussen SK (2008) Phytate: impact on environment and human nutrition. A challenge for molecular breeding. J. Zhejiang Univ. Sci. B 9(3):165-191.
Bradbury JH (1988). The chemical composition of tropical root crops. ASEAN Food J. 4:3-13.
Charles AL, Sriroth K, Huang TC (2005). Proximate composition, mineral contents, hydrogen cyanide and phytic acid of 5 cassava genotypes. Food Chem. 92: 615-620.
Chung KT, Wong TY, Wei CI, Huang YW, Lin Y (1998) Tannins and human health: a review. Crit. Rev. Food Sci. Nutr. 38(6): 421-64
Coursey DG' Aidoo A (1966) Ascorbic acid levels in Ghanaian yams. J. Sci. Food Agric. 17(10):446 - 449.
Coursey DG, Haynes PH (1970). Root crops and their potential as in the tropics. World Crops 22:261-265.
Eke-Ejiofor J, Owuno F (2012). Functional and Pasting Properties of Wheat/ Three-leaved Yam *(Dioscorea dumentorum)* composite flour blend. Global Res. J. Agric. Biol. Sci. 3(4):330 -335.
Ezeocha VC, Ojimelukwe PC (2012). The impact of cooking on the proximate composition and anti-nutritional factors of water yam (*Dioscorea alata*). J. Stored Prod. Postharvest Res. 3(13):172 - 176.
FAOSTAT (2013). FAO Statistical databases http://www.faostat.fao.org (Accessed 17/07/2013)
Farombi EO, Britton G, Emerole GO (2000). Evaluation of the antioxidant activity and partial characterization of extracts from browned yam flour diet. Food Res. Int. 33:493-499.
Food and Agricultural Organization (FAO) (2008). Statistical data. http://www.fao.org
Foss Analytical AB (2003). Manual for Kjeltec System 2300 Distilling and Titration Unit.
Hikino H, Konno C, Takahashi M, Murakami M, Kato Y, Karikura M, Hayashi T (1986). Isolation and hypoglycemic activity of dioscorans A, B, C, D, E, and F; glycans of Dioscorea japonica rhizophors. Planta Med. 3:168-171.
Hu K, Dong A, Yao X, Kobayashi H, Iwasaki S (1996). Antineopastic agents: I. Three spirostanol glycosides from rhizomes of Dioscorea collettii var. hypoglauca. Planta Med. 62:573-575
Hu K, Dong A, Yao X, Kobayashi H, Iwasaki S, Jing YK (1999). A new pregnane from Dioscorea collettii var. hypoglauca. J. Nat. Prod. 62: 299-301.
IITA (2009). Yam. www.iita.org (Accessed 16/07/13).
Izekor OB, Olumese MI (2010). Determinant of yam production and profitability in Edo State, Nigeria. Afr. J. Gen. Agric. 6 (4):205-210.
Joslyn MA (1970). Tannins and related phenolics. In: Methods in food analysis. pp. 701-725.
Kelmanson JE, Jager AK, Van Staden J (2000). Zulu medicinal plants with antibacterial activity. J. Ethnopharmacol. 69:241-246.
Lasztity R, Hidvegi M, Bata A (1998). Saponins in food. Food Rev. Int. 14(4):371-390.
Leach HW, McGowen LD, Schoch TJ (1959). Structure of the starch granule I: Swelling and solubility patterns of various starches. Cereal Chem. 36: 534-544.
Liang X, King JM (2003). Pasting and crystalline property Differences of Commercial and Isolated Rice Starch with Added Amino Acids. J. Food Sci. 68:3.
Makover RU (1970) Extraction and determination of phytic acid in beans. Cereal Chem. 47:288-296.
Mali S, Karam LB, Ramos LP, Grossmann MV (2004). Relationships among the composition and physicochemical properties of starches with the characteristics of their films. J. Agric. Food Chem. 52(25):7720-7725.
Marcus DL, Thomas C, Rodriguez C, Simberkoff K, Tasi JS, Strfaci JA, Freedman ML (1998). Increased peroxideratoin and reduced antioxidant enzyme activity in Alzheimer's disease. Exp. Neurol. 150: 40-44.
Medicalf DG, Gilles KA (1965). Wheat Starches I: Comparison of physicochemical properties. Cereal Chemistry 42:558-568.
Okorie PA, Okolie EC, Ndie EC (2011). Functional and Pasting Properties of Lesser Known Nigerian Yams as a Function of Blanching Time and Particle Size. Adv. J. Food Sci. Technol. 3(6):404-409.
Onitilo AA, Sanni LO, Daniel I, Maziya-Dixon B, Dixon A (2007). Physicochemical and functional properties of native starches from cassava varieties in southwest Nigeria. J. Food Agric. Environ. 5(3,4):108-114
Opara LU (1999). Yam storage. In: Bakker-Arkema et al., (eds). CIGR Handbook of Agricultural Engineering Volume IV Agro Processing. The American Society of Agricultural Engineers, St. Joseph, MI. pp. 182-214.
Opara LU (2003). Yams: Post-Harvest Operation. http://www.fao.org/fileadmin/user_upload/inpho/doc/Post_Harvest_Compendium_-_Yams.pdf
Osagie AU (1992). The yam tuber in storage. Post Harvest Research Unit, University of Benin, Nigeria. pp. 107-173.
Osunde ZD (2008) Minimizing Postharvest Losses in Yam (*Dioscorea spp.*): Treatments and Techniques. International Union of Food Science & Technology.
Polycarp D, Afoakwa EO, Budu AS, Otoo E (2012). Characterization of chemical composition and anti-nutritional factors in seven species within the Ghanaian yam (*Dioscorea*) germplasm. Int. Food Res. J. 19 (3):985-992
Richard JR, Asaoka MA, Blanshard JM (1991). The physico-chemical properties of cassava starch. Trop. Sci. 31:11189-207
Sanusi WA, Salimonu KK (2006). Food security among households: evidence from yam production economics in Oyo state, Nigeria. Agric. J. 1: 249 - 253
Shanthakumari S, Mohan VR, De Britto J (2008) Nutritional evaluation and elimination of toxic principles in wild yam (*Dioscorea* spp.) Trop. Subtrop. Agroecosystems 8(3):319-325
Undie AS, Akubue PI (1986). Pharmacological evaluation of Dioscorea *dumentorumdumentorum* tuber used in traditional antidiabetic therapy. J. Ethnopharmacol. 15:133-144.
Waitt AW (1963). Yam: Dioscorea species. Field Crop Abstract 16(3): 145-157.
Wheeler EL, Ferrel RE (1971). A method for Phytic acid determination in wheat fractions. Cereal chem. 48:312-316.
Williams VR, Wu W, Tsai HY, Bates HG (1985) Varietal differences in amylose content of rice starch. J. Agric. Food Chem. 6:47-48.

16

Antioxidant capacity of extracts from hawthorn (*Crataegus mexicana*) skin

Méndez-Iturbide Daniel[1], Banderas-Tarabay José Antonio[1], Nieto-Camacho Antonio[2], Rojas-Chávez Agustina[3] and García-Meza Maria G[3]

[1]Laboratorio de Química de Alimentos, Escuela de Nutrición, Facultad de Ciencias Médicas, Universidad Autónoma de Tlaxcala. Av. Xicohtencatl No. 466, Barrio de San Nicolás, San Pablo del Monte 90920, Tlaxcala, México.
[2]Instituto de Química, Universidad Nacional Autónoma de México, Circuito Exterior s/n, Ciudad Universitaria, Coyoacán, México, DF, C.P. 04510, México.
[3]Centro de Investigación en Biotecnología Aplicada, Instituto Politécnico Nacional, Ex-Hacienda San Juan Molino Carretera Estatal Tecuexcomac-Tepetitla Km 1.5, Tlaxcala C.P. 90700, México.

The hawthorn *Crataegus mexicana* is a traditional Mexican fruit with properties that make this fruit useful for the treatment of many ailments, including diseases of the respiratory and urinary tract. This paper reports the antioxidant capacity of the *n*-hexane, dichloromethane, ethyl acetate, acetone, ethanol and methanol extracts of *C. mexicana*. Samples were evaluated for total phenolic and carotenoid contents, 2,2-diphenyl-1-picryl-hydrazyl (DPPH) radical scavenging, the inhibition of the formation of thiobarbituric acid reactive species (TBARS) and the neutralization of the cation-radical 2,2´-azino-bis(3-ethyl benzothiazoline-6-sulfonic acid) diammonium salt (ABTS). The total phenolic content was 2.65 ± 0.23 mg of gallic acid equivalents per gram, and the carotenoid content was 26.4 ± 0.02 µg/g in dry hawthorn skin. The most active extract in scavenging DPPH radicals and inhibiting TBARS formation was the acetone extract, with activities of 21.9 ± 0.15 and 13.27 ± 0.70%, respectively, at 10 mg/L. The extracts were compared for activity against ascorbic acid, caffeic acid, α- tocopherol and quercetin. The acetone extract was the most active, with an IC_{50} value of 15.2 mg/L in DPPH and 17.7 mg/L in TBARS. A high correlation was observed between the results for TBARS and DPPH. These results demonstrate the potential nutritional and antioxidant value of this Mexican fruit.

Key words: Antioxidant capacity, *Crataegus mexicana,* fruit skin, hawthorn, organic extract.

INTRODUCTION

The worldwide increase in human deaths caused by several chronic degenerative diseases has been well documented (WHO/FAO, 2003). The development of complications from these diseases increases the economic costs of health services. Therefore, it is important to consider how the quality of life in these subjects changes due to a decrease in their working productivity (Biedrzycka and Amarowicz, 2008).

The pathophysiology of many diseases is associated with an increase of free radicals derived from reactive oxygen species (ROS) in the cells (Singh et al., 2008). A higher production of ROS can cause oxidative damage in DNA (Spencer et al., 2012), as well as in proteins and lipids in the cell membranes, and is associated with several inflammatory processes (Dufour et al., 2007). From a physiological perspective, an equilibrium should exist between free radicals and antioxidants in the organism (Blumberg, 2004). Antioxidants play an important role in protecting cells against ROS (Jayakumar and Kanthimathi, 2012). The antioxidant network systems in the body include enzymatic defenses, such as the superoxide dismutase (SOD), catalase (CAT) and glutathione peroxidase (GSH) mechanisms. Other non-enzymatic defense systems utilize ascorbic acid, carotenoid, tocopherol, and phenolic substances (Guimaraes et al., 2010).

Several studies have shown certain clinical benefits associated with the intake of fruits (Crowe et al., 2011). Some fruits have protective properties and help prevent the complications of chronic degenerative diseases, such as atherosclerosis, chronic inflammation, diabetes mellitus, cataracts, coronary diseases and certain types of cancer (Havsteen, 2002).

These characteristics have been attributed to the antioxidant activity of many secondary metabolites in fruits. Many types of fruits also have nutraceutical properties (Andlauer and Fürst, 2002) and a nutritional capacity to prevent chronic degenerative diseases (Willis and Wians, 2003). Studies in fruits from Asia, Europe and America in the *Rosaceas* family and the genus *Crataegus* had shown that contain terpenoids, polyphenols (catechins mainly epicatechin, polymeric and oligomeric proanthocyanidins such as B_2 dimeric procyanidin, cyanidin-3-*O*-galactoside, idaein and chlorogenic acid) and flavonoids (flavonol-*O*-glycoside such as quercetin-3-*O*-galactoside, vitexin-2"-*O*-rhamnoside, acetylvitexin-2"-*O*-rhamnoside, flavone-*C*-glycosides, and hyperóside) (Froehlicher et al., 2009; Shao-Jiang et al., 2011; Calişkan et al., 2012; Edwards et al., 2012). The hawthorn *Crataegus mexicana,* also named "tejocote", is a traditional fruit in the Mexican diet and in herbal medicine (Cabrera, 1958). The wild type, "criollo", is found throughout most of Mexico and Northeastern Guatemala. However, there is no scientific evidence that supports the biological and antioxidant properties of *C. mexicana*. In this context, the objective of the present work was to assess the antioxidant activity present in the skin of the hawthorn *C. mexicana*.

MATERIAL AND METHODS

Chemicals and reagents

Polyphenol (quercetin), ascorbic acid, caffeic acid, α-tocopherol, Folin-Ciocalteu reagent, gallic acid, 2,2-diphenyl-1-picrylhydrazyl (DPPH), sodium carbonate, trichloroacetic acid, bovine serum albumin (BSA), potassium persulfate, ferrous sulfate, ethylenediaminetetraacetic acid (EDTA), sodium acetate trihydrate, phosphate buffered saline (PBS), 1,1,3,3-tetramethoxypropane, 6-hydroxy-2,5,7,8-tetramethylchroman-2-carboxylic acid (Trolox), 2,2′-azino-bis(3-ethyl benzothiazoline-6-sulfonic acid) diammonium salt (ABTS), and 2-thiobarbituric acid (TBA) were purchased from Sigma-Aldrich Co (St Louis, MO). Deionized (MilliQ) water was used in the experiments. All of the solvents were of analytical grade.

Plant materials

A total of 4 kg of hawthorn (*C. mexicana)* fruits were harvested 120 days after flowering in the State of Tlaxcala (19°19'20.55''N/98°22' 56.15''W; altitude of 2242 m above sea level) in the month of December. Fruits were identified by a biologist according to the guidelines of the botanical garden at the Benemerita Universidad Autonoma de Puebla (voucher 14648). All samples had the same morphological characteristics with respect to the size and color and were free of surface damage. Due to the fact that many fruits contain the highest levels of antioxidants in the skin, only the skin samples of the hawthorn samples were studied. Fruits were hand-washed, and the skin was manually peeled, frozen and lyophilized. A total of 36 g of skin was obtained, which was then processed using a mesh (mesh size 20). Samples were kept frozen at -20°C in a nitrogen atmosphere until use.

Preparation of hawthorn extracts

The hawthorn skin powder (~30 g) was homogenized at room temperature and sequentially extracted for 72 h into 300 mL of the following solvents: *n*-hexane, dichloromethane, ethyl acetate, acetone, ethanol, and methanol. The extracted fractions from each solvent were decanted using a separatory funnel, evaporated, and kept in a sealed amber flask at 5°C.

Quantification of antioxidants

Measurement of total phenolic compounds

The total phenol content of the hawthorn skin was determined using the Folin-Ciocalteu reagent method (Folin and Ciocalteu, 1927) with minor modifications (Singleton et al., 1999; Everette et al., 2010). Briefly, 500 mg of dried skin sample was added to 40 mL of methanol/water (50:50, v/v) for 1 h at 25°C, and the supernatant was then collected and stored (methanol extract). The residual sample was added to 40 mL of acetone/water (70/30, v/v) at 25°C for 1 h (acetone extract). The methanol and acetone extracts were mixed, evaporated and resolubilized in 10 mL of ethanol (ethanol extract). Finally, the residual sample was dissolved in 10 mL of H_2O (aqueous extract). 1 mL of each aqueous and ethanol extract was brought to a volume of 2.5 mL using 0.3% HCl. Subsequently, 50 µL of either this sample or a standard was added to 1 mL of 2% Na_2CO_3, followed by the addition of 50 µL of Folin-Ciocalteu reagent diluted in water (1:1). After a 30 min incubation at 25°C, the absorbance at 750 nm was recorded using a spectrophotometer (Genesys 10 UV, Thermo Electron Co. U.S.A.). The total phenol content was expressed as milligrams of gallic acid equivalent per gram of sample dry weight (GAE/g) using a standard curve (concentration 0.2 to 1.0 mg/mL) for a freshly prepared gallic acid solution (Cai et al., 2004).

Measurement of total carotenoids

Total carotenoids were determined according to Speek et al. (1988). Briefly, 2.0 g of dried skin sample was added into 10 mL of *n*-hexane/acetone/ethanol (50:25:25 v/v). The solution remained at room temperature for 1 h and was then centrifuged at 3000 x *g* for 5 min. The organic phase was dissolved in *n*-hexane to a final volume of 25 mL. A 1.0 mL aliquot was used to measure the absorbance at 450 nm. Concentration values were obtained using the extinction coefficient of β-carotene, ε = 2505, as a standard and were reported as µg/g of dried hawthorn skin according to the following equation:

$$x\ (\mu g) = \frac{A^{*}\ y\ (mL)\ 10^{6}}{A_{1cm}^{1\%}\ 100} \qquad (I)$$

$$x\ (\mu g/g) = \frac{x\ (\mu g}{g\ sample} \qquad (II)$$

Where, *x* is the sample concentration of carotenoids, *y* is the sample volume and A* is the extinction coefficient of β-carotene.

Antioxidant activities

Scavenging effect on DPPH

The effects of hawthorn extract antioxidants on free radicals were determined using DPPH (Kim et al., 2005). Briefly, 100 µL of the aforementioned extracts was prepared at a concentration of 20 mg/mL in DMSO. From each sample stock, 1000, 100 and 10 mg/L final dilutions in cool ethanol were prepared. DPPH was dissolved in ethanol (1 mM). A total of 50 µL of sample extract and 150 µL of DPPH were mixed and incubated for 30 min at 37°C in an orbital shaker. The absorbance at 517 nm was recorded using an enzyme linked immunosorbent assay (ELISA) microplate reader (Bio-Tek EL800, U.S.A.). The percent reduced activity of DPPH was calculated and expressed as the IC_{50} (Khattak et al., 2008). Ascorbic acid, caffeic acid, α-tocopherol and quercetin were used as standards.

Lipid peroxidation induced by $FeSO_4$ in rat brain homogenate (TBARS)

The antioxidant activity of the hawthorn skin was estimated using the spectrophotometric method described by Ohkawa et al. (1979). Extract samples at 1000, 100 and 10 mg/L were prepared to inhibit lipid peroxidation induced by Fe^{2+} in rat brain homogenate. Male adult Wistar rats (200 to 250 g) were provided by the Tlaxcala Research Center of Behavior Biology (CTBC) at Universidad Autónoma de Tlaxcala. The study and animal handling were approved by the ethics committee of the Faculty of Health Sciences and the School of Nutrition at the same university. Rats were sacrificed using CO_2 and decapitated as humanely as possible.

The brains were immediately dissected, washed and homogenized in 10 mL of 0.1 M PBS (0.2 g/L KCl, 0.2 g/L KH_2PO_4, 8 g/L NaCl and 2.16 g/L $NaHPO_4.7H_2O$; pH 7.4) at 5°C to produce a 1:10 homogenate (w/v) (Dominguez et al., 2005). The homogenate was centrifuged for 10 min at 3000 x *g*, and the supernatant was adjusted to 2.66 mg of protein/mL in PBS for total protein determination according to the method of Lowry et al. (1951). The analysis of BSA was used to generate the standard curve.

A total of 375 µL (1.0 mg protein content) of brain sample was incubated in an orbital shaker at 37°C for 30 min with 50 µL of 10 µM EDTA and 50 µL of each of the hawthorn extracts to obtain working concentrations of 3 to 50 mg/L. Lipid peroxidation was initiated by the addition of 50 µL of $FeSO_4$ (final concentration = 10 µM) (Ng et al., 2000). After 1 h, the samples were cooled on ice, and 0.5 mL of TBA (1% 2-thiobarbituric acid in 0.05 N NaOH and 30% trichloroacetic acid, 1:1) was added. The samples were centrifuged for 10 min at 3000 x *g*. The final products of the lipid peroxidation from rat brains were assayed by malondialdehyde (MDA) determination using the TBARS assay (Kibanova et al., 2009). The samples were heated at 94°C for 30 min in a water bath. After cooling on ice, 200 µL of each sample and sample blank was placed in an ELISA microplate reader, and the absorbance was recorded at 540 nm. The TBARS results were expressed as the equivalent of MDA relative to 1,1,3,3-tetramethoxypropane as the standard. The percent inhibition is defined as the decrease of TBARS formation due to inhibition caused by the extract (Esterbauer and Cheeseman, 1990).

The inhibition of the lipid peroxidation up to 50% (IC_{50} µg/mL) in each extract was plotted against the *log* of the concentration in each sample of *C. mexicana* and the percent of lipid peroxidation induced by $FeSO_4$ 25 µM (100%).

Neutralization of the ABTS radical cation

The procedure of Delgado et al. (2011) was used to determine the capability of the hawthorn extracts to trap ABTS free radicals. ABTS at 7 mM was mixed with KH_2PO_4 (2.45 mM final concentration) and incubated for 12 to 16 h at 25°C in the dark. The solution was diluted in ethanol to obtain an absorbance of 0.70 at 436 nm. A total of 10 µL of sample extract diluted to 100 mg/L was added to $ABTS^{\cdot+}$ to obtain a final volume of 1.0 mL. The absorbance was measured during 6 min periods to determine the trapping capacity of the extract. The trapping capacity was compared against that of the equivalent of Trolox and was expressed as µM Trolox/g sample extract.

Statistical analysis

Data were expressed as the means ± standard error of mean (SEM) unless otherwise indicated. A one-way analysis of variance (ANOVA) was used to compare the antioxidant capacity and phenolic contents of the hawthorn skin extracts. The Duncan, Tukey, and Pearson correlations were used to compare the extracts. In all cases, significance was accepted at the $p < 0.05$ level. Each experiment was run in triplicate with two control samples. The OrigenLab program (*v*.8) was used for all analyses.

RESULTS AND DISCUSSION

Phenol content of hawthorn skin

A total of 36.0 g of dry hawthorn skin was obtained as previously described. The amount of extract obtained using nonpolar to polar solvents was as follow: *n*-hexane (0.63 g), dichloromethane (0.60 g), ethyl acetate (0.70 g), acetone (0.61 g), ethanol (3.90 g), and methanol (4.10 g). The total phenol content was 2.65 ± 0.23 mg GAE/g of dry sample. These values are similar to the polyphenol contents of other *Crataegus* species such as *C. azarolus* (1.85 mg/g) in Asia and *Crataegus germanica* (3.5 mg/g) in Europe (Edwards et al., 2012), and also in fruits that have an antioxidant activity higher than certain apple varieties (0.46 to 15.1 mg of GAE/g dry sample), as reported in other studies (Cam and Aaby, 2010; Sudha et al., 2007; Garcia et al., 2009). Phenolic compounds can protect against free radicals in aerobic cell metabolic processes. The antiradical activity of the polyphenols is attributed to the interaction of structural parts catalyzing metal chelation, the activation of certain antioxidant enzymes and the inhibition of oxidase enzymes (Ji-Hyun and Kwang-Deog, 2011). Such biological processes could be present in the skin of *C. mexicana*.

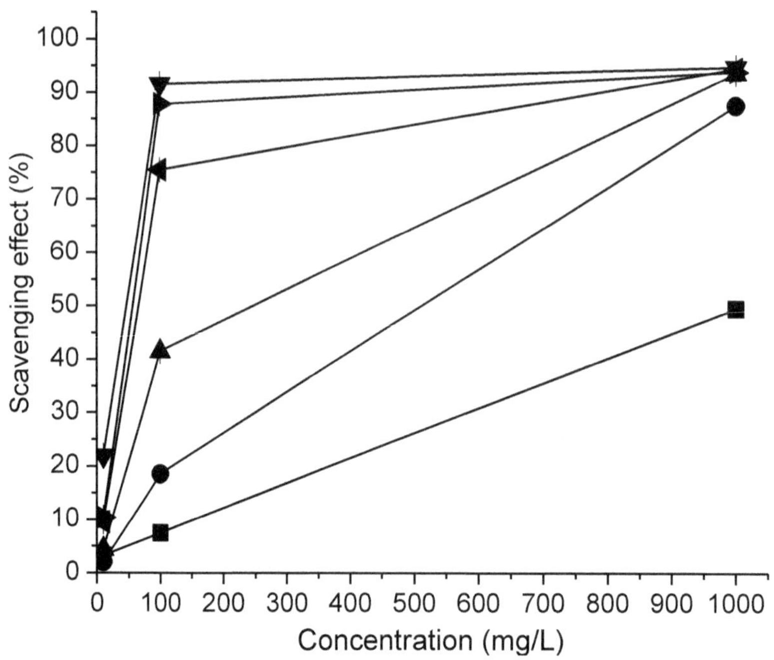

Figure 1. Scavenging of DPPH• radicals in the presence of increasing concentrations of *Crataegus mexicana* extracts at 10, 100 and 1000 mg/L. ■, *n*-Hexane; ●, dichloromethane; ▲, ethyl acetate; ◄, ethanol; ►, methanol; ▼, acetone (*p < 0.05).

Carotenoid content of hawthorn skin

The β-carotenoid content is associated with the intensity of color in fruits and leaves. *C. mexicana* has an intense yellow coloration in both the skin and pulp. This coloration may be associated with the carotenoid content of 26.4 ± 0.02 μg/g measured in dry hawthorn skin. Ying and Oey (2012), reported β-carotenoid contents of cherries, pears and plums of 20, 40 and 30 μg/g, respectively, and Monge-Rojas and Campos (2011) reported a carotenoid content of 27.6 μg/g in apples. A comparison of these results indicates that there is a greater quantity of carotenoids in dry hawthorn skin than in the aforementioned fruits. Therefore, the data demonstrate the large quantity of antioxidant metabolites in this fruit and its nutraceutical properties.

Free radical capture of DPPH

The capacity of *C. mexicana* extract to scavenge DPPH, $O_2^{\bullet -}$, $^{\bullet}OH$ and NO was measured (Figure 1). In this assay, antioxidants react with DPPH, a purple-colored stable free radical, and convert it to a colorless solution. The amount of DPPH reduced can be quantified by measuring the decrease in absorbance at 517 nm (Kaur et al, 2006). The DPPH scavenging ability of the extract may be attributed to its hydrogen donating ability.

The nonpolar extracts (*n*-hexane, dichloromethane and ethyl acetate) were less active in trapping the DPPH radical (7.33 to 41.46%) relative to the polar extracts (acetone, ethanol and methanol) (75.46 – 91.56%) at 100 mg/L. The hawthorn-acetone extract was more active at 10 mg/L, with a DPPH reduction of 21.9 ± 0.15%, compared with the *n*-hexane extract, with a DPPH reduction of 3.2 ± 0.23%. These results show that polar extracts have metabolites with a high capacity to donate hydrogen and electrons, as indicated by the effective neutralization of DPPH (Mishra et al., 2012).

MDA determination by TBARS

Increased lipoperoxide levels have been correlated with cancer, Alzheimer's disease, atherosclerosis and the aging process. The quantitative assessment of conjugated dienes, lipid hydroperoxides, alkanes, aldehydes and isoprostanes has been extensively studied. The effects of free radical damage have received increasing attention (Moore and Roberts, 1998). Therefore, during the last decade, substantial work has been conducted to refine the measurement of lipid peroxidation products.

Rat brain tissue contains both saturated and unsaturated lipids and fatty acids that are susceptible to peroxidation by ROS. Malondialdehyde (MDA) is one of the final products of cell membrane lipoperoxidation. MDA reacts with TBA to form a chromophore that is

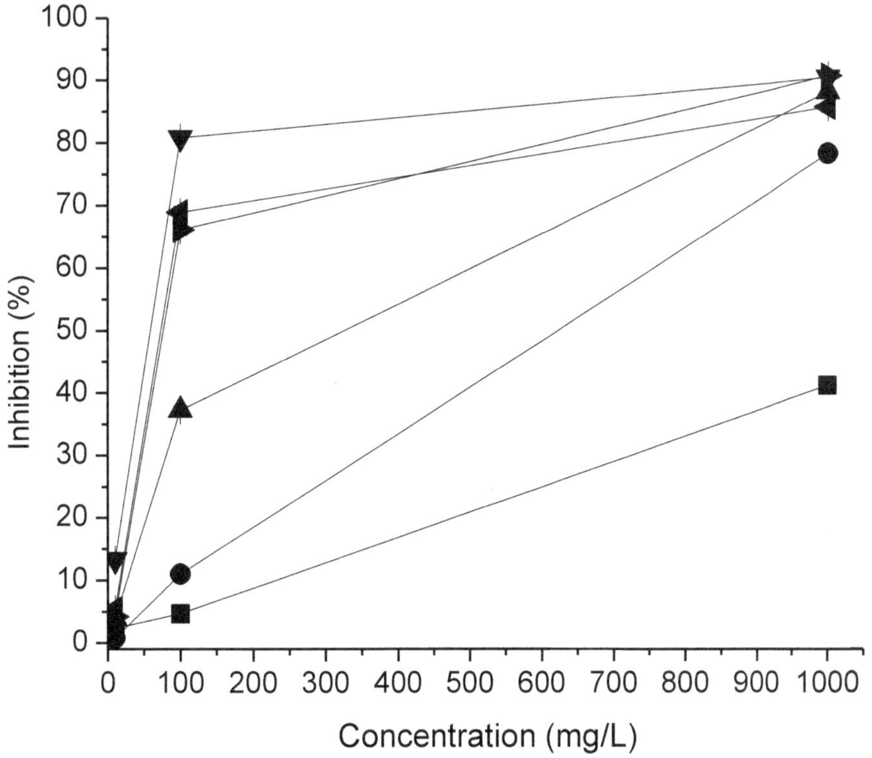

Figure 2. Effects of *Crataegus mexicana* extracts on the production of TBARS in rat brain homogenate. Increasing concentrations of *C. mexicana* extracts at 10, 100 and 1000 mg/L were used. ■, *n*-Hexane; ●, dichloromethane; ▲, ethyl acetate; ◄, ethanol; ►, methanol; ▼, acetone (*$p < 0.05$).

easily measured spectrophotometrically. To determine the antioxidant capacity of hawthorn extracts in rat brain tissue, $FeSO_4$ was used to induce the lipoperoxidation of rat brain cells. The ferrous ion (Fe^{2+}) can form lipid peroxides, which induce the formation of highly reactive substances, such as hydroxyl radicals, perferryl and ferryl species.

To evaluate the ability of the hawthorn extracts to protect the rat brain homogenate from being lipoperoxidated, different concentrations of extracts were used as described above. The polar extracts had a better performance in inhibiting lipid peroxidation, with the most active being the acetone extract (13.27 ± 0.70%) at 10 mg/L and the least active being the *n*-hexane extract (2.35 ± 0.27%) at 10 mg/L (Figure 2).

When compared with the data obtained from the DPPH assay, there was similar trend in the activity of the polar and nonpolar extracts in both experiments.

IC_{50} determination in DPPH and TBARS

To determine the IC_{50} in the DPPH and TBARS experiments, various concentrations of the hawthorn extracts were used according their inhibition response. For the most active extract, concentrations from 3.16 to 56.23 mg/L were used. For the extracts with medium activity, concentrations from 10 to 100 mg/L were used. For the extracts with a low activity, concentrations from 100 to 1000 mg/L were used.

The IC_{50} data for the experiments with DPPH showed that the hawthorn acetone extract was 3-times less active (15.26 ± 0.20 mg/L) than the relative potency of quercetin. The methanol extract ranked second in activity, and the *n*-hexane extract was the least active. Antioxidant activity has been shown to be related to polyphenol content in natural extracts.

The IC_{50} results from the TBARS experiments showed that the acetone extract demonstrated a strong inhibition of lipoperoxidation at a concentration of 16.78 ± 0.33 mg/L, followed by the extracts of methanol, ethanol and *n*-hexane, which was the least active. The acetone extract data were similar to those obtained for the IC_{50} of pure ascorbic acid, caffeic acid, α-tocopherol and quercetin antioxidants at concentrations from 1 to 40 mg/L in TBARS assays. When compared with quercetin (1.00), which had an IC_{50} of 3.56 ± 0.07 mg/L, the acetone extract was five-times less active, with a relative potency of 0.21 for the inhibition of lipoperoxidation (Table 1). In both the TBARS and the DPPH experimental models, the polar extracts were the most active, with a consistent antioxidant activity.

Table 1. [1]Hawthorn (*Crataegus mexicana*) IC_{50} antioxidant capacity compared with synthetic antioxidants determined using DPPH and TBARS assays.

Pure antioxidant	TBARS[1]	Relative potency[2]	DPPH[3]	Relative potency[2]
Ascorbic acid	12.95 ± 0.01	0.27	15.93 ± 0.37	0.33
Caffeic acid	13.55 ± 0.05	0.26	18.65 ± 0.73	0.28
α-tocopherol	33.17 ± 1.13	0.10	23.15 ± 0.67	0.23
Quercetin	3.56 ± 0.07	1.00	5.37 ± 0.14	1.00
Hawthorn extract				
n-Hexane	nd	nd	nd	nd
Dichloromethane	684.90 ± 1.58[a]	5.20×10^{-3}	724.05 ± 1.13[a]	7.41×10^{-3}
Ethyl acetate	550.63 ± 0.72[b]	6.46×10^{-3}	458.49 ± 0.78[b]	0.01
Acetone	16.78 ± 0.33[c]	0.21	15.26 ± 0.20[c]	0.35
Ethanol	54.49 ± 3.56[d]	0.06	54.20 ± 1.93[d]	0.10
Methanol	27.66 ± 0.73[e]	0.12	26.64 ± 0.66[e]	0.20

Means ± SEM of three different experiments. [2]Relative to quercetin as the unitary value. [3]IC_{50} for the neutralization of the radical DPPH in pure antioxidants and hawthorn skin (*C. mexicana*). nd, not determined. Values are expressed as μg/mL (mg/L). Different letters represent a significant difference at $^*p < 0.05$ (Duncan analysis).

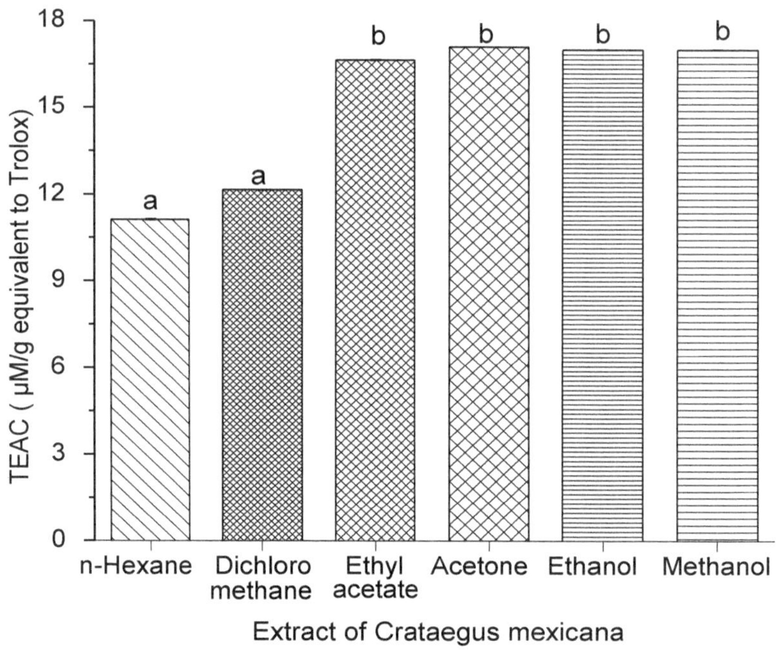

Figure 3. ABTS antioxidant capacity expressed as μM equivalent of Trolox/g of dry skin sample in the six hawthorn extracts of *Crataegus mexicana*. Means ± SEM of three different experiments. Different letters indicate a significant difference at $^*p < 0.05$ (Tukey analysis).

ABTS radical scavenging activity

The determination of ABTS radical scavenging activity in hawthorn extracts indicated a lower activity in the nonpolar extracts of 11 and 12 μM of Trolox/g of dry skin sample. In the more polar extracts, the results were between 16 and 17 μM of Trolox/g of dry skin sample (Figure 3). These results are similar to those reported in the skin of dehydrated exotic fruits, with values of 14.6 to 428 Trolox μmol/g (Contreras-Calderón et al., 2011). For comparison, Trolox equivalent antioxidant capacity (TEAC) values of 4.62, 3.87 and 15.18 μM trolox/g have been reported for apple *malus domestica* (red delicious), mango (hainan) and guava (Li et al., 2011), respectively.

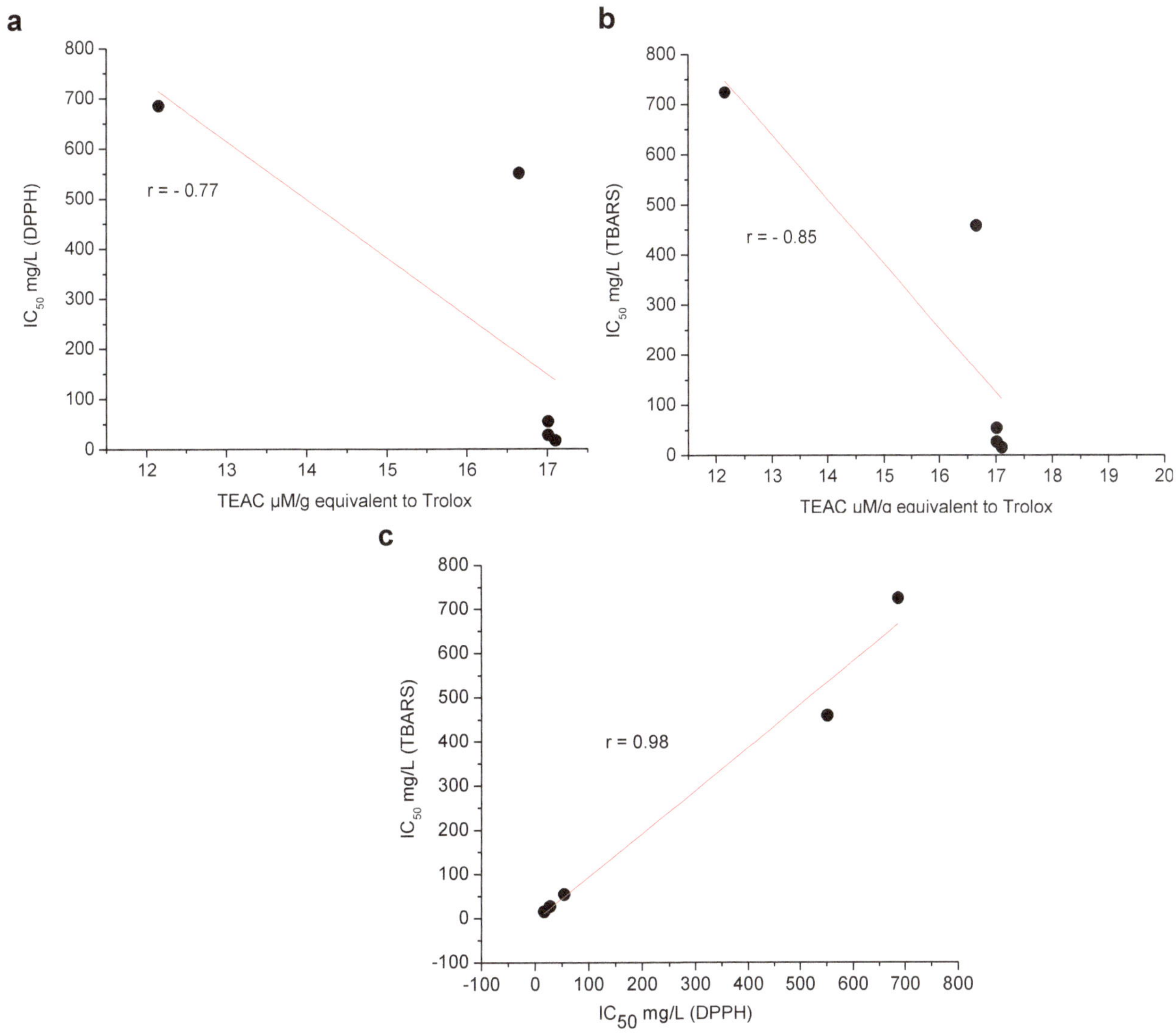

Figure 4. Pearson analyses of the six extracts from hawthorn skin (*Crataegus mexicana*). (a) DPPH *vs.* ABTS, (b) TBARS *vs.* ABTS, **(c)** TBARS *vs.* DPPH.

The data suggests a high polyphenol content in the polar extracts of hawthorn *C. mexicana* (Figure 3).

These results demonstrate that the dry hawthorn skin sample has a greater antioxidant capacity than those of apple, mango and guava. The results of the DPPH and TBARS activity assays using the polar extracts showed a higher correlation with the equivalent of TEAC. The Pearson statistical analysis shows a negative correlation between ABTS and DPPH (R = -0.77). These data show that extracts with low polarity have a low antioxidant capacity (Figure 4).

A lower correlation (R = -0.85) was observed for the TBARS and ABTS results because the methanol extract had a poor response in the TBARS assay and a good response in the ABTS assay. Finally, the correlation between the TBARS and DPPH data (R = 0.98) showed a similar behavior among the different extracts.

Conclusions

Fruits from the *rosaceas* family and the genus *crataegus* contain terpenoids and flavonoids (Shao-Jiang et al., 2011; Calişkan et al., 2012). However, there is insufficient data on the contents of phenolic and terpenic metabolites in these fruits. This work evaluated the antioxidant capacity of *C. mexicana* and revealed the major antioxidant activity of the acetone extract. This activity is likely the result of the action of medium- and high-polarity metabolites.

The total antioxidant capacity of hawthorn (*C. mexicana*) was evaluated *in vitro* using colorimetric techniques. The final data are similar to those reported in other *Crataegus* species in Asia (Calişkan et al., 2012) and Europe (Froehlicher et al., 2009; Zesheng et al., 2001), and also for the empire and golden delicious apple varieties, strawberry, black thorn, plum and cherry (Lata and Tomala, 2007). These results improve the knowledge of the potential nutritional and healthful effects of the bioactive compounds in this Mexican fruit and support its possible use as a supplement to increase the antioxidant activity of food products. Metabolite analysis of the polar content and protective properties in the red blood cell membranes of these extracts is in progress to identify the chemical structures of the bioactive components. The active components of *C. mexicana* could be used to prevent chronic degenerative diseases and to contribute to the maintenance of human health.

ACKNOWLEDGMENTS

We thank Karla Vega Flores for identifying fruit samples. This study was supported by grant 106653 from the Mexican Council of Science and Technology (CONACYT), and by grant P/PROMEP/103.5/11/6862.

REFERENCES

Andlauer W, Fürst P (2002). Nutraceuticals: a piece of history, present status and outlook. Food Res. Int. 35: 171–176.

Biedrzycka E, Amarowicz R (2008). Diet and health: Apple polyphenols as antioxidants. Food Rev. Int. 24: 235-251.

Blumberg J (2004). Use of Biomarkers of Oxidative Stress in Research Studies. J Nutr. 134: 3188S–3189S.

Cabrera GL (1958). Plantas curativas de Mexico: Propiedades medicinales de las más conocidas plantas de México. (2nd ed.). México: Ciceron.

Cai Y, Luo Q, Sun M, Corke H (2004). Antioxidant activity and phenolic compounds of 112 traditional Chinese medicinal plants associated with anticancer. Life Sci. 74: 2157-2184.

Calişkan O, Gündüz K, Serçe S, Toplu C, Kamiloğlu O, Şengul M, Ercişli S (2012). Phytochemical characterization of several hawthorn (*Crataegus* spp.) species sampled from the Eastern Mediterranean region of Turkey. Pharmacogn. Mag. 8(29): 16-21.

Cam M, Aaby K (2010). Optimization of extraction of apple pomace phenolics with water by response surface methodology. J. Agric. Food Chem. 58: 9103–9111.

Contreras-Calderón J, Calderón-Jaimes L, Guerra-Hernández E, García-Villanova B (2011). Antioxidant capacity, phenolic content and vitamin C in pulp, peel and seed from 24 exotic fruits from Colombia. Food Res. Int. 44: 2047–2053.

Crowe LF, Roddam WA, Key JT, Appleby NP, Overvad K, Jakobsen UM, Tjønneland A, Hansen L, Boeing H, Weikert C, Linseisen J, Kaaks R, Trichopoulou A, Misirli G, Lagiou P, Sacerdote C, Pala V, Palli D, Tumino R, Panico S, Bueno-de-Mesquita HB, Boer J, Gils HC, Beulens WJJ, Barricarte A, Rodríguez L, Larrañaga N, Sánchez JM, Tormo MS, Buckland G, Lund E, Hedblad B, Melander O, Jansson JH, Wennberg P, Wareham JN, Slimani N, Romieu I, Jenab M, Danesh J, Gallo V, Norat T, Riboli E (2011). Fruit and vegetable intake and mortality from ischaemic heart disease: Results from the European Prospective Investigation into cancer and nutrition (EPIC)-Heart study. Eur. Heart J. 32: 1235-1243.

Delgado OMC, Tironi AV, Añón MC (2011). Antioxidant activity of amaranth protein or their hydrolysates under simulated gastrointestinal digestion. LWT 44: 1752-1760.

Dominguez M, Nieto A, Marin JC, Jeffery E, Cespedes CL (2005). Antioxidant activities of extracts from *Barkleyanthus salicifolius* (Asteraceae) and *Penstemon gentianoides* (Scrophulariaceae). J Agric. Food Chem. 53: 58589-5895.

Dufour D, Pichette A, Mshvildadze V, Bradette-Hebert ME, Lavoie S, Longtin A (2007). Antioxidant, anti-inflammatory and anti-cancer activities of methanolic extracts from *Ledum groenlandicum* retzius. J. Ethnopharmacol. 111: 22–28.

Edwards JE, Brown PN, Talent N, Dickinson TA, Shipley PR (2012). A review of the chemistry of the genus *Crataegus*. Phytochemistry 79: 5–26.

Esterbauer H, Cheeseman KH (1990). Determination of aldehydic lipid peroxidation products. Meth. Enzymol. 186: 407-421.

Everette DJ, Bryant MQ, Green MA, Abbey AY, Wangila WG, Walker BR (2010). Thorough study of reactivity of various compound classes toward the Folin-Ciocalteu reagent. J. Agric. Food Chem. 58: 8139–8144.

Folin C, Ciocalteau V (1927). Tyrosine and trypthopan determination in proteins. J. Biol. Chem. 73: 627-650.

Froehlicher T, Hennebelle T, Martin-Nizard F, Cleenewerck P, Hilbert JL, Trotin F, Grec S (2009). Phenolic profiles and antioxidative effects of hawthorn cell suspensions, fresh fruits, and medicinal dried parts. Food Chem. 115: 897–903.

Garcia YD, Valles BS, Lobo AP (2009). Phenolic and antioxidant composition of by-products from the cider industry: Apple pomace. Food Chem. 117(4): 731–738.

Guimaraes R, Barros L, Carvalho AM, Ferreira ICFR (2010). Studies on chemical constituents and bioactivity of *rosa micrantha*: An alternative antioxidants source for food, pharmaceutical, or cosmetic applications. J. Agric. Food Chem. 58: 6277–6284.

Havsteen HB (2002). The biochemistry and medical significance of the flavonoids. Pharmacol. Ther. 96: 67–202.

Jayakumar R, Kanthimathi MS (2012). Dietary spices protect against hydrogen peroxide-induced DNA damage and inhibit nicotine-induced cancer cell migration. Food Chem. 134: 1580–1584.

Ji-Hyun J, Kwang-Deog M (2011). Inhibition of polyphenol oxidase and peroxidase activities on fresh-cut apple by simultaneous treatment of ultrasound and ascorbic acid. Food Chem. 124: 444–449.

Kaur G, Alam MS, Jabbar Z, Javed K, Athar M (2006). Evaluation of antioxidant activity of *Cassia siamea* flowers. J. Ethnopharmacol. 108: 340–348.

Khattak KF, Simpson TJ, Ihasnullah (2008). Effect of gamma irradiation on the extraction yield, total phenolic content and free radical-scavenging activity of *Nigella staiva* seed. Food Chem. 110: 967-972.

Kibanova D, Nieto-Camacho A, Cervini-Silva J (2009). Lipid peroxidation induced by expandable clay minerals. Environ. Sci. Technol. 43: 7550-7555.

Kim HJ, Chen F, Wang X, Chung YH, Jin Z (2005). Evaluation of antioxidant activity of vetiver (*Vetiveria zizanioides L*) oil and identification of its antioxidant constituents. J. Agric. Food Chem. 53: 7691-7695.

Lata B, Tomala K (2007). Apple peel as a contributor to whole fruit quantity of potentially healthful bioactive compounds. Cultivar and year implication. J. Agric. Food Chem. 55: 10795–802.

Li F, Bo-Tao X, Xiang-Rong X, Ren-You G, Yuan Z, En-Qin X, Hua-Bin L (2011). Antioxidant capacities and total phenolic contents of 62 fruits. Food Chem. 129: 345–350.

Lowry OH, Rosebrough NJ, Farr AL, Randall RJ (1951). Protein measurement with the Folin phenol reagent. J. Biol. Chem. 193: 265-275.

Mishra K, Ojha H, Chaudhury KN (2012). Estimation of antiradical properties of antioxidants using DPPH assay: A critical review and results. Food Chem. 130: 1036–1043.

Monge-Rojas R, Campos H (2011). Tocopherol and carotenoid content of foods commonly consumed in Costa Rica. J. Food Com. Anal. 24: 202–216.

Moore K, Roberts LJ (1998). Measurement of lipid peroxidation. Free Radic. Res. 28: 659–671.

Ng TB, Liu F, Wang ZT (2000). Antioxidative activity of natural products from plants. Life Sci. 66: 709-723.

Ohkawa H, Ohishi N, Yagi K (1979). Assay for lipid peroxides in animal tissues by thiobarbituric acid reaction. Anal. Biochem. 95: 351-358.

Shao-Jiang S, Ling-Zhi L, Pin-Yi G, Peng Y, Jing-Yu Y, Chun-Fu W (2011). Terpenoids and hexenes from the leaves of *Crataegus pinnatifida*. Food Chem. 129: 933–939.

Singh M, Madeleine A, Sanderson T, Murthy V, Ramassamy C (2008). Challenges for research on polyphenols from foods in Alzheimer's disease: bioavailability, metabolism, and cellular and molecular mechanisms. J. Agric. Food Chem. 56: 4855–4873.

Singleton V, Orthofer R, Lamuela-Reventos R (1999). Analysis of total phenols and other oxidation substrates and antioxidants by means of Folin-Ciocalteu reagent. Meth. Enzymol. 299: 152–178.

Speek AJ, Speek-Saichua S, Schreurs WHP (1988). Total carotenoid and β-carotene contents of Thai vegetables and the effect of processing. Food Chem. 27: 245-257.

Spencer AW, Vadhanam VM, Jeyabalan J, Gupta CR (2012). Oxidative DNA damage following microsome/Cu(II)-mediated activation of the estrogens, 17β-estradiol, equilenin, and equilin: role of reactive oxygen species. Chem. Res. Toxicol. 25: 305−314.

Sudha ML, Baskaran V, Leelavathi K (2007). Apple pomace as a source of dietary fiber and polyphenols and its effect on the rheological characteristics and cake making. Food Chem. 104(2): 686–692.

WHO/FAO (2003). Diet, nutrition and the prevention of chronic diseases. Report of a WHO/FAO consultation. Ginebra, World Health Organization (WHO Technical Report Series, No. 916).

Willis MS, Wians FH (2003). The role of nutrition in preventing prostate cancer: A review of the proposed mechanism of action of various dietary substances. Clin. Chim. Acta. 330: 57–83.

Ying LS, Oey I (2012). Effects of processing on anthocyanins, carotenoids and vitamin C in summer fruits and vegetables. Food Chem. 133: 1577–1587.

Zesheng Z, Qi C, Min Z, Yu H, Walter KKH, Zhen-Yu C (2001). Characterization of antioxidants present in hawthorn fruits. J. Nutr. Biochem. 12: 144–152.

Lipid composition of medicinal plants: *Portulaca oleracea* and *Ficus thonningi*

Ochulayi P. Ora[1]* and Anekwe G. E.[2]

[1]Department of Biochemistry, National Research Institute for Chemical Technology, Zaria Nigeria.
[2]Department of Biochemistry, University of Jos, Plateau state, Nigeria.

Lipid compositions of these plants *Portulaca oleracea* and *Ficus thonningi* were analyzed after extraction to determine the biochemical evidence for their therapeutic use as medicinal plants used in Nigeria. The lipids of *P. oleracea* and *F. thonningi* were extracted using chloroform and methanol (2:1), purified using Folch wash and analyzed using thin layer chromatography (TLC) and column chromatography to identify neutral and polar lipids and the various classes of lipids .The percentage composition of the various lipid fractions were calculated. The TLC result showed the presence of some neutral lipids, linoleic acid (66.7%) and triolene (26.7%) in *P. oleracea* and triolene (70%) in *F. thonnningi* and the polar lipids: Linolenic acid (50%),Oleic acid (35%), phospholipids (5%) and diacylglycerol in *P. olerecea* while in *F. thonningi* they were triolene (70%), triglycerol (30%), phosphatidate (26.7%), phospholipid (13.3), diacylglycerol (10%).Total lipid of *P. oleracea* is 35% and *F. thonningi* is 30%.Omega-3 fatty acids (50%) as well as omega-6 fatty acids (66.7%) presented in the biochemical evidence for the therapeutic use of purslane as a medicinal plant.

Key words: Lipids, chromatography, purslane, omega-3, omega-6 fatty acids.

INTRODUCTION

Medicinal properties have been attributed to many plants for thousands of years (Man et al., 2001; Jung et al., 2004).The plants extracts are widely sold as nutritional supplements or tonic and touted as beneficial for health (Brochers et al., 1999). Medicinal plants are useful against some kinds of cancer, lipid peroxidation, hypercholesterolemia, immune function, etc as known in different countries. Coronary heart diseases (CHD) continue to be a leading cause of mortality and morbidity in united state (Starfield, 2000; Smith, 2000; Jacobson, 2001; D'Agostino et al., 2000). It was well established that hypercholesterolemia is a major risk factor for coronary atherosclerosis and arteriosclerosis.

Hyperlipidemia particularly elevated serum cholesterol and low density lipoprotein (LDL) level is a risk factor in the development of atherosclerotic heart disease (Romerol-corral et al., 2006). Although several studies have reported that cholesterol lowering drugs like statins have been quite effective in lowering total cholesterol, LDL cholesterol prevent incidence of coronary heart disease and of ischemic stroke (Law et al., 2003; Wanner et al., 2005; Colivicchi et al., 2007). However, some of the drugs have been associated with side effects like elevated liver enzymes, muscle pain and joint aches, nausea, diarrhea, constipation (FDA MEDWATCH Reporting system, 2003; Chung et al., 2001; Scheen, 2001; Silva et al., 2006) as a result there is potential side effect from cholesterol lowering drugs, growing patient's demand for non-traditional or diet related approaches that could lower blood lipid cholesterol and prevent cardiovascular disease and stroke. Thus there is a renewed effort from researchers (Movahedian et al., 2007; Prasad, 2005) to identify natural sources of nutraceuticals and phytochemical with hypolipidemic and hypocholesterolemic properties.

*Corresponding author. E-mail: oraokibs@gmail.com or orachuks2008@yahoo.com.

Abbreviations: TLC, Thin layer chromatography; **CHD,** coronary heart diseases; **LDL,** low density lipoprotein.

Table 1. Weight of lipids of *P. oleracea* and *F. thonningi*.

Plant sample	Weight (g) of test tube	Weight (g) of test tube + lipid	Weight of lipids (g)
Portulaca oleracea	16.39	16.60	0.21
Ficus thonningi	14.43	14.50	0.14

Table 2. Percentage of total lipids.

Parameter	*Portulaca oleracea*	*Ficus thonningi*
Weight of plant leaves	20 (g)	20 (g)
Weight of total lipid extract	70 (mg)	60 (mg)
Percentage lipid in leaves	35%	30%
Weight of polar lipid	20 (mg)	30 (mg)
Percentage polar lipid	28.6%	50%
Weight of neutral lipid	30 (g)	10 (g)
Percentage neutral lipid	42.9%%	16.7%

It has been reported that medicinal plants like Puslane (*Portulaca Oleracea*) might provide therapeutic benefit to patientswithcardiovasculardiseaseriskfactor (Movahedian et al., 2007). Purslane is found growing in the wild or cultivated in many parts of the world (Minaiyan et al., 2005) and the leaves of purslane are commonly eaten extensively as soups and salads in the Mediterranean countries (Chan et al., 2000) *P. oleracea* is widely distributed in the tropical and sub tropical areas of the world where it is consumed as a nutritious vegetable and used for its pharmacological properties (Ezekwe et al., 2004). The aim of this work is to investigate the lipid compositions of the above mentioned medicinal plants to know the biochemical evidence of their therapeutic properties. Lipids are diverse group of fatty substances found in all living organisms which do not dissolve in water but are soluble in alcohol, ether and other organic solvents. Lipids are essential to life and that they are essential fats that play a very important role in the human body. They help our body to run smoothly in many ways, help the brain to function well, joint mobilization, energy production and also help the body absorb vitamins, coats the cells with lubrication therefore providing protective barrier, acts as messengers in the body (internal cellular communication), work to regulate steroidal hormones. *F. thonningi* is used to treat colds, sore throats, diarrhea and wounds and to stimulate lactation.

MATERIALS AND METHODS

The plant material used was obtained in the pharmacology department of University of Jos. Preparation of activated charcoal 50 ml of sulfuric acid and nitric acid was mixed in a beaker and the mixture was gently poured into a conical flask containing 100 g of charcoal powder with gentle stirring to ensure an even combination of charcoal powder with acid mixture. The next day the aqua region was decanted properly leaving behind the acid-soaked powder, and the charcoal was washed three times with chloroform and methanol to reduce the acidity and was further washed with distilled water. The charcoal was dried in an oven at 40°C for 1 h.

Lipid extraction

200 g of the leaves were washed, aerated and crushed in a mortar using chloroform and methanol for homogenization after which chloroform and methanol 2:1 was used to extract lipids from the leaves. using the method of Folch et al. (1957).

Purification

Lipid extract was purified to eliminate some contaminants by first pouring the extracts into a conical flask through a filter paper containing activated charcoal to remove coloring matters like chlorophyll. A clear supernatant was obtained which was then further purified using The "Folch" procedure as described by Folch et al. (1957)

Desolvenization of purified lipids

Purified lipids were transferred into dry, weighed test tubes and the weight of the lipids was determined as shown on Table 1.

Separation of neutral and polar lipids by silicic acid column chromatography

The separation of lipids was done using the method described by Anekwe et al (1999). The results are shown in Table 2.

Thin layer chromatography (TLC)

This method was used to identify different classes of lipids as described by Anekwe et al. (1999), the results are shown in Tables 2 to 7.

Table 3. Comparism of the standard RF values to the ones obtained for Portulaca oleracea and Ficus thonningi.

Fatty acids/standard RF value	*Portulaca oleracea*	*Ficus thonningi*
Linoleic acid (0.69-0.72)	0.69	-
Triolene (0.78)	0.78	0.78
Linolenic acid(0.31-0.43)	0.45	-
Oleic acid(0.26-0.32)	0.27	-
Phospholipids(0.0)	0.01	0.01
Diaclglycerol(0.04)	0.02	0.02
Triacylglycerol(0.23)	-	0.20
phosphatidate (0.60)	-	0.55
Phosphocholine(0.43)	-	-
Acetylcholine(0.75)	-	-

Table 4. *Portulaca oleracea* (neutral lipids).

Lipid component	Weight (mg)	Percentage (%)
Linoleic acid	20	66.7
Triolene	8	26.7

Table 5. *Portulaca oleracea* (polar lipid).

Lipid component	Weight (mg)	Percentage (%)
Linolenic acid	10	50
Oleic acid	7	35
Phospholipids	1	5
Diacylglycerol	1	5

Table 6. *Ficus thonningi* (neutral lipids).

Lipid component	Weight (mg)	Percentage (%)
Triolene	7	70%

Table 7. *Ficus thonningi* (polar lipids).

Lipid component	Weight(mg)	Percentage (%)
Triglycerol	9	30
phosphatidate	8	26.7
Phospholipid	4	13.3
Diacylglycerol	3	10

RESULTS AND DISCUSSION

The result shows that the percentage lipid content of *P. oleracea* is 35% while that of *F. thonningi* is 30%.The neutral lipids are 42.9% for *P. oleracea* and 16.7% for *F. thonningi* while the polar lipids are 28.6% for *P.oleracea* and 50% for *F. thonningi*. By comparing the RF values obtained to standard retention factor (RF) values. The result of the TLC and column chromatography shows that *P. oleracea* contains; Linoleic acid (66.7%), Triolene (26.7%), Linolenic acid (50%), Oleic acid (35%), Phospho-pholipids (5%), Diacyleglycerol (5%) while that of *F. thonningi* are triolene (70%), triglycerol (30%), phosphati-date (26.7%), Phospholipid (13.3%) diacyglycerol (10%). The predominant lipid of *P. oleracea* is linoleic acid (66.7%) while that of *F. thonningi* is triolene (70%). Both medicinal plants contained triolene, phospholipids and diacylglycerol. These results show that *P. oleracea* contains Essential fatty acids; linolenic acid (omega-3) and linoleic acid (omega-6) and *F. thonningi* contains oleic acids (omega-9) which are precursors of other important biochemical compounds. This work agree with several studies that shows that purslane have high content of omega-3 (Liu et el., 2002; Simopoulos et al., 2005; Omara-Awala et al., 1991).Polyunsaturated fatty acids are very important due to their numerous health benefit; it improves the health of the heart by reducing cholesterol level LDL whose accumulation blocks the arteries resulting to atherosclerosis, if it blocks the heart muscles it results to heart attack, if it blocks the arteries taking blood to the brain it results to stroke, its accumulation in the blood also results to high blood pressure. PUFA also prevents cancer, diabetes, and arthritis amongst other, effective against dermatitis, cystic fibrosis and have anti-inflammatory and acne reductive properties therefore the use of these plants *P. oleracea* and *F. thonningi* as a medicinal plant should be encouraged.

Conclusion

These essential fatty acids (omega-3 and -6) present in *P. oleracea* and *F. thonningi* is the biochemical evidence for the therapeutic and pharmaceutical properties of these plants therefore supporting their use as medicinal plants that can provide therapeutic benefits to patients with cardiovascular disease risk factors (Movahedian et al., 2007) and it also support that purslane offers better nourishment than the major cultivated vegetables due to its shoot that is a rich source of omega-3 fatty acids.

In conclusion, this work encourages the use of *P. oleracea* and *F. thonningi* as medicinal plants to combat the incidence of CHD and the treatment of patients with CHD risk factors as well as maintaining the overall health of the body due to their ability to lower cholesterol levels, lower high blood pressure are also due to their therapeutic and pharmacological properties.

REFERENCES

Anekwe GE (1999). Techniques in Lipid Biochemistry. Heinemann Educational Books Plc. Godman Printers Ibadan.12-23.

Brochers AT, Stem JS, Halkman RM, Keen CL, Gershwin ME (1999). Mushroom tumors and immunity. Proc. Soc. Exp. Boil. Med. (22):281-293.

Chan K, Islam MW, Kamil M ,Radhakrishnan R, Zakaria MN, Habibullah M, Attas A(2000).The Analgesic and Antiflammatory effects of Portulaca oleracea L. J. Ethanopharmacol. 73:445-451.

D'Adostino RB, Russell MW, Huse DM (2000).Primary and subsequent Coronary Risk Appraisal: New results from the Framingham study, Am. Heart J.139:273-281.

Ezekwe MO, Nyoka QE, Igbokwe PE, Besong SA (2004).Dietary supplements of Purslane leaves lowers serum Cholesterol in growing pigs. A867. Abstract, Experimental Biology.

Folch J, Lees M, Sloane-Stanley GH (1957). A Simple Method for the Isolation and Purification of Total Lipids from Animal Tissues, J. Biol. Chem. 226, 497–509.

Jacobson TA (2001).Clinical Context: current concepts of Coronary Heart Disease Management. Am. J. Med.110 (suppl.6a):3s-11s.

Jung K, Kim IH, Han D (2004): Effect of medicinal plant extracts on forced swimming capacity in mice. J. Ethnopharmocol. 93:75-81.

Law M, Wald NJ, Rudnicka AR (2003).Quantifying effect of Statins on low density lipoprotein cholesterol, ischeamic heart disease and stroke: systemic review and meta-analysis.BMJ, 28:326(7404);1423.

Liu L, Howe P, Zhou YF ,Xu ZQ ,Hocart C, Zhang R (2000).Fatty acids and B-carotene in Australian purslane (portulaca oleracea) varieties. J. Chromatogr. A., 893:207-213.

Man JL, Lim HC ,Chem CC (2001).Non volatile components of several medicinal mushrooms. Food Res. Int. 34.521-526.

Minaiyan M, Ahamadi A, Salehi E (2005).Antiulcerogenic effect of Zataria multiflora Boiss on Cysteamine induced duodenal ulcer in rats. Iran. J. Pharm. Sci.1:223-224.

Movahedian A, Ahannadi A, Vashimia M (2007).Hypocholesterolemic Effects of Purslane extract on serum lipids in rabbits fed with high cholesterol levels. Int. J. Pharmacol. 3(3):285-289.

Omara-Awala TR ,Mebrahtu T,Prior DE ,Ezekwe MO (1991).Omega-3 fatty acids in Purslane (Portulaca oleracea) Tissues. J. Am. Oil Chem. Soc. 68:198-199.

Prasad K (2005).Hypocholesterolemic and Antiatherosclerotic Effect of flax Lignin Complex isolated from flaxseed. Atherosclerosis 179:269-275.

Romero-Corral A, Somers VK , Korinek J, Sierra Johnson J, Thomas RJ, Alison TG ,lopez Jimenez F(2006).Update in prevention of Atherosclerotic Heart Disease :management of major Cardiovascular risk factors. Revista de investigation Clinica 58:237-244.

Scheen AJ (2001). Fatal Rhabdomyoly is caused by Cervastatin. Rev. Med. Liege 56(8):592-594.

Silva MA, Swanson AC, Gandhi PJ, Tataronis GR (2006). Statin-related adverse effects :a meta-analysis. Clin. Ther. 28(1):26-35.

Simopoulos AP, Tan DX , Manchester LC, Reiter RJ (2005).Purslane: A plant source of omega-3 fatty acids and Melatonin. J. Pineal Res. 39:331-332.

Smith D (2000). Cardiovascular Disease; a historic perspective.JPN J. Vet. Res. 48(2-3):147-156.

Starfield B (2008). Is US health really the best in the word? JAMA 28(2-3):483-485.

18

Phenolic content and antioxidant activity of selected Ugandan traditional medicinal foods

ANDABATI Brain and MUYONGA John

School of Food Technology, Nutrition and Bioengineering, Makerere University, P.O. Box 7062, Kampala, Uganda.

Twenty one (21) traditional food plants recognized as medicinal by communities in Kamuli and Gulu districts in Uganda were identified and analyzed for their phytochemical content and antioxidant activity. The total phenolic content (2.6 ± 0.1 to 184.2 ± 6.4 mg GAE gDW^{-1}) and flavonoid content (0.3 ± 0.1 to 162.2 ± 3.5 mg CE gDW^{-1}) as well as antioxidant activity (0.1 ± 0.1 to 57.8 ± 0.5 mg VCE gDW^{-1}) varied widely among the food plants. The seeds of *Tamarindus indica* L. and leaves of *Ipomea eriocarpa* R.Br., *Corchorus trilocularis* L., *Ocimum suave* L., *Corchorus olitorius* L., *Acalypha bipartita* Müll.Arg., and *Hibiscus acetosella* Welw. ex Hiern showed the highest phenolic contents and antioxidant activities. Juice prepared from tamarind pulp alone or mixed with tamarind seed powder (0.5, 1 and 3%) and analyzed for sensory acceptability, phenolic content and anti-oxidant activity were found to be highly acceptable, recording an overall hedonic acceptability scores of 7.40 ± 0.4, 6.65 ± 0.3, 7.01 ± 0.1 and 5.34 ± 0.2, respectively on a nine-point scale for juice containing 0, 0.5, 1 and 3% seed powder. Incorporation of tamarind seed powder into tamarind pulp juice resulted in a dose dependent increase in antioxidant activity from 8.5 ± 0.2 mg VCE 100 ml^{-1} for the control to 12.05 ± 0.3, 15.33 ± 0.7 and 17.22 ± 0.7 mg VCE 100 ml^{-1} for tamarind juice containing 0.5, 1 and 3% seed powder, respectively. A similar trend was recorded for the total phenolic and flavonoids.

Key words: Medicinal foods, nutraceuticals, antioxidant activity, phenolics, flavonoids.

INTRODUCTION

The consumption of significant amounts of certain particular foods, especially those of plant origin, including fruits, vegetables and whole grain cereals is associated with reduced risk of developing chronic degenerative diseases, such as cardiovascular diseases, cancers and diabetes mellitus (Espín et al., 2007). The health benefits of consuming such foods are ascribed partly to bioactive compounds found in the foods. For example, lycopene in tomatoes may protect against prostate cancers (Kucuk et al., 2002) and carotenoids in the various fruits and vegetables may also reduce the risk of prostate cancer (Jian et al., 2005) while flavonoids in tea act against cardiovascular diseases (Kris-Etherton and Keen, 2002). The phenolic and polyphenolic compounds comprise of important group of health promoting bioactive compounds, their protective effects are mainly attributed to

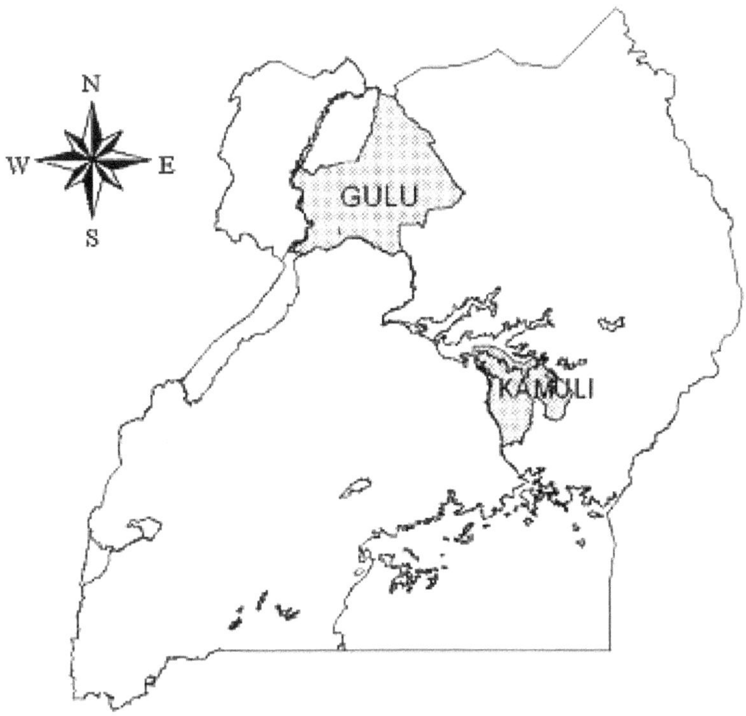

Figure 1. Map of Uganda showing the geographic location of Gulu and Kamuli districts, where a study was conducted to identify foods traditionally considered having medicinal properties. Source: http://en.wikipedia.org/wiki/Gulu_District;http://en.wikipedia.org/wiki/Kamuli_District.

their antioxidant activity (Shahidi, 1997).

Several traditional foods from different communities have been analyzed for presence of relevant bioactive compounds and biological activities (Buřičová and Réblová, 2008; D'Antuono et al., 2010; Heinrich et al., 2005, 2006) and such investigations have provided scientific evidence and validation for health claims of the different traditions. In Uganda, many foods are traditionally believed to provide medicinal benefits (Rubaihayo, 2002; Tabuti and Van Damme, 2012). However, there is inadequate scientific evidence for the ascribed medicinal properties of these traditional foods. Thus, there is need to validate the nutraceutical properties of traditional foods, given the rapidly increasing burden of chronic degenerative diseases in the country. It is anticipated that non-contagious chronic diseases will overtake infectious, maternal and nutritional diseases as the most common cause of death in Uganda by the year 2030 (WHO, 2011a, b).

In view of the above, this study was undertaken to analyze the phenolic content and antioxidant activity of selected traditional foods considered by local communities in Uganda to have health benefits. In addition, the potential of using certain traditional food additives in enhancing the natural antioxidant activity of foods was evaluated.

MATERIALS AND METHODS

Study area

A study to identify foods traditionally considered to have medicinal properties was conducted among the rural communities in Gulu and Kamuli districts, Uganda. Geographically, Gulu district is situated in the northern part of Uganda (02° 45'N, 32° 00'E) (Figure 1), and is home to the ethnic Acholi tribe who are estimated to be approximately 1.1 million people, representing 4.8% of the population of Uganda (UBS, 2002). Kamuli district on the other hand, is located in the eastern part of Uganda (00° 55'N, 33° 06'E), with the Basoga tribe as the main ethnic group. The Basoga are the third largest ethnic group in Uganda with population of 2.1 million people, representing about 8.6% of Uganda's population (UBS, 2002). The communities found in these two districts are mainly of rural setting (Lamordea et al., 2010), and consume rich diversity of foods originating from local plant or animal sources through gathering or harvesting (Oryema et al., 2013; Tabuti, 2007).

Data collection

A survey was conducted to identify foods considered to have medicinal value in Gulu and Kamuli districts in Uganda. Ethnobotanical information was obtained from informants through key informant interviews and focus group discussions. A total of 51 informants (27 in Gulu district and 24 in Kamuli district), consisting of traditional healers, health workers and elderly persons who are natives of the area were involved. Semi-structured questionnaires were used to collect information on local food plants considered to have medicinal value. Samples of identified foods reported to have medicinal value were collected and transported to the herbarium at Makerere University for identification.

Determination of the phenolic content and antioxidant activity of the identified foods

Plant material

The food plant materials collected were analyzed for total phenolic and flavonoid contents including total antioxidant activity.

Sample preparation

Different parts of the food plant (leaves and seeds) were cleaned, dried in shade at ambient temperature of about 25°C, were ground into fine powder using a stainless steel blender 7011S (Waring Commercial, Torrington, CT, USA) and sieved through a 500 µm sieve. The fine powder obtained was stored in airtight plastic containers at 4°C until further analysis, within 30 days.

Extraction of phenolic compounds

The extraction of phenolic compounds was done according to the method described by Makkar (2000) with minor modifications. Briefly, a small quantity (0.1 g) of the powdered sample was extracted in 5 ml of solvent mixture methanol : water (50:50 v/v) for 30 min at room temperature, with intermittent shaking. The extract was cooled in a refrigerator at 4°C for 10 min and then centrifuged at 3000 *xg* for 10 min using a centrifuge (Fisher Scientific 225, Pittsburgh, PA, USA). The supernatant was recovered and the pellet re-extracted for 45 min under the same conditions until a second supernatant was obtained. Finally, the two supernatants were pooled together and used in the determination of total phenolic and flavonoid contents and total antioxidant activity.

Determination of total phenolic content

The total phenolic content (TPC) was determined using spectrophotometry following Makkar (2000) method. To 100 µl of sample, 400 µl of distilled water was added to make the total volume to 0.5 ml. This was followed by addition of 0.25 ml of 1 N Folin-Ciocalteu reagent and 1.25 ml of 20% sodium carbonate solution. After 40 min incubation at room temperature, absorbance was read at 725 nm on a GENESYS spectrophotometer 10UV (Thermo Electron Corporation, Madison, WI, USA) against a blank solution containing methanol instead of the sample. A calibration curve was constructed within the concentration range 0.025 - 0.225 mg ml^{-1} (R^2 = 0.999). Mean values were calculated from three parallel analyses. The total phenolic contents were expressed in milligram gallic acid equivalents per gram of dry weight (mg GAE gDW^{-1}) of plant material.

Determination of total flavonoid content

The total flavonoid content (TFC) was measured using the colorimetric assay procedure reported by Muanda et al. (2011). Briefly, a 250 µl of a standard solution of catechin at different concentrations was added to 10 ml volumetric flask containing 1 ml of double distilled water (ddH_2O). At zero min, 75 µl of 5% $NaNO_2$ solution was added to the flask. After 5 min, 75 µl of 10% $AlCl_3$ solution was added. After 6 min, 500 µl solution of 1 N NaOH was added to the mixture. Immediately after diluting the solution by addition of 2.5 ml ddH_2O and thoroughly mixing, the absorbance of the pink solution mixture was measured at 510 nm in a GENESYS spectrophotometer 10UV (Thermo Electron Corporation, Madison, WI, USA) against a blank solution containing distilled water instead of sample. A calibration curve was constructed within the concentration range 0.025 - 0.225 mg ml^{-1} (R^2 = 0.999). Mean values were calculated from three parallel analyses. The total flavonoid contents of the samples were expressed as milligram catechin equivalents per gram dry weight (mg CE gDW^{-1}) of plant material.

Determination of total antioxidant activity

The total antioxidant activity (TAC) of the methanol extracts was estimated using the 1, 1-diphenyl-2-pycrylhydrazyl (DPPH) free radical scavenging assay (Kim et al., 2002). Briefly, to a 2.95 ml solution of freshly prepared absolute methanol, 100 µM solution of DPPH, 50 µl of the sample extract or control (50% [v/v] methanol) were added.

The mixture was shaken and allowed to stand in the dark at room temperature for 30 min. The absorbance of the resulting solution was measured at 517 nm in a GENESYS spectrophotometer 10UV (Thermo Electron Corporation, Madison, WI, USA) against a blank solution of absolute methanol. The free radical scavenging activity was calculated as:

$$Scavenging\ activity\ (\%) = [1 - \left(\frac{absorbance\ of\ sample}{absorbance\ of\ control}\right)] \times 100$$

A standard solution of ascorbic acid was run using several concentrations ranging from 0.002 to 0.1 mg ml^{-1}. A standard calibration curve of ascorbic acid was then prepared by plotting the percentage of free radical scavenging activity against concentration (R^2 = 0.999). The final value was expressed as milligram ascorbic acid equivalents per gram dry weight (mg VCE gDW^{-1}).

Production of juice enriched with *Tamarindus indica* L. seed powder

Preparation of tamarind seed powder (TSP)

Fruits of *T. indica* L. were purchased from a local market and transported to the Food Chemistry laboratory at Makerere University. The tamarind fruits were soaked in cold water (seed to water ratio of 1:2 (w/v) and the seeds removed manually from the pulp. The seeds were thereafter autoclaved at 121°C for 30 min to remove astringency (Legesse and Emire, 2012). The seeds were dried under shade at ambient temperature of about 25°C, milled into fine flour using the WonderMill grain mill (Grote Molen Inc, Pocatello, ID, USA), sieved through a 200 µm sieve size and stored in airtight plastic containers at 4°C until further use.

Preparation of mixture of tamarind pulp and seed powder juice

Tamarind pulp juice was prepared by first dissolving 295 g of tamarind fruit into one litre of cold water while stirring for about 30 min and discarding all the seeds and fibres. The resulting pulp juice (about 900 ml) was filtered through cheese cloth to obtain a volume of 600 ml of juice. The volume of juice was then diluted with an equal volume of water and 120 g of sugar added to sweeten the taste. Thereafter, the sweetened juice was further divided into four equal portions of 300 ml each. A preparation of tamarind seed powder (TSP) was added to levels of 0.5, 1 and 3% (w/v) to each of the respective first, second and third portions of the 300 ml of the sweetened juice. All the three portions were heated to boiling point to facilitate extraction of phenolic compounds from the seed powder and the mixture was filtered through cheese cloth to obtain the final tamarind pulp and seed powder juice. The remaining fourth 300 ml portion of the sweetened juice containing no tamarind seed powder (0% TSP) was used as a control.

Sensory evaluation of the tamarind pulp and seed powder juice

The prepared tamarind pulp and seed juice was evaluated by a panel of 30 randomly selected judges consisting of students of the Department of Food Technology and Nutrition, Makerere University Kampala. The sensory attributes evaluated included color, taste, odor, consistency and general acceptability. A nine-point hedonic scale (9 = like extremely, 8 = like very much, 7 = like moderately, 6 = like slightly, 5 = neither like nor dislike, 4 = dislike slightly, 3 = dislike moderately, 2 = dislike very much, 1 = dislike extremely) was used (Larmond, 1977). The sensory assessors were presented with 30 ml of each sample at room temperature under normal lighting conditions. Water was provided to the assessors for mouth or palate cleansing between each sample.

Analysis of phytochemicals in the prepared tamarind pulp and seed powder juice

Determination of total phenolic content

The phenolic content in the tamarind pulp and seed juice was estimated by Folin-Ciocalteu method as described by Thimmaiah (1999). Briefly, an aliquot (0.5 ml) of each juice was mixed with 2.5 ml of distilled water. To this, 0.5 ml of 1.0 N Folin-Ciocalteu reagent was added and incubated for 3 min. To each of the tube, 2 ml of 20% sodium carbonate solution was added and the tubes incubated in a boiling water bath for 1 min. The test tubes were cooled and the absorbance of each reaction mixture was read at 650 nm in a GENESYS spectrophotometer 10UV (Thermo Electron Corporation, Madison, WI, USA). A standard calibration curve of different concentrations of gallic acid (0.025 - 0.2 mg ml^{-1}) against 100 ml of juice were plotted (R^2= 0.999). The total phenolic content was expressed in milligram gallic acid equivalents per 100 ml of juice (mg GAE 100 ml^{-1}).

Determination of total flavonoid content

The total flavonoid content was determined using the method of Zhishen et al. (1999). In brief, 0.5 ml of standard solution of catechin or the sample diluted (1:2 with water) was mixed with 2 ml of deionized water and with 0.15 ml of 5% sodium nitrite. After incubation for 5 min at room temperature, 0.15 ml of 10% aluminum chloride was added, and after another 6 min, 1 ml of 1M sodium hydroxide solution was added. The total volume of the solution was adjusted to 5 ml with deionized and the absorbance was read at 510 nm in a GENESYS spectrophotometer 10UV (Thermo Electron Corporation, Madison, WI, USA). A standard calibration curve of different concentrations of catechin (0.025-0.2 mg ml^{-1}) versus 100 ml of juice was plotted (R^2 = 0.999). Total flavonoid content was expressed in milligram catechin equivalents per 100 ml of juice (mg CE 100 ml^{-1}).

Determination of total antioxidant activity

The antioxidant activity of the prepared juice sample was estimated using the 1,1-diphenyl-2-pycrylhydrazyl (DPPH) free radical scavenging assay as described by Shimada et al. (1992). In brief, to 2.9 ml of freshly prepared 100 μM of DPPH in 80% ethanol solution, 0.5 ml of the juice sample or control solution (DPPH alone) was added. The mixture was shaken and allowed to stand in the dark at room temperature for 30 min. The absorbance of the resulting solution was measured at 515 nm using a GENESYS UV-vis spectrophotometer 10UV (Thermo Electron Corporation, USA) against a blank solution containing only 80% ethanol. The free radical scavenging activity was calculated as:

$$Scavenging\ activity\ (\%) = [1 - \left(\frac{absorbance\ of\ sample}{absorbance\ of\ control}\right)] \times 100$$

A standard solution of ascorbic acid or vitamin C was similarly run using different concentrations ranging from 0.002 to 0.1 mg ml^{-1}. A calibration curve of the percentage free radical scavenging activity of ascorbic acid versus its concentration per 100 ml of juice was then prepared (R^2 = 0.999). The final result was expressed as milligram vitamin C equivalents per 100 ml of juice (mg VCE 100 ml^{-1}).

Statistical analysis

Data from three independent experiments were subjected to analysis of variance (ANOVA) at α = 0.05 and means were separated using Turkey's post hoc test. Regression analysis was performed to indicate the relationship between total phenolic or flavonoid contents and antioxidant activity. Statistical analysis was conducted using Statistical Package for the Social Sciences (SPSS), version 16 (SPSS Inc, Chicago, IL, USA).

RESULTS AND DISCUSSIONS

Phytochemical content and antioxidant activity of food materials

Total phenolic content

The total content of phenolics of the different foods ranged from 2.6 to 184.2 mg GAE gDW^{-1} and was highest for *T. indica* L. seeds and lowest for *Amaranthus hypochondriacus* L. (Table 1). Based on the classification of Lai and Lim (2011), samples with TPC > 20 mg GAE gDW^{-1} are classified as very high, those with TPC 10-20 mg GAE gDW^{-1} are considered high, those with TPC 5-10 mg GAE gDW^{-1} are moderate and samples with TPC < 5 mg GAE gDW^{-1} are considered low. All the foods regarded in the communities studied as having medicinal value showed very high TPCs (> 20 mg GAE gDW^{-1}), except *Hyptis spicigera* Lam. which showed a high TPC (18.5 ± 0.2 mg GAE gDW^{-1}) and foods of the Amaranthaceae family (*A. hypochondriacus* L. and *Amaranthus cruentus* L.) with low TPCs of 2.6 ± 0.1 and 2.9 ± 0.2 mg GAE gDW^{-1} respectively (Table 1).

Total flavonoid content

The result in Table 1 also showed variation in the total content of flavonoids from 0.3 to 162.2 mg CE gDW^{-1}. Total flavonoid content was highest for *T. indica* L. seeds (184.2 ± 0.2 mg CE gDW^{-1}) and lowest for *A. hypochondriacus* L. (2.6 ± 0.1 mg CE gDW^{-1}). Flavonoids

Table 1. Total phenolic content, total flavonoid content and total antioxidant activity of selected traditional medicinal foods identified in Kamuli and Gulu districts.

Plant scientific name	Local names	Plant parts analyzed	TPC (mg GAE gDW^{-1})	TFC (mg CE gDW^{-1})	TAC (mg VCE gDW^{-1})
Tamarindus indica L.	Enkooge	Seeds	184.2 ± 6.4^{a}	162.2 ± 3.5^{a}	57.8 ± 0.5^{a}
Ipomea eriocarpa R.Br.	Padowia kuri	Leaves	91.9 ± 2.9^{b}	78.9 ± 2.7^{b}	57.1 ± 0.6^{a}
Corchorus trilocularis L.	Otigo lum	Leaves	76.5 ± 0.2^{c}	46.0 ± 1.4^{c}	48.7 ± 1.6^{c}
Ocimum suave L.	Mujaaja	Leaves	74.5 ± 3.6^{c}	50.1 ± 4.3^{c}	$55.8 \pm 1.0^{a,b}$
Corchorus olitorius L.	Otigo Diri	Leaves	62.3 ± 0.9^{d}	36.2 ± 0.6^{c}	48.2 ± 0.8^{c}
Acalypha bipartita Müll.Arg.	Ayuyu	Leaves	62.0 ± 1.4^{d}	40.2 ± 0.3^{c}	49.7 ± 0.9^{b}
Hibiscus acetosella Welw. ex Hiern.	Gwanya	Leaves	$51.7 \pm 1.2^{d,e}$	24.4 ± 0.4^{e}	36.1 ± 2.0^{d}
Cassia obtusifolia L.	Oyado	Leaves	$48.1 \pm 1.7^{e,f}$	$12.0 \pm 0.2^{f,g}$	$20.0 \pm 1.2^{e,f}$
Solanum aethiopicum L.	Nakati	Leaves	$45.7 \pm 0.5^{e,f}$	$13.2 \pm 0.4^{f,g}$	$19.8 \pm 1.3^{e,f}$
Solanum nigrum L.	Ensugga	Leaves	$45.1 \pm 0.5^{e,f}$	$17.5 \pm 0.3^{e,f}$	25.6 ± 2.5^{e}
Vigna unguiculata L.	Eggobe	Leaves	$39.0 \pm 1.1^{f,g}$	$6.1 \pm 0.6^{g,h}$	$14.4 \pm 0.9^{f,g}$
Hibiscus cannabinus L.	Nyarogena	Leaves	$38.4 \pm 0.9^{f,g}$	$13.4 \pm 0.2^{f,g}$	22.2 ± 1.8^{e}
Crotalaria ochroleuca G.Don.	Lala	Leaves	$38.3 \pm 2.3^{f,g}$	$2.6 \pm 0.3^{h,i}$	$8.8 \pm 0.7^{g,h}$
Cleome gynandra L.	Ejjobyo	Leaves	$38.0 \pm 0.8^{f,g}$	$8.7 \pm 0.7^{g,h}$	$15.0 \pm 0.7^{f,g}$
Hibiscus sp. near H diversifolius Jacq.	Malakwang	Leaves	$37.7 \pm 2.4^{f,g}$	$12.9 \pm 1.0^{f,g}$	$19.2 \pm 2.1^{e,f}$
Solanum anguivi Lam.	Katunkuma	Fruits	$32.7 \pm 1.0^{f,g}$	$7.2 \pm 0.5^{g,h}$	$11.4 \pm 0.6^{g,h}$
Solanum gilo L.	Entuula	Leaves	$25.3 \pm 0.3^{h,g}$	$2.6 \pm 0.5^{h,i}$	$6.4 \pm 0.5^{f,i}$
Hyptis spicigera Lam	Lamola	Seeds	18.5 ± 0.2^{h}	0.4 ± 0.1^{i}	1.2 ± 0.6^{i}
Amaranthus cruentus L.	Doodo	Seeds	2.9 ± 0.2^{i}	0.5 ± 0.2^{i}	0.1 ± 0.0^{i}
Amaranthus hypochondriacus L.	Doodo	Seeds	2.6 ± 0.1^{i}	0.3 ± 0.1^{i}	0.1 ± 0.1^{i}

TPC: Total phenolic content; TFC: Total flavonoid content; and TAC: Total antioxidant activity. Data are expressed as mean ± standard error from three independent experiments (n=3). Mean values in the same column with different superscript letters are significantly different ($p < 0.05$).

are a large class of polyphenolic compounds that have been attributed to nutraceutical properties of several plants. Flavonoids are potent antioxidants and metal chelators (Tapas et al., 2008) and have long been recognized to be beneficial against many chronic diseases such as cardiovascular diseases (Kris-Etherton et al., 2002), cancer (Birt et al., 2001; Middleton et al., 2000), inflammation (Manthey et al., 2001) and neuro-degenerative disorders (Lu et al., 2010; Moosmann and Behl, 2002).

Total antioxidant activity

Antioxidant activity ranging from 0.1 to 57.8 mg VCE gDW^{-1} was recorded for the different food materials. *T. indica* L. seeds with 57.8 ± 0.5 mg VCE gDW^{-1}, *Ipomea eriocarpa* R.Br. (57.1 ± 0.6 mg VCE gDW^{-1}), *Ocimum suave* L. (55.8 ± 1.0 mg VCE gDW^{-1}), *Acalypha bipartite* Müll.Arg. (49.7 ± 0.9 mg VCE gDW^{-1}), *Corchorus trilocularis* L. (48.7 ± 1.6 mg VCE gDW^{-1}), *Corchorus olitorius* L. (48.2 ± 0.8 mg VCE gDW^{-1}) and *Hibiscus acetosella* Welw. ex Hiern. with 36.1 ± 2.0 mg VCE gDW^{-1} exhibited the highest antioxidant activities.

The total antioxidant activity of the traditional food plants significantly correlated to both the total phenolic content (R^2 = 0.883; $p < 0.05$) (Figure 2) and the total flavonoid content (R^2 = 0.965; $p < 0.05$) (Figure 3), suggesting that antioxidant activity may be attributed to the bioactivity of the phenolic and flavonoid compounds. Abu Bakar et al. (2009) reported a strong correlation between phenolic and flavonoid content and DPPH free radical scavenging ability.

Phytochemical content of the *Tamarindus indica* L. pulp and seed juice

Incorporation of tamarind seed powder (TSP) into tamarind pulp juice resulted into a proportionate increase in total phenolic content of the enriched juice from 24.68 ± 1.3 mg GAE 100 ml^{-1} for the control (juice without TSP) to 33.78 ± 2.2, 39.14 ± 2.0 and 53.34 ± 0.6 mg GAE 100 ml^{-1} for juice samples containing 0.5, 1 and 3% TSP, respectively (Table 2). The total flavonoid content (TFC) showed a similar pattern, increasing from 0.92 ± 0.0 mg CE 100 ml^{-1} for the control to 9.62 ± 0.1, 11.46 ± 0.3 and 17.5 ± 0.1 mg CE 100 ml^{-1} for juice samples enriched with 0.5, 1 and 3% (w/v) TSP, respectively. Similarly, total antioxidant activity increased from 8.50 ± 0.2 mg VCE ml^{-}

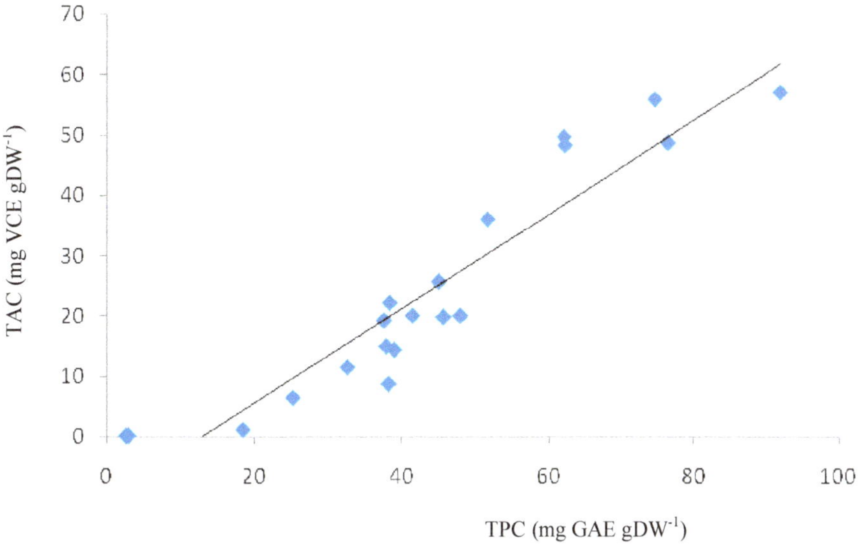

Figure 2. Relationship between antioxidant activities and the total phenolic contents in traditional food plants (R^2 = 0.883).

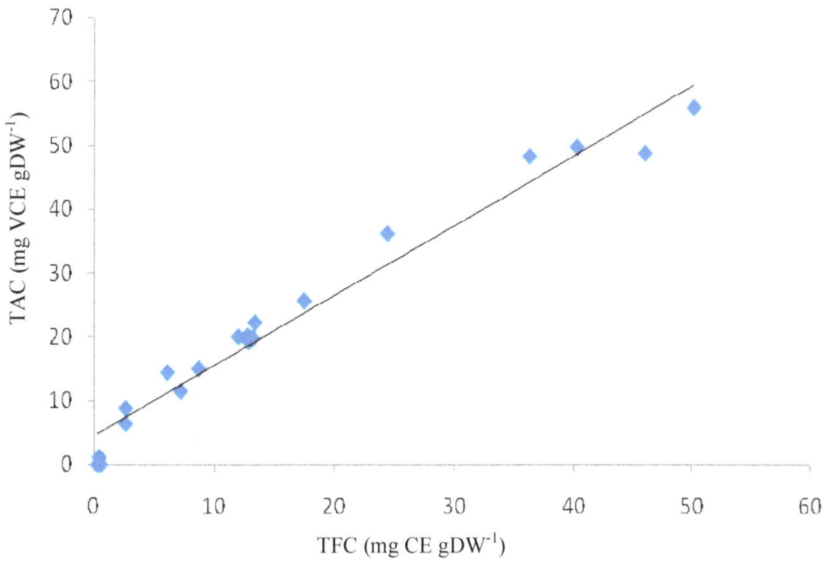

Figure 3. Relationship between antioxidant activities and the total flavonoid contents in traditional food plants (R^2 = 0.965).

Table 2. Total phenolic content, total flavonoid content and total antioxidant activity of tamarind pulp juices with different concentrations of tamarind seed powder.

Tamarind seed powder concentration (%)	TPC (mg GAE 100 ml^{-1})	TFC (mg CE 100 ml^{-1})	TAC (mg VCE 100 ml^{-1})
0	24.68 ± 1.3c	0.92 ± 0.0d	8.50 ± 0.2d
0.5	33.78 ± 2.2^{b}	9.62 ± 0.1c	12.05 ± 0.3c
1.0	39.14 ± 2.0^{b}	11.46 ± 0.3b	15.33 ± 0.7b
3.0	53.34 ± 0.6^{a}	17.50 ± 0.1a	17.22 ± 0.7a

TPC: Total phenolic content, TFC: Total flavonoid content, and TAC: Total antioxidant activity. Data are expressed as mean ± standard error from three independent experiments (n=3). Mean values in the same column with the same superscript letters are significantly different ($p < 0.05$).

Table 3. Sensory evaluation of juice prepared from tamarind pulp and seed powder.

Tamarind seed powder concentration (%)	Colour	Aroma	Taste/flavor	Mouth feel	General acceptability
0	6.70 ± 0.0^{a}	6.50 ± 0.2^{a}	6.50 ± 0.2^{ab}	6.90 ± 0.1^{a}	7.40 ± 0.4^{a}
0.5	6.45 ± 0.2^{ab}	6.80 ± 0.5^{a}	6.45 ± 0.4^{ab}	6.75 ± 0.1^{a}	6.65 ± 0.3^{ab}
1.0	6.23 ± 0.2^{ab}	5.34 ± 0.3^{ab}	6.80 ± 0.1^{a}	6.65 ± 0.2^{a}	7.01 ± 0.1^{a}
3.0	6.08 ± 0.1^{b}	4.65 ± 0.2^{b}	5.80 ± 0.1^{b}	5.30 ± 0.3^{b}	5.34 ± 0.2^{b}

Data are expressed as mean ± standard error from three independent experiments (n=3). Mean values in the same column with different superscript letters are significantly different ($p < 0.05$).

^{1} for the control to 12.05 ± 0.3, 15.33 ± 0.7 and 17.22 ± 0.7 mg VCE 100 ml^{-1} for the tamarind pulp juice sample containing 0.5, 1 and 3% (w/v) TSP, respectively. The phenolic contents of tamarind juice samples enriched with tamarind seed powder were comparatively higher than those of natural fresh juices made from apricot (23.75), pineapple (36.16), white grape (37.69), apple (45.38) and red grape juices (49.4), as reported by Mahdavi et al. (2010). The observed increments in total phenolic and flavonoid contents and in total antioxidant activity of the juice may be attributed to the high content of phenolic compounds and antioxidant activity in the *T. indica* L. seeds (Table 1). Previous studies have also reported high phenolic content and antioxidant activity in tamarind seeds (Siddhuraju, 2007). Overall, these results have demonstrated that incorporation of tamarind seed powder boosted the content of phenolic compounds and antioxidant activity of tamarind pulp juice. It further showed that the phenolic content and antioxidant activity increased proportionately with the amount of tamarind seed powder incorporated in the tamarind juice. However, the extent of incorporating tamarind seed powder in juice might be limited by the high content of tannins in tamarind seed coat and associated astringency (Pugalenthi et al., 2004). Thus, pre-treatment methods such as autoclaving of the seeds were found to reduce the astringency associated with tamarind seed (Legesse and Emire, 2012).

Sensory evaluation of tamarind pulp and seed juice

Addition of tamarind seed powder to tamarind juice to level of 1.0% (w/v) did not result in any significant difference in the perceived sensory attributes of tamarind pulp and seed juice ($p > 0.05$) as compared to the control (juice without tamarind seed powder) (Table 3). However, incorporation of tamarind seed powder at level of 3% (w/v) in the juice resulted in reduced scores in all the sensory attributes such as colour, aroma, taste and mouth feel. The largest difference in sensory scores between the control juice and juice containing 3% tamarind seed powder was in flavor and consistency, implying that the tamarind seeds disproportionately affected these sensory attributes. Tamarind seed kernels are known to contain some polysaccharides which when mixed with water form mucilaginous dispersions, leading to increase in viscosity of the juice. Furthermore, the presence of antinutritional factors such as tannins in tamarind seed testa renders the whole seed unsuitable for consumption (Caluwé et al., 2010). Overall, the colour, aroma, flavor and consistency of 0.5 and 1% tamarind pulp and seed powder juices were well accepted and the respective general acceptability scores for these were 6.65 ± 0.3 and 7.01 ± 0.1.

Conclusions

Biochemical analysis of traditional food plants identified to have health benefits by local communities in Gulu and Kamuli districts of Uganda showed that most exhibited high levels of total phenolic and flavonoid contents as well as high antioxidant activity. The present study therefore provides support for continued use of these investigated traditional foods for health promotion in Uganda. In particular, the seeds of *T. indica* L. and leaves of *I. eriocarpa* R.Br., *C. trilocularis* L., *O. suave* L., *C. olitorius* L., *A. bipartita* Müll.Arg. and *Hibiscus acetosella* Welw. ex Hiern. exhibited high levels of total phenolic and flavonoid contents including high antioxidant activity. The study has further demonstrated experimentally the potential use of *T. indica* L. seed powder in boosting antioxidant activity of juices as well as enhancing levels of phenolic and flavonoid compounds.

Conflict of interests

The authors did not declare any conflict of interests.

ACKNOWLEDGEMENT

This study was funded through a grant from the Swedish International Development Cooperation Agency, Department for Research Cooperation (Sida-SAREC) to Makerere University.

REFERENCES

Abu Bakar MF, Mohamed M, Rahmat A, Fry J (2009). Phytochemicals and antioxidant activity of different parts of bambangan (*Mangifera pajang*) and tarap (*Artocarpus odoratissimus*). Food Chem. 113:479-483.

Birt DF, Hendrich S, Wang W (2001). Dietary agents in cancer prevention: flavonoids and isoflavonoids. Pharmacol. Ther. 90:157-177.

Buřičová L, Réblová Z (2008). Czech medicinal plants as possible sources of antioxidants. Czech. J. Food Sci. 26:132-138.

Caluwé E, Halamová K, Van Damme P (2010). *Tamarindus indica L.* – A review of traditional uses, phytochemistry and pharmacology. Afrika Focus 23(1):153-83

D'Antuono F, Sanches-Silva A, Costa HS (2010). BaSeFood: sustainable exploitation of bioactive components from the Black Sea Area traditional foods. Nutr. Bull. 35:272-278.

Espín JC, García-Conesa MT, Tomás-Barberán FA (2007). Nutraceuticals: facts and fiction. Phytochemistry 68:2986-3008.

Heinrich M, Leonti M, Nebel S, W Peschel (2005). Local food-nutraceuticals: An example of a multidisciplinary research project on local knowledge. J. Physiol. Pharmacol. 56(1):5-22.

Heinrich M, Muller WE, Galli C (2006). *Local Mediterranean food plants and nutraceuticals*. Basel, Switzerland: Karger.

Jian L, Du CJ, Lee AH, Binns CW (2005). Do dietary lycopene and other carotenoids protect against prostate cancer? Int. J. Cancer. 113(6):1010-1014.

Kim JK, Noh JH, Lee S, Choi JS, Suh H, Chung HY, Song YO, Choi WC (2002). The first total synthesis of 2, 3, 6-tribromo-4,5-dihydroxybenzyl methyl ether (TDB) and its antioxidant activity. Bull. Korean Chem. Soc. 23(5):661-662.

Kris-Etherton PM, Hecker KD, Bonanome A, Coval SM, Binkoski AE, Hilpert KF, Griel AE, Etherton TD (2002). Bioactive compounds in foods: their role in the prevention of cardiovascular disease and cancer. Am. J. Med. 113:71S-88S.

Kris-Etherton PM, Keen CL (2002). Evidence that the antioxidant flavonoids in tea and cocoa are beneficial for cardiovascular health. Curr. Opin. Lipidol. 13(1):41-9.

Kucuk O, Sarkar FH, Sakr W, Khachick F, Djuric Z, Banerjee MM, Pollack MN, Bertram JS, Wood DP (2002). Lycopene in the treatment of prostate cancer. Pure Appl. Chem. 74:1443-1450.

Lai H, Lim Y (2011). Antioxidant properties of some Malaysian ferns. International Proceedings of Chemical, Biological and Environmental Engineering 20:8-12.

Lamordea MJ, Tabuti RS, Obua C, Kukunda-Byobona C, Lanyero H, Byakika-Kibwika P, Bosa GS, Lubega A, Ogwal-Okeng J, Ryan M, Waako PJ, Merry C (2010). Medicinal plants used by traditional medicine practitioners for the treatment of HIV/AIDS and related conditions in Uganda. J. Ethnopharmacol. 130:43-53.

Larmond E (1977). Laboratory methods for sensory evaluation of food. Ottawa, Ontario: Food Research Institute.

Legesse MB, Emire SA (2012). Functional and physicochemical properties of mango seed kernels and wheat flour and their blends for biscuit production. Afr. J. Food Sci. 3(9):193-203.

Lu YX, Zhang Q, Li J, Sun YX, Wang LY, Cheng WM, Hu XY (2010). Antidiabetic effects of total flavonoids from Litsea Coreana leve on fat-fed, streptozotocin-induced type 2 diabetic rats. Am. J. Chin. Med. 38:713-725.

Mahdavi R, Nikniaz Z, Rafraf M, Jouyban A (2010). Determination and comparison of total polyphenol and vitamin C ontents of natural fresh and commercial fruit juices. Pakistan J. Nutr. 9(10):968-972.

Makkar HPS (2000). Quantification of tannins in tree foliage: A laboratory manual for the FAO/IAEA co-ordinate research project on use of nuclear and related techniques to develop simple tannin assay for predicting and improving the safety and efficiency of feeding ruminants on the tanniniferous tree foliage. Vienna, Austria: A Joint FAO/IAEA Division of nuclear techniques in Food and Agriculture.

Manthey JA, Grohmann K, Guthrie N (2001). Biological properties of citrus flavonoids pertaining to cancer and inflammation. Curr. Med. Chem. 8:135-153.

Middleton E, Kandaswami C, Theoharides TC (2000). The effects of plant flavonoids on mammalian cells: Implications for inflammation, heart disease, and cancer. Pharmacol. Rev. 52:673-751.

Moosmann B, Behl C (2002). Antioxidants as treatment for neurodegenerative disorders. Exp. Opin. Investig. Drugs 11:1407-1435.

Muanda NF, Kon´e, D, Dicko A, Soulimani R, C Younos (2011). Phytochemical composition and antioxidant capacity of three Malian medicinal plant parts. J. Evid. Based Complementary Altern. Med. 1-8.

Oryema C, Oryem-Origa H, Roos N (2013). Edible wild fruit species of Gulu District, Uganda. Int. J. Biol. Sci. 2(4):068-082.

Pugalenthi M, Vadivel V, Gurumoorthi P, Janardhanan K (2004). Comparative nutritional evaluation of little known legumes, *Tamarindus indica* L., *Erythrina indica* and *Sesbania bispinosa*. Trop. Subtrop. Agroecosyst. 4:107-123.

Rubaihayo EB (2002). The contribution of indigenous vegetables to household food security (IK Notes No. 44). Washington, D.C: The World Bank.

Shahidi F (1997). Natural antioxidants: an overview. In: Natural antioxidants: Chemistry, health effects and applications, ed. F. Shahidi, 1-7. Champaign, IL: AOCS Press.

Shimada K, Fujikawa K, Yahara K, T Nakamura (1992). Antioxidative properties of xanthin on autoxidation of soybean oil in cyclodextrin emulsion. J. Agric. Food Chem. 40:945-948.

Siddhuraju P (2007). Antioxidant activity of polyphenolic compounds extracted from defatted raw and dry heated *Tamarindus indica* seed coat. Food Sci. Technol-LEB. 40:982-990.

Tabuti JRS (2007). Status of non-cultivated food plants in Bulamogi County, Uganda. Afr. J. Ecol. (Suppl. 1) 45:96-101.

Tabuti JRS, Van Damme P (2012). Review of indigenous knowledge in Uganda: implications for its promotion. Afrika Focus 25(1):29-38.

Tapas AR, Sakarkar DM, Kakde KB (2008). Flavonoids as nutraceuticals: A review. Trop. J. Pharm. Res. 7(3):1089-1099.

Thimmaiah SK (1999). Standard Methods of Biochemical Analysis. New Delhi, India: Kalyani Publishers.

Uganda Bureau of Statistics (UBS) (2002). Uganda population and housing census. Kampala, Uganda.

World Health Organization (2011b). Non-communicable diseases country profiles 2011. Geneva, Switzerland.

World Health Organization (WHO) (2011a). Global status report on non-communicable diseases 2010. Geneva, Switzerland.

Zhishen J, Mengcheng T, W Jianming (1999). The determination of flavonoid contents in mulberry and their scavenging effects on superoxide radicals. Food Chem. 64:555-559.

Dietary effects of two commercial feed additives on growth performance and body composition of African catfish *Clarias gariepinus* fingerlings

A. A. Dada and B. D. Olugbemi

Department of Fisheries and Aquaculture Technology, Federal University of Technology, Akure, Ondo State, Nigeria.

An experiment was conducted to evaluate the effect of two commercial feed additives on growth, nutrient utilization and body composition of African catfish, *Clarias gariepinus* fingerlings. Two commercial feed additives Aqua booster® and Aqua pro® were added to the basal diet at 0.5 g kg^{-1} feed and fed to the fish for 56 days at 3% body weight. Fish fed on diets with commercial feed additives showed significantly improved growth performance, feed utilization and body composition compared to fish fed on the control diet. The best growth was observed in the group fed with dietary Aqua pro®. The protein efficiency ratio (2.73 ± 0.58) and specific growth rate (9.13 ± 0.60) were significantly higher in groups fed with the Aqua pro® diets. The highest protein content (67.15%) was obtained in the fish fed with dietary Aqua pro® and it was significantly higher than in all other groups ($p<0.05$). Moisture content did not differ among treatments. Ash and lipid contents were significantly affected by feed additives. The results revealed that using Aqua pro® at level of 0.5 g kg^{-1} feed was the best in terms of growth performance and feed utilization.

Key words: Feed additives, catfish, growth performance, feed utilization.

INTRODUCTION

Over the years, the total world fishery production decreased slightly and the human consumption for aquatic product increased (FAO, 2010). The reduction in capture fisheries was partly compensated for the fast growth of aquaculture industry. The need for enhanced disease resistance, feed efficiency and growth performance of cultured organisms is substantial for various sectors of this industry (EL-Haroun et al., 2006). If growth performance and feed efficiency are increased in commercial aquaculture, the costs of productions are likely to be reduced. There is large number of feed additives available to improve fish growth performance in the Nigeria markets and some of these additives used in feed mill are chemical products especially hormones and antibiotics which may cause unfavourable side effects. World Health Organization encourages using of medicinal herbs and plants to substitute or minimize the use of chemicals through the global trend to go back to nature. Attempts to use the natural materials such as medicinal plants could be widely accepted as feed additives to enhance efficiency of feed utilization and aquaculture productive performance. Recently, medicinal plants and probiotics have been reported as potential alternatives, among other feed additives, to antibiotics in aquaculture diets. Much of this interest arises from increased public awareness and banned to the use of antibiotics as growth promoters in aquaculture diets. In the last decade, some studies show the positive effects of dietary medicinal

Table 1. Ingredients and proximate composition of the experimental diets.

Ingredient	Experimental diet		
	Diet 1 (control)	Diet 2 (Aqua booster®)	Diet 3 (Aqua pro®)
Fishmeal (65% CP)	250.00	250.00	250.00
Yellow maize (9% CP)	150.00	150.00	150.00
Soybean meal	350.00	350.00	350.00
Blood meal	100.00	100.00	100.00
Cod liver oil	60.00	60.00	60.00
Vegetable oil	40.00	40.00	40.00
Vitamin/mineral premix	30.00	30.00	30.00
Binder (starch)	20.00	20.00	20.00
Aqua booster® (0.5 g/kg)	0.00	0.50	0.00
Aqua pro® (0.5 g/kg)	0.00	0.00	0.50

Vitamin-mineral premix in the diet contains, *vitamin 8,000,000 IU, vitamin D_3 1,600,000 IU, vitamin E 6,000 IU, vitamin K 2,000 mg, thiamine B_1 1,5000 mg, riboflavin B_2 4,000 mg, pyriodoxine B_6 1,5000 mg, niacin 15,000 mg, vitamin B 1210 mg, pantothenic acid 5,000 mg, folic acid 500 mg, biotin 20 mg, choline chloride 200 g, antioxidant 125 g, manganese 80 g, Zinc 50 g, iron 20 g, copper 5 g, iodine 1.2 g, selenium 200 mg, cobalt 200 mg.

Table 2. Chemical analysis (% on DM basis) of the experimental diets.

Parameter	Experimental diet		
	Diet 1 (control)	Diet 2 (Aqua booster®)	Diet 3 (Aqua pro®)
Crude protein	42.53 ± 0.72	42.58 ± 0.72	42.253 ± 0.72
Ash	7.69 ± 0.56	6.48 ± 0.56	8.28 ± 0.56
Crude lipid	8.96 ± 0.43	11.43 ± 0.43	10.03 ± 0.43
Moisture	14.33 ± 1.02	12.50 ± 1.02	16.92 ± 1.02

plants and feed additives on growth and feed utilization in fish (Turan and Akyurt, 2005; Turan, 2006; Soosean et al., 2010; Deng et al., 2011; Farahi et al., 2011; Dada and Oviawe, 2011; Prasad and Mukthiraj, 2011; Cho and Lee, 2012).

The African catfish *Clarias gariepinus* is widely cultured in Nigeria under intensive management (Adewolu et al., 2009). Increased production of fry and fingerlings with attributes of faster growth rates and higher environmental tolerance is *sine qua non* to ensuring fish food security in Nigeria. Aquaculture requires high quality feeds which should contain not only necessary nutrients but also complementary feed additives to keep organism's healthy, faster growth and environmental friendly. The main objective of this study was to investigate the effect of using two commercial feed additives (Aqua booster® and Aqua pro) on growth, feed utilisation and body com-position in African catfish (*C. gariepinus*) fingerlings.

MATERIALS AND METHODS

Formulation of experimental diets

Commercial feed additives (Aqua booster® and Aqua pro®) were obtained from a chemical store in Akure, Nigeria. 0.5 gkg^{-1} of the feed additives were mixed with a basal feed comprising standard amounts of fish meal, yellow maize, soybean meal, blood meal, fish oil, vegetable oil, vitamin premix and starch, formulated using pearson square method (Table 1). All dietary ingredients were milled to a 3 mm particle size. The ingredients were thoroughly mixed in a Hobbart A-2007 pelleting and mixing machine (Hobart Ltd, London, UK) after which the experimental diets were pelleted with a laboratory pelleting machine with a 0.9 mm die and dried at ambient temperature (27 to 30°C). Experimental feeds were stored at deep-freezer (-20°C) until used. Proximate composition of experimental diets were determined according to AOAC methods (2005): dry matter (DM) after drying in an oven at 105°C until constant weight; crude protein (N = 6.25) by Kjeldahl digestion and distillation after acid digestion; crude lipid by petroleum ether extraction in a Soxtec apparatus; ash by incineration in a muffle furnace at 550°C for 8 to 12 h (Table 2).

Experimental procedure

Nine glass aquaria tanks (2 × 2 × 1.5 m), each half-filled with water were aerated continuously using an air compressor. Ninety farm-raised *C. gariepinus* fingerling (with an average initial weight of 7.27 to 7.52 g) were acclimated to laboratory conditions for 14 days before being distributed randomly into the nine tanks (10 fish $tank^{-1}$) representing two dietary treatments (0.5 gkg^{-1} of feed additives) respectively and a control (0 gkg^{-1} of feed additive). Fish were fed at 5% of their body weight (bw) per day in three equal meals, every 5 h between 08:00 and 18:00. All fish were weighed and counted fortnightly and feeding rates were adjusted accordingly. At the end of the experimental period, the following growth and feed utilization

Table 3. Chemical composition of whole body of *C. gariepinus* fingerlings fed experimental diets.

Parameter	Experimental diet		
	Diet 1 (control)	Diet 2 (Aqua booster®)	Diet 3(Aqua pro®)
Ash	11.27±0.24[a]	16.11±0.24[c]	12.24±0.24[b]
Crude lipid	8.49±0.26[a]	9.30±0.26[ab]	9.07±0.26[ab]
Moisture	12.92±0.68[a]	13.17±0.68[a]	14.22±0.68[a]
Crude protein	61.01±0.66[a]	62.86 ± 0.66[ab]	67.15 ±0.66[c]
NFE	0.19±0.20[a]	0.42±0.20[a]	1.63±0.20[b]

Mean within the row with different superscripts are significantly different (p<0.05).

indices were calculated: weight gain (WG), specific growth rate (SGR), food conversion ratio (FCR), feed efficiency (FE) and protein efficiency ratio (PER) using the following formulae:

WG = Final average weight (g) − initial average weight (g);

SGR (% d – 1) = 100 × (ln *Wt* − ln *W0*)/*t*

Where *Wt* and *W0* represent final and initial body weights of fish, respectively, and *t* represents the duration of the feeding trial;

FCR = Dry weight of feed (g) / wet weight gain by fish (g); and
PER = wet weight gain by fish (g) / protein intake (g)

Where protein intake (g) = Protein (%) in feed × total weight (g) of diet consumed / 100

Water temperature, pH and dissolved oxygen concentration were routinely monitored in all tanks. At the beginning and end of the feeding trial, pooled samples of 15 fingerlings were analyzed for carcass composition according to AOAC methods (2005): dry matter (DM) after drying in an oven at 105°C until constant weight; crude protein (N = 6.25) by Kjeldahl digestion and distillation after acid digestion; crude lipid by petroleum ether extraction in a Soxtec apparatus; ash by incineration in a muffle furnace at 550°C for 8 to 12 h (Table 3).

Statistical analyses

Analysis of variance (ANOVA) was used at 95% significance level to test for significant differences between the various treatment means obtained for the growth, feed utilization and carcass composition using the SAS package. Tukeys multiple range test was used to determine which pairs of the treatment means differed significantly.

RESULTS

Mean water quality parameters during the experiment were: dissolved oxygen 6.40 ± 0.19 mg l^{-1}, pH 7.24 ± 0.13 and temperature 27.10 ± 0.33°C. There were improvements in the growth responses of fish fed on dietary feed additives. The fastest growth responses were obtained in the fish fed on dietary Aqua pro® (9.13 ± 0.60) while the slowest growth was obtained in the fish fed on the control diet (8.01 ± 0.60) (Table 4). There were significant differences in growth performance across the different dietary feed additives. There were greater improvements in the feed conversion ratio (FCR) of fish fed on dietary feed additive than the control fish. The average FCRs were 0.86 and 0.83 for diets Aqua booster® and Aqua pro®, respectively. The PER was 2.53 and 2.73 for the fish fed diets Aqua booster® and Aqua pro®, respectively. Fish fed on dietary feed additives had a significantly ($p < 0.05$) higher protein content than fish fed the control diet. The body composition values are given in Table 3.

DISCUSSION

Water quality parameters were not significantly different between treatments and were within the recommended ranges for the culture of *C. gariepinus* (Viveen et al., 1986). The results suggest that dietary feed additive promoted the growth of *C. gariepinus* fingerlings. These results showed that feed additive enhances nutrient utilization, which is reflected in improved weight gain, FCR, PER, PE and SGR. Generally, better feed conversion ratio values were obtained in all treatments, but the poorest occurred in control (Table 4). There were significant differences in the FCR among the treatments ($p < 0.05$). Aqua superliv® in diets also promoted growth and feed conversion efficiency in Nile tilapia fingerlings (Dada, 2012). Similar results were reported for using two different commercial feed additives for Nile tilapia fingerlings (Abdelhamid and Mohammed, 2008). EL-Haroun (2007) also reported that African catfish fingerlings fed on diets supplemented by commercial feed additive Biogen® exhibited faster growth than those fed with the control diet. Similarly, in catfish *C. gariepinus* (Turan and Akyurt, 2005), tilapia *Oreochromis niloticus* (Khattab et al., 2004; Felicitta et al., 2013), olive flounder *Paralichthys olivaceus* (Cho and Lee, 2012) and shrimp *Peneaus indicus* (Olmedo Sanchez et al., 2009) feed additives in diets promoted growth and feed efficiency. From a proximate composition point of view, feed additives increased the level of protein in *C. gariepinus*.

The body composition values obtained in this study were similar to those reported by Diab et al. (2002), Lara-Flores et al. (2003) and Abdelhamid and Mohamed (2008) using commercial feed additives. Interestingly, we

Table 4. Mean growth performance and feed utilisation of *C. gariepinus* fingerlings fed experimental diets for 56 days.

Parameter	Experimental diet		
	Diet 1 (control)	Diet 2 (Aqua booster®)	Diet 3 (Aquapro®)
Initial weight (g)	7.27±0.93^{a}	7.52±0.93^{a}	7.43±0.93^{a}
Final weight (g)	22.58±2.19^{a}	25.20±2.19^{b}	26.79±2.19^{b}
Weight gained (g)	15.31±2.16^{a}	17.68±2.16^{b}	19.36±2.16^{b}
FCR	0.86 ± 0.12^{ab}	0.83 ± 0.12^{a}	0.74 ± 0.12^{b}
Survival (%)	66.67	76.67	90.00
SGR (% day^{-1})	8.01±0.60^{a}	8.48±0.60^{a}	9.13±0.60^{b}
PER	2.14±0.58^{a}	2.53±0.58^{ab}	2.73 ± 0.58^{b}

Means in a given column with the same superscript letter were not significantly different at $p < 0.05$. SGR, Specific growth rate; PER, protein efficiency ratio; FCR, feed conversion ratio.

found that the survival of *C. gariepinus* can be improved by feed additives supplementation. Therefore, we expect that this result will stimulate a series of studies on the utilization of these two commercial feed additives in diets for fishes.

Conclusion

This study established the efficacy of Aqua booster® and Aqua pro® feed additives as a growth promoter in *C. gariepinus* fingerlings and fish farmers should be encouraged to supplement these feed additives in fish diet. Future research should focus on the improvement of rearing technologies for different species of fish reared using feed additive as a feed supplement.

REFERENCES

Abdelhamid E, Mohamed KA (2008). Effect of using probiotics as growth promoters in commercial diets for monosex Nile tilapia (*Oreochromis niloticus*) fingerlings. In: Elghobashy H, Fltzsimmons K, Dlab AS (eds), Proceedings of the Eighth International Smposium on Tilapia in Aquaculture, 12–14 October 2008, Cairo, Egypt. pp. 241–252.

.Adewolu MA, Akintola SL, Akinwunmi OO (2009). Growth performance and survival of hybrid African catfish larvae (*Clarias gariepinus X Heterobranchus bidorsalis*) fed different diets. Zoologists 7(1) 45-51.

AOAC International Official Methods of Analysis (2005). 18th ed. AOAC International, Gaithersburg, MD.

Cho SH, Lee S (2012). Onion powder in the diet of the olive flounder, *Paralichthys olivaceus*: Effects on the growth, body composition, and lysozyme activity. J. World Aquac. Soc. 43(1): 30 – 38.

Dada AA (2012). Effects of herbal growth promoter feed additive in fish meal on the performance of nile tilapia (*Oreochromis niloticus* L.). Egypt. Acad. J. Biolog. Sci. 4(1): 111 – 117.

Dada AA, Oviawe NE (2011). The use of bitter kola *Garcinia kola* dry seed powder as a natural growth- promoting agent for African sharptooth catfish *Clarias gariepinus* fingerlings. Afr. J. Aquat. Sci. 36(1): 97 – 100.

Deng J, An Q, Bi B, Wang Q, Kong bL, Tao L Zhang X (2011). Effect of ethanolic extract of proplis on growth performance and plasma biochemical parameters of rainbow trout (*Oncorhynchus mykiss*). Fish Physiol. Biochem. 37(4): 959 – 967.

Diab AS, El-Nagar OG, Abd-El-Hady MY (2002). Evaluation of *Nigella sativa* L. (black seeds, Baraka), *Allium sativum* (garlic) and Biogen as feed additives on growth performance and immunostimulants of *Oreochromis niloticus* fingerlings. Suez Canal Vet. Med. J. 2: 745–753.

EL-Haroun ER (2007). Improved growth rate and feed utilization in farmed African catfish Clarias catfish Clarias gariepinus (Burchell 1822) through a growth promoter Biogen® supplementation. J. Fish. Aquat. Sci. 2(5):319-327.

EL-Haroun ER, Ma A, Goda S, Kabir Chowdhury MA (2006). Effect of dietary probiotic Biogens supplementation as growth promoter on growth performance and feed utilization of Nile tilapia, *Oreochromis niloticus* (L.). Aquac. Res. 37:1473-1480.

FAO (2010). The State of World Fisheries and Aquaculture. pp.197.

Farahi A, Kasiri M, Sudagar M, Soleimani Irael M, Zorriehzahra SMJ (2011). Effec t of dietary supplementation of Melissa officinalis and Aloe vera on hematological traits, lipid oxidation of carcass and performance in rainbow trout (*Oncorhynchus mykiss*). Online J. Anim. Feed Res. 2 (1): 1-5.

Felicitta J, Arti Manju R, Ronald J,Sakthika T, Nagarajan R, Chelladurai G (2013). Effect of different concentrations of somephytoadditives (Allium sativa and Allium cepa) On growth, survival and hematological parameters in tilapia (Oreochromis mossambicus) juveniles. The Israeli Journal of Aquaculture- Bamidgeh, 64: 8p. (Accepted paper).

Khattab YAE, Shalaby AME, Sharaf Saffa M, El-Marakby H, RiziAlla EH (2004). The physiological changes and growth performance of the nile tilapia Oreochromis niloticus after feeding with Biogen® as growth promoter. Egypt. J. Aquat. Biol. Fish. 8(2): 145 – 158.

Lara-Flores M, Olvera-Novoa MA, Guzman-Mendez BE, López-Madrid W (2003). Use of the bacteria *Streptococcus faecium* and *Lactobacillus acidophilus*, and the yeast *Saccharomyces cerevisiae* as growth promoters in Nile tilapia (*Oreochromis niloticus*). Aquaculture 216: 193–201.

Olmedo Sanchez JA, Curiel Flores A, Orozco JR (2009). The effect of a herbal growth promoter feed additive on shrimp performance. Res. J. Biol. Sci. 4: 1022- 1024.

Prasad G, Mukthiraj S (2011). Effect of methanolic extract of Andrgraphis paniculata (Nees) on growth and haematology of Oreochromis mossambicus (Peters). World J. Fish Mar. Sci. 3(6): 473 – 479.

Soosean C, Marimuthu K, Sudhakaran S, Xavier R (2010). Effect of mangosteen (*Garcinia mangostana* L.) extracts as a feed additive on growth and hematological parameters of African catfish (*Clarias gariepinus*) fingerlings. Eur. Rev. Med. Pharmacol. Sci. 14: 605 – 611.

Turan F (2006). Improvement of growth performance in Tilapia (*Oreochromis niloticus* Linnaeus) by supplementation of red clover (*Trifolium pretense*) in diets. Isr. J. Aquac. - Bamidgeh 58(1): 34–38.

Turan F, Akyurt I (2005). Effects of red clover extract on growth performance and body composition of African catfish *Clarias gariepinus* (Burchell, 1822). Fish. Sci. 71: 618 – 620.

Viveen WJAR, Richter CJJ, Van Oordt PG, Janssen JAL, Huisman EA (1986). Practical manual for the culture of the African catfish, *Clarias gariepinus*. The Hague, Netherlands: Section for Research and Technology, Agricultural University of Wageningen.

Production and evaluation of specialty bread from sprouted mixed-grains

Onyeka E. U. and Obeleagu O. S.

Federal University of Technology, Owerri, Imo State, Nigeria.

Specialty bread was made from sprouted mixed-grains using adaptable technology. The grains were sprouted at atmospheric conditions (30°C, 75% RH) over a period of 6, 12, 18 and 24 h. Paste obtained from the sprouts was mixed at different ratio together with other ingredients to obtain five bread samples. Nutritional, functional and sensory properties of the resulting loaves were carried out using standard analytical methods. Conventional bread served as a control. Sprouting improved ($p<0.05$) the nutritional and functional properties of mixed-grain bread. Dietary fiber was 5.36 and 20.91 g/100 g^{-1}, acidity 1.23 and 5.73% and protein energy 29.13 and 108.72 KJg^{-1} for control and 24 h sprouted mixed-grain bread, respectively. The developed bread may qualify as a health food judging from the data generated in this research. The breads were equally acceptable to the panel judges. Twenty-four hour sprouting is considered best for production of mixed-grain bread.

Key words: Bread, consumer, mixed grains, sprouting.

INTRODUCTION

People all over the world depend on bread for their regular and convenient source of energy and other nutrients. In fact, bread has become a food staple in many countries of the world including Nigeria. Judging from the culinary role of bread, it is necessary that bread is not only safe but nutrient-dense. Unfortunately conventional bread in Nigeria and some other parts of the world is low in important nutrients such as dietary fiber and protein (Jideani and Onwubali, 2009; Edema et al., 2005). A high proportion of bread is baked with refined (white) flour and using a technology that does not permit full generation of metabolites in the dough (AOAC, 1990; Ostman et al., 2005). Ancient breads were prepared with coarse milled flour and using a sour-dough process (Ostman et al., 2005). Increased incidence of some diseases including diabetes, cancer, celiac disease, intestinal dysbiosis, constipation, heart-burn, decreased pancreatic function and sensitized intestinal tract have been associated with diets consisting of refined foods such as white bread. There is enough evidence that these diseases result directly or indirectly from the consumption of wheat (gluten) in bread (Moroni et al., 2009; Dewettinck et al., 2008; Crockett et al., 2011; http://www.foodforlife.com). Some people avoid bread in order to remain healthy, while some uninformed individuals continue to suffer diverse ailments associated with bread meal without knowing the root cause of their ill-health.

Over the years, man has searched for healthy bread and has come up with many types of bread namely white bread, brown bread, whole-meal bread, whole- germ bread, granary bread and composite-flour breads.

Table 1. Sprouting scheme for selected grains.

Total time of sprouting (h)	Steps involved in sprouting
6	No sprouting, just soaked for 6 h and grind
12	6 h soaking + 30 min initial air-rest + 5 h soaking + 30 min final air-rest
18	6 h soaking + 1 h initial air-rest + 5 h soaking + 5 h soaking + 1 h final air-rest
24	6 h soaking + 1½ h initial air-rest + 5 h soaking + 5 h soaking + 5 h soaking + 1½ h final air-rest

One of the latest bread products is the sprouted grain bread that is fondly called *Ezekiel* bread (Brenna and Cleary, 2005). *Ezekiel* bread is made from mixtures of sprouted grains and legumes (Brenna and Cleary, 2005) and is claimed to be richer in protein and vitamins than the conventional white bread (Kent, 1984). As reported by Shipard (2005), bread from sprouted mixed grains could contain all the nine essential amino acids and is low in fat, has no trans-fat or cholesterol and is relatively low in sodium. Shipard (2005) claimed that sprouting of grains will take care of the anti-nutritional factors and also circumvent the use of additives and chemicals to achieve improved nutrient content. This according to him is because during germination and sprouting of grains and seeds, many enzyme inhibitors are effectively neutralized, whilst at the same time, the activity of beneficial plant digestive enzymes is greatly enhanced. The digestible energy, bio-available vitamins, minerals, amino acids, proteins and phytochemicals necessary for a germinating plant to grow are also essential for human health (Moroni et al., 2009; http://www.foodforlife.com). The objectives of this present research are to: (a) investigate the potential of sprouting locally available grains/legume for use in preparation of specialty bread (b) to determine the functional and nutritional content of the sprouted mixed-grain bread and (c) to determine the consumer acceptability of the sprouted mixed-grain bread.

MATERIALS AND METHODS

All the consumables excluding chemicals were bought from a local market in Owerri, Imo State, Nigeria. The grain samples were identified at the Crop Science Laboratory, Federal University of Technology, Owerri, Nigeria. The grains included wheat (*Triticum aestivum*), maize (*Zea mays*), sorghum (*Sorghum vulgare*), millet (*Eleusine caracana*), lentil (*Digitaria excelis staph)*, soybean (*Glycine max*) and sesame seed (Sesamum *indiacum L).* They were chosen from locally available cereals/legumes with full consideration to ease of sprouting. Chemicals were obtained from Food Science and Technology Laboratory, Federal University of Technology, Owerri, Nigeria.

Sprout production

The method of Eneje and Orji (2002) with slight modifications was used for the sprouting of the grains. The grains were sprouted under ambient conditions (30°C, 75% RH) for 6, 12, 18 and 24 h before milling. The period of sprouting was decided following a trial sprouting experiment with the various grains. Half a kilogram of each grain (maize, sorghum, millet, wheat, sesame seed, lentil and soybean) was used for each sprouting period. The scheme for the sprouting experiment is illustrated in Table 1. The grains were sorted and washed with ethanol/water (70/30 v/v) to remove stones and dirt before sprouting. Then, the seeds were first soaked in water (50/50 w/w) for 30 min to help bring out the seeds out of quiescence (Glennie et al., 1983). Soaking was followed by draining and rinsing at regular intervals until the seeds sprouted. The sprouted grains were milled using a manual attrition milling machine (corona corn mill, manufactured by YCIA, S.A. Colombia). Samples of the sprouted grains are shown in Plate 1a.

Product formulation

The combinations of the sprouted grain dough were based on nutrient composition of the grains. The quantity of other ingredients namely honey (20 ml), salt (0.01), eggwhite (22 ml), yeast (0.3 g) and water (10 ml) per 100 g of the dough remained constant. Production panel judges (PPJ) were used for the trial testing of samples to select the best four formulations as shown in Table 2.

Production of sprouted mixed-grain bread

The production of sprouted mixed-grain bread was according to the Biblical Ezekiel, chapter four verse nine, cited from Brenna and Cleary (2005). One hundred grams of paste was weighed out from the four batches of mixed paste generated from the sprouting scheme, and was mixed with the other ingredients. Yeast was dissolved in 10 ml of water (40°C) and 20 ml of honey, and was allowed for 10 min at 45°C for yeast reaction to take place. After 10 min, the dissolved yeast, honey, salt and egg were mixed with the paste and stirred continuously for 5 min. The paste was allowed to ferment for another 10 min at 45°C and to proof before it was placed in an aluminium baking pan and was baked at 250 to 300°C for 45 min. A stove heated sand-box was used as the baking oven. The pictorial view of the whole process is shown in Plate 1a to d.

Analysis of bread samples

Nutritional composition

The determination of proximate composition (moisture, carbohydrate, crude protein, fat and ash content) of the bread samples was carried out according to the method of Association of Official Analytical Chemists (1990). Energy was calculated using Atwater formula (FAO, 1973) using values of 9, 4, and 3.8 $kcalg^{-1}$ for fat, protein and carbohydrate, respectively. pH and acidity of the samples were determined using the method described by AOAC

Plate 1. A to d: Pictorial view of Ezekiel bread process methods.

(1995). The total dietary fiber was determined according to the method of the AOAC (1995) with slight modifications. One gram of each sample (dry basis) was put into a beaker, and 10 ml of distilled water was added and the mixture was stirred. The beaker with the sample was then put into a water bath containing boiling water and was gelatinized over a hot plate while stirring continuously. When the sample had gelatinized, the pH of the gelatinized sample was adjusted to 6.0. Termamyl was added and incubated at 100°C for 30 min. This was followed by protease incubation and lastly by amyloglucosidase enzyme incubation at 60°C for 30 min. At the end of amyloglucosidase incubation, the sample was precipitated with four volumes of ethanol and acetone. The sample was then dried in a drying oven at 100°C. Total dietary fiber was calculated as:

$$\text{Total dietary fiber} = \frac{\text{Weight of residue after drying}}{\text{Weight of sample}} \times \frac{100}{1}$$

Determination of functional properties

Water absorption capacity (WAC) and foaming capacity (FC) were determined by a combination of the AOAC (1990) methods. For WAC, 2 g of each dough sample was dispersed in 20 ml of distilled water, stirred using a magnetic stirrer at 1000 rpm and centrifuged at 4600 rpm for 30 min. The supernatant was carefully decanted. The WAC of the protein was expressed as percentage of water bound. For FC determination, 5 g of dough samples was dispersed in 20 ml of distilled water, homogenized using a metabolic shaker for 30 min and centrifuged (Hettich Universal Bayern, Germany) at 3000 rpm for 15 min at 30°C. The supernatant obtained was mixed for 5 min using a magnetic stirrer at 1200 rpm, poured into a 100 ml measuring cylinder and its volume was immediately noted. Foaming capacity (FC) was then expressed using the equation:

$$FC = \frac{\text{Vol. of foam after blending - Vol. of foam befiire blending}}{\text{Vol. of foam before blending}} \times \frac{100}{1}$$

Sensory evaluation

Sensory evaluation of the bread samples was carried out using 20-member panel selected randomly from the university community. The panelists were instructed to rate the samples in terms of aroma, texture, taste, crust colour, crumb colour and overall acceptability based on 9-point hedonic scale ranging from 9 (like extremely) to 1 (dislike extremely). The panelists worked under controlled condition to avoid biased results.

Experimental design and statistical analysis

This research is a two factor experiment in a randomized complete block design (RCBD) with 4 sprouting periods and 5 grain proportions as sources of variation. Therefore, the experimental design is thus, 4 x 5 factorial designs.

Table 2. Formulations for preparation of sprouted mixed-grain bread.

Product code	Grains ($g\ 100\ g^{-1}$)							Total weight
	Maize	Soybean	Lentil	Sesame	Sorghum	Millet	Wheat	
PD1	0	0	0	0	0	0	100	100
PD2	20	10	10	10	10	20	20	100
PD3	5	10	5	10	5	5	60	100
PD4	10	10	10	10	10	10	40	100
PD5	Control (unsprouted commercial white wheat bread)							

RESULTS

Nutritional composition of sprouted mixed-grain bread

The effect of sprouting on proximate composition of mixed grain bread is shown in Table 3. There was a steady increase in protein content, protein energy and dietary fiber content of bread samples as sprouting period increased while fat, energy value and carbohydrate content decreased. Protein increased from 6.42 to 16.55 gg^{-1}, protein energy from 43.76 to 94.03 KJg^{-1}, and total dietary fiber from 8.33 to 21.65 $g\ 100\ g^{-1}$ for unsprouted and 24 h sprouted mixed-grain bread, respectively. There were variations in proximate composition of the bread samples as a result of the grain proportion (Table 4). Expectedly, samples with low wheat inclusion had low carbohydrate content. Generally, control sample had more carbohydrate content but low protein energy and dietary fiber than the specialty bread. Bread samples made from sprouted wheat grain (PD1) had the lowest fat content of 6.01 g/100 g, while the commercial loaf had 9.53 g/100 g. Higher ($p = 0.05$) moisture content (14.15 to 17.01 g/100 g) occurred in sprouted mixed-grain bread samples when compared with the commercial loaf (9.49 g/100 g). Interaction of sprouting period and formulation showed that 24 h sprouting and uniform (even) distribution (except for wheat) of the grains gives bread better parameters (Table 5). This explains why sample PD4 had better nutritional parameters than others.

pH and acidity of bread samples

The effects of sprouting period and grain proportions on pH and acidity of sprouted specialty bread are shown in Table 6. The average acidity of bread increased ($P < 0.05$) from 3.59 to 4.67% as a result of sprouting, while the values for pH decreased from 5.44 to 5.13.

Functional properties of mixed grain bread

Water binding capacity (WBC) decreased with sprouting period, while foaming capacity increased (Table 3). Commercial bread sample had higher water absorption capacity (WAC) when compared with the specialty breads, but their foaming capacity measured up with the specialty bread samples.

Product acceptability

The result of one-way analysis of variance apparently showed that sprouting up to 24 h produced bread with better sensory qualities (Table 7). Bread samples with closer distribution of the grain exhibited better sensory parameters than others. It was observed that mixed-grain bread ($PD4^{24}$) sprouted for 24 h with grain proportions of maize 10 g; soybean 10 g; lentil 10 g; sesame 10 g; sorghum 10 g; millet 10 g and wheat 40 g compared more favourably with the commercial bread in all the sensory parameters tested.

DISCUSSION

The increase in protein content following grain sprouting could be as a result of solubilization of nitrogen part of protein/amino acids, making them more available for digestion. Increase in dietary fiber contents as observed in this work could be an indication of the hydrolysis of the complex starch molecules during sprouting and the decrease in fat content could be as a result of hydrolysis of triacylglycerols to glycerol and constituent fatty acids, and part of the lipids contained inside the germ of the grains being used for germination. The observed decrease in carbohydrate content and energy value of specialty bread is in agreement with previous report of Akpapunam and Dedeh (1995) that at 24 h sprouting, the growth process and metabolic activities suspended at maturation, dehydration and storage of the grain resumed under favourable conditions. These observations also tally with the report of Chavan and Kadam (1989) who stated that the desirable nutritional changes (such as increase in protein and dietary fiber content) that occur during sprouting are mainly due to the breakdown of complex compounds into simpler form. Okun et al. (1980) found that in sprouted wheat at 30°C the free amino acids did not change significantly for two days of sprouting but significant increases in free amino acids were noticed at 40°C for 24 to 48 h of sprouting. Protein synthesis has

Table 3. Effect of sprouting period on nutrient and functional properties of mixed grain bread[a]

SP	Proximate (g 100 g^{-1}DM)					TDF	Energy (KJg^{-1} DM)		PEP	Functional properties	
	Carb	Protein	Fat	Ash	Moisture	(g 100 g^{-1})	Total	Protein		WBC (g g^{-1})	FC (g g^{-1})
6 h	60.17±13.01	6.42±1.25	9.01±2.22	20.50±4.22	13.48±4.41	8.33±4.31	1452.04	43.76	0.0303	4.99 ±0.55	2.81±1.08
12 h	44.51±10.71	14.10±2.21	8.18±1.28	20.82±1.22	16.57 ±5.21	15.07±4.65	1323.50	68.30	0.0532	4.49± 0.51	2.96±1.07
18 h	39.04±12.00	15.51±2.05	7.16±2.27	21.62±1.32	15.01 ±4.30	18.16±5.52	1210.49	82.77	0.0721	3.19± 0.52	4.81±0.9
24 h	37.17±13.14	16.55±1.15	2.85±3.21	15.78±2.21	17.87 ±5.47	21.65±7.09	1148.62	94.03	0.0879	2.43± 0.99	4.91±0.6
Control	67.67±7.21	5.44±0.5	9.55±0.22	11.13±0.51	9.49±1.51	5.36±0.4	1452.79	24.93	0.0012	4.11± 0.05	4.19±0.11
LSD	1.03	0.09	0.35	0.21	1.04	0.57	121.12	1.96	0.024	1.21	1.32

[a]SP = sprouting period; Carb = carbohydrate; TDF = total dietary fiber; Control, commercial white bread; WBC, water binding capacity; FC, foaming capacity; PEP, proportion of total energy derived from protein; LSD, least significant difference; DM, dry matter.

Table 4. Effect of grain formulation on nutrient and functional properties of mixed-grain bread[a].

Products	Proximate (g 100 g^{-1} DM)					TDF	Energy (KJg^{-1}DM)		PEP	Functional properties	
	Carb	Protein	Fat	Ash	Moisture	(g 100 g^{-1})	Total	Protein		WBC (g g^{-1})	FC (g g^{-1})
PD1	41.99±1.9	14.67±1.5	6.01±3.1	20.48±2.1	17.01±1.5	16.69±3.7	1226.03	81.33	0.0683	3.46 ± 1.14	4.70± 0.97
PD2	38.76±1.2	15.66±1.1	7.33±4.1	18.93±1.4	16.16±1.5	18.58±4.5	1229.91	84.30	0.0724	3.29 ± 1.51	4.63± 1.49
PD3	40.37±9.5	15.07±1.3	6.83±4.7	18.72±3.1	16.06±1.5	19.53±1.5	1208.65	81.33	0.0695	3.22± 1.56	4.13± 1.99
PD4	39.93±3.5	16.17±1.4	6.22±4.5	18.47±3.5	16.15±1.5	18.36±3.5	1214.67	87.29	0.0753	2.76± 1.05	3.49± 1.48
Control	67.67±7.2	5.44±0.5	9.55±0.2	11.13±0.5	9.49±1.5	5.36±0.4	1539.98	29.13	0.0189	4.11 ± 0.05	4.19± 0.01
LSD	1.03	0.09	0.35	0.21	1.04	0.57	135.28	18.63	0.022	1.21	1.71

*SP = Sprouting period; Carb= carbohydrate; TDF = total dietary fiber; WBC, water binding capacity; FC, foaming capacity; PEP, proportion of total energy derived from protein; LSD, least significant difference; DM, dry matter; Control, commercial white bread. Decoding of product code is as shown in Table 2.

been proven to be one of the essential processes associated with sprouting (Adeyemo et al., 1992; Wang and Fields, 1978). An increase in proteolytic activity during sprouting is desirable for nutritional improvement of cereals because it leads to hydrolysis of prolamins and liberates amino acids such as glutamic, and conversion of storage proteins into albumins and globulins, and also proline are converted to limiting amino acids such as lysine. Cuddeford (1989) stated that crude fibre, a major constituent of cell walls, increases both in percentage and real terms, with the synthesis of structural carbohydrates such as cellulose and hemicelluloses. The decrease in ash content only on the 24 h sprouting might be as a result of leaching out of minerals into the soaking medium (Shipard, 2005; Cuddeford, 1989) otherwise ash content increased with sprouting. Shipard (2005) claims that when seeds are sprouted, minerals chelate or merge with protein in a way that increases their function, and sprouts are considered a significant source (plant) of digestive enzymes. Solubilization of protein, increase in dietary fiber and decrease in fat and carbohydrate content of *Ezekiel* bread may explain the increased digestibility of the bread as reported by Brenna and Cleary (2005) and Dewettinck et al. (2008). According to Akpapunam and Dedeh (1995), improvements in amino acid composition, B-group vitamins, sugars, protein and starch digestibility and decrease in phytates and protease inhibitors are the metabolic effects of sprouting process. The result of the proximate

Table 5. Interaction of sprouting and formulation on nutrient and functional properties of mixed-grain bread[a].

SP (h)	PDT	Proximate (g 100 g^{-1} DM)					TDF	Energy (KJg^{-1} DM)		PEP	Functional properties	
		Carb	Protein	Fat	Ash	Moisture	(g 100 g^{-1})	Total	Protein		WBC (gg^{-1})	FC (gg^{-1})
6	PD1	60.24	9.05	9.85	20.44	12.04	7.15	1478.73	42.77	0.0289	4.91	3.67
6	PD2	55.21	10.18	9.91	21.32	11.22	8.33	1419.98	50.10	0.0353	4.83	2.66
6	PD3	56.88	9.45	9.45	19.45	13.52	7.45	1416.99	46.61	0.0329	4.87	2.51
6	PD4	50.14	10.11	11.77	20.03	14.01	10.05	1408.25	50.17	0.0356	4.71	3.51
Mean		55.62[a]	9.70	10.25[a]	20.31[a]	12.70	8.25	1430.99[a]	47.41	0.0332	4.83[a]	2.59
12	PD1	43.63	13.06	9.71	21.17	18.98	12.58	1268.46	72.02	0.0568	4.76	3.55
12	PD2	38.64	15.17	11.90	18.69	13.92	13.65	1337.43	79.57	0.0595	4.92	2.98
12	PD3	37.53	13.43	9.40	19.90	15.65	19.14	1170.61	81.40	0.0695	4.90	1.92
12	PD4	36.35	14.75	12.19	20.33	17.78	14.94	1301.02	79.38	0.0610	3.80	1.85
Mean		39.04[b]	14.10	10.8[a]0	20.0[a]2	16.58[a]	15.08	1269.38	78.09	0.0617	4.60[a]	2.58
18	PD1	39.82	14.44	5.80	22.46	16.88	16.38	1112.24	90.68	0.0815	3.05	4.50
18	PD2	38.60	14.78	7.64	20.68	11.93	18.07	1159.22	89.06	0.0768	3.01	5.90
18	PD3	36.89	16.88	10.52	21.89	16.08	20.79	1155.82	102.04	0.0883	2.98	5.80
18	PD4	35.27	15.95	4.68	21.48	15.17	17.41	1085.15	102.97	0.0949	2.80	3.92
Mean		37.65	15.51	7.16	21.63[a]	15.02	18.16	1128.11	96.19	0.0854	2.96	5.03 [a]
24	PD1	40.11	16.51	2.46	17.82	19.05	21.10	1044.71	110.66	0.1059	2.58	5.76
24	PD2	37.02	15.03	2.56	17.42	17.60	17.41	1003.02	118.47	0.1181	1.96	5.01
24	PD3	37.22	17.88	2.59	14.36	16.88	21.10	1166.44	89.13	0.0764	1.80	4.69
24	PD4	34.51	17.79	1.79	13.52	17.98	24.01	1064.28	116.63	0.1096	1.70	4.72
Mean		37.22	16.80[a]	2.35	15.78	17.88[a]	20.91[a]	1069.61	108.72[a]	0.1025[a]	2.01	5.05[a]
Commercial		67.67	5.44	9.55	11.13	9.49	5.36	1539.98	29.13	0.0189	4.10	4.19

[a]SP = Sprouting period; ; Carb carbohydrate; TDF = total dietary fiber; WBC, water binding capacity; FC, foaming capacity; PEP, proportion of total energy derived from protein; Control, commercial white bread. Decoding of product code is as shown on Table 2.

composition of the bread samples implies that the commercial bread will offer more calories but fewer nutrients. High dietary fibre is considered an advantage since dietary fibre reduces bowel problems including constipation, which is basic to numerous diseases such as appendicitis, diabetes, large bowel cancer, obesity and gallstone (Moroni et al., 2009; Dewettinck et al., 2008; Onyeka, 2007). The dietary status of the conventional bread as observed in this research is alarming, and calls for the attention of health stakeholders. This is particularly important as bread has become one of the staple foods in Nigeria and for the fact that low intake of dietary fiber is associated with a number of ailments (Moroni et al., 2009; Onyeka, 2007; Montonen et al., 2003; FAO/WHO, 1997). Since sprouted grain bread had low fat content, an increase in the fat content of the dough through the use of grains rich in oil may be considered. The commercial bread had high fat content because the manufactures normally use shortening as part of the major ingredients. The high residual moisture in the specialty bread could explain the short shelf-life (≤24 h) noticed in the present specialty bread samples. According to Kuye and Sanni (1999), high residual moisture content is a

Table 6. Interaction of sprouting and formulation on pH and acidity of mixed grain bread.

Products	Period of sprouting in hours							
	6 h		12 h		18 h		24 h	
	pH	Acidity	pH	Acidity	pH	Acidity	pH	Acidity
PD1	5.18	4.01	5.01	4.64	5.33	5.18	4.95	5.43
PD2	5.15	4.12	5.15	4.72	4.97	5.19	4.62	5.46
PD3	5.00	4.24	4.99	4.84	4.89	5.29	4.74	5.54
PD4	5.04	4.41	4.84	4.91	4.71	5.30	4.56	5.73
Control	6.85	1.21	6.85	1.21	6.85	1.21	6.85	1.21
Mean	5.44	3.59	5.37	4.06	5.35	4.43	5.14	4.67

LSD ($p<0.05$); pH = 0.21; acidity = 0.33. Decoding of product code is as shown in Table 2.

Table 7. Comparative sensory evaluation scores (1 to 9 scale) for mixed-grain specialty bread and white wheat bread (control)[a].

Parameter	Bread samples				
	Control	PD1[24]	PD2[24]	PD3[24]	PD4[24]
Aroma	$8.50^{a}\pm0.60$	$7.20^{ab}\pm0.02$	$6.45^{b}\pm0.75$	$6.80^{b}\pm0.89$	$7.55^{ab}\pm0.60$
Texture	$8.50^{a}\pm0.68$	$6.80^{b}\pm0.83$	$6.70^{b}\pm0.57$	$6.85^{b}\pm0.81$	$7.45^{ab}\pm0.60$
Taste	$8.45^{a}\pm0.51$	$6.75^{b}\pm0.78$	$6.45^{b}\pm0.68$	$6.90^{ab}\pm0.78$	$7.45^{ab}\pm0.60$
Crust colour	$8.00^{a}\pm0.50$	$6.40^{b}\pm0.59$	$6.85^{ab}\pm0.81$	$6.95^{ab}\pm0.68$	$7.20^{ab}\pm0.69$
Crum colour	$8.00^{a}\pm0.50$	$6.90^{ab}\pm0.91$	$7.05^{ab}\pm0.82$	$7.25^{ab}\pm0.71$	$8.50^{a}\pm0.51$
Acceptability	$8.00^{a}\pm0.50$	$6.00^{c}\pm0.72$	$6.70^{b}\pm0.82$	$7.15^{ba}\pm0.47$	$7.40^{ab}\pm0.50$

[a]Mean scores with the same letter in a row do not differ significantly ($p \leq 0.05$) from each other. Only bread from 24 h sprouted grains were used for sensory evaluation because production panel judges (PPJ) scored them higher than others; Control, commercial white bread. Decoding of product code is as shown in Table 2.

disadvantage in that microbial proliferation is enhanced and storage life is reduced. However, moisture content alone cannot be responsible for the relatively short shelf-life of the samples. The activities of the inherent enzymes could also be part of the reason.

The results for pH and acidity obtained in this research are in agreement with the finding of Chavan and Kadam (1989) who stated that increase in sprouting period decreases pH of bread. Unsprouted (low acid) wheat is among a small number of foods that contain measurable amounts of oxalate and when oxalates become too concentrated in body fluids, they crystallize, complex with calcium, and cause health problems (Chandrasheka and Sadyanaranyana, 2006).

Thus, commercial bread from unsprouted wheat which has low acidity (1.21%) and high pH (6.85) may not be good for excellent body health. As reported by Dewettinck et al. (2008), sprouted grains become more alkaline in the process of sprouting, changing it from an acid forming food, pre soaked, to an alkaline forming food source, post soaked. Increase in acidity (organic acid) is an advantage because such foods are alkaline forming which help the body use less energy in its never ending battle to maintain acid-base (pH) balance (Brenna and Cleary, 2005; Ostman et al., 2005). Moreover, foods high in organic acids tend to offer good glycemic response and control of satiety (Ostman et al., 2005; Yoom et al., 1983).

The decrease in WBC following sprouting is because of the disappearance of starch due to hydrolysis. The increased foaming capacity in the specialty bread samples may be attributed to the increased protein content especially albumin and globulins during grain sprouting (Okaka and Potter, 1979; Chavan and Kadam, 1989; Fennema's Food Chemistry, 2008). The foaming capacity observed in the commercial breads could be because the breads were prepared using chemical emulsifiers. The functional properties of the specialty bread samples might have been enhanced by the egg (22 ml) and honey (20 ml) added during processing.

In the findings of Brenna and Cleary (2005), an increase in sprouting period, from 12 to 48 h, improved the colour, taste, aroma and texture of bread. A complementary role of each grain might have impacted positively on the sensory attributes of the bread samples. This is similar to the observation of Olaoye et al. (1983) that maize, sorghum, soybean and wheat grains impacted acceptable sensory properties on bread.

Conclusions

This study has indicated that specialty bread could be prepared from a mixture of sprouted grains. Inclusion of grains other than wheat helped to improve the nutritional value of bread. Functional, nutritional as well as acceptability of the specialty bread was achieved through sprouting of the grains. The combination of cereals and legumes guaranteed proper balance of amino acids to provide a complete protein. However, long-term study is required to extend and explore the beneficial metabolic effects of using mixed grains for bread production. The acceptability of the product by about 82% of the judges shows that the product would be easily accepted by consumers. Therefore, this bread could be utilized to improve the nutritional status of people, especially the aged since the product has low sugar content and a claim of high digestibility (Dewettinck et al., 2008). It would be particularly good in developing countries where consumption of white bread is significant. And the technology used in this research will not be difficult even at local (village) set-up. Grain sprouting is a viable option towards improving nutritional value of locally made bread.

REFERENCES

Adeyemo SO, Olayode OO, Odutuga AA (1992). Biochemical analysis of germinating white maize (*Zea mays*). Niger. J. Nutr. Sci. 13(4):24-28.

Akpapunam M, Dedeh SS (1995). Traditional lactic acid fermentation, malt addition and quality development in maize –cowpea weaning blends. Food Nutr. Bull. 16(4):75-80.

AOAC (1990). Official Methods of Analysis Ed.15, 2. Association of Official Analytical Chemists, Washington D.C.

AOAC (1995). Enzymatic Grammetric Method of Analysis. http://www.Fao.org/docrep/006/y5022eo5.htm_10k (accessed June 30, 2004).

Brenna and Cleary (2005). Cereal complex carbohydrates and their contribution to human health. J. Cereal Sci. 42(1):1-13.

Chandrasheka A, Sadyanaranyana KV (2006). Disease and pest resistance in grams of sorghum and millets J. Cereal Sci. 44(3):2873-3304.

Chavan J, Kadam S (1989). "Nutritional improvement of cereal by sprouting. Crit. Rev. Food Sci. Nutr. 28(5):401-437.

Crockett R, Rachel L, Pauline VY (2011). Effects of soy protein isolate and egg white solids on the physiochemical properties of gluten-free bread. Food Chem. 129(1):84-90.

Cuddeford D (1989) "Hydroponic" In practice 11(5):211-214.

Dewettinck K, VanBockstacle F, Kuhne B, Vande Walle D, Courtens TM, Gellynck X (2008). Nutritional value of bread: Influence of processing, food interacting and consumer perception. J. Cereal Sci. 48(2):243-249.

Edema MO, Sanni LO, Sanni AI (2005). Evaluation of maize –soybean flour blends for sour maize bread production in Nigeria. Afr. J. Biotechnol. 4(2):911-918.

Eneje LO, Orji OJ (2002). Technology of Alcoholic and Non Alcoholic Beverages 2nd Ed, pp. 10-12.

FAO (1973). Report of joint FAO/WHO ad hoc Expert Committee on Energy and Protein Requirement. FAO Nutrition Meeting Report Series No 52. Food and Agricutural Organization of the United Nations, Rome.

FAO/WHO (1997). Carbohydrate in Human Nutrition. http://www.fao.org/docrep/8079e/w8079eOa.htm (accessed June 28, 2004).

Fennema's Food Chemistry (2008). 4th edn, CRC press London pp. 781-793.

Glennie CW, Carr JG (1983). Birth, T Endosperm modification in sorghum grain during germination. Cereal Chem. 61(2): 285-289. http://www.foodforlife.com/ (accessed 18/02/2010)

Jideani VA, Onwubali F (2009). Optimization of Wheat sprouted soybean bread using Response Surface Methodology. Afr. J. Biotechnol. 8(12):6364-6373.

Kent NL (1984). Technology of cereals 3rd Edition Pergamon press, Oxford U.K. 152 pp.

Kuye A, Sanni LO (1999). Industrialization of fermented Food process. How far in Nigeria? J. Sci. Ind. Res. 58(11):837-843.

Montonen J, Knekt P., Jarvinen R, Ramaen A (2003). Wholegrain and fiber intake and the incidence of type 2 diabetics. Am. J. Clin. Nutr. 77:622-629.

Moroni AV, DalBello F, Arendt E (2009). Sour Dough in gluten –free bread-making: An ancient technology to solve a novel issue? Food Microbiol. 26(7):676-680.

Okaka JC, Potter NN (1979). Physico-chemical and functional properties of cowpea flours. J. Food Sci. 44(5):1235-1237.

Okun CW, Wang YYD, Masham AC (1980). Wheat and its characteristics. Blackwell scientific publications. California U.S.A. 84-89 pp.

Olaoye OA Onilude AA, Idowu OA (2006). Quality characteristics of bread produced from composite flours of wheat, plantain and soybean. Afr. J. Biotechnol. 5(3):1102-1106.

Onyeka EU (2007). Glycemic and physiochemical properties of five common cowpea (vigna unguiculata) cultivars in Nigeria. J. Food Proc. Preserv. 31:618-631.

Ostman E, Lijeberg H, Bjork I (2005). Impact of organic acid on the glycemic and insuliaemic responses to whole grain bread product in healthy humans www.int.th.se/research/posters/elim.pdf. Accessed march, 2007.

Shipard I (2005). "How can I Grow and use sprouts as living Food? Stewart publishing, Australia. pp. 10-20.

Wang YYD, Fields MD (1978). Germination of corn and sorghum in the home to improve nutritive value. J. Food Sci. 43(5):1113-1115.

Yoom JH, Thompson LU, Jenkens I, David JA (1983). "The effect of phytic acid on in vitro rate of starch digestibility and blood glucose response". Am. J. Clin. Nutr. 38(6):835-842.

Assessment of structure and conduct of yam wholesale market in Efon Alaaye Local Government Area of Ekiti State, Nigeria

Folayan, J. A

Department of Agricultural Economics and Extension Service, Joseph Ayo Babalola University, Ikeji-Arakeji, P.M.B. 5006 Ilesa, Osun State, Nigeria. E-mail: foldio2000@yahoo.com.

This study assessed the structure and conduct of yam wholesale market in Efon Alaaye Local Government Area of Ekiti State of Nigeria. The primary data was collected with structured questionnaire and administered to yam wholesalers in the study area between 1995 and 2008. The study employed the use of descriptive statistics of frequency, and percentages. The study revealed that the capital required to enter the yam marketing industry in 2008 ranged between ₦25,000 and ₦100,000. There was no evidence of dominant firms, price collision or predatory pricing among the wholesalers and most traders sold their product according to the prevailing price in the market irrespective of the costs incurred. 100% each of the respondents expressed inadequate capital, high cost of labour, and high cost of transport while 85.7% expressed lack of storage facilities, and 71.1% competition from other firms as the problems encountered in the industry. Both formal and informal financial arrangement should be encouraged by the government and nongovernmental organization while the wholesalers too should endeavour to form themselves into cooperatives, they should reduce the quantity of yam kept in storage per unit of time, while research into techniques and economics of establishment and use of yam storage should be conducted. Government should improve rural road conditions all in an attempt to enhance marketing of yam and other agricultural products.

Key words: Wholesalers, market structure, conduct.

INTRODUCTION

Yam (*Dioscorea* species) is a premium crop in Nigeria food system and Nigeria is the world's largest producer with an aggregate annual output in excess of 50% of total world production of the 47.5 million tonnes of yam produced worldwide in 2000. Nigeria alone accounted for 26 million tonnes followed by Ghana and Coted'Ivoire with 3 million and 2.9 million tonnes respectively (IITA, 2006).

According to Net industries (2009), there are two centres of yam cultivation worldwide. The first is the high rainfall region of Western Africa from ivory Coast to Cameroon. The secondary areas of yam cultivation are West Indies Pacific Island and South Westhern United States. As reported by IITA (2007), yams are produced on 5 million hectares in tropical and subtropical regions of the world.

This study was embarked upon for the purpose of assessing the structure and conduct of yam wholesale market in Efon-Alaaye Local Government Area of Ekiti State with a view to have in depth knowledge of elements of market structure and conduct as it affect the market of study. The outcome of the study will go a long way to assist in providing recommendations that would improve marketing of yam wholesale in Efon Local Government in particular and serve as a benchmark for further research into yam marketing in general.

More and Walsh (1966) made efforts to formulate elicit criteria of adequacy or optimality of markets. They viewed a market as being adequate if its actual performance is not extremely detrimental with respect to several firm. A market is considered adequate if its market structure conducts and performance are as favorable in all respect

Table 1. The elements of structure and conducts.

S/N	Structure	Conduct
1	A brief history of the industry interest,	Non unfair, exclusionary, predatory or coercive tactics
2	Barrier to entry,	Non misleading sales promotion,
3	Product differentiation	Non tacit or exposed price collusion
4	The industry's growth rate of demand,	Non shielding of inefficient firms
5	Size and number of buyers and sellers,	Non pricing practices which discourage uniform high quality and uniform seasoned output
6	Elasticity of demand	
7	Ratio of fixed costs to variable costs in the short run.	

Source: Olufokunbi (1984).

as unavoidable circumstances permits. They recommended criteria of adequacy of structure conduct and performance of the market.

Structure has been defined as the significant economic variables that characterize an organization while conduct can be defined as the behavior of the industry of interest with reference to its activities (Olufokunbi, 1984). The elements of structure include those shown in Table 1.

The elements of structure and conduct (Table 1) of yam wholesale market have been used by Akinboro (1981) to evaluate the wholesale marketing of tomato in Ibadan, by Folayan (1987) to analyse marketing of yam in Itawure, Ekiti West Local Government Area of Ondo State, by Akinola (2003) to appraise gender factor in the marketing of Cocoa in Ondo State and by Folayan (2005) to analyse the economics, of Cocoa marketing in Ondo State and Ekiti States of Nigeria.

A wholesaler is defined as one who sells not less than 50% of his/her yam to yam retailers. A retailer on the other hand was classified as one who sold not less than 50% of his/her total yam to those he/she strongly believes will not re-sell the commodity in the form it was purchased even if some reselling will eventually take place.

The focus of the detail and rigorous analysis was at the wholesaling stage, as this stage serves as the focal point for supply and demand. Major sources of information for marketing decision have been know to spring from this level (Shepherd et al., 1976) and prices serving as guidelines for yam marketing appear to be generated from this level.

MATERIALS AND METHODS

The study was carried out in Efon-Alaaye Local Government Area of Ekiti State, Nigeria. The Local Government was bounded in the South and East by Oriade Local Government of Osun State and in the North and West by Ekiti West Local Government of Ekiti State. It is a major distribution center for Agricultural products, in general and yam in particular.

Sampling size, technique and data

Primary data were collected with structured questionnaire while secondary data was collected from published and unpublished materials. The questionnaires of 35 of these wholesalers were processed for adequate information. The 35 cases would satisfy the assumption of central limit theorem for hypothesis testing should the need arises. Also, wholesalers so interviewed have consistently been marketing significant quantity of yam for over 10 years. Data collected on market structure included age in business, capital requirement, sources of capital, ownership structure, volume of operation and barrier to entry while that of conduct included information on opening duration, functional acts employment, buying practices, discount, pricing, mark-up, marketing outlets, selling practices, credit, advertising, spoilage, other issues and problems encountered in the industry.

Weight determination of yams handled per annum

The total weight of yam tubers handled in the market per annum was estimated by use of the formular expressed in Equation 1:

$$2.27NRnXY \quad (1)$$

Where, 2.27 kg = Average weight of tuber; N = number of weeks of operation by yam marketers, Rn = 35 = number of marketers/respondents, X = Number of yam tubers marketed per week and Y= number of days of operation by yam marketers.

RESULTS AND DISCUSSION

The findings on market structure and conduct of firm in Efon-Alaaye Local Government based yam wholesale industry are discussed as follows.

Age in business

As contained in Table 2, the oldest industry reported that she has been in business for 51-60 years. The mode of age of firms in the industry was 31-40 years while the median was 21-30 years and the arithmetic mean of age

Table 2. Age in Business Efon Alaaye Local Government based on yam wholesale industry 2008.

Years	Frequency	Percentage
1-10	10	28.6
11-20	6	17.1
21-30	5	14.3
31-40	11	31.4
41-50	2	5.7
51-60	1	2.9

Table 3. Capital required to enter the yam wholesaling industry in, Efon Alaaye Local Government Area, 2008.

Capital required (₦)	Number of firms	% of firm
25,000	7	20.0
30,000	4	11.4
40,000	2	5.7
50,000	9	25.7
60,000	6	17.1
80,000	4	11.4
100,000	3	8.6

Mode = ₦50.000; Median = ₦50, 000; Mean = ₦515, 70; Minimum = ₦100, 000.

Table 4. Barriers to entry for Efon Local Government wholesale yam industry, 2008.

Factor	Number of firms that indicated the factors	%
Control over suppliers	0	0
Profit potential	20	57
Capital required	35	100

in business was 21-30 years. The number indicated that the business had been on the increase over the years. Although data on the actual firm that exited from the industry over the years were not available.

Ownership structure

All the 35 firms surveyed consisted of sole proprietorships. All the firms were separate business units and these were no evidence of organizational combination. The owners of the yam wholesaling firms assumed the responsibility of management. They did not hire individuals as managers of the firms.

Capital required to enter the industry

The detail of the capital required to enter the industry is summarized in Table 3 The mode of the capital required was fifty thousand naira (₦50,000), the mean was about five thousand one hundred and sixty Naira only (₦5,160) and the medium was fifty thousand Naira only (₦50,000). The above capital amounts required to enter the business are small relative to capital requirement on marketing of some other agricultural products.

Barrier to entry

Table 4 revealed that 100% of the respondents indicated that capital required to enter the industry could discourage a potential entrant. However, the fact that the numbers of firms that enter the market over the year's increases suggest that this requirement had not exerted too much barriers to entry.

About 57% respondent reported that potential entrant into the yam wholesale business could be scared away

Table 5. Distribution of firms by Opening Duration per day: Efon Local Government Area based yam wholesale Industry, 2008.

Opening hours per day (h)	Number of firms	% of firms
9	2	5.71
10	9	25.71
11	8	22.86
12	7	20.00
13	9	25.71

Mode = 10 and13 h; median = 11 h; mean = 11.34 h; minimum = 9 h; maximum = 13 h.

due to profit potential which they described as very low. None of the respondents indicated control over suppliers and sales as a barrier to entry.

Sources of capital

All the wholesalers reported that they did not have financial assistance whatsoever; this they alleged limited the size of their marketing activities and volume of trade.

Opening duration

The minimum opening duration per day for all the firms was 9 h while the maximum opening duration was 13 h per day. The modes were 10 and 13 h per day, the medium was 11 h and the mean was 11.34 h per day (Table 5).

Opening days per week

The arithmetic mean of the opening days per week for the industry was six, while the mode was six days and the median six days. The maximum was seven days while the minimum was five days.

Week on vacation

The minimum number of week on vacation per year was two weeks, while the maximum was 7 weeks. The mean number of weeks on vacation was 3.57 weeks while the median was 3 weeks per year and the mode number of weeks on vacation per year was 2 weeks.

All the wholesalers stated that they go on special vacation each year. It was found from this study that the reasons given for being absent from the market are often identical. Such reasons are itemized as: (a) bereavement or death of a family member or of a friend; (b) marriage of a son or daughter; (c) religious and other ceremonial festivals such as Egungun and Ogun festivals; (d)naming and house warming ceremonies; (e) ill-health; f)attending a family meeting, visiting friends and relatives.

Function acts

The principal functions performed by the yam wholesalers were:

Buying

(a) Wholesalers bought yam from local grower in and around Efon Alaaye Local Government Area.
(b) They purchased yams from other wholesalers within the market.

Selling

The wholesalers of yam sold through many outlets such as:

(a) Other wholesalers in Efon Alaaye Local Government Area
(b) Retailers from Aramoko, Erio, Ikoro, Igede, Ijero, Ilesa, Ijebu-Ijesa, Efon-Alaaye, Ido-Ajinare, Imesi and Ado-Ekiti.
(c) Direct consumers and other users of yams

Physical function

The wholesalers performed the following physical functions;

(a) Handled yams
(b) Transported yam to the market
(c) Sort out yams of different sizes
(d) Assembled the yam purchased from different sources.
(e) Stored unsold yams before sale.

Table 6. Duration of storage of yam in Efon Local Government yam wholesale market, 2008.

Number of days	Number of respondent firms	% of respondent firms
1-7	14	40
1-15	11	31.43
1-21	10	28.57

Table 7. Tasks performed by the Efon Local Government based yam wholesalers, 2008.

Tasks performed	Number of firm	% of firms
Grading	0	0
Pricing	35	100
Wholesale buying from middlemen	15	42.9
Wholesale buying from producers	35	100
Sorting	35	100

Storage

The yam wholesalers had no special storage facilities and so were not able to keep their yams for too long a time. The storage period as contained in Table 6 ranged between 1 and 21 days.

Risk bearing

The wholesalers bore the changes in yam prices, both in transit and market.

Merchandising

This entailed haggling with potential buyers. No particular attempt is made to promote sale or advertise. The attention of purchases is caught by calling the potential customers, saying "customer, come and buy yam of good quality". In some cases, it is just the exposure of the displayed yam that draws the attention of the potential customers.

Task performed

The distribution of wholesalers by tasks performed are summarized in Table 7.

Employment

About 11.4% (4 firms) of the firms employed hired full-time labour and each worker was paid ₦4, 500 per month. 18 firms that is 51% of the wholesalers employed full-time family labour; the remaining 37.2% employed no labour. There were no special fringe benefits given to labour.

Buying practices

The wholesalers in this market purchased their yam from farmers, and other wholesalers within the market. Table 8 shows the summary of the institutions from which the wholesalers obtained their supply and the proportion of the wholesalers that got supply from specified institutions.

Almost all the wholesalers bought their yam from the farmers while none of them bought from cooperative societies. The preference for yam obtained directly from farmers was claimed to be due to the fact that it enable the wholesalers to make cheaper and bulkier purchases. The wholesalers did not have problems in getting good quality product from the local growers and other wholesalers. Purchases by contract were not common among the wholesalers.

Discount obtained

The wholesalers in Efon-Alaaye Local Government yam wholesale market did obtain discounts for immediate cash purchase, bulk buying and pre- payment. The wholesalers however reported that cases of pre-payment are uncommon. Table 9 summaries the detail while Table 10 shows the distribution and the statistics of the amount of the discount in naira granted per 100 tubers of yam by the producers to the wholesalers.

Price quoting conduct and outcome of pricing

Findings obtained from wholesaler concerning price

Table 8. Institution from which wholesalers obtain their supply and the proportion of wholesalers that mentioned the institution: Efon Alaaye Local Government based yam wholesale industry in 2008.

% Of yam purchased from	Number of firm	% of firm
Farmers		
50 – 100	33	94.3
30 – 49	2	5.7
Other wholesales		
1 – 20	13	37.1
21 – 40	1	2.9
Cooperatives		

Table 9. Obtaining discounts on immediate cash purchases, bulk purchase and prepayment: Efon local government yam wholesale industry in 2008.

Condition for which discount is obtained	Yes	%	No	%
Immediate Cash purchases	30	85.7	5	14.3
Bulk purchases	27	77.1	8	22.9
Prepayment	15	42.9	20	57.1

Table 10. Discount in naira per 100 tubers of yam granted by yam producers to the wholesalers: Efon Local Government based yam wholesale industry in 2008.

Discount (₦) granted per 100 tubers (227 kg) of yam	Discount as % of purchase price	Number of firms that granted the discount	% Of firms that granted the discount
159	1.79	7	20
298	2.98	10	28.6
300	3.57	5	14.3
350	4.17	5	14.3
400	4.76	3	8.6
500	5.95	5	14.3

Mean = ₦300 (3.57%); Median = ₦300 (3.57%); Mode = ₦250 (2.98%); Minimum = ₦ 150 (1.79%); Maximum = ₦500(5.95%).

Table 11. Different means of price quoting and factors that determined quoted and final prices of yam of Efon Local government based yam wholesale in 2008.

Means of quoting prices	Number of firms	% Of firms
Follow the lead of dominant seller	0	0
Attempt to meet competitors price	35	100
Place mark-up on cost	35	100
Emergency demand for yam	31	88.6
Usual large deliveries or failure	31	88.6

quoting conduct and outcome of pricing are summarized in Table 11.

In quoting prices for their products, the wholesalers used a variety of ways. After quoting a price for the customers, the final price, which the customer pays, was often determined through haggling. Here, the seller starts

Table 12. Reported mark-up charged on purchase price of 100 tubers (227kg) of yam by Efon local Government yam wholesale industry in 2008.

1 Markup on sale (%)	2 mid point	3 Number of firms	4 % Of Respondent firms	5 Col.2xCol.4	6 Expected value of mark-up
10.01-12	11.01	1	2.89	0.318	
12.01-14	13.01	1	2.89	0.376	
14.01-16	15.01	2	5.71	0.857	
16.01-18	17.01	2	5.71	0.971	
18.01-20	19.01	5	14.29	2.716	
20.01-22	21.01	5	14.29	3.002	
22.01-24	23.01	5	14.29	3.288	
24.01-26	25.01	6	17.14	4.287	
26.01-28	27.01	6	17.14	4.623	
28.01-30	29.01	2	5.71	1.656	
					22.094

Mean mark-up = 22.094; Mode mark-up =18.21%; Median mark-up = 22.81% Minimum mark-up = 10.50%; Maximum mark-up = 30%.

Table 13. Yam Marketing Outlets: Efon Alaaye Local Government Based Yam Wholesale Industry in 2008.

% Sold to different outlet	Number of firm	% of firm
% sold to Retailers		
60 – 100	28	80.0
35 – 59	7	20.0
% Sold to other wholesalers		
20 – 4o	10	28.6
5 – 25	8	22.9
% sold to the direct consumers		
5-15	16	45.7

with a price which she knows is excessively high, but which he knows may produce some sales. The sellers might subjectively evaluate the customers before quoting a price on a given quantity and quality of yam. An experienced buyer however knows that this is merely an offering price. He therefore begins to lower it in gradual steps noting on each occasion the reaction of the seller as he reduces the price. This continues until a price is reached which is acceptable to both.

It was observed that the final price the buyer pays is always influenced by some none-price factors as whether the customer is a male or female, how well dressed he or she is, whether he or she has a car, his or her skill at bargaining, and his or her familiarity with the market. Thus, the yam wholesaling industry was found to be imperfectly competitive in conduct.

In the industry, there was no evidence of price collusion or predatory pricing among the wholesalers. The forces of supply and demand also affected how the wholesalers quoted prices for their product.

All the wholesalers also indicated that meeting competitor's prices, placing of mark-upon cost of product and consideration for the need for profit were some of the factors that determined the final prices for their product. Table 12 summarizes the percentage mark-up placed on the cost of yam by the wholesalers.

Marketing outlets

The marketing outlets for the product of the wholesalers were retailers, other wholesalers and direct consumers. Tables 13 and 14 summarize the marketing outlets for the wholesalers in order of priority respectively.

Almost all the wholesalers chose retailers as the most preferred outlet. The choice of the retailers as the most preferred outlet may be due to the fact that while it enabled them (wholesalers) quicker and bulkier sales than when the sale is made through the direct consumers, it may also afford them (wholesalers) higher level of profit margin than when sale is made through other wholesalers.

Selling practices

As shown in Table 15, majority of the wholesalers granted discount privileges to buyers for immediate cash

Table 14. Outlets preferred by wholesalers in order of priority.

Outlet	1		2		3		4	
	Number	%	Number	%	Number	%	Number	%
Retailers	33	94	2	5.7	-	-	-	-
Other wholesalers	2	5.7	18	51.4	5	14.3	-	-
Direct consumers	-	-	15	42.9	13	37.1	-	-

Table 15. Purchase - Payment strategies by yam wholesale industries, Efon Local Government.

Condition	Number of firm	% of firm
Immediate cash purchases	35	100
Bulk buying	35	100
Advance payment	25	71.4

Table 16. Discount in Naira per 100 tubers (227 kg) of yam granted by wholesalers to buyers: Efon Alaaye Local Government based wholesale industry in 2008.

Discount in N/100 tubers (227 kg)	Discount as of selling price (N)	% of firms that granted the discount	% of firm
00	0.64	4	11.43
150	0.96	4	11.43
200	1.28	10	28.57
250	1.60	5	14.28
300	1.92	5	14.28
350	1.23	3	11.43
400	2.55	4	11.43

Mean = ₦240 (1.53%); Median = ₦200 (1.28%); Mode = ₦ 200 (1.28%); Minimum = ₦100 (0.64%); Maximum = ₦ 400 (2.55%).

Table 17. Granting of credit to customers: Efon Local Government yam wholesale industry in 2008.

Credit repayment duration	Number of firm	% of firm
Below 7 days	15	46.7
8 – 15 days	11	34.4
16 – 30 days	6	18.7

purchase, bulk purchase and advance payment. Most of them however reported that the issue of advance payment by the customers was uncommon. Table 16 shows the distribution and statistics of the discount amount in Naira granted to the customers for purchase of 100 tubers of yam (227 kg).

Extension of credit to customers

Credit extension to the customers, especially during the glut period was a common practice among the wholesalers of yam in Efon Alaaye Local Government Area.

About 91% of the wholesalers sold on credit to customers, while the remaining 9% do not sell on credit. Table 17 summarizes the duration of the credit sales in the Efon Alaaye Local Government Area in 2008.

There were no formal procedures that are usually followed in granting credit. Credit extension depends on wholesalers personal judgment which ranges from degree of relationship with the potential debtor,

Table 18. Distribution of credit extension in Efon Local Government based yam wholesale firms in 2008.

Distribution	% Of sales on credit	Amount sold on credit at a time (₦)	Bad debt (₦)	% Of credit on bad debt
Mean	5.6	24,550	3,800	15.9
Mode	4	20,000	3,000, 4,000	20
Median	5	20,000	3,500	16.4
Minimum	2	12,000	2,000	4
Maximum	12	50,000	7,000	24

Table 19. Wholesalers' anticipation over credit sales: Efon Local Government Area based yam wholesale industry in 2008.

Reaction	Number of firm	% Of firm
Quilting Credit Extension	3	9.4
Reducing credit extension	4	12.5
Increasing credit extension	0	0
No change in the present policy	25	78.1

recommendation by a third party that is well known to the wholesaler: to a good credit record by the customers.

About 97% of the firms that extend credit had between 2 to 10% of their sales on credit (see detail distribution of percentage sales on credit and amount of credit granted per time in Table 18)

The above distribution of bad debt loss is small relative to bad loss incurred in some other agricultural products. For example, Utomakili (1981) reported mean percentage of bad debt of 32% on rice wholesaling in Bendel state in 1981.

Because of the minimal effect of bad debt losses on the wholesalers of yam in Efon Alaaye Local Government based yam wholesale industry in 2008, about 78.1% of them anticipated no change in the present credit extension policy. None of the wholesalers however planned to increase credit extension; about 12.5 and 9.4% to reduce credit extension and quit credit extension respectively (Table 19).

Advertising practices

There are some forms of advertising and sales promotion of which none of the wholesale expended part of their capital in Efon Local Government Area yam wholesale industry in 2008. It takes the form of the wholesalers calling the attention of the potential customers to the product. It also takes the form of displaying the product in a way that the potential customers can easily see it.

Trade association

The name of the trade association in Efon Local Government Area is "Egbe Onisu Efon" that is Efon yam sellers association. The membership of this trade association was not compulsory. About 66% of the wholesalers were members of the association. The association had no impact on market structure, and conduct.

Visiting and donating on naming and house warming ceremonies, marriages and funerals are some of the social functions of the association. The influence of the trade association may have been conditioned by the awareness and aspiration of the members, their customs and tradition, education and enlightenment, and financial constraint. The extent to which these factors actually influence the roles of trade associations in primary trade in Nigeria could be suggested for future research.

Issues on which the wholesalers expressed thoughts

The issues on which the wholesalers expressed thought include the exit and entry of firms. Table 20 shows the summary. The general prospects seem to be that of the entering of new firms and expansion of supply quantity of yam in the market. The wholesalers also reported that they did not foresee any merging or vertical integration in the industry.

Problems encountered in the industry

The problems encountered in the industry as reported by the wholesalers were high cost of transport, lack of storage facilities, inadequate capital, high cost of labour and competition from other firms. None of them indicated insufficient demand for product as one of the problems

Table 20. Industrial prospects of yam marketing: Efon Alaaye Local Government based wholesaling industry in 2008.

Prospect	Number of firms with view	% of firm
New firms entering	24	68.6
Existing firms folding up	3	8.6
Existing firms expanding	17	48.6
Existing firms merging	0	0
No foreseeable changes	5	14.3

Table 21. Problems faced by yam wholesaling firms in Efon Local Government 2008.

Problems encountered	Number of firm	% of firm
Inadequate capital	35	100
Competition from other firms	25	71.1
High cost of labour	35	100
Lack of special storage facilities	30	85.7
Insufficient demand for the product	0	0
High cost of transport	35	100

encountered in the industry. Table 21 shows the summary of the views of the wholesalers on the problems encountered in the industry.

They indicated that the most serious problem that limits volume of operations and therefore income is lack of capital, insufficient credit for establishment, keeping the concern going and expansion.

High transportation cost is another crucial factor that limits marketing volume. These costs are probably related to the type of roads over which yams are transported given other conditions. High labor cost was also mentioned as a limiting factor to the quantity of yam marketed; lack of special storage facilities and high spoilage rates were also mentioned as limiting factors.

All the wholesalers indicated that insufficient demand for their produce was not a problem at all. In fact, they felt that they were able to provide between 50 and 60% of the quantity demanded of their produce. All the wholesalers could not identify which of these problems they themselves could solve. They however pointed out that the government could come to their aid in extending credit facilities.

The operationalisation of this recommendation will be militated against by the fact that none of the wholesalers indicated that she is keeping a bank account and no bank manager would grant any loan to somebody unless the character and credit worthiness of the borrower and the probability of the applicant's business being successful are known. Because many food traders do not use banks' facilities, it is difficult for banks to assess their credit worthiness. In addition to this, most food sellers are small-scale middlemen who are often unable to offer any security or collateral acceptable to the bank. Also, banks loath to extend loans to food traders because of the perishable nature of food products, the tendency towards frequent fluctuation of food prices, and the cost of collecting small loans from a large member of middlemen.

Conclusion

This study contemplated at assessing the structure and conduct of yam wholesale market in Efon Local Government Area of Ekiti State, Nigeria. The market site was chosen on the basis of its being a major center for distributing foodstuffs particularly in Ekiti State of Nigeria. The research focused mainly on the wholesalers of yam because of the important roles that have been identified by previous studies as played by this institution in the distribution of food crops from the producer to the consumers. During the study, 35 industries (about 83% of the whole) were interviewed in the market site.

The capital that is required to enter the yam marketing industry in Efon Alaaye Local Government of Ekiti State in 2008 cannot be said to be too high. The amount ranged between ₦25,000 and ₦100,000. There also was no indication of any restriction to entering the industry. There was no evidence of dominant firms, price conclusion or predatory pricing among the wholesalers. Most traders sell their product according to the routing price in the market irrespective of the costs incurred.

The functional acts of the yam wholesalers included buying, selling, storing, risk bearing and merchandising. Also, they obtained discount, quoted prices, put mark up charges on purchase price, granted discount to buyers;

extended credit to customers. They expressed their views on issues bothering on their trade and problems encountered in the industry.

RECOMMENDATIONS

In order to solve the problem of financial inadequacy facing the wholesalers, both formal and informal financial institutions should be encouraged by the Government to provide financial assistance to them so that they can expand and improve their existing business volume. This can be operationalised and facilitated if they form cooperative organizations so that the pecuniary and non pecuniary cost of borrowing per wholesaler will all be reduced and in order to alleviate the problem of high transport costs reputed as limiting the size of operations of wholesale firms, the government should make conscious efforts to organize yam wholesaling cooperatives. These cooperatives can buy in bulk from the sources of supply and can even buy their vehicles, which they can use and maintain in transporting purchased yam; an arrangement that can considerably reduce transport cost and therefore marketing costs.

The government and nongovernmental organization should make efforts towards improving rural road conditions in Nigeria, and also construct rural feeder roads that will go a long way at reducing marketing costs, and therefore at enhancement the marketing of not only yams, but also many agricultural products.

REFERENCES

Akinboro LO (1981). Structure, Conduct and Operational and Pricing Efficiencies of wholesale Marketing of Tomato in Ibadan Unpublished M.Phl. Thesis, Department of Agricultural Economics, University of Ife, Nigeria.

Akinola GO (2003). Gender factor in the marketing of cocoa in Ondo State. unpublished M.Sc. thesis Department of Accounting and Administration, Obafemi Awolowo University, Ile- Ife, Nigeria. pp. 35-45.

Folayan JA (1987). Analysis of yam wholesales marketing in Itawure, Ekiti- west Local Government of Ondo State, Nigeria. Unpublished M.sc Thesis, Department of Agricultural Economics, Obafemi Awolowo University, Ile-Ife, Nigeria.

Folayan JA (2005). Economic Analysis of Cocoa Marketing in Ondo and Ekiti States of Nigeria. Unpublished PhD Thesis, Department of Agricultural Economics Federal University of Technology Akure, Nigeria.

IITA (2006). International Institute for Tropical Crop www.iita.org/crop/yam.htmt+ymt prod. Assessed on the 27th July, 2007.

IITA (2007). International Institute for Tropical Crop www.iita.org/crop/yam.htmt+ymt prod. Assessed on the 13th September, 2010.

Net Industries (2009). yam http://Science.jrank.or/page/7437/yam.html. Assessed on 13th August, 2010.

Olufokunbi B (1984). Evaluation of Agricultural Policies in Nigeria proceeding of an agricultural policy. A paper presented at the MISER workshop.

Shepherd GS, Futrell G, Strain RJ (1976). Marketing Farm Products, Iowa University Press, 448 pp.

Utomakili JB (1981). An Analysis of Rice Marketing in Bendel State of Nigeria, Unpublished M.Sc Thesis, Department of Agricultural Economics University of Ife, Ile-Ife, Nigeria.

Simultaneous high performance liquid chromatography analysis of water soluble vitamins in some foods prepared by a Nigerian eatery

Israel Olusegun Otemuyiwa and Steve Adeniyi Adewusi

Department of Chemistry,Obafemi Awolowo University,Ile-Ife, Nigeria.

The present study carried out simultaneous analysis of water soluble vitamins (thiamine, pyridoxine, niacin, folic acid and ascorbic acid) in some foods prepared by a Nigerian eatery using a reverse phase high performance liquid chromatography. The results (mg /100 g) indicate that pyridoxine ranged from 0.16 in amala to 1.14 in fried rice, niacin from 0.18 in doughnut to 0.70 in meat pie, thiamine from 0.30 in doughnut to 2.40 in cowpea. Ascorbic acid was lowest in doughnut and highest in *eba* composite meal while folic acid ranged from 30 µg /100 g in doughnut to 195 µg /100 g in meat pie. Thiamine was significantly higher in village kitchen meals and below detection limit in meal of rice alone. The gram per serving of the village kitchen would give 108-200% and 32-48 % of the recommended dietary allowance (RDA) for thiamine and folic acid, respectively, whereas the level of ascorbic acid was too low to meet the RDA. The study shows that fast foods have low content of water soluble vitamins and its consumption could compromise intake of some vitamins.

Key words: Thiamine, pyridoxine, niacin, folic acid, ascorbic acid, eatery.

INTRODUCTION

Micronutrient deficiency diseases (MDD) are widespread and affect large numbers of people in developing countries. It is estimated that one in every three people in the world is at risk for one or more micronutrient deficiencies, while approximately 2 billion people world-wide suffer from some kind of micronutrient deficiency, causing a wide array of disorders and increasing the risk of death, disease and disability (WHO, 2005). Vitamin (an important micronutrient), is essential for most of the body's functions and is required by the body in small amounts for metabolism, to promote health and proper growth especially among children (Olaniyi, 2000). Vitamin deficiency diseases do not only occur in poverty stricken communities; in affluent societies, deficiency diseases also occur often as the result of poor choice of food or unhealthy eating habits. With fast food eateries springing up in nooks and crannies of Nigeria, consumption of fast food is gradually becoming a habit among Nigerian youths and elite. The eateries probably because of pre-ference for taste, apart from conventional fried or baked fast foods, incorporated village kitchen meals (village kitchen meals are local meals of the people where the eatery is located).

Fast food operations involved non-conventional coo-king methods like deep frying, microwave and pressure cooking in order to save time. From general point of view, thermal processing employed in food preparation has been found to lead to loss of water soluble vitamins.

According to Berechet and Segal (2007) water soluble vitamins may be lost during cooking in two ways: first, by degradation which can occur by destruction or by other chemical changes such as oxidation; and second by leaching into the cooking medium.

The increasing interest in good eating habit has provided a greater awareness of the vital role that micronutrients play in growth and health. Diets which do not contain adequate amounts of these vitamins often result in deficiency diseases; for instance with thiamine deficiency, a number of important body functions can be disturbed. They include brain function, nerve function (especially of the legs), and heart function; these three impairments are called Wernicke-Korsakoffs psychosis', dry beriberi and wet beriberi (Ozawa et al., 2001). Pyridoxine deficiency may lead to depression, convulsions, abnormal nerve functions (especially in the limbs), dermatitis, cracking of skin at the corner of the mouth and the lips, a smooth tongue, and anaemia (Anyakora et al., 2008).

Pellagra is a condition resulting from niacin deficiency, in which there are symptoms of dermatitis in skin exposed to the sun, diarrhoea and dementia (Berger, 1985). The consequences of deficiency of folic acid include anaemia and defective lining of the gut, adversely affecting absorption of many nutrients. Since the number of blood platelets (which play a part in blood clotting) can be low with folic acid deficiency, a tendency to prolonged bleeding can also occur (Bailey et al., 2010), while vitamin C have been reported to reduce the damage by free radicals and check degenerative disease its deficiency, leads to scurvy (Rickman et al., 2007).

The National Agency of Food and Drug Administration and Control (NAFDAC), in conjunction with the Micronutrient Initiative (MI), of Nigeria, have contributed immensely to the creation of this awareness of the impact of eating foods deficient in micronutrients especially vitamins (Anyakora et al., 2008). With the increase in establishment of fast food eateries and popularity of consumption of fast foods, it is therefore of utmost importance to know if conventional fast foods and village kitchen meals prepared from eateries will contribute significantly to the dietary vitamin intake. The objective of this study was to determine the content of some water soluble vitamins (niacin, pyridoxine, folic acid, thiamine and ascorbic acid) in selected foods prepared by a Nigerian eatery.

MATERIALS AND METHODS

Samples

Samples were collected as consumed from five outlets of a popular fast eatery located in Lagos, Nigeria. The meals were weighed as served and pooled together, then those served with soup were mixed together with the soup before being blend using warring blender ((National model MX-795N, Matsushita, Malaysia). The samples were kept in a freezer and the extraction of vitamins was initiated within 12 hof sample collection.

Analysis of vitamins

Analysis of water soluble vitamins was carried out simultaneously using High Performance Liquid Chromatography with UV detector (Agilent Technologies Model 1200, Germany) following the method of Khor and Tee (1996) with slight modifications.

Reagent and chemicals

Vitamin standards (pyridoxine, niacinamide, folic acid, ascorbic acid and thiamine) and HPLC grade methanol were purchased from Sigma (Sigma-Aldrich, Germany). Sodium monohydrogen phosphate was purchased from BDH.

Chromatographic conditions

The HPLC used was a reversed phase type. The chromatographic separation column consisted of a stainless steel (4.6 × 150 mm) Eclipse XDD, and 5 µm mbondapack C_{18} column. The HPLC had an integrated UV detector which was set at 254 nm wavelength to monitor the column effluent. The mobile phase comprised of 90% of 0.01 M sodium mono-hydrogen phosphate and 10% HPLC grade methanol. The mobile phase was sonicated for 30 min before use. The elution was isocratic and the flow rate was 0.60 mL / min.

Extraction and analysis of vitamins from samples

The samples (2.0 g) were weighed into 200 mL Erlenmeyer flask, 30 mL of 0.1 M Hydrochloric acid was added and the mixture was heated at 40°C for 5 min, the mixture was filtered through ashless filter paper and centrifuged for 10 min at 1800 rpm. The supernatant was sonicated and 20 µL were injected into the column and separated. Peaks were identified by comparison of the retention time to that of the known standards and the value of vitamins was obtained from computer printout (Chemstation software).

Preparation of standards

The stock standard solutions of thiamine, pyridoxine and niacin were prepared by dissolving 50 mg of the standard in 0.1 M hydrochloric acid in 25 mL standard flask and made up to mark with the acid. For preparation of folic acid and ascorbic acid, 0.1 M sodium hydroxide and deionized water was used. The working standards were prepared from the stock standard solutions by taking 0.1, 0.2, 0.3, 0.4, 0.5 mg/mL of the stock into 10 mL standard flask and made up to mark with appropriate solvent. The solutions were sonicated for 30 min and 20 µL of various concentrations of vitamin standards were injected into the HPLC column separately and the retention times were noted. The solutions of the standards were mixed in equal proportion and 20 µL was injected into the column. The retention time of each vitamin in the complex mixture was noted and used to identify the vitamins in the sample. The detection limit for the vitamins was 0.05 mg/ml.

Statistical analysis

Statistical analysis was carried out by the use of Microsoft Excel Statistical Packages (Microsoft Corporations, USA) and Graph-Pad Instat-3 Packages (Graph Pad software Inc, USA). All analyses were carried out in triplicate and the results presented as mean and

Table 1. Characteristics of the standard curve.

Vitamin	Retention Time (min)	Regression Equation (y = mx+b)	Correlation coefficient (r)
Ascorbic acid	2.3	y = 92511.36027x + 1048.70238	0.999
Pyridoxine	3.2	y = 18536.68580x + 16.31764	0.999
Thiamine	3.9	y = 14213.42610x + (-284.46318)	0.995
Folic Acid	4.8	y = 25719.56704x + (-327.04686)	0.998
Niacin	5.9	y = 44725.448x + (-586.167)	0.996

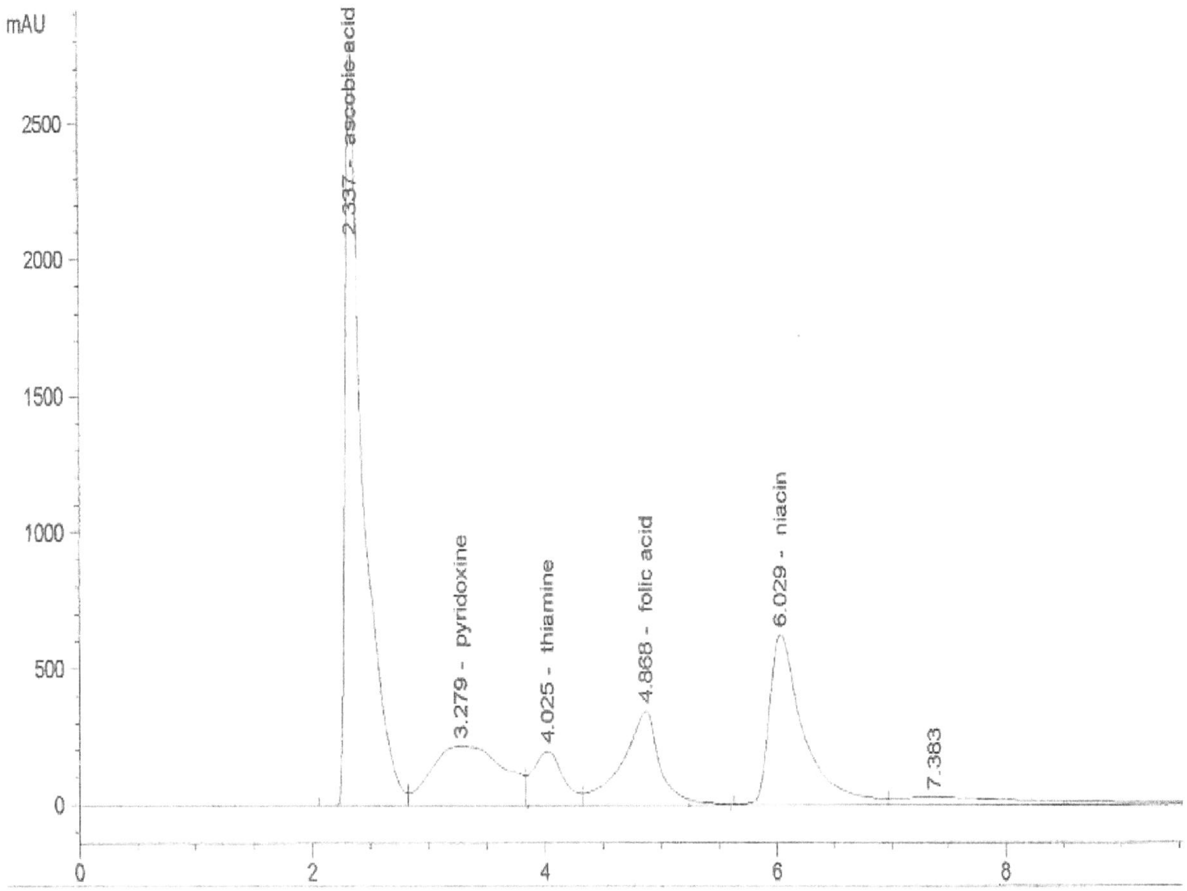

Figure 1. Chromatogram of the vitamin standards.

standard deviation. Analysis of variance was used to assess and compare results.

RESULTS AND DISCUSSION

The chromatographic characteristics and column is presented in Table 1 and Figure 1, while the mode of preparation of the samples is presented in Table 2. The results of analysis indicated that the highest moisture content (Table 3) was reported for cowpea meal (40 %), and it was observed that moisture was significantly higher in village kitchen meals than fast foods. Preparation of village kitchen meals involved the use of large volume of water which by implication could mean more dilution or less nutrients per volume of food compared to fast food fried and baked products. The village kitchen meals were served with soup and meat and/ or fish which apart from large portion size per serving could significantly affect the vitamin status of the meals compared with fast foods fries. While young adolescent and adult would require munching at least two serving size doughnut, beef roll, meat pie to satisfy appetite, a serving size of village kitchen meals and fast food rice kitchen meals and fast

Table 2. Sample description and mode of preparation.

Sample	Description and Method of Preparation
Conventional fast foods	
Jollof rice:	Rice cooked (or baked) with ground tomatoes, peppers, sometimes with meat and other spices.
Fried rice:	Rice cooked in vegetable oils, liver chop, meat, carrot, green beans and spices are added to taste
Meat pie	Snack made from wheat flour dough containing pastry of seasoned meat, potatoes and other vegetables
Doughnut:	Snack made from wheat flour dough mixed into a circular ball fried in deep frying oven and then robbed with sugar or honey
Hotdog	Snack made from wheat flour dough containing cooked sausage made from meat slurry, rolled up in a pastry and baked in the oven
Beef roll	Snack made from wheat flour dough containing cooked sausage is rolled up in a pastry and baked in the oven
Village kitchen meals	
Eba	Prepared from gari (peeled, grated cassava tuber, fermented, partially dehydrated and baked or fried) mixed with hot water to thick paste served with vegetable soup and fish
Amala	Prepared by mixing dried yam flour powder in boiled water into a thick paste, served with vegetable soup and fish
Pounded yam	Prepared by pounding boiled yam in a mortal with pestle to smooth paste served with vegetable soup and fish
Cowpea (Mashed Beans)	Prepared by cooking cowpea (*Vigna unguiculata L. Walp*), pepper, palm oil and other ingredients are added to the broth and mashed to paste.

food rice (dry weight presented) (Table 3) could satisfy appetite.

The result of pyridoxine content indicated that village kitchen meals recorded significantly lower value than fast foods. The least value was reported for *amala* (0.16 mg) and highest for fried rice (1.14 mg). In fast foods, pyridoxine value followed this decreasing order: fried rice > meat pie > jollof rice > beef roll. The high pyridoxine recorded in meals with lower moisture content is consistent with observation that pyridoxine in food is susceptible to light induced degradation and exposure to water which can cause leaching and consequent losses (Perveen et al., 2009), thus low level of pyridoxine observed in village kitchen meals (*amala, eba* and pounded yam) could be also adduced to high volume of water employed in preparation that could cause leaching of the vitamin into the cooking broth. The range of 0.34 to 0.60 mg /100 g was reported for soy product and 0.416 mg /100 g for buckwheat flour (Labiedzinka and Szefer, 2006).

The level of thiamine unlike pyridoxine was significantly higher in village kitchen meals than fast foods. Thiamine in village kitchen meals appear in this decreasing order; cowpea > amala > pounded yam, while baked hot dog and meat pie recorded higher values than doughnut which is processed by deep fat frying. Several studies have shown significant decreases in thiamine content during thermal processing although the extent of degradation depends on the commodity. Baking, pasteurization, or boiling of foods fortified with thiamine has been reported to decrease its content by up to 50% (Prodanov et al., 2004). The high content of thiamine recorded for village kitchen meals is consistent with the report that deficiency disease associated with thiamine is rare in developing countries except for those who suffer eating disturbance or alcoholism (Gordon, 2003). The impact of heat in village kitchen meals would be highly reduced due to volume of water and this could explain in part the reason for higher thiamine compared to fast food fries and baked foods.

Analysis of rice alone (fried/ Jollof) without meat or fish indicated that thiamine was below detection limit. This observation is in agreement with the report that rice (especially polished rice) is a poor source of thiamine, which was the main reason why "beriberi" a noticeable deficiency disease of thiamine has been reported to be endemic in areas where rice was the staple diet (Ozawa et al., 2001). The result presented for thiamine in rice in this study reflects the thiamine in composite meals and not rice alone, therefore, in Nigeria, where polished rice is becoming staple diet in homes, incorporation of complementary source of thiamine becomes inevitable if occurrence of thiamin deficiency diseases is to be averted.

The highest level of niacin (0.70 mg) was recorded in meat pie and least for beef roll and *amala* (0.18 mg). There was no significant difference between village kitchen meals and fast food rice but doughnut and beef roll recorded significantly lower values. Most data suggest that niacin is stable to processing, retention rates of 93% or higher were found after canning and subsequent storage of green peas, green beans, peaches and sweet potatoes on both wet and dry weight bases (Rickman et al., 2007).

Table 3. Hydro- soluble vitamin content of some foods prepared by a Nigerian eatery (mg / 100 g) and % realization of RDA (per serving).

Sample/ Vitamin	Moisture %	weight / per serving (dry)	Pyridoxine	% RDA	Thiamine	% RDA	Niacin	% RDA	Ascorbic acid	% RDA	Folate**	% RDA
Jollof rice	24 ± 0.1^{b}	239* ± 10^{b}	0.75 ± 0.02^{c}	44	0.29 ± 0.01^{d}	54	0.27 ± 0.02^{b}	19	0.94 ± 0.04^{b}	0.65	80 ± 0.9^{d}	41
Fried rice	19 ± 0.2^{c}	362* ± 13^{a}	1.14 ± 0.0^{a}	66	0.18 ± 0.02^{e}	53	0.53 ± 0.1^{b}	37	0.53 ± 0.01^{d}	0.75	53 ± 1.0^{e}	13
Meat pie	14 ± 0.1^{c}	116 ± 5.0^{c}	1.04 ± 0.01^{b}	61	0.30 ± 0.11^{d}	25	2.7 ± 0.3^{a}	50	0.5 ± 0.03^{d}	0.9	195 ± 5.0^{a}	48
Beef roll	16 ± 0.2^{c}	75 ± 0.8^{g}	0.67 ± 0.01^{d}	39	0.08 ± 0.0^{e}	6	0.18 ± 0.0^{c}	12	0.23 ± 0.01^{e}	0.2	30 ± 3.0^{f}	8
Doughnut	15 ± 3.1^{c}	60 ± 2.0^{h}	0.34 ± 0.03^{f}	20	0.11 ± 0.0^{e}	9	0.2 ± 0.0^{c}	14	0.2 ± 0.02^{e}	0.2	30 ± 3.0^{f}	8
Hot dog	15 ± 1.9^{c}	60 ± 4.2^{h}	0.5 ± 0.01^{e}	29	0.33± 0.01^{d}	27	0.38± 0.01^{b}	27	1.0 ± 0.05^{b}	1.1	130 ± 3.0^{a}	32
Eba	35 ± 2.8^{a}	148* ± 13^{d}	0.30 ± 0.02^{f}	17	3.3 ± 0.05^{a}	25	0.3 ± 0.0^{b}	21	1.1 ± 0.06^{a}	1.1	37 ± 2.0^{f}	9
Amala	38 ± 4.0^{a}	124* ± 8.0^{e}	0.16 ± 0.02^{g}	09	1.4 ± 0.03^{c}	116	0.18± 0.04^{c}	12	0.27 ± 0.04^{e}	0.3	58 ± 4.0^{e}	14
Pounded Yam	36 ± 5.0^{a}	158* ± 10^{d}	0.25 ± 0.0^{g}	14	1.3 ± 0.1^{c}	108	0.24 ± 0.0^{b}	17	0.2 ± 0.0^{e}	0.2	46 ± 1.0^{e}	11
Cowpea meal	40 ± 2.8^{a}	180* ± 15^{c}	0.29 ± 0.0^{f}	17	2.4 ± 0.02^{b}	200	0.44± 0.05^{b}	31	0.76 ± 0.0^{c}	0.8	169 ± 3.0^{b}	43

Values are mean ± standard deviation of triplicate analysis. Values with the same superscripts within the column are not significantly different ($p < 0.001$). *Foods served with meat or fish; **folic acid reported as µg /100g.

The content of folate (Table 3) ranged between 30 to 195 µg/100g; the least value was reported for doughnut and the highest was recorded for meat pie. The trend in the level of folate is in the order: meat pie > cowpea meal > hot dog. Legumes have been reported to contain high level of folate; this is evident in that cowpea meal recorded high folate whereas the relatively high folate content recorded in meat pie and hot dog could be as a result of fortification of the flour with folate. Leichter et al. (1978) reported that folate is relatively stable to heat and humidity hence the loss of folate in foods has been adduced to leaching and not to degradation of the molecule. One of the major results of a folate deficiency is that in the early phase of red blood cell formation, the cell cannot divide because they cannot form new DNA. The cells remain in immature form known as megaloblasts. The bone marrow of a folate deficient person produces mostly immature megaloblastic cells; few mature red blood cells arrive in the blood stream, thus the blood capacity to carry oxygen decreases causing a form of anaemia known as megaloblastic anaemia (Lieberman and Bruning, 1990).

The content of ascorbic acid (Table 3) ranged from 0.2 to 1.1 mg / 100 g with the highest value reported for eba. The content of ascorbic acid was generally low in all samples compared to fruits and vegetables; for instance Fox and Cameron (1989), reported 15 - 50 mg / 100 g for raw fruits. It is a common knowledge that ascorbic acid is heat labile and is by far the least stable nutrient during processing; it is highly sensitive to oxidation and leaching into water-soluble media hence is easily destroyed during food processing. Ascorbic acid is one of the ligands that promote absorption of iron but its action on copper is dose dependent (Hazell and Johnson, 1986). The action of ascorbic acid on food iron is usually related to its reducing power and metal chelating activities thereby forming soluble complexes.

For consumption of conventional fast food fries, on the average, a consumer can munch two servings of doughnut, hotdog or meat pie in order to satisfy hunger. Assuming the grams per serving (applying food weight on dry weight basis) of any of these food products was consumed per day, the percentage realization of the recommended dietary allowances for vitamins would indicate that jollof rice and meat pie would provide between 44 and 61% of pyridoxine whereas village kitchen meals would provide below 20%; for niacin all meals would provide a range of 12-40% though niacin is a semi-essential vitamin in man since niacin could be spared by tryptophan, from which it could be synthesized at the rate of 60 mg of the amino acid for every one mg of niacin (Gordon, 2003). This way, the body could synthesize up to 50% of the needed niacin. The use of tryptophan in this manner could however jeopardize the protein nutriture of the diets, making tryptophan a major limiting amino acid. On the other hand, the percentage realization of the RDA revealed that village kitchen meals would give 108-200%, whereas doughnut and beef roll could provide less than 10% of the RDA for thiamine, also cowpea, meat pie and hot dog could give 32-48 % while doughnut and beef roll would provide less than

10% of the RDA for folate. The level of ascorbic acid observed in this study is too low and could not meet the RDA of 90 mg hence the meals must be complemented with fruits and vegetables.

The study reveals that village kitchen meals contain higher level of some water soluble vitamins than conventional fast foods; it is envisaged that consumption of fast foods could lead to poor intake of the vitamins.

REFERENCES

Anyakora C, Afolami I, Ehianeta T, Onwumere F (2008). HPLC selected food products in Nigeria. Afr. J. Pharm. Pharmacol. 2 (2): 29-36.

Berger NA (1985). Poly (ADP-ribose) in the cellular response to DNA damage Radiat. Res.101:4-15.

Bailey RL, Dodd KW, Gahche JJ (2010). Total folate and folic acid intake from foods and dietary supplements in the United States 2003–2006: Am. J. Clin. Nutr. 91:231–7.

Berechet G, Segal R (2007). Vitamins Retention in some Microwave Dishes. The Annals of the University Dunarea de-Jos of Galati Fascicle IV. Food Technol, 27-32.

Fox, BA, Cameron AG (1989). Fruits and Vegetables Food Science. Nutrition and Health 5th Ed. Edward Arnoid London Melbourne Awekland pp. 273-280.

Gordon M (2003). Contemporary Nutrition; Issues and Insights 5th Edition Mc Graw Hill London.pp 250 -550.

Hazell, T. and Johnson, I.T. (1986). Effect of Food Processing and Fruit Juices on *in-vitro* estimated Iron Availability from Cereal, Vegetables and Fruits. J. Sci. Food Agric. 36, 73-82.

Khor S, Tee E (1996): Development of a HPLC method for the simultaneous determination of several B-vitamins and Ascorbic Acid. Mal. J. Nutr. 2:49–65.

Labiedzinka A, Szefer P (2006). Vitamin B in Grain and Cereal-Grain Food, Soy products and Seeds. Food Chem. 56:112-118.

Leichter J, Switzer VP and Landymore AF (1978). Effect of Cooking on Folate Content of Vegetables. Nutr. Rep. Int. 18:475–479.

Lieberman S, Bruning N, (1990). The Real Vitamin and Mineral Book NY Avery Group, pp. 3-5.

Olaniyi AA (2000). Essential Medicinal Chemistry, 2nd Edition, Shaneson, Ibadan, p. 324.

Ozawa H, Homma Y, Arisawa H, Fukuuchi F, Handa S. (2001). Severe metabolic acidosis and heart failure due to thiamine deficiency. Nutrition. 17:351–352.

Prodanov M, Sierra I, Vidal-Valverde C. (2004). Influence of Soaking and Cooking on theThiamin, Riboflavin and Niacin contents of Legumes. Food Chem. 84:271–277.

Perveen S, Arfa YA, Khanb KM (2009). Quantitative Simultaneous Estimation of Water Soluble Vitamins, Riboflavin, Pyridoxine, Cyanocobalamin and Folic Acid in Neutraceutical Products by HPLC. The Open Anal. Chem. J. 3:1-5

Rickman JC, Diane M, Barrett DM (2007). Nutritional comparison of fresh, frozen and canned fruits and vegetables. Part1. Vitamins C and B and phenolic compounds. J. Sci. Food. Agric 87: 930–944.

WHO (2005). Diet, Nutrition and Prevention of chronic diseases A joint WHO/FAO Expert Consultation report series 916.

Study of electrical properties of vegetable oils for the purpose of an application in electrical engineering

Khouloud LAKRARI[1] , Mouloud EL MOUDANE[2], Imane HASSANAIN[1], Imane ELLOUZI[1], Said KITANE[3] and Mohamed ALAOUI El BELGHITI[1]

[1]Laboratoire de Chimie Physique Générale, Département de Chimie, Université Mohammed V-Agdal, Faculté des Sciences, Avenue Ibn Batouta, BP 1014 Rabat.
[2]Laboratoire de Matériaux, Nanotechnologies et Environnement (LMNE), Département de Chimie, Université Mohammed V-Agdal, Faculté des Sciences, Avenue Ibn Batouta, BP 1014 Rabat.
[3]Laboratoires de chimie appliquée à l'Ecole Nationale de l'Industrie Minérale, Avenue Hadj Ahmed Cherkaoui BP 753 Agdal Rabat.

The growing importance of the issues for sustainable development requires the consideration of environmental criteria in the development of all new materials and equipment. In the case of dielectric liquid, a better balance must be found between technical performance within the filled equipment and the impact on the environment always possible leaks. In transformers, a stable liquid, inert, with good electrical and thermal properties is necessary; outside, the liquid must be non-toxic to the environment and readily biodegradable. The intrinsic properties of natural vegetable oils in terms of fire resistance, environmental performance and electrical and thermal characteristics of dielectric compositions are particularly useful products in the field of electrical engineering. The objective of our work was to make a comparative study of different vegetable oils: Argan, Rapeseed, Sunflower and analyze the behavior of the electrical resistivity and viscosity as a function of temperature. This study may be useful for a possible application of these oils in technology.

Key words: Electrical resistivity, viscosity, vegetable oil, electrotechnical.

INTRODUCTION

The environmental impact of human activity causes increasing concern: the situation is worst between climate problems, pollution, overexploitation of resources and individual consumption, which are continually increasing. These developments make it necessary to consider practical and compatible solutions with sustainable development (Bozell and Patel, 2006).

The concept of green chemistry (Anastas and Warner, 1998) was created to "support the design of products and processes that reduce or eliminate the use and formation of hazardous substances." These eco-design rules sometimes seem restrictive, but they can be beneficial, especially for chemists and manufacturers. The principles of green chemistry tend to more environmentally friendly processes for the synthesis of the environment and the use of renewable raw materials (Baumann et al., 1988). Industrials (and financials) have logically turned to the substitution of fossil fuels with renewable raw materials, such as vegetable oils for biofuel sector. However, we must ensure that the consumption of vegetable oils for these new applications does not disturb the balance of existing networks, in particular those related to human

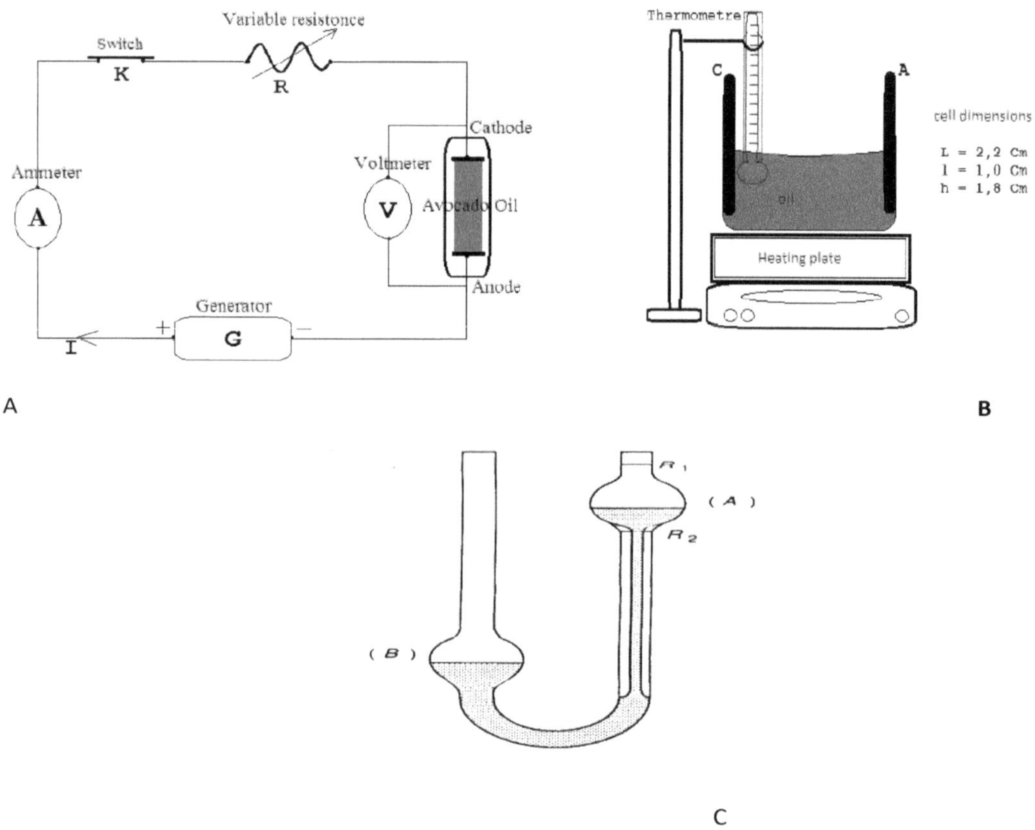

Figure 1. The diagram of the cell used to measure the electrical resistivityand viscosity. a. Construction of the used apparatus. b. Measurement of cell resistivity. C, Osswald viscometer.

consumption. Vegetable oils according to their fatty acid composition, have non-food applications. For example, azelaic acid and Pelargonic can be industrially produced by ozonolysis of oleic acid (Goebel et al., 1957), the majority fatty acid of olive oil, rapeseed and sunflower. Azelaic acid has pharmacological properties and is used in cosmetics and dermatology. The pelargonic acid is itself regarded as a by product of this reaction between and in the formulation of lubricants.

The oils are the main class of lubricants as well as the number of applications that the volume of lubricant used in the world (remember that fat is composed of 80 to 90% oil). Due to their natural affinity for metal surfaces and their viscosity, oil seep between the surfaces in relative motion and reduce friction, heat and wear.

Because of their chemical inertness when subjected to electric fields, oils are often used as electrical insulation for certain applications, including transformers, circuit breakers, cables, and capacitors.

For electrotechnical applications, in addition to electrical properties (rigidity, resistivity, and specific power inductor) are also sought good thermal properties.

Mineral oils, especially used at the beginning of electrical engineering, are now abandoned in favor of synthetic oils whose properties are much better for this type of application (Berger, 2002).

There are prototypes of biodegradable insulating oil for transformers based on derivatives of vegetable oils (fatty acids) and with unmatched performance at low temperatures. The biodegradability and non-toxicity of these insulating oils are substitutes choice for those that are commonly used in full based petrochemical derivatives, and therefore more prone to price fluctuations of petroleum products (Vitilingon et al., 1998) The application of vegetable oils in technology, requires a thorough study of their electrical and physical cleanliness (resistivity, viscosity etc.)

MATERIALS AND METHODS

Vegetable oils are generally very low toxics and have excellent biodegradability. These qualities are due in particular to a low resistance to oxidation and hydrolysis. Because of these two characteristics that are conducive to eco-toxicological aspects, plant oils are already used in distribution transformers and attempts are being made to expand their use in power transformers. (Darwin et al., 2007)

The vegetable oils: Argan, Rapeseed, Sunflower commercially available were purchased form the market.

The diagram of the cell used to measure the electrical resistivity is shown in Figure 1a and b. We used the method of resistivity measurement called "colon": the electrical resistance of the oil was determined by measuring the current and the potential difference

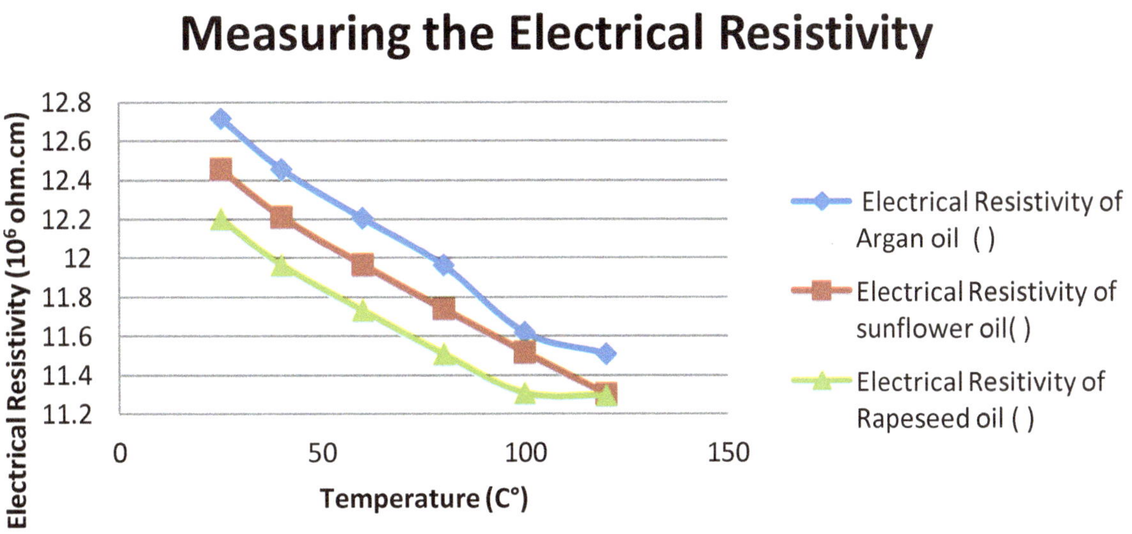

Figure 2. represents measurements of dynamic viscosity of vegetable oils: argan, rapeseed, sunflower.

(ddp) between the two electrodes of the cell. The viscosity WAS measured by a viscometer Osswald (Figure 1c):

Measurement of the electrical resistivity of vegetable oils

We used the following formula to measure the electrical resistivity (ρ):

$$\rho = R \times \frac{S}{L}$$

Where ρ is the resistivity (Ω.cm); S is the section (cm²); L is the length (cm).

S= l x L; l = 1.1 cm; L= 2.2 cm (distance between the two electrodes).

Measurement of the dynamic viscosity of vegetable oils

Measuring the time of a flow of a volume V of fluid through a capillary tube, the kinematic viscosity is proportional to the flow time:

$$\nu = K \cdot \Delta t$$

The constant K of the device is given by the manufacturer of the viscometer.

RESULTS AND DISCUSSION

We studied the variation of the electrical resistivity and the dynamic viscosity as a function of the temperature of vegetable oils: Argan, rapeseed and sunflower, the results obtained are represented in Figures 2 and 3.

We observe that the electrical resistivity decreased with increasing temperature, this decrease can be explained by: 1) By the chemical various changes occurring in the oil when heated; 2) the orientation of the molecules when the temperature increases (decrease in viscosity with increasing temperature) which promotes the flow of current in the oil. Tekin and Hammond (2000) studied the influence of temperature on the measurements of the electrical resistivity of soybean oil. These authors observed the same changes in resistivity as a function of temperature. Our results are in good agreement with those of Tekin and Hammond (2000).

The dependence of the viscosity of the fluid temperature is a manifestation of the behavior of cohesive energies and heat, with a rise in temperature, the attraction between polar molecules decreased while their thermal energy increased and the viscosity of the oil decreased.

Our measurements are in agreement with the measurements of Dilip et al. (2013) because the estimated coefficient by the author is very small so we can neglect it.

Measuring the electrical resistivity and the viscosity is a method for the control of the quality of oil and as an indicator of changes in new or oil in service resulting from contamination or damage.

We can conclude that the temperature facilitates when it increases the electrical conductivity and viscosity of the oils we studied. This study has allowed us to make a comparison of our results on the behavior of the resistivity and viscosity as a function of temperature with those of other researchers working on the same research topic (Pace et al., 1968; Risman and Bengtsson, 1971). Electrical measurements of oils that depend on the temperature can be used as a strong indicator of the deteriorating food quality oils at high temperatures.

The study of the viscosity and electrical resistivity of these oils can be useful for application in the field of technology (insulation, transformer etc). In perspective, we plan to complete our work by a study of the thermal resistivity as a function of temperature.

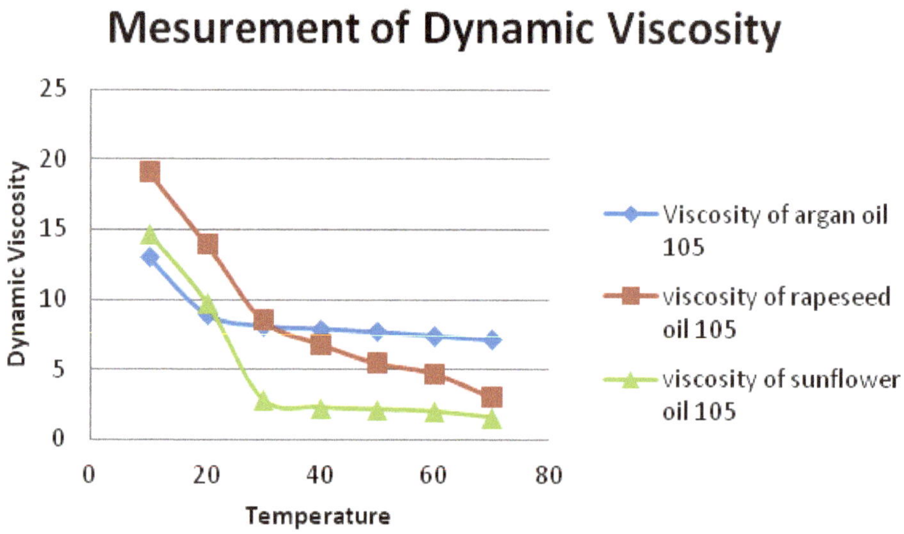

Figure 3. Measurements of dynamic viscosity of vegetable oils: argan, rapeseed, sunflower.

REFERENCES

Anastas PT, Warner JC (1998). Green chemistry theory and practice, Edition Oxford University press.

Baumann H, Bühler M, Fochem H, Hirsinger F, Zoebelein H, Falbe J, Angew (1988). "Natural Fats and Oils –Renewable Raw Materials for the Chemical Industry. Chem. Int. Ed. Eng. 27:41-62.

Berger N (2002)."liquids isolants en électrothechnique-caractéristiques des produits",techniques de l'ingénieur, Traité D2471, Novembre 2002, p. 24.

Bozell JJ, Patel MK (2006). Feedstocks for the Future – Renewable for the Production of Chemicals and Materials, Edition American Chemical Society.

Darwin A, Perrier C, Foliot P (2007). "The use of natural ester fluids in transformer", Proceedings of MATPOST conference, Lyon (France), Paper 0036: 15-16.

Dilip K, Amarjit S, Paramjit ST (2013). Interrelationship between viscosity and electrical properties for edible oils. J. Food Sci. Technol.

Goebel CG, Brown AC, Oehlschlaeger HF, Rolfes RP (1958). Method of making azelaic acid, Patent US 2813113, 1957. Chem. Abstr. 52:2431.

Pace WE, Westphal WB, Goldbith SA (1968). Dielectric properties of commercial cooking oil. J. Food Sci. 33:30-36.

Risman PO, Bengtsson NE (1971). Dielectric properties of foods at 3 GHz determined by cavity perturbation technique. J. Microw. Power. 6:101-106.

Tekin A, Earl G. Hammond. (2000). J. Am. oil Chemists' Society. 77 (3): 281-283.

Vaitilingom G, Perilhon Ch, Liennard A, Gandon M (1998). Development of rapeseed oil burners for drying and heating. Indus. Crops Prod. 7, 273-279.

Screening, identification and antagonistic activity of halo stable *Bacillus* sp. Mk22 used as probiotic in *Penaeus monodon* Fabricius, 1798

Sekar Ashokkumar [1] and Packyam Mayavu[2]

[1]International Maritime Academy, Chennai, India.
[2]Faculty of Marine Sciences, Annamalai University, Parangipettai, India.

The research of probiotics for aquatic animals is increasing with the demand for environmental friendly aquaculture. Most attempts to propose probiotic have been undertaken by isolating and selecting strains from saltpan environment. 16S rRNA gene sequencing showed that strain *Bacillus* sp. Mk22 (accession number: JF794553) was gram positive, had rod shape and was 937 nm in size. The isolated strain that was applied to *Penaeus monodon* culture under laboratory condition revealed that maximum survival rate was 92.7 ± 1.53%, wet weight was 6.9 ± 0.15 g, production was 378 ± 1.0 g and feed conversion ratio was 0.9 ± 0.03 in experiment-III compared to those of control experiment. The present study also showed that the halophilic (*Bacillus* sp.) bacterium was able to colonize both the culture in water and shrimp digestive tract. The minimum total bacterial counts (6.0 ± 1.0×10^5 Cfu ml^{-1}) and maximum *Bacillus* counts (5.6 ± 0.75×10^4 Cfu g^{-1}) were recorded in experiment-III and not in the control.

Key words: *Bacillus* sp., polymerase chain reaction, *Penaeus monodon*, probiotic, Saltpan.

INTRODUCTION

The UN FAO estimates that half of the world's seafood demand will be met by aquaculture in 2020, as wild capture fisheries are overexploited and are in decline. Shrimp (or prawn) culture is wide spread throughout the tropical world. It is in an industry set for a period with strong growing demand and is currently worth around US$10 billion. In most of the world, aquaculture industry is beset by disease outbreak, caused mostly by bacteria (especially for genome *Vibrio*) and viruses. The high density of animals in hatchery tanks and ponds is conducive for the spread of pathogens; and the aquatic environment as well as regular applications of protein-rich feed is ideal for culturing bacteria. Mostly, *Vibrio* spp. (*V. paraheamolyticus, V. harveii, V. alginolyticus*) cause major problems in aquaculture, resulting in reduced growth rate, poor feed consumption, loss of body weight and ultimately mass mortality.

A number of alternative strategies for the prevention and control of diseases have been proposed and have already been applied successfully in aquaculture, such as antibiotics, vaccines and immunostimulants. Antibiotics, one of the feed additives, were commonly used in the early 1950s (Ahilan et al., 2004). The problem of antibiotic contamination of aquaculture facilities and livestock and the indiscriminate worldwide use of antibiotics in aquaculture have led to the development of drug-resistant bacteria which are becoming increasingly difficult to control and eradicate (Hayashi et al., 1993; DePaola, 1995; Bruun et al., 2000; Sahul and Balasubramanian, 2000; Van der Waaij and Nord, 2000; Miranda and Zemelman, 2002). Subsequently, certain antibiotics such as chloramphenicol have been banned in many countries (Robert et al., 1995; FAO website – URL in reference list). As a result of resistant bacterial strains becoming more prevalent and difficult to treat, alternative methods of controlling the microbial environment are being investigated. One of the methods gaining recognition for controlling pathogens within the aquaculture industry is the use of beneficial or probiotic bacteria (Verschuere et al., 2000; Irianto and Austin, 2002a).

Some halophilic bacteria against aquculture pathogens are used in aquaculture. These bacteria both live in high saline and medium saline environment. Halophiles can be classified as slight, moderate and extreme microorganisms depending on their NaCl requirement. Slight halophiles require 0.2 – 0.85 M (2-5%) NaCl, moderate halophiles require 0.85 – 3.4 M (5-20%) NaCl and extreme halophilic microorganisms grow optimally above 3.4 – 5.1 M (20-30%) NaCl concentrations. Hence, in the present study, probiotic microorganisms were isolated from the saltpan environment and applied.

MATERIALS AND METHODS

Sediment samples were collected monthly from Marakkanam saltpan environment (Lat. 12° 14' 29N; Long. 79° 56' 28E) at Tamilnadu, India. Isolation of probiotic bacteria was done using well plate method in antagonistic activity. Physico-chemical character was not included in the results.

Isolation of halophilic bacteria

Isolation of halophilic bacterial was done using the halophilic agar (Hi-Media, Mumbai) (Mellado et al., 1998). Totally, 25 strains were isolated from the saltpan environment and kept at 4°C for further analysis.

Identification

16S rDNA sequencing was used to identify the beneficial bacteria. Once the DNA sequence was edited and assembled, identification was done using Microseq Analysis Software and Sequence Database and universal primers. Identification was based on the pairwise alignment algorithms and phylogenetic tree. The two shrimp pathogenic *Vibrio* strains namely, *V. parahaemolyticus* and *V. harveyi* were used to screen the antagonistic activity of the isolated halophilic bacteria. The beneficial bacteria were sequenced using the forward and reverse primer; 16S rRNA forward primer: *AGA GTT TGA TCC TGG CTC AG* and reverse primer: *ACG GCT ACC TTG TTA CGA CTT.* The sequences were assembled using Clustal W software version 1.82 (Thompson et al., 1994) available at http://www.ebi.ac.uk. Strains' morphology and their surface were observed under the scanning electron microscope (Hitachi-s-1500X-SEM).

Rearing of shrimp

Separate experiment was conducted to examine the effect of probiotic administrated to black tiger shrimp (*P. monodon*). The stocking (*P. monodon*), post larval-15 were purchased from commercial hatcheries and screened for WSSV using 2 step PCR test based on the methods of Lo et al. (1996). WSSV-negative PL was brought to the laboratory and stocked in 50 L capacity culture tanks having 35 L chlorinated filtered estuarine water. The animals were provided with proper aeration and feeding was given at 7.5% of the body weight initially; subsequent feeding was adjusted to 5 to 3.5% of the body weight per day according to the left-out, unutilized feed and increasing body weight of animals. The feed was given twice a day; 60% at dawn (6.00 a.m.) and 40% at dusk (6.00 p.m.). In the experimental tanks, the water quality parameters were maintained at the optimum range; the bottom water in the tank along with excess feed and fecal matter was siphoned out using 2 cm dia plastic hose to enhance the survival of the animals.

Experiment I

The effect of without probiotic on juveniles (*P. monodon*) was examined in experiment I. Shrimp was stocked in 50 L plastic tanks at density of 20 per tank in triplicate. The commercial pellet feed was used as a control. Water was exchanged weekly and animal behavior was observed every day.

Experiment II

The effect of commercial probiotic on juveniles (*P. monodon)* was examined in experiment II. Shrimps were stocked in 50 L plastic tank at density of 20 per tank in triplicate. The commercial pellet feed mixed with commercial probiotic contained spore of two species of *Bacillus* sp. at a concentration of 1.2×10^6 CFU/ml. Water was exchanged weekly and animal behavior was observed every day.

Experiment III

The effect of halophilic bacteria (*Bacillus* sp.) on juveniles (*P. monodon)* was examined in experiment III. Shrimps were stocked in 50 L plastic tanks at a density of 20 per tank in triplicate. The commercial pellet feed was mixed with halophilic bacteria (*Bacillus* sp.) at a concentration of 1.2×10^6 CFU/ml. Water was exchanged weekly and animal behavior was observed every day.

Survival and growth rate

At the end of each experiment, the percentage survival was determined and 10 shrimps were sampled randomly from each tank to determine wet weight, feed conversion ratio (FCR), production and survival rate with the following equation.

Production

The production rate was calculated by the following formula:

Production = Initial weight (g) / Final weight (g)

Feed Conversion Ratio (FCR)

The feed conversion was calculated using the formula:
FCR = Feed taken in (dry weight in g) / Weight gain (wet weight g)

Monitoring of bacteria

Water samples and digestive tract samples from shrimp were used to determine the counts of total bacteria and counts of the halophilic probiont bacteria (*Bacillus*). Prior to dissection or homogenation, the shrimps were rinsed with sterilized distilled water, washed with 0.1% benzalkonium chloride according to the method of Gatesoupe (1999) and then rinsed again with sterilized distilled water to remove all external bacteria. In all the experiment, the digestive tract was dissected out using sterile technique and then was homogenized. All samples were diluted serially with sterilized normal saline solution (0.85% w/v NaCl). Total counts of bacteria were determined by plating on zobal marine agar (with 1% w/v NaCl). *Bacillus* bacteria in water and digestive tract samples were cultured

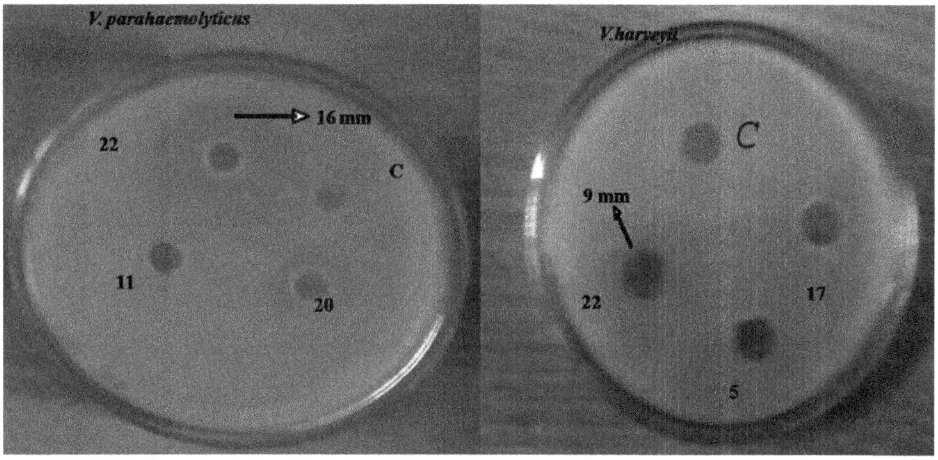

Figure 1. Experimental studies showed antagonistic activities of halophilic strain (*Bacillus* sp.)

according to the method recommended by Probiotics International Ltd. (Protexin Aquatech, Registration Dossier, unpublished pamphlet). The number of colonies on each plate was counted after incubation for 68 h at 25°C for water samples and for 72 h at 37°C for digestive tract samples. Most of the *Bacillus* spp. (Bacillus sp. MK22) cultured in fresh or marine water grow well in high and moderate salt concentration. *Bacillus* sp. had maximum cell count inside the digestive tract of shrimp for 54 h. All statistical analysis was significance; the level of '$P < 0.05$ was used for all tests. Data are reported as means ± standard deviation.

RESULTS

The present study showed that every month, the total bacterial population increased between 12 x10^6 CFUg^{-1} to 45 x10^6 in saltpan environment.

A total of 25 halophilic bacterial strains were preliminary screen based on the zone of inhibition; five strains showed acceptable activities. Among these, a strain, *Bacillus* sp. Mk22 showed large clear zones around the bacterial colonies and it was selected for further studies.

The initial screening is shown in Figure 1. The strain showed maximum zone of inhibition against *V. paraheamolyticus* (16 mm) and the minimum was observed against *V. harveyii* (9 mm) (Figure 1).

The SEM images were taken to verify the size and morphology of bacterial cells. Figure 2 shows the SEM image of *Bacillus* sp: rod shape, smooth and had approximate size of 3.0 µm. The isolate, *Bacillus* sp. Mk22 was subjected to molecular level identifications.

In the Phylogenetic analysis based on a comparison of the 16S rDNA sequence, data showed that the genus *Bacillus* is phylogenetically homogeneous, forming a distinct lineage within the radiation of the bacteria.The dendrogram placed the strain *Bacillus* sp. Mk22 in a separate line of descent within the genus *Bacillus*, representing a distinct phylogenetic lineage (Figure 3).

Further, representatives of other species like *Bacillus baekryungensis*, *B. marisflavi*, *Cloacibacterium normanense*, *B. ferrariarum*, *B. aquimaris*, *Bacillus* sp., *Bacillus* sp., and *Bacillus* sp. Mk22 (JF794553.1) with 99% similarity in the genus levels formed a distinct cluster.

The maximum survival (92.7 ± 1.53%) was recorded in experiment-III and the minimum (60 ± 2.53%) was observed in experiment I (Table 1). Administration of the probiotic significantly increased survival in all treatments (generally by 20–30%) over the controls, except in experiment III, where survival was significantly ($P < 0.05$) different from that in the controls (Table 2).The maximum wet weight (6.9 ± 0.15 g) was recorded in experiment III and minimum (3.5 ± 0.10 g) was observed in experiment I (Table 1). Wet weight was significantly ($P < 0.05$) greater in treatments than in the control. The production rate was found to be higher (378 ± 1.0 g) in experiment III followed by experiment II (301 ± 1.0 g); in the control experiment, it was low (159.3 ± 1.53 g) (Table 1). Production was more ($P < 0.05$) significant in experiment III than in the control. The maximum (1.2 ± 0.28) FCR was recorded in experiment I and minimum (0.9 ± 0.03) FCR was recorded in experiment III (Table 1). There was 5% level of significance in experiment-III compared to the control.

Monitoring of bacteria

In all the experiments, Halophilc (*Bacillus* sp.) bacteria successfully colonized both the culture water and the digestive tract of the shrimp (Table 2). The maximum total count (24 ± 0.79×10^5 Cfu ml l^{-1}) was recorded in experiment I and minimum (6.0 ± 1.0×10^5 Cfu ml l^{-1}) was observed in experiment III; and the maximum *Bacillus* count (3.7 ± 0.58×10^4 Cfu g^{-1}) was recorded in experiment III and minimum (0×10^4 Cfu g^{-1}) was observed in experiment I in water (Table 2).

The maximum total *Bacillus* count (35.4 ± 0.59×10^5 Cfu ml l^{-1}) was recorded in experiment I and minimum (4.1 ± 0.47×10^5 Cfu ml l^{-1}) was observed in experiment III; and the maximum *Bacillus* count (5.6 ± 0.75×10^4 Cfu g^{-1}) was recorded in experiment III and minimum (0×10^4 Cfu g^{-1}) was observed in experiment I in the digestive tract (Table

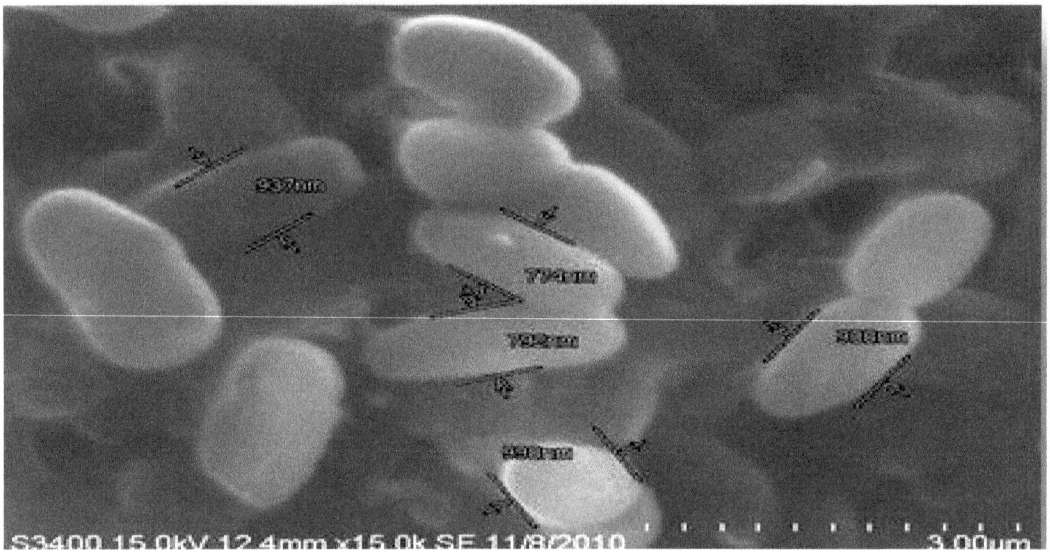

Figure 2. Scanning electron microscopy (SEM) image of *Bacillus* sp. Mk22.

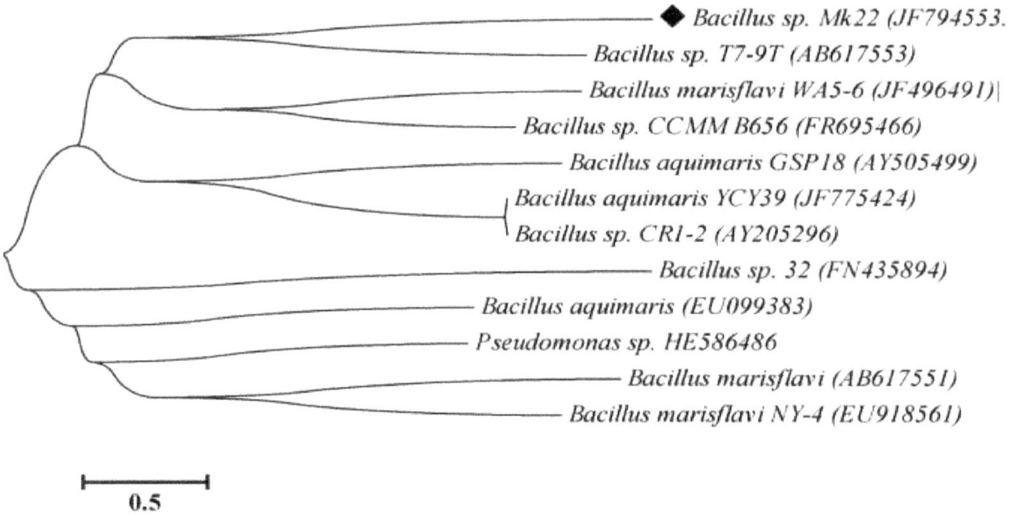

Figure 3. 16S rRNA gene sequence-based phylogenetic relationships of *Bacillus* sp. Mk22 (830 nucleotides) and closely related members of the genus *Bacillus*. The tree was constructed using the neighbourjoining algorithm. GenBank accession numbers are given in parentheses. Only bootstrap values above 50% are shown (1000 replications). Bar 0.5 substitutions per 100 nucleotides. Triangle indicates the present study strain.

Table 1. Growth and survival parameters of *P. monodon* reared with and without *Bacillus* probiotic added to water.

Experiment	Stocking density (*P. monodon*)	Survival (%)	Wet weight (g)	Final production(g)	FCR
I	60	60 ± 2.52	3.5 ± 0.10	159.3 ± 1.53	1.2 ± 0.28
II	60	83 ± 2.0	5.5 ± 0.36	301 ± 1.0	1.0 ± 0.04
III	60	92.7 ± 1.53	6.9 ± 0.15	378 ± 1.0	0.9 ± 0.03

Mean ± S.D. Exp-I (Control no probiotic provided); Exp-II- Commercial probiotic (Superbiotic) provided; Exp- III *Bacillus* sp. Mk22 provided.

Table 2. Total bacterial count and *Bacillus* count in water and in digestive tracts of *P. monodon* reared with and without *Bacillus* probiotic added to water.

Experiment	Water		Digestive tract	
	Total counts (10^5CFU ml l^{-1})	*Bacillus* count (10^4 CFU ml l^{-1})	Total counts (10^5 CFU g^{-1})	*Bacillus* count (10^4 CFU g^{-1})
I	24 ± 0.79	0	35.4 ± 0.59	0
II	8 ± 1.00	5 ± 1.00	8.7 ± 0.85	4.4 ± 0.42
III	6 ± 1.00	3.7 ± 0.58	4.1 ± 0.47	5.6 ± 0.75

Mean ± S.D. Exp-I (Control no probiotic provided); Exp-II- Commercial probiotic (Superbiotic) provided; Exp- III *Bacillus* sp. Mk22 provided.

2). The ANOVA showed 5% level of significance ($P < 0.05$) between the water and digestive tract.

DISCUSSION

Aquaculture is badly affected by parasites, fungi, bacteria, viruses and non-infectious diseases cause problems. Various toxins and other water quality stressors can also affect crustacean health. With the increasing importance of crustacean aquaculture as well as mounting pressures on fisheries, there is a need for reliable accurate means for developing the health of crustaceans (Sindermann, 1990). Particularly, vibriosis is one of the major diseases in aquatic organisms. This is usually caused by the species belonging to the genus *Vibrio* which are the natural inhabitants of estuarine and marine environments, well known for causing vibriosis in fish worldwide (Schaperclaus, 1986). Several species are known to be pathogenic to aquatic animals as well as humans. They are highly abundant in aquatic environments, including estuaries, marine coastal waters and sediments. Even though there are several remedies in controlling *Vibrio,* there exists several constraints such as the prevalence of the antibiotic resistance bacteria. In this present study we aimed to find the potential of the application of halophilic bacteria for controlling the crisis of vibriosis in the culture of *P. monodon*.

The total bacterial count varied between 12 x10^6 and 45 x10^6 CFUg^{-1}. 16S rDNA gene sequencing is a powerful tool that has been used to trace phylogenetic relationships between microorganisms and to identify bacteria from various sources, such as environmental or specimens. This technology is used today in laboratories for routine identifications, not only for slow-growing, unusual or fastidious bacteria but also for bacteria that are poorly differentiated by conventional methods.

Among the five isolates, *Bacillus* sp. Mk22 was selected for molecular level identification and it was subjected to further probiotic study. In phylogenetic analysis based on a comparison of the 16S rDNA sequencing, data showed that the genus, *Bacillus* is phylogenetically homogeneous, forming a distinct lineage within the radiation of the bacteria. The dendrogram placed the strain *Bacillus* sp. Mk22 in a separate line of descent within the genus *Bacillus*, representing a distinct phylogenetic lineage and forming a distinct cluster. Gracia et al. (1987) reported an endose spore forming gram positive *Bacillus* sp from saltpan in USA. Even though Garabito et al. (1997) proposed the new species *Bacillus salexigens* in hypersaline soils located in different soils located in different geographical areas of Spain, Yoon et al. (2004) proposed the transfer of *Bacillus* halodenitrificans a gram variable, endospore forming moderately halophilic rod isolated from a marine solar saltern of the yellow sea in Korea.

Bacillus sp. Mk22 is closely related to the *Bacillus* sp., 32 JD-2009 (FN435894); there is detection of halophilic bacteria and archaea on the extreme salt environment that attacks monuments. There was identification of halophilic bacteria from a salt marsh and two salterns in the protected ecosystem of lower Loukkos (Larache, Morocco). *Bacillus* sp T7-9T (AB617553) and *Bacillus* sp., CCMM B656 (FR695466) are all the strains isolated from the saltpan environment in South Korea (Na et al., 2011); and other species related to *Bacillus aquimaris,* with accession number EU099383, have 99% similarity in the species level. *B. aquimaris* is not only present in the saltpan environment but also in the marine environment. *B. aquimaris* is isolated from the Kumta Coast of Karnataka, India; this bacterium grows at pH 7.5- 9.5 and 40°C (Pooja and Jayaraman, 2009). *B. aquimaris* is a relatively novel marine bacterium. Recent reports are available on the culture characteristics, sequence information and phylogeny of the organism (Gontang et al., 2007; Liu and Shao, 2007). Yoon et al. (2004) reported the isolation of endospore-forming, rod shaped moderately halophilic, *B. aquimaris* (TF-12T) from seawater.

Halophilic bacterial forms produce several secondary metabolites such as bacteriocins, bacteriocin like substances and antibacterial lipopeptides and these metabolites have received considerable attention as biological control agents in pharmaceutical industry because they are generally recognized as safe (GRAS), have low toxicity, high biodegradability, and are environmental friendly (Oliveira and Pijoan, 2004). In the present study, the halophilic bacterial strain *Bacillus* sp. showed antagonistic activties against all shrimp pathogens (*V. paraheamolyticus* (16 mm) and *V. harveii* (9 mm)). Lee et al. (2010) reported a novel analysis of *Bacillus subtilis* SC-8 antagonistic to *Bacillus cereus.* Liu et al. (2008) proved antagonistic activities of volatiles from four strains of *Bacillus* spp. and *Paenibacillus* spp. against soil-borne plant pathogens.

Several early reports are available for mode of action and application of probiotics in ponds (Sugita et al., 1998). In the present investigation, halophilic bacterial strain *Bacillus* sp. was used as probiotic for *P. monodon.* Under laboratory conditions, the result showed excellent production rate (378 ± 1.0 g), wet weight (6.9 ± 0.15g) and FCR (0.9 ± 0.03). Rengpipat et al. (2000) reported that *Bacillus* sp. will activate both cellular and humoral immune defense against shrimp pathogens in tiger shrimp (*P. monodon*). Balcazar et al. (2006) stated that the administration of a mixture of bacterial strains (*Bacillus* and *Vibrio* sp) positively influence the growth and survival of juvenile of white shrimp. The present study also revealed that *Bacillus* sp. was able to colonize both the culture water and the shrimp digestive tract. Based on the above observations, it was concluded that the strain of halophilic *Bacilus* sp. was effective in inhibiting the shrimp pathogens, like *V. paraheamolyticus* and *V. harveyi* both *in vitro* and *in vivo.* The probionts significantly reduce mortality and also do not have any pathogenic effect on the shrimp larvae. Halophilic (*Bacillus*) bacteria were administered to *P. monodon* under laboratory condition; final production, FCR, wet weight and survival were significantly higher in experiments I-III than in the control. Therefore, these bacterial probionts can be used effectively to control the shrimp pathogens and enhance production that may substitute the use of antibiotics in aquaculture.

ACKNOWLEDGEMENTS

We are grateful to Prof. T. Balasubramanian, Dean, Faculty of Marine Sciences, CAS in Marine Biology, Annamalai University, Parangipettai and also Prof. S. Balasubramaniam, Dean International Maritime Academy, Chennai for providing the materials and for the encouragement.

REFERENCES

Ahilan B, Shine G, Santhanam R (2004). Influence of probiotics on the growth and gut microflora load of juvenile Gold fish (*Carassius auratus*). Asian Fish. Sci. 17: 271–278.

Balcazar JL, Blas I, Ruiz-Zarzuela I, Cunningham D, Vendrell D, Muzquiz JL (2006). The role of probiotics in aquaculture. Vet. Microbiol. 114: 173–186.

Bruun MS, Schmidt AS, Madsen L, Dalsgaard I (2000). Antimicrobial resistance patterns in Danish isolates of Flavobacterium *psychrophilum*. Aquaculture 187:201-212.

DePaola A (1995). Tetracycline resistance by bacteria in response to oxytetracycline contaminated catfish feed. J. Aquat. Anim. Health 7: 155-160.

Garabito MJ, Arahal DR, Mellado E, Marquez MC, Ventosa A (1997). *Bacillus salexigens* sp. nov., a new moderately halophilic *Bacillus* species. Int. J. Syst. Bacteriol. 47: 735-741.

Gatesoupe FJ (1999). The use of probiotics in aquaculture. Aquaculture 180:147- 165.

Gontang EA, Fenical W, Jensen PR (2007). Phylogenetic diversity of Gram positive cultures from marine sediments. Appl. Environ. Microbiol. 73: 3272–82.

Hayashi F, Fuse A, Inoue M, Ishii H, Barcs I, Mitsuhashi S (1993). Frequency of drug resistance in *Vibrio* spp. in Japan. Lett. Appl. Microbiol. 16: 28-31.

Irianto A, Austin B (2002). Probiotics in aquaculture Review. J. Fish Dis. 25:633-642.

Lee NK, Yeo IC, Park JW, Kang BS, Hahm YT (2010). Isolation and characterization of a novel analyte from *Bacillus subtilis* SC-8 antagonistic to *Bacillus cereus*. J. Biosci. Bioeng. 110: 298-303

Liu W, Wei M, Zhu B, You-chen D, Liu F (2008). Antagonistic Activities of Volatiles from Four Strains of *Bacillus* spp. and *Paenibacillus* spp. Against Soil-Borne Plant Pathogens. Agric. Sci. China 7(9):1104-1114.

Liu Z, Shao ZZ (2007). The diversity of alkane degrading bacteria in the enrichments with deep sea sediment of the South China Sea. Pub. Med. 47: 869-73.

Lo CF, Leu JH, Chen CH, Peng SE, Chen YT, Chou CM, Wang CH, Kou GH (1996). Detection of B*aculovirus* associated with White Spot Syndrome (WSBV) in penaeid shrimp using polymerase chain reaction. Dis. Aquat. Org. 25:133–141

Mellado E, Garcia MT, Roldan E, Nieto JJ, Ventosa A (1998). Analysis of the genome of the Gram-negative moderate halophiles Halomonas and Chromohalobacter by using pulsed-field gel electrophoresis. Extremophiles 2: 435-438.

Miranda CD, Zemelman R (2002). Bacterial resistance to oxytetracycline in Chilean salmon farming. Aquaculture 212(1-4):31-47.

Na J-M, Kang M-S, Kim J-H, Jin Y-X, Je J-H, Kim J-B, Cho Y-S, Kim J-H, Kim S-Y (2011). Distribution and Identification of Halophilic Bacteria in Solar Salts Produced during Entire Manufacturing Process. Korean J. Microbiol. Biotechnol. 39:(2) 133–139

Oliveira S, Pijoan C (2004). Haemophilus parasuis: new trends on diagnosis, epidermiology and control. Vet. Microbiol. 99:1-12.

Pooja S, Jayaraman G (2009). Production of extracellular protease from halotolerant bacterium, *Bacillus aquimaris* strain VITP4 isolated from Kumta coast. Process Biochem. 44:1088-1094.

Rengpipat S, Pukpratanporn S, Piyatitivorakul S, Menasaveta P (2000). Immunity enhancement in black tiger shrimp (*Penaeus monodon*) by a probiont bacterium (*Bacillus* S11). Aquaculture 191: 271-288.

Robert R, Nicolas JL, Miner P (1995). Mortality control of Pecten maximus larvae in hatchery. In Abstracts. 10th. Pectinid Workshop, Cork Ireland. pp. 51-52.

Sahul Hameed AS, Balasubramanian G (2000). Antibiotic resistance in bacteria isolated from Artemia nauplii and efficacy of formaldehyde to control bacterial load. Aquaculture 183:195-205.

Schaperclaus W (1986). Fish Diseases. Akademie-Verlag, Berlin. 2: 665-666 .

Sindermann CJ (1990). Principal diseases of marine fish and shellfish, Vol. 2. Gulf Professional Publishing, Academic Press, San Diego, CA

Sugita H, Hirose Y, Matsuo N, Deguchi Y (1998). Production of the antibacterial substance by *Bacillus* sp. strain NM12, an intestinal bacterium of Japanese coastal fish. Aquaculture 165:269-280.

Thompson JD, Higgins DG, Gibson TJ (1994). CLUSTAL W: Improving the sensitivity of progressive multiple sequence alignment through sequence weighting, position specific gap penalties and weight matrix choice. Nucleic Acids Res. 22: 4673-4680.

Van der Waaij D, Nord CE (2000). Development and persistence of multi-resistance to antibiotics in bacteria; an analysis and a new approach to this urgent problem. Aquaculture 16:191-197.

Verschuere L, Rombaut G, Sorgeloos P, Verstraete W (2000). Probiotic bacteria as biological control agents in aquaculture. Microbiol. Mol. Biol. Rev. 64: 655-671.

Yoon JH, Oh TK, Park YH (2004). Transfer of *Bacillus* halodenitrificans Denariaz et al., 1989 to the genus *Virgibacillus* as *Virgibacillus* halodenitrificans comb. nov. Int. J. Syst. Evol. Microbiol. 54: 2163-2167.

Sensory evaluation and sugars contents of coconut (*Cocos nucifera* l.) water during nuts ripening

ASSA Rebecca Rachel[1,4] , PRADES Alexia[2], KONAN Amoin Georgette[1], NEMLIN Jean[3] and KONAN Jean-Louis[4]

[1]Unité de Formation et de Recherche (UFR) Biosciences, Laboratoire de Biochimie et Sciences des Aliments; Université Félix Houphouët Boigny-Abidjan, (*) 21 BP 1126 Abidjan 21, Côte d'Ivoire.
[2]Centre de Coopération Internationale en Recherche Agronomique pour le Développement (CIRAD), Département PERSYST "Performances des systèmes de production et de transformation tropicaux", Unité Mixte de Recherche "QUALISUD", TA B 95/16. 73, rue JF Breton 34398 Montpellier Cedex 5 France.
[3]Centre National de Recherche Agronomique (CNRA) – Station sur la Recherche Technologique, 08 BP 881 Abidjan 08, Côte d'Ivoire.
[4]CNRA- Station Marc Delorme pour la recherche sur le cocotier de Port- Boüet, 07 BP 13 Abidjan 07, Côte d'Ivoire.

Sensory characteristics of coconut water from four coconut cultivars (*Cocos nucifera* L.) were investigated during nuts ripening. The study was carried out on the West African Tall, the Malaysian Yellow Dwarf, the Equatorial Guinea Green Dwarf and the improved hybrid PB121. Tested parameters were tastes (sweet, salty and sour), gustatory preferences and sugars contents. Obtained results showed significant interactions between cultivars and maturity stages for analyzed parameters. So, during the ripening of nuts, the sweet taste of coconut water predominated according to the maturity stages. It was maximal at the rank 21 of dwarf cultivars with perception equivalent from 5.80 to 6.10% sugars contents. At full maturity, the coconut water sweet taste lowers because its sugars enhance the formation of the kernel. The results of hedonic tests led to the preference of the coconut water according to its sweet taste. So, the water of dwarf cultivars immature nuts, more sweetened, was the most appreciated. Sugars contents are specific for cultivars and maturity stages. Their values were maximal at ranks 19 and 21 of dwarf cultivars. Moreover, gustative data were closed to those of the sweet taste perception. This study allowed to determinate the sensory characteristics of the coconut water in order to give indicators for its efficient technological promotion.

Key words: Coconut water, sensory tests, sugars contents, cultivars, ripening.

INTRODUCTION

The coconut tree (*Cocos nucifera* L.) is a permanent culture of tropical countries. It is also called 'life's tree' due to its numerous uses (Manisha and Shyamapada, 2011). The coconut tree plays a socio economic primordial role in all humid inter tropical zones. Indeed, about 10 millions families, draw their resources with coconuts speculations (Moore and Batugal, 2004). So, in Ivory Coast, the coconuts farms cover 50000 ha among which more than 90% are located along the coastal region (Amrizal, 2003). In this region, the coconut tree represents the main source of incomes for most peasants (Konan et al., 2008). Indeed, the water and the kernel are edible and economically profitable parts of the coconut. However, technologies transformation is not fully

Table 1. Concentrations (%) of the coconut water tasting witnesses substances.

Witnesses solutions codes	Sweet (sucrose)	Salty (sodium chloride)	Acid (citric acid)
1	0	0	0
2	1.75	0.05	0.01
3	3.5	0.1	0.02
4	4.375	0.15	0.03
5	6.125	0.2	0.04

developed there. They are limited to the manufacture of the copra (dried kernel at 6% of humidity) and oil extraction, at small scale, with mature nuts kernels. Besides, an accomplished inquiry into coastal region of Ivory Coast showed that 90% of populations use the coconut water to quench thirst or for its therapeutic virtues (Assa et al., 2006). Moreover, the physic-chemical characteristics of the coconut water were previously studied. So, the immature coconuts water is a delicious and nutritious drink (Jackson et al., 2004). It can be used also for the manufacture of syrups or fermented drinks (Enonuya, 1988). At maturity, coconut water loses its delicious taste to the advantage of the kernel (Jayalekshmy et al., 1986). However, in spite of coconut water, incentive prices, therapeutic virtues and high consumption, its sensory characteristics are not much studied, with few exceptions (Lapitan and Mabesa, 1983). Besides, these previous studies do not always treat cultivars intra and inter variations.

Also, they are generally centered on hedonic characteristics of fruits, without indications on their sensory profiles. Thus, sensory methods have an important place among analytical techniques because they use human as instrument measure (Raoux, 1998). Moreover, in Ivory Coast, sensory data of coconuts are non existent. So, the present document studies the sensory precisely gustatory parameters of coconuts water during nuts ripening. Values will be compared to sugars contents in order to characterize specifically coconut water and give indicators for their efficient promotion. Then, this work is interesting for the field because it will permit a consensus about analytical and sensorial methods for coconuts characterization.

MATERIALS AND METHODS

The biological material was constituted of coconut cultivars mostly used by their actors. It was about the West African Tall (WAT), the Malaysian Yellow Dwarf (MYD), the Equatorial Guinea Green Dwarf (EGD) and the improved hybrid PB121 (PB121+); obtained by a crossing between MYD and WAT (Bourdeix et al., 1992). Fruits were harvested on healthy grown-up coconut trees plots located on the station Marc Delorme of the National Centre of Agronomical Research (CNRA) in Abidjan, Ivory Coast.

Sampling

Six coconut palms were haphazardly chosen per cultivar, giving 24 palms samples. Among these, six bunches at six different stages of maturity (ranks) were simultaneously harvested. Maturation ranks were as follows: 17, 19, 21, 23, 25 and 26, corresponding respectively to fruits of 5, 7, 9, 11, 13 and 14 months old after conception. Five fruits from each bunch of a coconut palm were sampled. Their waters were homogenized to form a representative sample for analysis. This study was repeated over three successive harvest campaigns. Harvested fruits were sheltered to prevent the effects of sun and rain. Then, coconut water samples were analyzed less than 24 h after the harvest. Experiments were conducted using the same palms over two successive years; in 2007 and 2008 from February to April.

Sensory analyses

Sensory methods included descriptive tests, relating to the realization of sensory profiles of coconut water tastes. It was besides; on hedonic tests intended to assess the preferred level of the coconut water samples. Contrary to hedonic analyses, the descriptive tests required skilled tasters. So, a jury of 13 tasters was formed on the Marc Delorme station after the selection tests. These were consisted on the determination of gustatory acuteness according to the norms NF V09-002 of AFNOR (1984). They included identification tests of basic tastes (sour, bitter, salty and sweet) and determination of their perceptibility thresholds from substances witnessed. A room was adjusted for the tests according to norms NF V09-105 (AFNOR, 1984).

Descriptive tests

Panelists performed descriptive tests of the coconut water on a scale of notation in 5 points according to norm NF V09-16 (AFNOR, 1984). Tested parameters were sweet, salty and sour tastes. Concentrations of the tasting witnesses were prepared by addition of the witness substance in the mineral water (Table 1). Their maximum values were determined according to bibliographic data (Campos et al., 1996). For each tested taste, every subject savored successively the sample and the tasting witnesses. The intensity of the taste was immediately noted on the form. Gathered numerical informations were converted into corresponding concentrations of the substances witnesses. Indeed, for a given taste, 5 points on the scale represented respectively 5 successive concentrations of the tasting witness. These data allowed the sensory profiles establishment of the coconut water according to cultivars and maturity stages.

Hedonic tests

These tests were performed to assess the degree of samples gustatory evaluation. They required 35 not trained tasters, on a 5 points notation scale. Each taster registered on a form, a numerical value characteristic of the savored sample, according to his own

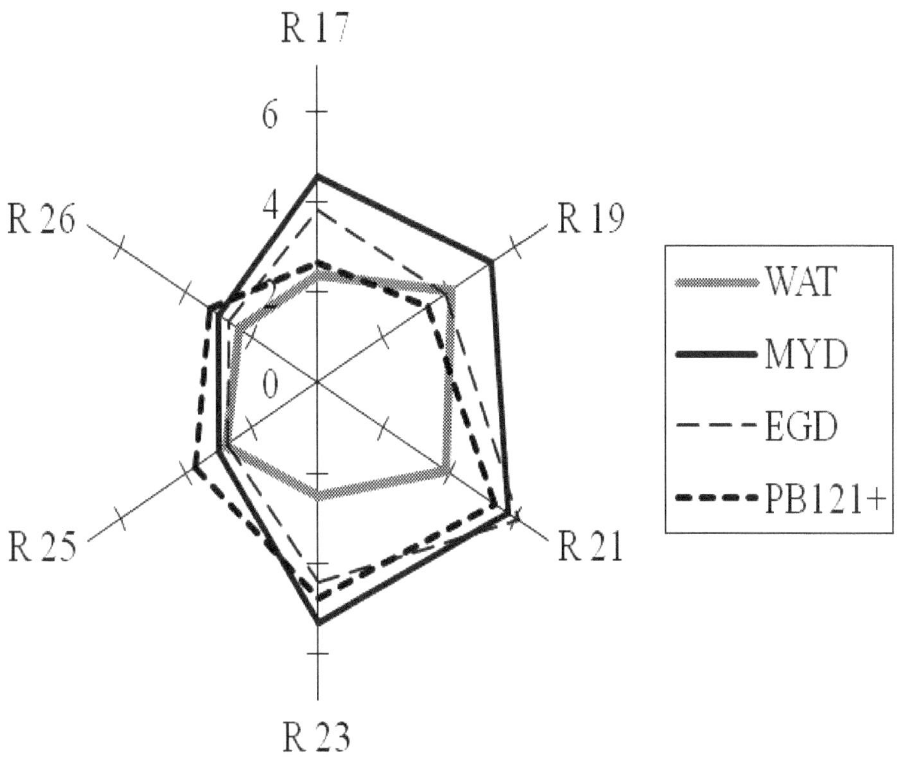

Figure 1. Sweet sensory profile of the coconut water according to cultivars and nuts maturity stages (LDS: 0.74).

evaluation. So, value 1 meaned no preference, 2 points indicated a weak preference, 3 translated a medium preference, 4 meaned a strong preference and 5 a very strong preference.

Sugars determination

The coconut water sugars were determined with spectrophotometer (Spectronic Genesis 5) according to Dubois et al. (1965) methods. It was about the determination of total soluble sugars with the sulfuric-phenol method and reducer sugars with 2-hydroxy-3, 5-dinitrobenzoic acid (DNS).

Statistical analyses

Data were treated with statistical software GenStat (GenStat Discovery, Edition 2). So, an analysis of variances (ANOVA) with two classification criteria was performed at 1 and 5% significant level for treatments of the coconut water sensory profiles and sugar contents. Criteria of classification were cultivars and maturity stages, Campaigns constituted blocks. Averages values were compared by the least significant difference (LSD). Results of hedonic tests were compared by Duncan test.

RESULTS

Statistical analysis showed significant variations during nuts ripening with a highly significant interaction ($p < 0.01$) between cultivars and ranks for each tested parameters.

Sensory profiles

Sweet taste predominated in the water of especially immature nuts whoever were the studied cultivars. It was maximal for dwarf cultivars MYD and EGD at rank 21 and equivalent to respective contents of 5.80 and 6.10% (Figure 1). The sweet taste of the coconut water lowered, whoever was the cultivar, beyond the rank 21 up to nuts full maturity. So, it was minimal at WAT rank 26 with a perception equivalent to 2.39%. But, at full maturity, PB121+ nuts water had the strongest sweet taste. The coconut water had lesser salty and sour (acid) tastes in comparison with its sweet taste. So, the salty taste, expressed in percentage of sodium chloride (NaCl), decreased from rank 17 ($11.90.10^{-2}$%) to 26 ($8.90.10^{-2}$%) for WAT. On the contrary, salty tastes of MYD, EGD and PB121+ nuts water, went down at the rank 21. They increased later up to the rank 26 with a maximum value equivalent to $12.6.10^{-2}$% content for EGD (Figure 2). The sour taste of the coconut water, expressed in percentage of citric acid, evolved during the ripening of nuts. So, for WAT cultivar, it augmented from the rank 21 ($1.9.10^{-2}$%) to 26 ($1.2.10^{-2}$%). The sour tastes went down up to 21 where they represented a minimal content of $1.1.10^{-2}$% for EGD and PB121+. Values increased then until $1.8.10^{-2}$ at 26 for MYD (Figure 3).

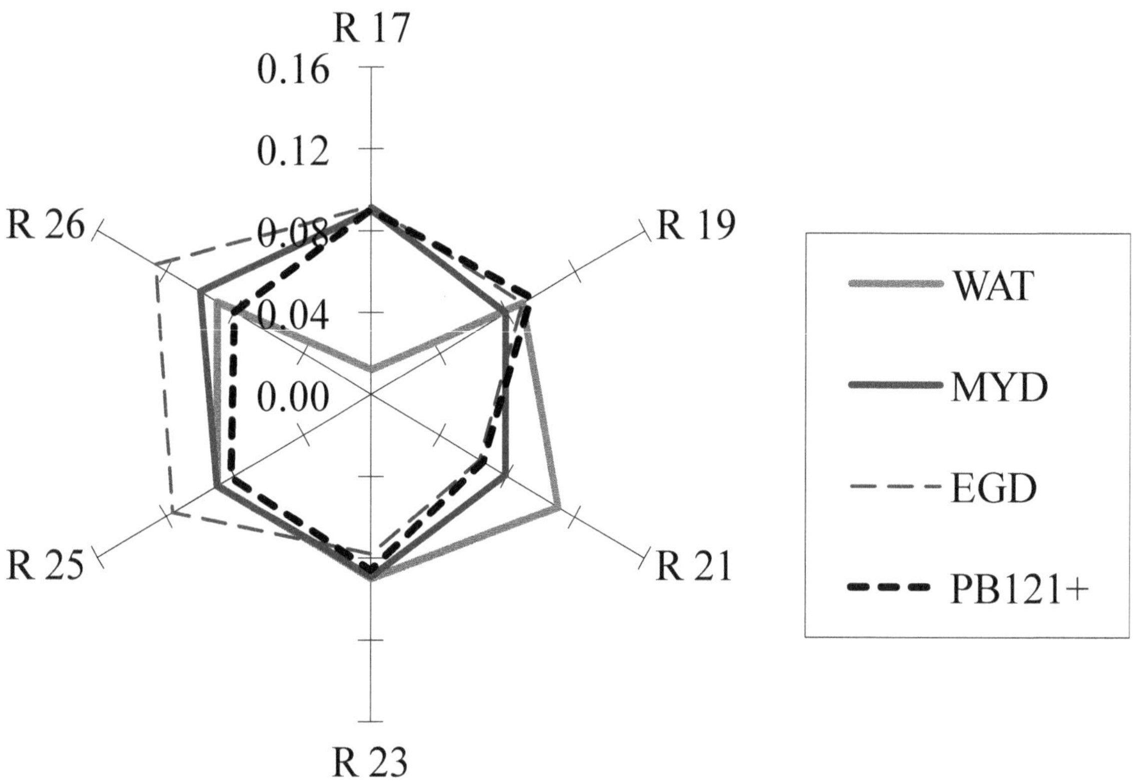

Figure 2. Salty sensory profile of the coconut water according to cultivars and nuts maturity stages. LDS: 0.02.

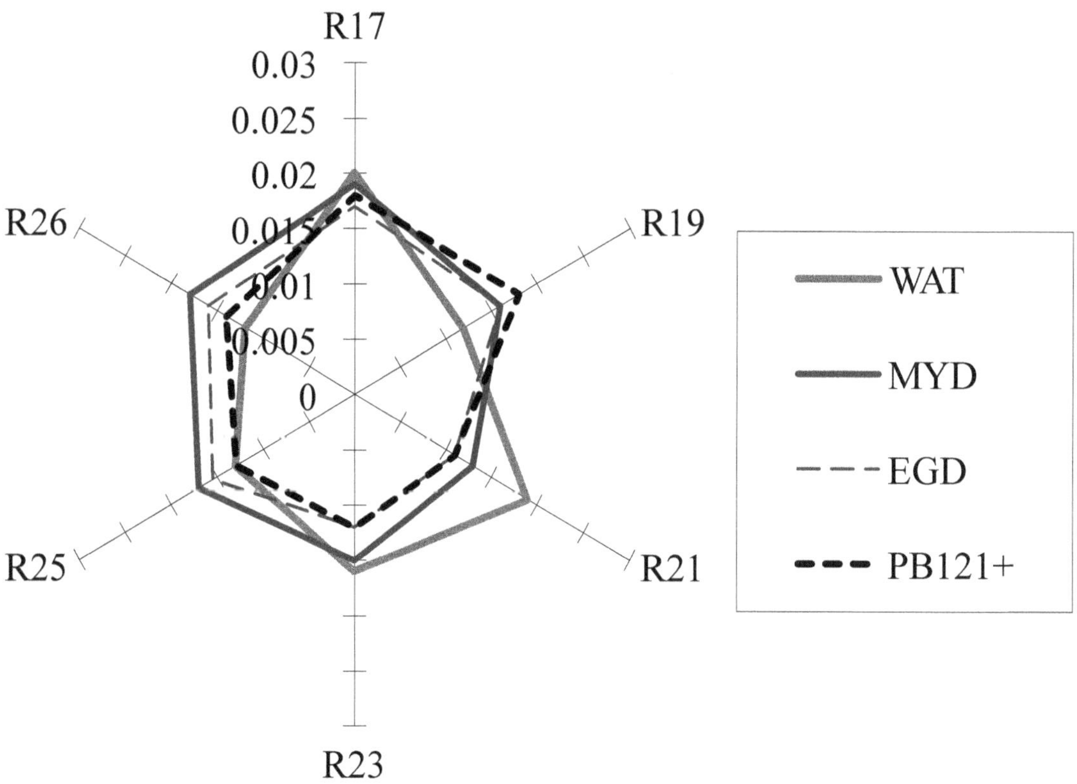

Figure 3. Sour sensory profile of coconut water according to cultivars and nuts maturity stages. LDS: 0.005.

Table 2. Gustatory preferences for the coconut water on 5 points scale according to cultivars and maturity stages.

Cultivars	Maturity stages						Score / cultivar
	R17	R19	R21	R23	R25	R26	
WAT	2.91[d]	4.2[b]	3.12[d]	2.73[d]	3.03[d]	1.76[e]	2.96[d]
MYD	2.78[d]	4.08[b]	4.61[a]	3.50[bc]	2.42[e]	2.22[e]	3.27[c]
EDG	2.00[e]	4.81[a]	3.50[bc]	3.38[c]	3.03[d]	2.69[d]	3.23[c]
PB121+	1.77[e]	3.17[d]	4.37[b]	3.57[c]	3.14[d]	3.11[d]	3.19[d]
Score / rank	2.36[e]	4.08[a]	3.90[b]	3.29[c]	2.91[d]	2.45[d]	3.16[d]

Scores followed by the same letter (from a to e) are not significantly different.

Gustatory preference

The preference of coconut water augmented until the rank 19 or 21 for all tested cultivars with maximal scores for EGD (4.81) and MYD (4.61). Beyond the rank 21, preferences decreased until the rank 26 where they were 2.22 for MYD and minimal (1.76) for WAT. Coconut water of EGD, MYD and PB121+ at ranks 21 and 23 were moderately preferred with notes wobbling between 3.38 and 3.50 (Table 2). However, coconut water of ranks 25 and 26 was not generally appreciated whoever was the cultivar. These samples had average notes of 3 at rank 25 and 2 at rank 26.

Sugars contents

The evolution of total sugars (TS) and reducer sugars (RS) contents during nuts ripening included an increasing stage followed by a decreasing phase (Table 3). So, for WAT, total sugars (3.98 g.100 ml^{-1}) and reducer sugars (3.88 g.100 ml^{-1}) contents were maximal at the rank 19. For MYD, the maximum contents of total sugars (5.13 g.100 ml^{-1}) were obtained at the rank 19. However, they were 4.82 and 3.83 g.100 ml^{-1} in the rank 21 respectively for EGD and PB121 +. Not reducers sugars (NRS) contents had an evolution opposite to those of reducer and total sugars. Indeed, for WAT, MYD and EGD cultivars, respective maximum values of 1.74, 1.88 and 1.53 g.100 ml^{-1} were obtained at the rank 26. As for the PB121 +, its maximum content (1.31 g.100 ml^{-1}) was obtained at the rank 25.

DISCUSSION

The results of our study showed gustatory characteristics differences of coconut water according to varieties and maturiy stages. The EGD coconut water of rank 26 had the most salty taste and that of WAT rank 21, the most sour. Sweet, salty and sour tastes of the studied coconuts water corresponded to those of Campos et al. (1996) on the immature nuts. So, sweet taste predominated in coconut water whatever were cultivars and the maturity stages. This justified its evaluation according to this sensory criterion into the investigated populations, according to Assa et al. (2006). Specifically, the coconut water from MYD and EGD cultivars fruits of ranks 19 (7 months) and 21 (9 months) had the most sweet taste. These samples were more appreciated by most of testers. They are moreover, most appreciating by the consumers. Indeed, the water of immature coconuts is more used thanks to these virtues and especially its sweet taste (Nadanasabapathy and Kumar, 1999). This justifies its uses in the manufacture of energetic and refreshing drinks (Jackson et al., 2004). The immature coconut water from WAT and $PB121^{+}$ was not appreciated by the consumers because it is not much sweetened. So, nuts water of these cultivars could be promoted in drinks only after addition of sweetening compounds. Moreover, previous results (Assa et al., 2007) revealed sugars chromatographic profiles of WAT, MYD, EGD and PB121+ coconut water. The results showed qualitatively, identical sugars, for these cultivars. Indeed, fructose, sucrose, glucose and sorbitol are majority sugars with different concentrations according to varieties. These results showed that coconut water from MYD had total sugars content statistically identical to that of EGD at the rank 21. It confirms the sensory profiles obtained in our study. So, the coconut water sweet tastes perception has the same evolutions as those of their total sugars contents. The consumer could therefore act efficiently as mean of the coconut water sugars contents valuation. However, at the rank 19, water of EGD nuts with little sweetness than that of MYD, is the most appreciated by the tasters. It means that, except the total sugars contents of the coconut water, other factors would intervene for its preference.

Indeed, according to Lemordant (1988), the main sweet constituents have different sweetening powers. So, sucrose is used as reference with a sweetening power of 1. Fructose and sorbitol have respective sweetening powers of 1.7 and 0.7; while those of glucose, glycerol and galactose are 0.5. This difference of sweetening powers suggests distances between coconut sugars contents and their gustatory preferences. So, the maximum proportions in sucrose and fructose of the coconut water from EGD at ranks 19 and 21 would justify

Table 3. Sugars contents (g.100 g^{-1}) of coconut water of different cultivars at different maturity stages.

Cultivars	Ranks (R)	TS	RS	NRS	P intra cultivar
WAT	R17	3.53	3.5	0.03	P<0.01
	R19	3.98^{b}	3.88	0.11	
	R21	3.66	2.94	0.74	
	R23	2.22	1.08	1.14	
	R25	1.87	0.45	1.42	
	R26	2.01	0.25	1.74	
MYD	R17	4.33^{b}	4.15	0.18	P<0.01
	R19	5.03^{a}	4.42	0.39	
	R21	5.13^{a}	3.76	1.37	
	R23	3.52	1.82	1.71	
	R25	2.51	0.70	1.82	
	R26	2.19	0.29	1.88	
EGD	R17	3.80	3.42	0.38	P<0.01
	R19	4.75^{a}	4.41	0.08	
	R21	4.82^{a}	4.38	0.46	
	R23	3.05	1.71	1.34	
	R25	2.22	0.89	1.33	
	R26	2.26	0.71	1.53	
PB121 +	R17	2.50	2.38	0.12	P<0.01
	R19	3.41	3.22	0.18	
	R21	3.83	3.59	0.24	
	R23	3.62	2.85	0.76	
	R25	2.76	1.45	1.31	
	R26	2.51	1.57	0.92	
P inter cultivar		P<0.01	P<0.01	P = 0.05	
LSD		0.44	0.52	0.26	

TS: Total sugars. RS: reducer sugars. NRS: not reducer sugars; p: significant level. LSD: little significant difference.

their preference in comparison with that of MYD (Assa et al., 2007). Besides, according to some oral information, at rank 19; the coconut aroma of EGD would be more pleasant, preferable to that of MYD. Moreover, there is a biochemical explanation of the gustatory characteristics evolution of coconut water during the ripening of nuts. Indeed, at the early immature stages; immediately after fecundation, inflorescence sap sugars enter the fruit across the peduncle. They lead to a complete conversion of sucrose in glucose and especially in fructose, which has high sweetening power. It is effective under the effect of an enzyme (the invertase) exiting in the peduncle (Delrot, 2000). However, from the rank 19 or 21, the glucose of coconut water is progressively redeployed into kernel's sucrose and polysaccharides. This is due to hydrolysis reactions implicating alcohols functions (Omotosho and Odeyemi, 2012). At genetic level, the coconut water gustatory characteristics do not seem to be heritable parameters. Indeed, the coconut water sensory profiles from hybrid $PB121^{+}$, MYD and WAT (Bourdeix et al., 1992) shows a sweet taste very close to that of the male parent GOA, between ranks 17 and 19. On the contrary, it gets closer to the female parent; MYD between ranks 21 and 23. It is beyond the two parent's values between ranks 25 and 26.

The salty and sour tastes perception of hybrid PB121+ nuts water is close to those of the male parent especially from the rank 23. So, inventory and study on genes responsible of coconut water taste would allow to more understanding their expression and their regulation following Delseny et al. (2002) previous studies on *Arabidopsis thaliana*.

Conclusion

The coconut water sensory characterization during nuts ripening was performed to give indicators for an efficient technological promotion. This study underlines compliance between the sweet sensory profiles of coconut water and its sugars contents obtained by physic-chemical tests. Our results show that the immature coconut water from Dwarf cultivars (MYD and EGD) at ranks 19 and 21 are more appreciated. Indeed, their sensory characteristics, with sweetness predominance, better meet needs of consumers. Specifically, these dwarf cultivars could be recommended to the peasants and other coconut actors of coastal regions where there is a big consumption of immature nuts water. Moreover, the coconut water of MYD and EDG cultivars can be specifically promoted as sweet and nutritious drinks in order to augment the profitability of the coconut tree.

REFERENCES

AFNOR (1984). Recueil de normes françaises d'agro- alimentaire : analyse sensorielle. Paris, la Défense (France). pp.159.

Amrizal I (2003). Coconut Statistical Yearbook, APCC, Jakarta (India), p. 233.

Assa R, Konan JL, Nemlin J, Prades A, Agbo N, Sie R (2006). Diagnostic de la cocoteraie paysanne du littoral ivoirien. Sci. Nat. 3(2):113-120.

Assa RR, Konan JL, Agbo N, Prades A, Nemlin J (2007). Caractéristiques physico- chimiques de l'eau des fruits de quatre cultivars de cocotier (*Cocos nucifera* L.) en Côte d'ivoire. Agron. Afr. 19(1):41-51.

Bourdeix R, N'cho Y, Sangaré A, Baudoin, De Nucé (1992). L'hybride de cocotier PB 121 amélioré, croisement du Nain jaune malais et de géniteur Grand Ouest Africain sélectionnés. Oléagineux 47(11):619-633.

Campos C, Souza P, Virgilio J, Beatriz M, Gloria A (1996). Chemical composition, enzyme activity and effect of enzyme inactivation on flavour quality of green coconut water. J. Food Process. Preserv. 20(6):487-500.

Delrot S (2000). Les sèves et leur circulation. Pour la Science. Dossier hors série: 28-33.

Delseny M, Verdoucq L, Maisonneuve S, Roscoe T (2002). Génomique et métabolisme des lipides des plantes. OCL 9(2):130-134.

Dubois M, Gilles K, Hamilton J, Rebers P, Smith F (1965). Colorimetric methods for determination of sugars and related substances. Anal. Chem. 28: 350-356.

Enonuya DO (1988). High performance Liquid Chromatographic analysis of nutwater syrup fractions from two varieties of Nigerian coconuts. Niger. J. Palms Oils Seeds 9:48-58.

Jackson JC, Gordon AG, Kayanne MC, Rolle R (2004). Changes in the chemical composition of coconut (*Cocos nucifera*) water during maturation of the fruit. J. Sci. Food Agric. 84(9):1049-1052.

Jayalekshmy A, Arumughan C, Narayanan M (1986). Changes in the chemical composition of coconut water during maturation. J. Food Sci. Technol. (23): 203-207.

Konan JL, Konan BR, Assa RR (2008). Caractéristiques physico-chimiques de l'amande mature des hybrides de « cocotiers grands » améliorés. Agro. Afr. 20(3):331-344.

Lapitan OB, Mabesa RC (1983). Chemical and sensory characteristics of Laguna and Golden coconuts (*Cocos Nucifera* L.). Philipp. Agric. (66):144-150.

Lemordant D (1988). Le sucre, les sucres, les édulcorants : leurs diverses utilisations. J. Agric. Trad. Bota. Appliq. 35:147-157.

Manisha D, Shyamapada M (2011). Coconut (*Cocos nucifera* L.: Arecaceae): In health promotion and disease prevention. Asian Pacific J. Trop. Med. 241-247.

Moore C, Batugal P (2004). Banking on the tree of life. In: *Geneflow*. A Publication about the Earth's Genetic Ressources. Ruth D. Raymond (eds.). Rome, Italy. p. 35.

Nadanasabapathy S, Kumar R (1999). Physico-chemical constituents of tender coconut. Indian J. Agric. Sci. 69(10):750-751.

Omotosho I, Odeyemi FA (2012). Bio-nutritional constituents of coconut fruit and its possible medicinal applications. Afr. J. Plant Sci. 6(12):309-313.

Raoux R (1998). Approche de l'analyse sensorielle : évolutions récentes. OCL (5)4:305-307.

Effect of concentration of debittering agent on the mineral, vitamin and phytochemical contents of *Lasianthera africana* leafy vegetable

Ani, J. C.[1], Inyang, U. E.[2] and Udoidem, I.[2]

[1]Department of Food Science and Technology, University of Nigeria, Nsukka, Enugu State, Nigeria.
[2]Department of Food Science and Technology, University of Uyo, Uyo, Akwa Ibom State, Nigeria.

White variety of *Lasianthera africana* leaves were blanched in hot water (control) and in different concentrations (0.25, 0.50, 0,75, 1.00 and 1.25%) of unripe plantain peel ash solution for 3 min at 100°C, cooled, drained, oven dried (50°C) for 36 h. Then, raw leaf and all oven dried samples were analyzed for minerals, vitamins and phytochemicals. The raw leaf contained potassium (78.98±0.78 mg/100 g), calcium (190.25±0.44 mg/100 g), sodium (75.69±0.95 mg/100 g), magnesium (14.68±0.74 mg/100 g), iron (3.96±0.55 mg/100 g), zinc (5.95±0.52 mg/100 g), phosphorus (17.79±0.81 mg/100 g), ascorbic acid (109.64±0.08 mg/100 g), beta-carotene (2.86±0.04 mg/100 g), riboflavin (0.22±0.03 mg/100 g), thiamine (1.01±0.06 mg/100 g), alkaloids (2.67±0.33 g/100 g), flavonoids (0.32±0.03 g/100 g), saponins (3.09±0.04 g/100 g) and tannins (0.28±0.01 g/100 g). Blanching the leaves either in hot water or in different concentrations of unripe plantain peel ash solution led to varying losses of the minerals, vitamins and phytochemicals. Samples blanched in different levels of ash solution retained higher mineral content than hot water blanched samples. Percentage minerals retained increase with increased levels of ash in the blanching solution. Conversely, percentage retention of vitamins and phytochemicals decreased with increase in the levels of ash in the blanching solution. Ash concentration had no significant (p>0.05) effect on the levels of magnesium, zinc, beta-carotene and tannins retained in the blanched samples. For higher retention of vitamins and health benefiting phytochemicals, lower concentration of unripe plantain peel ash solution (0.50% to 0.75%) should be used to debitter *L. africana* leaf.

Keywords: *Lasianthera africana* leaf, debittering agent, minerals, vitamins, phytochemicals.

INTRODUCTION

Green leafy vegetables (wild or cultivated) are important items of diets in many Nigerian home as valuable sources of nutrients especially in rural areas where they contribute substantially to micronutrients which are usually in short supply in most diets. Leaves in general are important sources of iron, potassium, calcium, magnesium, zinc, provitamin A, thiamine, riboflavin, ascorbic acid and folic acid (Uwaegbute, 1989; Fasuyi,

2006). The dietary fibre in vegetables increases bulk and reduces the incidence of constipation and other related diseases (Jalili et al., 2000). Green leafy vegetables also contain bioactive compounds (phytochemicals) that have potentials in helping to reduce the risk of several deadly diseases in man (Sofowora, 1989; Williamson et al., 1997; Chung et al., 1998; Agte et al., 2000). Epidemiological studies have shown that consumption of vegetables can protect humans against oxidative damage by inhibiting or quenching free radicals and reactive oxygen species (Ames et al., 1981). High consumption of green leafy vegetables therefore plays vital role in human nutrition and health (Odukoya et al., 2007). Various traditional post harvest processing treatments such as soaking in hot water, boiling, squeeze washing with water, and hot water blanching may however lead to losses of some of the characteristics which initially made green leafy vegetables consumers delight (Adeboye and Babajide, 2007).

Lasianthera africana is one of the top six commonly consumed green leafy vegetables by Efik and Ibibio ethnic groups of Nigeria (Williams *et al.,* 2009). It belongs to the family *Icacinaceae.* It is called "editan" in Efik and Ibibio local dialects of Nigeria. It is a perennial, glabrous, shrub that reaches a height of 61-136 cm (Hutchinson and Dalziel, 1973). Among the Ibibios, four local varieties distinguished by their taste, leaf colour and ecological distribution are known (Bassey *et al.*, 2006). The varieties are "afia" (white variety), "obubit" (black variety), "idim" (riverine variety) and "akai" (forest variety). The leaf has been used since pre-historic time for preparing soup and in many traditional concoctions for the treatment of various ailments (Sofowora, 1989). Ebana et al. (1996) reported that the leaves of *L. africana* are rich in chemical compounds of nutritional and medicinal importance. Preliminary screening of the leaves for phytochemicals indicated the presence of alkaloids, flavonoids, saponins, anthraquinones, glycosides and tannins in all the four ethnovarieties (Bassey et al., 2006).

One unique characteristic of *L. africana* leaf is that it has bitter taste that requires debittering prior to culinary use. Debittering helps to enhance palatability and acceptability of the soup prepared with the leaf. Traditionally, the leaf is usually debittered by squeeze washing with water or treatment with aqueous extract from unripe plantain peel ash. The use of higher concentration of unripe plantain peel ash usually gives objectionable flavour to the soup prepared with the debittered leaf and may have negative effect on other beneficial constituents in the leaf. Presently, nothing is known about the effect of debittering treatments on the nutrients, especially the water soluble and heat sensitive nutrients as well as constituents of medicinal value. This study was therefore conducted to assess the effects of debittering *L. africana* leaves in different concentrations of unripe plantain peel ash solution on the nutrients and phytochemical content in the leaves.

MATERIALS AND METHODS

Sample collection and preparation

Twigs of *L. africana* (white variety) were harvested from a garden at Aka Offot in Uyo Local Government Area of Akwa Ibom State and authenticated at the taxonomy unit of the Department of Botany and Ecological Sciences, University of Uyo. The harvested twigs were destalked, washed in potable water, spread out under a shade to air dry. The leaves were cut (2 mm width), shared into six portions (1 kg each) and blanched in 0.00, 0.25, 0.50, 0.75, 1.00 and 1.25% solution of unripe plantain peel ash (1:3 w/v) at 100°C for 3 min. The blanched samples were drained immediately, cooled, dried (50°C) for 36 h in a conventional air oven (Model P.P. 22 US, Genlab, England), milled, packaged in plastic containers, labeled and stored in a refrigerator (4°C) for analyses. The contents of minerals, vitamins and phytochemicals in the cut raw leaf were also determined.

Mineral analysis

The mineral elements (K, Na, Ca, Mg, Zn, Fe and P) were determined using atomic absorption spectrophotometer (UNICAM, Model, 939, UK) as described by AOAC (2000).

Vitamin analysis

Ascorbic acid, beta-carotene, thiamine and riboflavin were determined using the method described by AOAC (2000).

Phytochemical analysis

Alkaloid and flavonoid were determined using the method of Harborne (1973). Saponin and tannin were determined by AOAC (2000) method.

Statistical analysis

Data obtained were subjected to one way analysis of variance (ANOVA) using SPSS version 18 statistical package (SPSS, Inc., USA) to determine variation between treatments. Means of data generated were separated using Duncan multiple range test (DMRT). Results were expressed as mean ± SD (standard deviation) of triplicate determinations. Significant variation was accepted at $p<0.05$.

RESULTS AND DISCUSSION

Table 1 shows the mineral content of *L. africana* leaves blanched in hot water (control) and in different concentrations of unripe plantain peel ash solution. The raw leaves had higher mineral content than the blanched and dried samples. Hot water blanched and dried samples exhibited higher losses of minerals than ash solution blanched and dried samples.

The potassium content in the raw (unblanched) leaf was 78.98±0.78 mg/100 g. The level of potassium in the hot water blanched and dried samples was 70.43±0.10 mg/100 g, whereas the values for samples blanched in different concentrations of unripe plantain peel ash

Table 1. Effect of concentration (%) of unripe plantain peel ash blanching solution on the mineral content of *L. africana* leaf (mg/100 g).

Mineral	Ash concentration (%)					
	0.00	0.25	0.50	0.75	1.00	1.25
K	$70.43^{d}\pm0.14$	$72.20^{c}\pm0.33$	$72.95^{c}\pm0.31$	$73.92^{b}\pm0.09$	$74.31^{a}\pm0.14$	$74.88^{a}\pm0.27$
Na	$61.09^{c}\pm0.13$	$61.47^{c}\pm0.21$	$62.35^{b}\pm0.28$	$62.79^{ab}\pm0.29$	$63.04^{ab}\pm0.04$	$63.57^{a}\pm0.06$
Ca	$177.91^{d}\pm0.16$	$178.62^{c}\pm0.16$	$179.00^{c}\pm0.04$	$179.53^{b}\pm0.16$	$179.82^{ab}\pm0.08$	$180.06^{a}\pm0.08$
Mg	$12.23^{a}\pm0.17$	$12.24^{a}\pm0.19$	$12.30^{a}\pm0.14$	$12.35^{a}\pm0.14$	$12.38^{a}\pm0.07$	$12.42^{a}\pm0.16$
Zn	$5.010^{a}\pm0.13$	$5.14^{a}\pm0.04$	$5.17^{a}\pm0.03$	$5.18^{a}\pm0.06$	$5.20^{a}\pm0.03$	$5.21^{a}\pm0.01$
Fe	$3.45^{a}\pm0.44$	$3.49^{a}\pm0.21$	$3.53^{a}\pm0.08$	$3.56^{a}\pm0.00$	$3.57^{a}\pm0.13$	$3.61^{a}\pm0.29$
P	$15.34^{b}\pm0.19$	$15.93^{b}\pm0.41$	$16.22^{b}\pm0.39$	$16.79^{a}\pm0.45$	$16.86^{a}\pm0.23$	$17.01^{a}\pm0.41$

Values are means ± SD of triplicate determinations. Means on the same row with different superscripts are significantly different at $p<0.05$.

solution ranged from 72.20±0.33 mg/100 g in 0.25% ash solution blanched and dried samples to 74.88±0.27 mg/100 g in 1.25% ash solution blanched and dried samples (Table 1). Significantly, $p<0.05$ higher retention of potassium in ash solution blanched and dried samples relative to hot water blanched and dried samples could be due to differences in the soluble solute concentration of the blanching medium.

Ejoh et al. (2007) similarly reported that leafy vegetables blanched in 2.5 and 5.0% "kanwa" solution exhibited lower mineral losses than hot water blanched samples.

The sodium content in the raw leaf was 75.69±0.95 mg/100 g. The hot water blanched and dried samples had lower sodium content (61.09±0.14 mg/100 g). The ash solution blanched and dried samples had higher sodium content than hot water blanched and dried samples. The values for ash solutions blanched and dried samples ranged from 61.47±0.21 mg/100 g for sample blanched in 0.25% ash solution to 63.57±0.66 mg/100 g for samples blanched in 1.25% ash solution. This shows that sodium retention (%) increased with increase in ash concentration (Table 1).

The result suggests that the use of high concentration of unripe plantain peel ash to debitter *L. africana* leaves should not be encouraged as high sodium intake is an identified risk factor for the development of hypertension in susceptible individuals (Tobian, 1997; Campbell-Platt, 2009). Salt restriction and low salt diets have been shown to lower blood pressure in both normotensive and hypertensive subjects (Ebuehi et al., 2003).

Calcium was the most predominant mineral element in raw *L. africana* leaf (190.25±0.44 mg/100 g). The high calcium content in the leaf is significant because the cells need calcium and more than 99% of calcium in the body is used as a structural component of bones and teeth. This represents about 40% of all minerals present in the body (Grosvenor and Smolin, 2002). The blanched and dried samples had lower calcium content than the raw leaf. This could be due to leaching into the blanching medium. Samples blanched in ash solutions had significantly ($p<0.05$) higher calcium content than hot water blanched samples (Table 1). This could be due to differences in soluble solute concentration of the blanching medium. Percentage calcium retention increased with increase in the ash concentration (%) in the blanching medium ranging from 93.89% for samples blanched in 0.25% ash solution to 94.64% for samples blanched in 1.25% ash solution. Similar observation was made by Ejoh et al. (2007) who reported calcium content of 0.83±0.02 g/100 g, 1.04±0.11 g/100 g and 1.28±0.01 g/100 g for *Vernonia calvoana* blanched in hot water, 2.5% "kanwa" and 5.0% "kanwa" solutions, respectively.

The contents of magnesium, zinc, iron and phosphorus in the raw leaf were 14.68±0.74 mg/100 g, 5.95±0.52 mg/100 g, 3.96±0.55 mg/100 g and 17.79±0.81 mg/100 g, respectively. Their values in the blanched and dried samples were lower. This could be due to leaching of these minerals into the blanching medium. The hot water blanched and dried samples had lower content of these minerals than ash solutions blanched and dried samples. There were systematic increases in these minerals with increases in the concentration of ash in the blanching solutions. This could be due to possible contributions from the unripe plantain peel ash in the solutions used for blanching the leaves. Similar observation was reported by Ejoh et al. (2007) for *Vernonia calvoana* leaf blanched in hot water and in "kanwa" solutions.

Data on vitamin analysis (Table 2) indicate that there was inverse relationship between the increase in the concentrations of unripe plantain peel ash in the blanching solution and the level of vitamins retained in the blanched and dried samples. There were consistent reductions in the vitamins analyzed with increase in the concentration of ash in the solutions used for blanching the leaves.

The ascorbic acid content in the raw leaf was 109.64±0.08 mg/100 g. The human body cannot produce ascorbic acid, so it must be obtained entirely through one's diet. Hot water blanched and dried samples had the ascorbic acid content reduced to 65.12±0.14 mg/100 g.

Table 2. Effect of concentration (%) of blanching ash solutions on some vitamins in *L. africana* leaf (mg/100 g).

Ash concentration (%)	Ascorbic acid	Beta-carotene	Riboflavin	Thiamine
0.00	65.12[a] ±0.14 (59.39)	1.89[a] ±0.04 (66.08)	0.19[a] ±0.03 (86.36)	0.61[a] ±0.01 (60.40)
0.25	61.43[b]±0.04 (56.03)	1.87[a]±0.06 (65.39)	0.17[ab]±0.01 (77.27)	0.57[ab]±0.03 (56.44)
0.50	59.72[c]±0.17 (54.45)	1.86[a]±0.04 (65.04)	0.16[ab]±0.00 (72.73)	0.54[ab]±0.03 (53.47)
0.75	55.90[d]±0.09 (51.90)	1.86±0.02 (65.04)	0.14[ab]±0.03 (63.64)	0.52[ab]±0.05 (51.49)
1.00	55.03[e]±0.04 (50.19)	1.85±0.07 (64.69)	0.13[ab]±0.01 (59.09)	0.51[ab]±0.01 (50.50)
1.25	53.79[f]±0.04 (49.06)	1.85±0.00 (64.69)	0.11[b]±0.01 (50.50)	0.49[b]±0.04 (48.51)

Values are means ± SD of triplicate determinations. Mean on the same column with different superscripts are significantly different at p<0.05. Values in parenthesis indicate percentage retention of vitamins.

This could be attributed to leaching, thermal decomposition and to a lesser extents, oxidation (Fellows, 2000).

Ascorbic acid contents of unripe plantain peel ash solution blanched and dried samples were significantly ($p<0.05$) lower than the content in hot water blanched and dries samples (Table 2). The values for samples blanched in different concentrations of ash solution ranged from 53.79±0.04 mg/100 g for samples blanched in 1.25% ash solution to 61.43±0.94 mg/100 g for samples blanched in 0.25% ash solution. Ascorbic acid is not only soluble in water but is also the most heat labile of the vitamins (Grosvenor and Smolin, 2002; Bolarin, 2006).

The lower levels of ascorbic acid in samples blanched in solutions of unripe plantain peel ash relative to hot water blanched samples could be due to the fact that the alkaline media predisposed the vitamin to decomposition which enhanced the level of losses. Ihekoronye and Ngoddy (1985) noted that when leafy vegetables are cooked with baking soda to retain the greenness, they lose much of their ascorbic acid. Grosvenor and Smolin (2002) also stated that the loss of ascorbic acid is accelerated by low acid conditions.

The beta-carotene content in the raw leaf was 2.86±0.04 mg/100 g. Hot water blanched and dried samples had lower content of beta-carotene (1.89±0.06 mg/100 g) than the value in the raw leaf. This reduction in beta-carotene could be attributed to possible destruction by heat and oxidation reaction. According to Meyer (1978), McDowell (1989) and Roche (1990), beta-carotene is especially prone to oxidative destruction in the presence of heat, light, oxygen and metal ions. Blanching the leaves in various concentrations of ash solution did not have any significant ($p>0.05$) effect on the beta-carotene content of the blanched samples (Table 2). There were however, slight reductions in beta-carotene of the blanched leaves with increase in the concentration of ash in the solution ranging from 1.87±0.06 mg/100 g for samples blanched in 0.25% ash solution to 1.85±0.00 mg/100 g for samples blanched in 1.25% ash solution.

The non-significant difference ($p>0.05$) in the beta-carotene content of samples blanched in different concentrations of ash solution and hot water confirms the fact that beta-carotene is not affected by changes in pH (Meyer, 1978). Ihekoronye and Ngoddy (1985) also stated that dilute acid, bases and water have little effect on carotenoids.

The riboflavin content in the raw leaf was 0.22±0.03 mg/100 g. Samples blanched in either hot water or various concentrations of unripe plantain peel ash solution and dried had lower riboflavin contents than the value in raw leaf. Sample blanched in hot water had higher riboflavin content (0.19±0.03 mg/100 g) than samples that were blanched in different concentrations of ash solution whose values ranged from 0.17±0.01 mg/100 g for samples blanched in 0.25% ash solution to 0.11±0.01 mg/100 g for samples blanched in 1.25% ash solution.

The observed higher retention of riboflavin in hot water blanched and dried samples (86.36%) when compared with ash solutions blanched and dried samples (77.22 – 59.00%) could be due to the fact that riboflavin is more soluble in alkaline solution than in water (Jacob, 1999), and is destroyed under alkaline condition, light and excessive heat (Fellows, 2000).

The thiamine content in the raw leaf was 1.01±0.06 mg/100 g. Hot water blanched and dried samples and ash solutions blanched and dried samples had lower thiamine content than the raw leaf. Thiamine is known to

Table 3. Effect of concentration (%) of blanching ash solutions on the phytochemical content in *L. africana* leaf (g/100 g).

Ash concentration (%)	Alkaloid	Flavonoid	Saponins	Tannins
0.00	$2.14^{a} \pm 0.06$ (19.85)	$0.21^{a} \pm 0.01$ (34.37)	$2.98^{a} \pm 0.06$ (3.56)	$0.24^{a} \pm 0.04$ (14.29)
0.25	$1.96^{ab} \pm 0.04$ (26.59)	$0.16^{ab} \pm 0.03$ (50.00)	$2.33^{b} \pm 0.04$ (24.60)	$0.19^{a} \pm 0.00$ (32.14)
0.50	$1.84^{bc} \pm 0.06$ (31.08)	$0.15^{bc} \pm 0.00$ (53.12)	$2.31^{b} \pm 0.03$ (25.24)	$0.19^{a} \pm 0.01$ (32.14)
0.75	$1.76^{bc} \pm 0.07$ (34.09)	$0.13^{bc} \pm 0.01$ (59.37)	$2.27^{b} \pm 0.04$ (26.54)	$0.16^{a} \pm 0.04$ (42.86)
1.00	$1.69^{cd} \pm 0.56$ (36.70)	$0.10^{cd} \pm 0.00$ (68.75)	$2.25^{b} \pm 0.07$ (27.18)	$0.15^{a} \pm 0.01$ (46.43)
1.25	$1.60^{cd} \pm 0.00$ (40.07)	$0.09^{cd} \pm 0.01$ (71.87)	$2.20^{b} \pm 0.00$ (28.80)	$0.14^{a} \pm 0.03$ (50.00)

Values are means ± SD of triplicate determinations. Mean on the same column with different superscripts are significantly different at $p<0.05$ values in parenthesis indicate percentage reduction of phytochemicals.

decompose readily in alkaline and neutral solutions if heated (Jacob, 1999; Fellows, 2000; Grosvenor and Smolin, 2002). Hot water blanched and dried samples had higher thiamine content (0.61±0.03 mg/100 g) than ash solutions blanched and dried samples whose values ranged from 0.57±0.03 mg/100 g for samples blanched in 0.25% ash solution to 0.49±0.04 mg/100 g for samples blanched in 1.25% ash solution (Table 2). The value for hot water blanched sample was only significantly ($p<0.05$) higher than that of 1.25% ash solution blanched sample, whereas the rest of the values were not significantly ($p>0.05$) different from each other.

The contents of alkaloid, flavonoid, saponin and tannins in the raw *L. africana* leaf were 2.67±0.33 g/100 g, 0.32±0.03 g/100 g, 3.09±0.04 g/100 g and 0.28±0.01 g/100 g, respectively. The presence of these phytochemicals in the leaf is significant because of their health protecting and promotion potentials. Blanching the leaves in either hot water or different concentrations of unripe plantain peel ash solution had negative effect on these phytochemicals as their values in the blanched and dried samples were lower than those in the raw leaf (Table 3). This could be due to leaching of these phytochemicals into the blanching media. Similar reduction of phytochemical constituents in vegetables as a result of blanching had been reported by Adeboye and Babajide (2007), Onyeka et al. (2010) and Nkafamiya et al. (2010).

The unripe plantain peel ash solutions blanched and dried samples had lower alkaloids, flavonoids, saponins and tannins than hot water blanched and dried samples. This result corresponded with the report of Adeboye and Babajide (2007) who reported that leafy vegetables blanched in hot water contained higher phytochemicals than those blanched in 1% potash solution for five minutes. The reduction of alkaloids, flavonoids, saponins and tannins consistently increased with increase in the level of unripe plantain peel ash in the blanching solution. Percentage reduction of alkaloids, flavonoids, saponins and tannins ranged from 40.07, 71.87, 28.80 and 50.00% for 1.25% ash solution blanched and dried samples to 26.59, 50.00, 24.60 and 32.14% for 0.25% ash solution blanched and dried samples.

It is important to minimize losses of these phytochemicals as they are implicated with various health benefits. Alkaloids and flavonoids for instance are listed among the plant chemical constituents responsible for hypoglycemic effect in humans (Akah et al., 2002, Li et al., 2004). Flavonoids have strong anti-cancer properties and protection against cardiovascular disease by inhibiting the oxidation of low density lipoprotein (Grosvenor and Smolin, 2002, Okwu, 2004).

Saponin in food decreases the cholesterol absorption from the gastrointestinal tract and therefore lower blood cholesterol, a major risk factor for cardiovascular disease (Topping et al., 1980; Grosvenor and Smolin, 2002). Tannins on the other hand can form complexes with protein and iron thereby inhibiting their bioavailability (Grosvenor and Smolin, 2002). Besides this deleterious effect of tannins, they exhibit antioxidant properties and may inhibit activation of carcinogens and cancer promotion (Grosvenor and Smolin, 2002). This implies that tannin can serve dual purpose of reducing some essential nutrients and protecting the body against cancer.

Conclusion

In conclusion, debittering *L. africana* leaf in hot water and different concentrations of unripe plantain peel ash solution led to varying losses of minerals, vitamins and phytochemicals in the leaf. Mineral losses decreased with increase in the levels of ash concentration in the blanching solution. Conversely, losses of vitamins and phytochemicals increased with increase in the level of ash in the blanching solution. For higher retention of vitamins and health benefiting phytochemicals, lower ash concentration (0.50% to 0.75%) should be used for debittering the leaf.

Conflict of interests

The authors did not declare any conflict of interest.

ACKNOWLEDGEMENT

The authors gratefully acknowledge Mr. E. Umoh of the Biochemistry Department, University of Uyo, Akwa Ibom State, Nigeria for the technical assistance.

REFERENCES

Adeboye AS, Babajide JM (2007). Effect of processing methods on antinutrients in selected leaf. Nig. Food J. 25(2):77-87.

Agte VV, Tarwadi KV, Mengale S, Chiplonkar SA (2000). Potential of indigenous green vegetables as natural sources of fortification of eight micronutrients. J. Food Comp. Anal. 13:885-891.

Akah PA, Okoli CO, Nwafor SV (2002). Phytotherapy in the management of diabetes mellitus. J. Nat. Rem. 2:1-10

Ames BM, Moyer TP, Kavash, JS, Moertel CG, Rubin J (1981). Pharmacology of *amygdalin* (*Laetrile*) in cancer patients. Cancer Chemother. Pharmacol. 6:51-57.

AOAC(2000). Association of Official Analytical Chemists Official Methods of Analysis (17th edn.), Washington D.C.

Bassey ME, Etuk EUI, Ubom R, Obop IE (2006). Chemotaxonomic study of *Lasianthera africana* (*Icacinaceae*) in Akwa Ibom State of Nigeria. Nig. J. Bot. 19:99-102.

Bolarin D (2006). Hand book on clinical chemistry. Spectrum Books Ltd. Spectrum House, Ibadan. pp. 395-403.

Campbell-Platt G (2009). Food science and technology. Wiley-Blackwell Ltd., USA. pp. 207-246.

Chung AT, Weichang I, Johnson MG (1998). Are tannins a double edged sword inbiology and health? Trends Food Sci. Technol. 4:168-175.

Ebana RU, Essien AI, Ekpo OD (1996). Potential medicinal value of leaves of *Lasianthera africana* P. Bear. Glob. J. Pure Appl. Sci. 1:2-7.

Ebuehi OAT, Adegunloye BJ, Mojiminiyi FBO (2003). Effect of dietary salt/loading on blood pressure and erythrocyte Na^+K^+ - ATPase in rats. Afr. J. Med. Pharm. Sci. 7:27-35.

Ejoh RA, Nkonga DV, Inocent G, Moses MS (2007). Effect of the method of processing and preservation on some quality parameters of three non-conventional leafy vegetables. Pak. J. Nutr. 6(2):128-133.

Fasuyi AO (2006). Nutritional potentials of some tropical vegetable meals: chemical characterization and functional properties. Afr. J. Biotechnol. 5(1):49-53.

Fellows PJ (2000). Food processing technology: Principles and practice. CRS Press LLC., Boca Raton, USA, pp. 233-240; 312-352.

Grosvenor MB, Smolin LA (2002). Nutrition: From science to life. Harcourt College Publishers, New York, pp. 288-371; 404-469.

Harborne JB (1973). Phytochemical methods. Chapman and Hall Ltd., London. pp. 50-76.

Hutchinson J, Dalziel JM (1973). Flora of West Tropical Africa (2nd edn.). Crown Agent, London. p. 638.

Ihekoronye AI, Ngoddy PO (1985). Integrated food science and technology for the tropics. MacMillan Edu. Pub., London, pp. 28-91.

Jacob MB (1999). The chemical analysis of foods and food products. (3rd edn.), S. K. Jain, New Delhi, India, pp. 693-783.

Jalili T, Wildman RE, Medeiros DM (2000). Nutraceutical roles of dietary fibre. J. Nutra. Funct. Med. Foods 2(4): 19-34

Li WL, Zheng HC, Bukuru J, De Kimpe N (2004). Natural medicines used in the traditional Chinese medical system for therapy of diabetes mellitus. J. Ethnopharmacol. 92: 1-21.

McDowell RL (1989). Vitamins in animal nutrition. Comparative aspects to human nutrition. Academic Press, Inc., New York, pp. 365-427.

Meyer KH (1978). Food chemistry (3rd edn.). The AVI Pub. Coy. Inc. Westport, Connecticut. pp. 118-146.

Nkafamiya II, Oseameahon SA, Modibbo UU, Haggai D (2010). Vitamins and effect of blanching on nutritional and anti-nutritional values of non-conventional leafy vegetables Afr. J. Food Sci. 4(6):335-341.

Odukoya OA, Inya-Agha SI, Segun FI, Sofidiya MO, Ilori OO (2007). Antioxidant activity of selected Nigerian green leafy vegetables. Am. J. Food Technol. 2(3):169-175.

Okwu DE (2004). Phytochemical and vitamins content of indigenous spices of South Eastern Nigeria. J. Sustain. Agric. Environ. 6: 30-34.

Onyeka EU, Agha JC, Nwanjo AC (2010). Phytochemical, nutrient and anti-nutrient content of *Cnidosscolus oconitifolius,* a lesser known green leafy vegetable. Nig. Food J. 26(2): 14-24.

Roche (1990). Beta-carotene: Nutrients and colour. F. Hoffman-La Roche Ltd., Switzerland. pp. 1-28.

Sofowora A (1989). Medicinal plants and traditional medicine in Africa. John and Sons Ltd., London, pp. 12-146.

Tobian L (1997). Diatary sodium chloride and potassium have effects on the pathophysiology of hypertension in humans and animals. Am. J. Clin. Nutr. 65:606-611.

Topping DL, Storer GB, Calvert GD, Meller RA (1980). Effect of dietary saponins on fecall bile acid and neutrall sterols, plasma lipids and lipoprotein turnover in the pig. Am. J. Clin. Nutr. 33:783-786.

Uwaegbute AC (1989). Vegetables nutrition and utilization. In: B. N. Mba and D. O. Nnanyelugo (eds). Food crops production, utilization and nutrition. Dotam Publ. Ltd. Ibadan, pp. 145-153.

Williams IO, Parker RS, Swanson J (2009). Vitamin A content of South Eastern Nigerian vegetable dishes, their consumption pattern and contribution to Vitamin A requirement. Pak. J. Nutr. 8(7):1000-1004.

Williamson G, Dupont MS, Heaney RK, Roger G, Rhodes MJ (1997). Induction of slutathione Stransferase activity in hep G2 cells by extract of fruits and vegetables. Food Chem. 2:157-160.

Sensory evaluation of wheat-cassava-soybean composite flour (WCS) bread by the mixture experiment design

Udofia P. G., Udoudo P. J. and Eyen N. O.

Department of Hotel and Catering Management, Akwa Ibom State Polytechnic, Ikot Osurua, Akwa Ibom State, Nigeria.

Wheat flour is unique for bread baking, but due to high cost of wheat, its continuous use in a developing economy is no longer encouraged. Cassava is a high carbohydrate bearing crop while soybean has a good history of health-giving nutrients. Both plants are indigenous to the tropics, they are cheap and available, but are grossly underutilized in the bread industry. To study sensory effect of supplementation of non-wheat flour in bread, the Simplex-Centroid Design of the mixture experimental design {3,3} of the response surface methodology (RSM) was adopted. Fifteen (15) samples of bread samples were baked with each blend of the WCS composite flour. The bread samples were subjected to sensory evaluation by a panel of semi-trained panelists who had eaten bread for a long time. Optimization analysis on the data obtained from the sensory session showed that blend of 0.67:0.17:0.17 of wheat, cassava and soybean composite flour, respectively produced the best composite bread followed by pure blends of wheat bread in terms of taste, colour, texture, aroma, acceptability and preference. More work should be advanced on other factors of using composite flour for bread manufacturing as well as the promotion of the products on account of health claims of the product.

Key word: Sensory evaluation, composite flour, mixture experimental design, simplex-centroid design, response surface methodology (RSM), optimization.

INTRODUCTION

Most Nigerians are consuming bread as a traditional food item. Wheat flour is the major component of bread. Since wheat is not a tropical crop, it is imported at high cost to meet the rising trend in demand. According to David (2006), Nigeria is the world's largest importer of American red winter wheat. The rising trend of wheat importation may continue due to the relatively stable exchange rate, competitive prices of wheat, high crude oil prices and modest economic growth, besides, close substitute of wheat is yet to be identified among carbohydrate bearing crop (David, 2006). Efforts of governments to embark on local production of wheat have been failing; the situation is not likely to improve on account of unfavorable agronomic conditions of the crop in Nigeria.

The presence of glutein and glutenin confers wheat with unique baking properties. Wheat also contains other nutrients though in marginal amounts which may not support good health all on its own. Wheat may contain some anti -nutritional factors like phytate and tannins especially if not properly processed (Jayacopal, 2012).

Cassava is a tropical crop with high content of carbohydrate; most cassava varieties exhibit high level of

Table 1. Experimental layout of the Simplex Centroid Design showing un-randomized runs.

Run	A	B	C
1	0.33	0.33	0.33
2	0.00	0.50	0.50
3	0.00	0.00	1.00
4	0.17	0.17	0.67
5	1.00	0.00	0.00
6	0.50	0.50	0.00
7	0.00	1.00	0.00
8	0.00	0.00	1.00
9	0.00	1.00	0.00
10	1.00	0.00	0.00
11	0.50	0.50	0.00
12	0.17	0.67	0.17
13	0.50	0.00	0.50
14	0.67	0.17	0.17

Run = product from the respective mixture; A = Wheat flour, B = cassava flour, C = soybeans flour.

tolerance for poor agronomic conditions.Though some cassava varieties contain cyanide, however genetic engineering and improved processing methods are alleviating cyanide toxicity problems. Cassava does not contain the two proteins which makes wheat a preferred major baking ingredient. Physiochemical properties of cassava starch are suitable for supplementation of wheat flour in bread-making without compromising its sensory attributes (Eduardo et al., 2013).

Soybean is another major crop of the tropics, apart from its ability to strive in the tropical climate; it is a health plant due to good content of protein and other phytochemicals (Aleke et al., 2000). The FDA says that 25 g of soy protein per day can reduce the risk of coronary heart disease. The UK Joint Health Claims Initiative (JHCI) re-confirmed the assertion with a similar claim in 2002.

Sensory evaluation evokes, measures, analyses and interprets responses to products as perceived by senses of sight, smell, touch, taste and hearing (Stone and Sidel, 1993). The consumer buys nutrition, convenience, image and functionality in food. Therefore, new products must provide all the responses like or near the older product.

This study was devoted to the determination of sensory properties, preference of bread samples baked with blends of different proportions of wheat, cassava and soy bean flours. The sensory session used a taste panel of staff and students of the Department of Science Technology, Akwa Ibom State Polytechnic, Ikot Ekpene, Nigeria using the mixture experimental design of the response surface methodology (RSM).

MATERIALS AND METHODS

One hundred percent (100%) American hard red winter wheat flour was obtained from Niger Mills, Calabar. TMS (4) 1425 cassava variety was obtained from the International Institute of Tropical Agriculture (IITA) through the Root Crop Research Institute, Umudike. Other baking ingredients: yeast, fat, baking powder, sugar and salt, were obtained from a baking shop, all in Ikot Ekpene, Nigeria.

Preparation of cassava flour

The cassava tuber was peeled about 10 hours post-harvest, washed, sliced thinly with a papaya shredder (Qlink, China), removing the fibre part. The shredded cassava was washed with ample of clean water and allowed to drain, it was thinly spread on a straw mat spread on a clean concrete surface and sun dried for about 11 h until it was water drip-free before it was transferred to an hot air oven (Therma Grosskuchen, Le Chef, Sweden) at about 60°C until the moisture content was reduced to about 13% dry basis.

The cassava flake was ground through a corn meal grinder to pass through a sieve DIN 4188 (0.12 mm) aperture. The cassava flour was packed in plastic buckets.

Preparation of soy bean flour

Soy bean flour was screened of foreign bodies, washed with ample of clean water, soaked in boiling water for 24 h. Seed coat was removed from the soaked seed and sun dried for 12 h to reduce water content to about 13%. It was then fried to brittleness before ground with a Waring blender (Model 5011, Waring Products Corp. New Hartford, CT, USA) into powder and sieved through DIN 4188 (0.12 mm) aperture sifter and stored for experiments.

Experimental design

Blends (mixtures) are combinations of components that together produce an end product having one or more properties of interest (Cornell and Piepel, 2008). Response in mixture experiment is a function of proportions and not amounts of the mixture components. Given these two conditions and fixing the total (an equality constraint) facilitates modeling of the response as a function of component proportions (Huguo, 2002).

Mixture experimental design assumes Equation 1,

$$Y_i = \beta_1 x_1 + \beta_2 x_2 + \beta_3 x_3 + \beta_{12} x_1 x_2 + \beta_{13} x_1 x_3 + \beta_{23} x_2 x_3 + \beta_{ij} x_{ij}^2 \ldots 1$$

where Y_i is the response, the β symbols represent coefficient to be fitted via regression analysis, x_1, x_1x_2, xij^2 are the linear, interaction and quadratic effect of the components respectively.

A simplex-centroid design consists of all points that are equally weighted mixtures of 1 to q components. The components are distributed as presented in Table 1. The small option of the design was used in the allocation of the mixture components resulting in 14 bread samples (Table 1).

Preparation of composite flour

Composite flour was prepared by intimate mixture of proportions of the three component flours according to Table 1 (Huguo, 2002). Blends of the mixture were the pure (A, B, C), binary AB, AC, BC) and overall centroid (ABC).

Table 2. Experimental layout and responses.

Run	A	B	C	Taste	Colour	Aroma	Texture	Pref.	Accpt.
1	0.33	0.33	0.33	3.00	7.00	8.00	2.00	2.00	2.00
2	0.00	0.50	0.50	7.30	4.00	7.00	2.00	3.00	4.00
3	0.00	0.00	1.00	2.00	2.00	7.00	2.00	3.00	2.00
4	0.17	0.17	0.67	3.00	6.00	6.00	4.00	5.00	6.00
5	1.00	0.00	0.00	9.00	8.50	5.00	9.00	9.0	8.00
6	0.50	0.50	0.00	6.00	7.80	8.00	6.00	7.00	8.00
7	0.00	1.00	0.00	5.00	6.00	3.00	3.00	2.00	2.00
8	0.00	0.00	1.00	2.30	3.00	2.00	2.00	3.00	2.00
9	0.00	1.00	0.00	2.00	5.00	3.00	2.00	2.00	4.00
10	1.00	0.00	0.00	8.60	9.00	6.00	8.50	9.00	8.00
11	0.50	0.50	0.00	6.00	7.00	7.00	6.00	7.00	8.00
12	0.17	0.67	0.17	5.00	8.00	6.00	6.00	6.00	5.00
13	0.50	0.00	0.50	6.00	5.00	7.00	5.00	7.00	6.00
14	0.67	0.17	0.17	8.00	9.00	8.00	8.00	9.00	8.00

A, wheat flour, B = Cassava flour, C = soybean flour, run = experimental run, pref. = preference, accpt. = acceptability.

Table 3. p-Values of model, R^2 regression analysis of the models and mean score of hedonic tests.

Analysis	Taste	Colour	Aroma	Texture	Accept.	Prefer.
Model	0.0002	0.0225	0.0452	0.0018	0.0286	NS
Linear	0.0012	0.0057	NS	0.0002	0.0059	00014
AB	NS	NS	0.0217	NS	NS	NS
AC	NS	NS	NS	NS	NS	NS
BC	NS	NS	NS	NS	NS	NS
ABC	0.0331	S	S	S	S	S
R^2	0.9254	0.7577	0.7069	0.8759	0.7411	0.8095
Mean	6.24	5.23	5.93	4.74	5.21	5.29

S = significant, NS = not significant.

Preparation of dough and baking

The composite flour dough was prepared and baked according to the method specified by Eddy et al., (2007) and Edwardo et al. (2013).

All the ingredient; yeast 3 g, salt 4 g sugar 5 g, margarine 9 g, water, according to the absorption capacity of blend were initially mixed in a mixing machine, (KitchenAid, USA). Different levels of water were mixed for 2.5 min at the low machine speed and for 8 min at the medium machine speed with the original dry mix of each blend until soft dough resulted. After the first fermentation, the dough was divided into 40 g portions rounded, and placed into bread baking pans and proofed for another 40 min in an improvised fermentation cabin at about 35°C and 80% relative humidity. The proofed dough of blends were baked in a hot air oven (Therma Grosskuchen, Le Chef, Sweden) at 210°C for 8 min. The bread samples were allowed to attain room temperature (28.0±1.0°C) before used for sensory evaluation.

Sensory evaluation

Samples of the composite bread were served to a 10 member semi-trained panelists who were familiar with the major sensory attributes of good quality bread: taste, aroma, texture, colour of the samples. A 9-point hedonic scale of 9 (like extremely) to 1 (dislike extremely), was designed to measure the degree of preference of the samples. The samples were presented in identical containers, coded with 3-digit random numbers and served simultaneously. The categorical scores of the sample parameters were converted to numerical scores ranging from 1 to 9, with 1 as the highest and 9 as the lowest level of the quality attribute (Iwe, 2002).

RESULTS

Table 2 shows the experimental runs and the respective responses, Tables 3 and 4 show values of 2nd polynomial fit, analysis of variance (ANOVA) and coefficient of estimates respectively for the data obtained from sensory evaluation.

Equations 2 to 7 are models (mathematical equations) of the parameters of composite bread samples which resulted from the experimental runs. The graphical

Table 4. Coefficients of estimates.

Parameter	A	B	C	AB	AC	BC	ABC
Taste	8.7859	5.5940	2.5601	1.3606	-1.9268	0.9303	66.2963
Colour	9.1669	3.6288	2.0757	-3.6774	-3.0859	11.2856	NS
Aroma	5.6174	2.9507	4.4042	12.7761	7.4093	11.4093	NS
Texture	8.8823	2.7274	1.9964	2.3025	-0.6808	0.7478	NS
Accept.	9.1968	2.2920	3.1042	4.7989	1.4418	0.0132	NS
Prefer.	8.1665	3.0713	2.3833	7.3115	0.9710	2.3995	NS

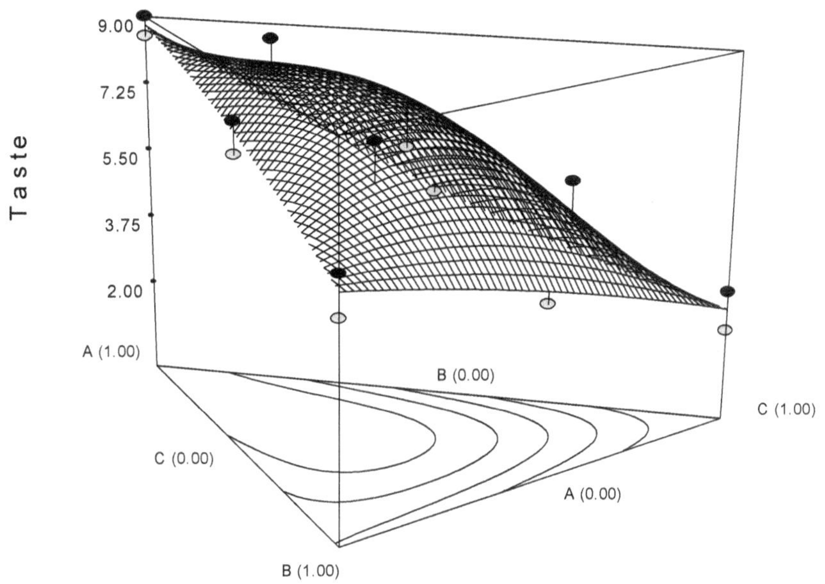

Figure 1. Response surface plot of bread on taste of bread samples;Taste = 8.78A + 5.59B + 2.56C **+1.36AB -1.92AC** + **0.93BC** + 66.29ABC……………1

presentation (Figures 1 to 4) plots visualize the fitted models in triangular surface plots.

DISCUSSION

Taste

Taste is an important sensory attribute of any food. Intake of bread is often enhanced by taste (Sim and Tam, 2001). Table 2 shows a reduction in hedonic score of samples of bread with high proportions of cassava and soy bean flour. Table 2 also shows that bread samples from pure blend and blend of 0.67:0.17:0.17 wheat, cassava and soy bean composite flour, respectively, showed high sensory scores on taste. However, some panelists still expressed liking for the ‘bad’ samples. This observation may be attributed to personal choice or an influence of the experimental conditions. Soy bean flour has beany flavor and poor aroma (Janet and Nyla, 2001; Dendy, 2013), addition of high proportions of the flour in the composite flour may introduce objectionable characteristics which overwhelmed the traditional taste attribute of the pure wheat flour blend samples and affected the choice of their taste. Table 3 shows that the taste model was significant ($p<0.05$) and reliable, ($R^2 = 0.9245$) with a mean hedonic score of 6.24.

Model of taste (Equation 1) shows that pure blend of the components was significant ($p<0.05$), their cross products were not significant ($p>0.05$) (bold type in Equation 1). Overall centroid of the components of mixture was significant ($p<0.05$). The response surface plot (Figure 2) eluciates this observation.

Colour

Colour of bread crust is an important sensory attribute, which can enhance acceptability. The local population thinks that pale coloured bread crust is indicative of

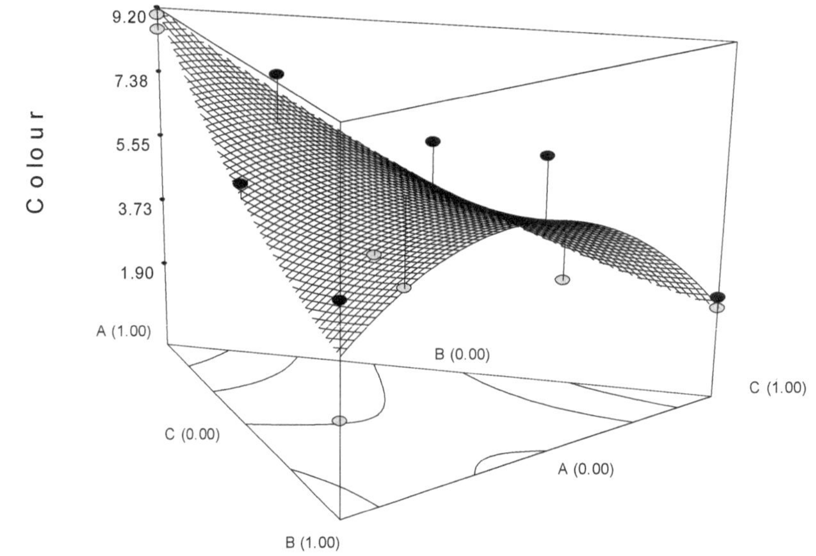

Figure 2. Response surface plot of bread on colour of bread samples; Colour = 9.17A +3.63B + 2.08C - 3.08AB - 3.09AC + 11.29BC.......... (2)

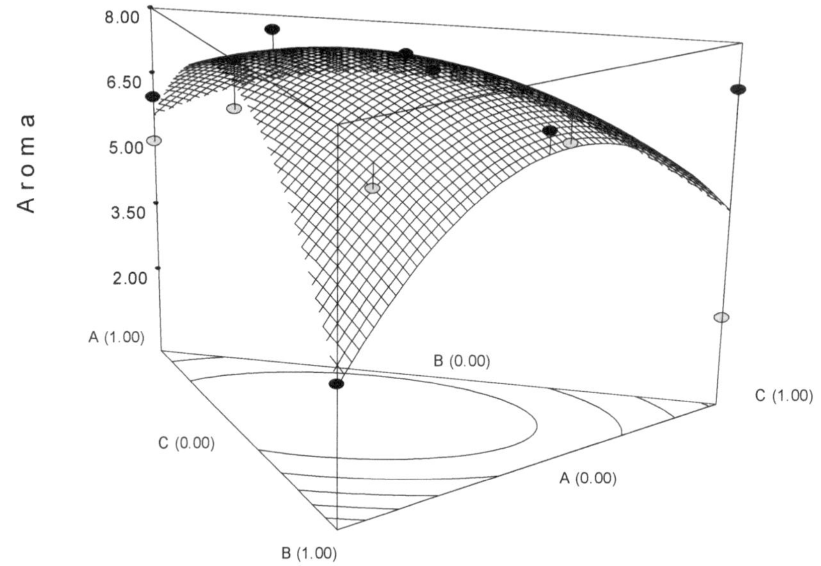

Figure 3. Response surface plot of bread on aroma of bread sample; Aroma = 5.61A + 2.95B + 4.40C +12.77+ 615AB + 7.41AC + 11.41BC + 1.22ABC..... (3)

improper baking. Besides it is assumed that the brown colour is what imparts nutrients, especially iron on the product. Browning of bread crust is an origin of Millard reactions during baking in the presence of amino acids, reducing sugars, temperature, time of baking and moisture levels of the fermented dough (Deny, 2013). Table 2 shows that browning was highest in samples with high supplementation of soy bean and cassava flours. This observation may be attributed to the high content of amino acids in soy bean and high content of sugars in cassava flours. However, all other blend samples showed some levels of brownness. Table 3 indicates that the models of the parameter was significant, ($p<0.05$), showing an adequate quadratic fit (R^2 = 75.77%) and an overall mean hedonic score of 5.23.

Model 1 shows that all pure blends were significant, their cross products: AB, AC and BC were not significant, $p<0.05$, while their overall centroid was significant,

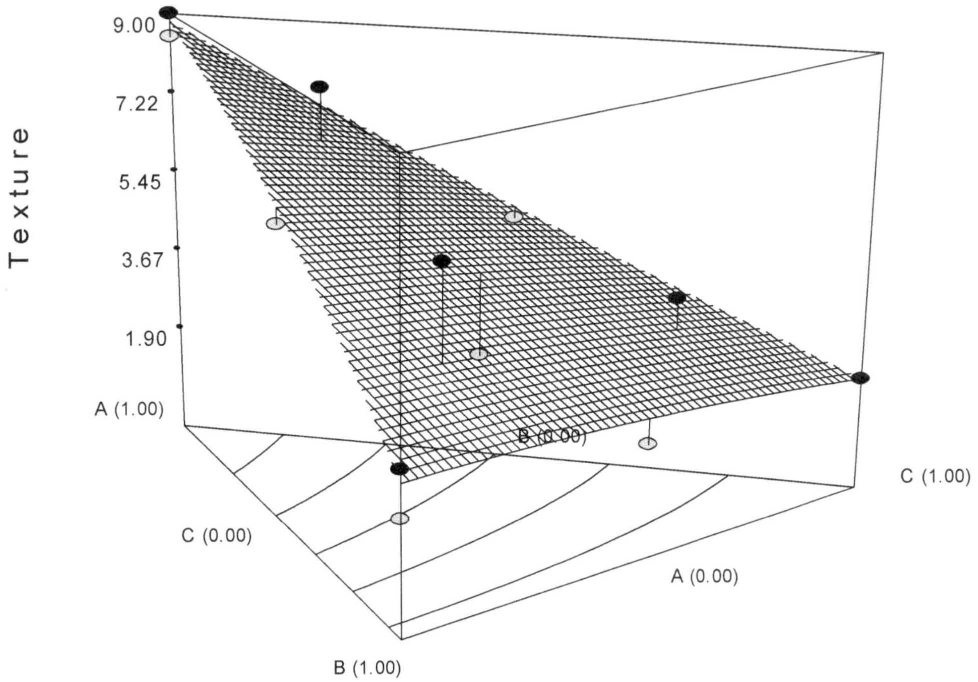

Figure 4. Response surface plot of bread on texture of bread samples; Texture = 8.824A + 2.72B + 1.99C + 2.30AB - **0.68AC** + 0.75BC………… (4)

$p<0.05$. The negative coefficient of estimate of cross-product of AC suggested that pure components may produce higher coloured product than in combination. This observation was supported by Figure 1. Statistically, the model should exclude the non significant estimates of coefficient, but in practice the non-significant estimates are experimented for any slight positive effect they might introduce.

Aroma

Aroma is an important parameter of food (Iwe, 2002). 'Good' aroma from food excites the taste buds, making the system ready to accept the product. 'Poor' aroma may cause outright rejection of food before they are tasted. The model of aroma was not very significant ($p = 0.0452$), while showing seemingly adequate quadratic fit ($R^2 = 70.69\%$).

The two cross-product terms in Table 4 (AC and BC) were not significant ($p>0.05$) while the overall centroid were significant ($p<0.05$), which was although included in the model, for further experiment-tation. The observation may be attributed to the beany aroma of raw soy bean, it may also be attributed to the strangeness of the product. Consumers attitudes may be tuned to accept new product if health claim, or social status is attached. In fact, there was the tendency for composite bread samples with high supplementation of wheat flour to consistency show high sensory scores.

Texture

Food texture sometime embraces appearance (Eduordo et al. 2013). Texture of composite bread samples were rated on a scale from 1 (heavy) to 9 (light). The mean texture scores of samples ranged from sample 8.5 for pure blends. Table 3 shows that samples of pure blends were significant ($p<0.05$), the model showing adequate quadratic fit ($R^2 = 87.59\%$) and a mean hedonic score of 4.74.

Terms in the model are coefficient of estimates (Table 4) and how they contribute to the parameter of the bread samples. Pure blend sample of A, AB and ABC were significant ($p<0.05$). The observation indicates that high supplementation of non-wheat flour showed low scores on texture. High supplementation reduces elastic property of wheat flour dough rendering the dough incapable of retaining the gas emanating from fermentations (Giami et al., 2004).

Optimization

Preference and acceptability

Both acceptance and preference are primarily economic concept. Acceptance of food varies with standards of living and cultural background, whereas preference refers to selection when presented with choice (Huguo, 2002). Preference is often influenced by prejudices, religious principles, group conformance, 'status value' and snobbery, in addition to the quality of the food. People have preferences, no matter how illogical they may appear. Therefore, the parameters are difficult parame-ters to determine in a new product development (Sim and Tam, 2001). In Table 2, hedonic scale scores for both parameters on the parameters showed a similar trend. The general trend is that the pure blend and low non-wheat composite flour samples showed strong likeness to colour, and texture also have strong positive influence on acceptability and preference. Models of both parameters are significant ($p<<0.005$), predictable ($R^2 = 0.7411$ and $R^2 = 0.8095$) respectively with mean scores of 5.21 and 5.29.

Generally, optimization analysis showed that blend samples of 0.67:0.17:0.17 of wheat flour, cassava and soybean flours respectively showed the highest hedonic scores after samples of pure blend of wheat flour.

Conclusion

In the study, it was observed that all samples of highly supplemented bread samples failed to meet the important parameters of pure bread; the failure hindered general acceptability and preference of the familiar bread sample. The high hedonic scores obtained in some of the new products was attributed to personal psychology of choice. However, samples of overall centroid blend showed encouraging scores on product acceptability. It is believed that regular supply, publicity and introduction of health claims on the products may enhance improved production strategy and quality. Besides, the product should be presented as the 'food for the rich'.

REFERENCES

Aleke DL, Germain AS et al. (2000). Isoflavone-rich soy protein isolate attenuates bone loss in the lumbar spine of perimenopausal women. Am. J. Clin. Nutr. 72(3):844-52.

David MO (2006). Nigeria, No. 1 market for U. S. wheat; potential for other grains and feeds, USAD Foreign Agricultural Service bulletin. pp. 1-2.

Dendy DAV (2013) Review of composite flour technology in the context of Tanzania. National Resources Institute, Chatham, UK. pp. 1-23.

Eddy NO, Udofia PG, Eyo D (2007). Sensory evaluation of wheat/cassava composite bread and effect of label information on acceptance and preference. Afr. J. Biotechnol. 6(16):123-134

Eduardo M, Svanberg U, Oliveira J, Ahrne L (2013). Effect of cassava flour characteristics on properties of cassava-wheat-maize composite bread types. Int. J. Food Sci. 1-10.

Giami GY, Amasisi T, Ekiyor G (2004). Comparison of bread making properties of composite flour from kernels of roasted and boiled African bread fruit (*Treculia africana*) seed. J. Mat. Res. 1 (1): 16-25.

Iwe MO (2002). Handbook of sensory methods and analysis. PROJOINT Communications Services Ltd, Enugu. pp. 70-72.

Sim J, Nya T (2001). Eating qualities of muffins prepared with 10% and 30% soy flour. J. Nutr. Recipe Menu Dev. 3:2.

Stone H, Sidel JL (1993). Sensory evaluation practices, Califonia: Academic Press.

Performance characteristics, biochemical and haematologcal profiles of broiler chickens fed synthetic and herbal methionine supplemented diets

F. A. Igbasan and O. A. Olugosi

Department of Animal and Health, School of Agriculture and Agricultural Technology, Federal University of Technology, P. M. B. 704, Akure, Ondo State, Nigeria.

This study was conducted to compare the efficacy of herbal methionine (HM) to synthetic methionine (SM) in broiler chickens. The herbal methionine (Meth-o-Tasr®) was supplied by Intas Pharmaceutical Limited, India. The HM and SM were added to standard diets at 0, 0.5, 1.0 and 1.5 kg/ton of finished feed and fed to 168 Arbor Acre broiler chickens. After seven days pre-experimental period, the birds were randomly allocated to seven isonitrogenous and isocaloric dietary treatments in a completely randomized experimental design with factorial arrangement of treatments. Each treatment was replicated four times with six birds per replicate. The trial lasted for 56 days and was divided into two phases; the starter phase and the finisher phase. The final body weight, body weight gain, feed intake and feed conversion ratio (FCR) of the birds were determined on a weekly basis. At the end of the feeding trial, two birds per replicate were randomly selected and slaughtered for the determination of carcass quality traits and biochemical and haematological profiles. The average final body weight, body weight gain and average feed intake of the birds fed on diets supplemented with SM were significantly different ($P \leq 0.05$) from those fed on diets supplemented with HM at both starter and finisher phases. The birds on SM supplemented diets gained more body weight and consumed more feed than those on HM supplemented diets. Feed conversion efficiency was not affected ($P \geq 0.05$) by the dietary methionine source. Birds on SM supplemented diets had better ($P \leq 0.05$) carcass yield than those on HM and control diets. Dietary HM supplementation did no significant ($P \geq 0.05$) effect on any of the biochemical or haematological variables determined. It can be concluded from this study that, though no detrimental health effects, were detected, dietary HM is not an effective substitute for synthetic methionine for optimum production performance.

Key words: broiler chickens, herbal methionine, performance, synthetic methionine.

INTRODUCTION

Despite the advances made in poultry nutrition in the last few decades, a lot of nutritional problems still remain unresolved. One of the most critical areas is amino acid nutrition. Of the essential amino acids required for poultry, methionine is usually the first to be limiting in diets based on maize and soybean meal (Fancher and Jensen, 1989). Methionine plays a significant role in energy production and protein synthesis. It enhances production and the size of the eggs, overall growth of the birds, feed conversion efficiency, and livability of broilers and layers (Binder, 2003 and Aerni et al., 2005). Methionine is a potent donor of methyl groups, which contributes to the synthesis of many important substances including epinephrine, choline, and creatinine (Bender, 1975). The increase in demand for cheap meat has given rise to the use of synthetic compounds in animal feeds. Recently, the safety of such practice has been questioned and their use is becoming restricted in many regions of the world. Therefore, there is renewed interest in developing natural alternative supplements to maintain animal performance and well being. Herbal methionine premixes, manufactured in India, have recently found their way into the animal feed industry. Methiorep® and Meth-o-Tas® are herbal methionine premixes that are available in the animal feed market. They are phyto-additives containing herbal ingredients that mimic the activity of methionine rations (Chattopadhyay et al., 2006). Under Indian conditions, herbal methionine (Methiorep®) has been found to replace DL-methionine very effectively when used in broiler rations (Chattopadhyay et al., 2006; Kalbande et al., 2009). Using Methiorep®, within the Nigerian environment, Itoe et al. (2010) reported that broilers fed herbal methionine performed below optimal levels and concluded that herbal methionine is an ineffective substitute for synthetic methionine for optimum production performance. Our recent study (Igbasan et al., 2012) with laying hens also revealed that birds fed on diets supplemented with herbal methionine (Meth-o-Tas®) produced less eggs, had a lower egg mass output and final body weight, and poorer feed conversion efficiency compared with those fed on synthetic methionine. The present study was conducted to compare the efficacy of herbal methionine (Meth-o-Tas®) with synthetic methionine in the diets of broiler chickens.

MATERIALS AND METHODS

The experiment was carried out at the Poultry Unit of the Teaching and Research Farm of the Federal University of Technology, Akure, Nigeria. The herbal methionine (HM) premix (Meth-o-Tas®) used for the study was supplied by Intas Pharmaceutical Ltd. India. The composition of the HM as specified by the manufacturer includes the following: *Andrographic paniculata*, 35%; *Zea mays*, 25%; *Ocimum sanctum*, 30%; *Asparagus racemosus*, 10%. The recommended inclusion rate is 1 kg/ton of finished feed.

Two corn-soybean meal basal diets were formulated to meet the nutrient requirements (NRC, 1994). The compositions of the diets are presented in Table 1. The methionine sources (SM and HM) were included in the basal diets at 0, 0.5, 1.0 and 1.5 kg/ton of finished feed. Therefore, there were seven isonitrogenous and isocaloric diets at the starter or finisher phase of the experiment.

A total of 200 day-old Arbor Acre male broiler chicks were purchased from a reputable commercial hatchery (Chi Hatchery, Ajanla Farms, Ibadan, Nigeria). The birds were fed on a commercial broiler starter ration for seven days and starved for 4 h before they were randomly allocated to seven treatment groups based on body weight. Out of these 200 chicks, 168 healthy ones were randomly assigned to seven dietary treatment groups with 24 birds per treatment group and six birds per replicate. There were four replicates per treatment group and the experiment lasted for eight weeks, four weeks for starter phase and another four weeks for finisher phase. The experimental design was completely randomized with factorial arrangement of treatments.

The growth performance data were determined for both starter and finisher phases. During each phase, weekly body weight and feed consumption were recorded. Feed conversion ratio was calculated from the body weight and feed consumption data.

At the end of the fattening period (56 days), eight birds were randomly chosen per treatment group, weighed, and slaughtered. Slaughtering was performed by severing the jugular vein with sharp surgical knife without anaesthetizing. The slaughtered birds were de-feathered using hot water, and the dressing weights were recorded. After removal of viscera, the eviscerated weights were recorded and organs (liver, heart, kidney, spleen, lung, bursa, gizzard, and pancreas) were removed and weighed. Carcasses were cut into basic parts (breasts, thighs, drumsticks, wings, sharks, heads, chests, backs, and necks) and weighed. Blood samples were also collected from the birds for the biochemical and haematological analyses. For haematological determinations, blood samples were collected into sample tubes containing anticoagulant as described by Lamb (1981). Plasma was harvested by centrifuging the blood samples at 3000 rpm for 15 min in centrifuge machine. The heparinized plasma samples were stored at -20°C in sample tubes until further analysis. Haematological parameters determined to include erythrocyte sedimentation rate, packed cell volume, red blood cell, haemoglobin, leucocyte, neutrophill, monocyte, basophil and eosinophil.

Plasma samples were analyzed for proteins (total proteins and albumin) and enzymes {alanine transferase (ALT), aspartate transferase (AST). Total protein, albumin, ALT and AST activities were determined in the Autoanalyzer, Microlab 200 using commercial kits (Randox Laboratories Ltd., Ardmore Diamond Road, Crumlin, Co. Antrim, United Kingdom, BT29 4QY). The globulin fraction was calculated by subtraction of albumin level from total protein level.

The proximate composition of the basal diets was determined according to the method of AOAC (2005). The data collected were subjected to one way analysis of variance (ANOVA) according to the General Linear Model Procedures of SAS (2008). When analysis of variance indicated a significant treatment effect, Duncan's multiple range test (Duncan, 1955) was used to compare treatment means. The model included main effects of methionine source, methionine level, and their interaction.

RESULTS

The initial body weight, final body weight, body weight gain, feed intake and feed conversion ratio for both starter and finisher phases are presented in Tables 2 and 3. There was a significant ($P \leq 0.05$) effect of methionine source on average final body weight, average body weight gain and average feed intake of birds at both starter and finisher phases. Dietary methionine level had no significant ($P \geq 0.05$) effect on any of the performance parameters determined. A source by level interaction was observed in average final body weight, average body weight gain, and average feed intake. There was no significant ($P \geq 0.05$) effect of methionine source, level or source by level interaction on the feed conversion ratio of the birds. Average body weight decreased by about 10% while that of feed intake decreased by 8% when HM was included in broiler starter diets up to 1.5 kg/ton of finished feed. The same trend was also followed for average body weight (11%) and average feed intake (10%) at the finisher phase. A closer look at the performance data (Table 2) also revealed that average body weights decreased with increasing the level of HM in broiler

Table 1. Composition of the basal diets.

Ingredient	Composition (Kg/ton)	
	Starter diet	Finisher diet
Maize	600.0	600.0
Wheat offal	-	108.5
Groundnut cake	180.0	120.0
Soybean meal	150.0	146.0
Fish meal	40.0	-
Bone meal	20.0	22.0
Lysine	2.5	0.5
Methionine[a]	0.5	0.5
Premix[b]	4.0	4.0
Salt	3.5	3.5
Total	1000.0	1000.0

	Calculated analysis (%)		Proximate analysis (%)	
	Starter	Finisher	Starter	Finisher
Crude protein	23.0	18.8	22.8	18.9
Crude fibre	3.63	3.63	2.15	3.06
Calcium	1.02	0.84	-	-
Crude Fat	-	-	15.5	18.7
Avail. P.	0.54	0.46	-	-
Moisture	-	-	6.5	6.8
Ash	-	-	9.6	6.8
Lysine	1.29	0.88	-	-
Methionine	0.35	0.43	-	-
Metabolizable energy (Kcal/kg)	3011.8	2917.7	-	-

[a]Methionine content varies depending on the inclusion level of the herbal and synthetic methionine. [b]Broiler premix per 1 kg used in the composition of the basal diets consist of the following: vit A (500,000 iu), vit D3 (1000,000 iu), vit E (1,400 iu), vit K (0.8 g), thiamine B1 (0.8 g) riboflavin B2 (2 g), nicacin B3 (16 g), D-calpan B5 (4.4 g), pyridoxine B6 (1.6 g), biotin (0.04 g), folic acid (vit B12) (0.006 g), manganese (28 g), copper (2.4 g), Iron (16 g), Iodine (0.4g), cobalt (0.1 g), selenium (0.06 g), sodium chloride (200 g), BHT (200g, BHT (50g) per kg.

starter diets and these were not significantly different (P ≥ 0.05) from the control. However, at the finisher phase (Table 3), though not significantly different (P ≥ 0.05) from the control, average body weights increased with increasing level of HM in the diets. Although not significant (P ≥ 0.05), those birds fed on SM supplemented diets utilized feed better than their counterparts fed to the control and HM supplemented diets.

The carcass characteristics of broiler chickens fed on diets supplemented with SM and HM are presented in Table 4. There was a significant (P ≤ 0.05) effect of methionine source on the dressed and eviscerated weights. The head, shank, back, neck, chest, and wing, as gram per kg live weight, were also influenced (P ≤ 0.05) by methionine source. Birds fed on diets supplemented with SM had a better carcass yield than their counterparts on HM and control diets. Carcass yield decreased by 8.8 and 11.8% for dressing and eviscerated weights respectively, for birds fed HM supplemented diets compared with those birds fed on SM supplemented diets. There was no effect (P ≥ 0.05) of methionine level, and source by level interaction on any of the carcass traits except for eviscerated weights. Statistical analysis of the organs (Table 5) showed that there was no significant (P ≥ 0.05) effect of methionine source, level, and source by level interaction on any of the parameters determined except for gizzard which was affected (P ≤ 0.05). The gizzards of broilers fed the SM supplemented were heavier than those of broilers fed HM supplemented diets (2.01±0.11 versus 1.63±0.07 g/kg live weight).

Table 6 shows the plasma biochemical profiles of the broiler chickens as influenced by SM and HM supplementation. Plasma concentrations of proteins and enzymes were similar (P ≥ 0.05) between the dietary treatments. Feeding diets containing HM did not significantly (P ≥ 0.05)alterthehaematologicalvariables(Table7)determined.

DISCUSSION

The results of the present experiment showed that

Table 2. Performance characteristics of broiler chickens at the starter phase fed diets supplemented with synthetic and herbal methionine.

Source	Level	Initial live weight (g/bird)	Final live weight (g/bird)	Total weight gain (g/bird)	Daily weight gain (g/bird)	Total feed intake (g/bird)	Daily feed intake (g/bird)	Feed conversion ratio
Control	0	128.33±0.00	1154.33±63.28^{b}	1030.0±63.28^{b}	36.79±2.26^{b}	2082.5±74.49^{ab}	74.38±2.66^{ab}	2.04±0.08
SM	0.5	128.33±0.00	1262.50±39.89^{ab}	1134.2±39.89^{ab}	40.51±1.43^{ab}	2155.8±80.28^{ab}	76.99±2.87^{ab}	1.90±0.03
SM	1.0	128.33±0.00	1329.20±23.93^{a}	1200.8±23.93^{a}	42.89±0.85^{a}	2221.8±37.92^{ab}	79.35±1.35^{ab}	1.85±0.01
SM	1.5	128.33±0.00	1350.00±48.59^{a}	1221.7±48.59^{a}	43.63±1.73^{a}	2295.8±107.11^{a}	81.99±3.82^{a}	1.88±0.02
HM	0.5	128.75±3.07	1229.20±25.80^{ab}	1100.4±22.93^{ab}	39.30±0.82^{ab}	2108.5±59.57^{ab}	75.31±2.13^{ab}	1.92±0.07
HM	1.0	128.75±3.07	1162.50±43.77^{b}	1033.8±44.23^{b}	36.92±1.58^{b}	2004.0±60.83^{b}	71.57±2.17^{b}	1.94±0.05
HM	1.5	128.75±3.07	1158.30±47.87^{b}	1030.0±47.87^{b}	36.79±1.29^{b}	2026.5±57.09^{b}	72.38±2.04^{b}	1.99±0.15
Source								
SM		128.33±0.0	1313.9±23.19^{a}	1185.6±23.19^{a}	42.34±0.83^{a}	2224.4±45.35^{a}	79.44±1.62^{a}	1.88±0.01
HM		128.61±1.31	1183.3±23.21^{b}	1054.7±23.0^{b}	37.67±0.82^{b}	2046.3±33.74^{b}	73.09±1.21^{b}	1.95±0.05
Level								
0.5		128.54±1.42	1245.8±22.88	1117.3±22.23	39.90±0.79	2131.1±47.13	76.15±1.68	1.91±0.04
1.0		128.54±1.42	1245.8±39.05	1117.3±39.23	39.91±1.40	2112.0±52.86	75.46±1.89	1.90±0.03
1.5		128.33±0.0	1254.2±40.05	1125.8±48.05	40.21±1.72	2161.0±75.80	77.18±2.71	1.93±0.07
Significance								
Source		NS	**	**	**	**	**	NS
Level		NS	NS	NS	NS	NS	NS	NS
Source*Level		NS	*	*	*	*	*	NS

[abc]Means with different superscript are significantly. SM, synthetic methionine; HM, herbal methionine; NS, not significant at $P \geq 0.05$; *significant at $P \leq 0.05$; **significant at $P \leq 0.01$.

Table 3. Performance characteristics of broiler chickens at the finisher phase fed diets supplemented with synthetic and herbal methionine.

Source	Level	Final weight (g/bird)	Total weight gain (g/bird)	Daily weight gain (g/bird)	Total feed intake (g/bird)	Daily feed intake (g/bird)	Feed conversion ratio
Control	0	2227.5±48.85^{bc}	1069.17±48.08^{ab}	38.18±1.72^{ab}	4064.3±129.91^{ab}	145.15±4.64^{ab}	3.82±0.13
SM	0.5	2370.8±72.13^{ab}	1108.30±43.30^{a}	39.59±1.55^{a}	4265.0±137.28^{a}	152.32±4.90^{a}	3.88±0.19
SM	1.0	2375.0±79.79^{ab}	1045.80±73.71^{ab}	37.35±2.63^{ab}	4192.5±168.52^{a}	149.73±6.02^{a}	4.07±0.30
SM	1.5	2508.3±122.95^{a}	1158.30±75.92^{a}	41.37±2.71^{a}	4256.3±144.15^{a}	152.01±5.15^{a}	3.70±0.14
HM	0.5	2107.5±61.02^{c}	878.34±45.71^{b}	31.37±1.63^{b}	3591.5±196.25^{b}	128.27±7.01^{b}	4.11±0.23
HM	1.0	2172.9±27.51^{bc}	1010.40±50.16^{ab}	36.09±1.79^{ab}	3719.5±111.64^{ab}	132.84±3.99^{ab}	3.71±0.23

Table 3. Contd.

HM	1.5	2191.7±98.48^{bc}	1033.30±89.24^{ab}	36.91±3.19^{ab}	4135.8±246.89^{a}	147.71±8.82^{a}	4.04±0.14
Source							
SM		2418.1±52.88^{a}	1104.2±37.16^{a}	39.44±1.33^{a}	4237.9±79.24^{a}	151.35±2.83^{a}	3.88±0.12
HM		2157.4±37.51^{b}	974.03±39.58^{b}	34.79±1.41^{b}	3815.6±122.8^{b}	136.27±4.39^{b}	3.95±0.12
Level							
0.5		2239.2±66.25	993.33±52.33	35.48±1.89	3928.3±168.79	140.29±6.03	3.99±0.15
1.0		2274.0±54.63	1028.1±41.81	36.72±1.49	3956.0±129.41	141.29±4.62	3.89±0.19
1.5		2350.0±94.33	1095.8±59.16	39.14±2.11	4196.0±134.29	149.86±4.8	3.87±0.11
Significance							
Source		***	*	*	*	**	NS
Level		NS	NS	NS	NS	NS	NS
Source * Level		*	NS	NS	*	*	NS

[abc]Means with different superscript are significantly (p≤0.05) different. SM, Synthetic methionine; HM, herbal methionine; NS, not significant at P p≤0.05. *Significant at P≤0.05; **significant at P ≤0.01.

Table 4. Carcass characteristics of broiler chickens fed diets supplemented with synthetic and herbal methionine (g/kg live weight).

Source	Level	Dressed weight	Eviscerated weight	Head	Shank	Drum stick	Chest	Breast muscle	Back	Neck	Thigh	Wings
Control	0	2003±64.50	1645±11.2	2.96±0.19	4.90±0.46	12.28±0.95	23.79±1.46	8.420.38	19.50±1.37	5.64±0.50	14.23±1.20	9.83±0.85
SM	0.5	2250±101.77	1925±88.14	2.80±0.06	4.89±0.16	14.82±1.45	24.84±0.48	8.90±0.34	17.50±0.68	5.90±0.13	14.12±0.18	10.02±0.21
SM	1.0	2300±118.02	1937.5±90.51	2.95±0.15	4.78±0.27	12.56±0.3	25.47±1.02	8.94±0.40	19.24±0.71	6.29±0.51	13.95±0.48	9.85±0.28
SM	1.5	2437.5±178.02	2051±146.39	2.70±0.12	4.57±0.22	12.86±0.37	25.73±0.71	9.01±0.61	19.44±0.45	5.53±0.26	14.90±0.83	9.73±0.32
HM	0.5	2150±84.32	1762.5±84.38	2.97±0.12	4.92±0.24	13.33±0.33	23.86±0.71	8.70±0.50	20.00±0.45	5.08±0.34	15.12±0.42	10.76±0.26
HM	1.0	2162.5±56.5	1775±70.08	3.09±0.07	5.59±0.21	13.52±0.30	23.08±0.64	7.78±0.34	19.80±0.36	5.27±0.3	14.40±0.98	11.32±0.29
HM	1.5	2062.5±154.62	1675±122.11	2.94±0.13	5.15±0.30	13.00±0.37	23.74±0.92	8.58±0.56	19.69±0.99	5.28±0.35	13.31±0.37	10.56±0.32
Source												
SM		2329.2±77.17	1970.8±62.69	2.82±0.07^{a}	4.74±0.13^{b}	13.41±0.53	25.35±0.43^{a}	8.95±0.26	18.72±0.34^{b}	5.91±0.20^{a}	14.32±0.32	9.87±0.15^{b}
HM		2125±59.68	1737.5±53.1	3.00±0.06^{a}	5.22±0.15^{a}	13.29±0.19	23.54±0.43^{b}	8.35±0.28	19.83±0.32^{a}	5.21±0.18^{b}	14.28±0.39	10.88±0.17^{a}
Level												
0.5		2200±65.19	1843.8±62.56	2.89±0.07	4.90±0.14	14.07±0.74	24.35±0.43	8.80±0.29	18.76±0.51	5.49±0.21	14.62±0.23	10.39±0.19

Table 4. Contd.

1.0	2231.3±65.65	1856.3±57.14	3.02±0.08	5.18±0.19	13.04±0.24	24.25±0.66	8.36±0.30	19.52±0.39	5.78±0.31	14.17±0.53	10.58±0.27
1.5	2250±123.83	1862.5±104.03	2.82±0.09	4.86±0.20	12.93±0.25	24.74±0.61	8.30±0.41	19.55±0.53	5.41±0.21	14.10±0.49	10.15±
Significance											
Source	**	**	*	*	NS	**	NS	*	*	NS	***
Level	NS	NS	NS	NS	NS	NS	NS	NS	NS	NS	NS
Source*Level	NS	**	NS	NS	NS	NS	NS	NS	NS	NS	NS

[abc]Means with different superscript are significantly (P ≤ 0.05) different. SM, Synthetic methionine; HM, herbal methionine. NS, not significant at P ≤ 0.05; *significant P ≤0.05; **significant at P ≤ 0.01.

Table 5. Relative organ weight of broiler chickens fed diets supplemented with synthetic and herbal methionine (g/kg live weight).

Source	Level	Liver	Heart	Kidney	Spleen	Lung	Bursa	Gizzard	Pancreas
Control	0	1.77±0.15	0.43±0.03	0.37±0.04	0.10±0.01	0.57±0.05	0.87±0.09	2.03±0.13	0.18±0.02
SM	0.5	1.89±0.09	0.43±0.02	0.45±0.06	0.1±0.02	0.53±0.03	0.86±0.04	2.06±0.17	0.16±0.02
SM	1.0	1.72±0.12	0.39±0.01	0.48±0.06	0.09±0.01	0.47±0.04	0.83±0.04	2.22±0.18	0.24±0.04
SM	1.5	1.81±0.10	0.46±0.05	0.48±0.05	0.11±0.02	0.58±0.02	0.96±0.06	1.74±0.2	0.17±0.02
HM	0.5	1.92±0.15	0.43±0.02	0.47±0.05	0.08±0.01	0.55±0.03	0.9±0.09	1.65±0.11	0.18±0.02
HM	1.0	1.74±0.11	0.44±0.03	0.47±0.07	0.08±0.01	0.54±0.04	0.91±0.08	1.72±0.13	0.16±0.01
HM	1.5	1.61±0.15	0.46±0.03	0.48±0.05	0.09±0.01	0.58±0.06	0.89±0.05	1.52±0.1	0.19±0.02
Source									
SM		1.80±0.06	0.43±0.02	0.47±0.03	0.10±0.01	0.53±0.02	0.88±0.03	$2.01±0.11^{a}$	0.19±0.02
HM		1.76±0.08	0.44±0.02	0.47±0.03	0.09±0.01	0.56±0.02	0.90±0.04	$1.63±0.07^{b}$	0.18±0.01
Level									
0.5		1.90±0.09	0.43±0.01	0.46±0.04	0.09±0.01	0.54±0.02	0.88±0.05	$1.85±0.11^{ab}$	0.17±0.01
1.0		1.73±0.08	0.42±0.02	0.48±0.04	0.09±0.01	0.50±0.03	0.87±0.04	$1.97±0.13^{a}$	0.20±0.02
1.5		1.71±0.09	0.46±0.03	0.48±0.03	0.10±0.01	0.58±0.03	0.92±0.04	$1.63±0.11^{b}$	0.18±0.01
Significance									
Source		NS	NS	NS	NS	NS	NS	**	NS
Level		NS	NS	NS	NS	NS	NS	NS	NS
Source*Level		NS	NS	NS	NS	NS	NS	*	NS

[abc]Means with different superscript are significantly (P ≤ 0.01) different. SM, Synthetic methionine; HM, herbal methionine. NS, not significant at P ≥ 0.05; *significant at P ≤ 0.05; **significant at P ≤ 0.01.

Table 6. Haematoloicgal Profiles of broiler chickens fed diets supplemented with synthetic and herbal methionine.

Source	Level	ESR (mm/hr)	PCV (%)	RBC (10^6mm^3)	HB (g/100ml)	Lymphocyte (%)	NEU	Monocyte (%)	Basophil (%)	EOS
Control	0	3.75±0.62	27.13±1.13	207.13±12.69	9.04±0.37	60.13±0.95	24.75±0.80	11.38±0.60	2.50±0.27	1.25±0.25
SM	0.5	3.25±0.25	27.38±0.63	198.75±9.52	9.14±0.20	59.13±0.97	26.50±0.46	10.75±0.75	2.63±0.26	1.00±0.27
SM	1.0	2.75±0.59	29.25±1.18	225.13±13.90	9.74±0.39	58.38±0.71	26.13±0.52	11.63±0.75	2.50±0.19	1.38±0.26
SM	1.5	2.88±0.35	28.25±0.80	210.13±11.07	9.46±0.27	59.38±1.28	25.75±0.82	10.63±0.53	3.13±0.30	1.13±0.23
HM	0.5	3.00±0.33	28.25±0.92	212.63±11.27	9.41±0.30	58.38±1.08	25.75±0.75	12.00±0.96	2.50±0.27	1.38±0.26
HM	1.0	3.13±0.48	28.25±0.94	211.25±12.70	9.43±0.31	60.25±0.88	25.25±0.65	11.00±0.71	2.63±0.26	0.88±0.23
HM	1.5	2.88±0.13	28.00±0.63	207.63±9.54	9.39±0.20	59.63±0.42	25.25±0.67	11.63±0.86	2.5±0.27	1.00±0.19
Source										
SM		2.94±0.24	28.29±0.52	211.33±6.80	9.45±0.17	58.96±0.57	26.13±0.35	11.00±0.39	2.75±0.15	1.17±0.14
HM		3.00±0.19	28.17±0.46	210.50±6.22	9.41±0.15	59.42±0.49	25.42±0.39	11.54±0.48	2.54±0.15	1.08±0.13
Level										
0.5		3.13±0.20	27.81±0.55	205.69±7.35	9.28±0.18	58.75±0.71	26.13±0.44	11.38±0.61	2.56±0.18	1.19±0.19
1.0		2.94±0.37	28.75±0.74	218.19±9.27	9.58±0.24	59.31±0.60	25.69±0.42	11.31±0.51	2.56±0.16	1.13±0.18
1.5		2.88±0.18	28.13±0.49	208.88±7.07	9.43±0.16	59.50±0.65	25.50±0.52	11.13±0.51	2.81±0.21	1.06±0.14
Significance										
Source		NS	NS	NS	NS	NS	NS	NS	NS	NS
Level		NS	NS	NS	NS	NS	NS	NS	NS	NS
Source*Level		NS	NS	NS	NS	NS	NS	NS	NS	NS

Mean ± standard deviation; SM, synthetic methionine; HM, herbal methionine; NS, not significant ($P > 0.05$); ESR, erythrocyte sedimentation; PCV, packed cell volume; RBC, red blood cell; HB, haemoglobin; NEU, neutrophils; EOS, eosinophils.

broiler chickens fed diets supplemented with synthetic methionine had better performance in terms of weight gain and feed consumption than those fed herbal methionine supplemented diets. Earlier studies by Chattopadhyay et al. (2006), Halder and Roy (2007) and Kalbande et al. (2009) demonstrated that broiler chickens fed diets supplemented with herbal methionine (Methiorep®) had similar performance in terms of body weight gain and feed conversion efficiency compared with those birds fed DL-methionine. In fact, the feed conversion ratio of broilers fed 15 g herbal methionine/kg diet was significantly better than that of broilers fed on 10 g DL-methionine/kg diet (Chattopadhyay et al. 2006). However, using the same herbal methionine (Methiorep®) and dietary inclusion levels, Itoe et al. (2010) showed that broiler chickens fed supplemental herbal methionine were inferior in weight gain and feed conversion efficiency to those birds fed supplemental DL-methionine.

A post-mortem measurement of the birds in the treatments shows that average carcass yields were significantly influenced by the dietary methionine source. Those birds fed on DL-methionine had better carcass yields than their counterparts on herbal methionine supplemented diets. This finding is not in agreement with that of Chattopadhyay et al. (2006) who reported that broilers given herbal methionine had a similar dressing yield but higher breast and thigh yields at herbal 15 g herbal methionine per kg feed than 10 g DL-methionine/kg diet.

Little information has been published on the effects of herbal methionine supplements on the biochemical profiles of domestic chickens (Prabhakaran et al., 1996). Non-significant differences were observed in virtually all the haematological variables determined for birds fed

Table 7. Biochemical profiles of broiler chickens fed diets supplemented with synthetic and herbal amethionine.

Source	Level	Total protein (mg/100ml)	Albumin (mg/100ml)	Globulin (mg/100ml)	AST (IU/L)	ALT (IU/L)
Control	0	92.59 ±9.31	33.02±3.63^{b}	59.58±8.10^{ab}	28.63±3.74^{ab}	43.45±12.04^{ab}
SM	0.5	96.77 ±13.35	39.88±3.00^{ab}	56.89±12.18^{ab}	33.78±3.07^{a}	48.95±12.73^{a}
SM	1.0	96.77 ±8.74	30.97±2.71^{b}	65.80±7.42^{ab}	21.30±3.02^{b}	16.20±8.94^{b}
SM	1.5	90.33 ±4.71	46.46±6.18^{a}	43.87±9.69^{b}	23.51±2.90^{ab}	30.20±7.69^{ab}
HM	0.5	107.56±9.02	39.74±3.37^{ab}	67.82±6.78^{ab}	25.03±4.33^{ab}	30.45±13.19^{ab}
HM	1.0	112.26±9.68	38.28±1.75^{ab}	73.98±8.63^{a}	32.14±2.84^{ab}	49.17±7.21^{a}
HM	1.5	93.99 ±6.88	34.62±4.53^{ab}	59.36±6.26^{ab}	28.60±3.87^{ab}	14.2±7.57^{b}
Source						
SM		94.62±5.34	39.10±2.70	55.52±5.80	26.20±2.0	31.78±6.19
HM		104.60±5.02	37.55±1.94	67.06±4.21	28.59±2.15	31.45±6.16
Level						
0.5		102.17±7.91	39.81±2.18	62.36±6.88^{ab}	29.40±2.80	39.70±9.17
1.0		104.52±6.61	34.62±1.82	69.89±5.60^{a}	26.72±2.44	32.95±7.03
1.5		92.16 ±.4.05	40.54±4.0	51.62±5.92^{b}	26.05±2.43	22.20±5.61
Significance						
Source		NS	NS	NS	NS	NS
Level		NS	NS	NS	NS	NS
Source*Level		NS	NS	NS	NS	NS

Mean ± standard deviation; SM, synthetic methionine; HM, herbal methionine; AST, aspartate transferase; ALT, alanine transferase.

diets supplemented with both SM and HM. This is in agreement with the results of Rekhatel et al. (2010) who showed that herbal methionine had no significant effect on haematological profiles of broiler chickens. Also, our recent study (Igbasan et al., 2012) revealed that herbal methionine did not affect the haematological profiles of domestic laying hens. The implication is that dietary herbal methionine has no detrimental effect on surviva-bility of chickens. Rajurker et al. (2009) also reported that herbal methio-nine supplement (Methiorep®) is totally safe and has no adverse effect even when used at the highest limit dose of 5 g/kg body weight of male Wistar rats. Total plasma protein, albumin, and globulin concentrations observed are within the normal range reported for chickens (Prabhakaran et al., 1996). There were no differences in total plasma protein and albumin concentrations. These results are in agreement with those of Halder and Roy (2007) and Rekhatel et al. (2010) who did not observe any significant effect of HM supplementation on total protein and albumin concentrations of broiler chickens.

Generally, the activities of plasma AST and ALT observed are within the normal range for chickens (Marjanovic et al., 1975). This suggests that supplemen-tation of broiler chicken diets with herbal methionine did not inhibit the activities of plasma enzymes or cause any liver dystrophy or other vital organs abnormalities where these enzymes are secreted. These results corroborate those of Halder and Roy (2007), Kalbande et al. (2009), Rekhatel et al. (2010), and Igbasan et al. (2012) who reported that the plasma concentrations of ALT, AST, and ALP were not affected by supplementation of broiler and laying hen diets with herbal methionine.

The current experiment shows that broiler chickens fed diets supplemented with synthetic methionine had better performance in terms of weight gain, feed consumption and carcass yield than those fed herbal methionine sup-plemented diets. Since the primary objective of poultry farmers is to maximize profit, it can be concluded that based on the results of this trial, that herbal methionine is not a substitute for DL-methionine for optimum perfor-mance of broiler chickens.

REFERENCES

Aerni V, Brinkhof MWG, Oester WH, Frohlich E (2005). Productivity and mortality of laying hens in aviaries: a systematic review. World Poult. Sci. J. 61: 130-142.

AOAC International Official Methods of Analysis (2005). 18th ed. AOAC International, Gaithersburg, MD.

Bender DA (1975). Amino Acid Metabolism. 1st ed. John Wiley and Sons Ltd. New York, USA. pp. 112-142.

Binder M (2003). Life cycle analysis of DL-methionine in broiler meat production. In: Information for the feed industry. Degussa feed additives, Hanau-Wolfgang, Germany. pp. 1-8.

Chattopadhyay K, Mondal MK, Roy B (2006). Comparative efficacy of DL-methionine and herbal methionine on performance of broiler chicken. Int. J. Poult. Sci. 5: 1034-1039.

Duncan DB (1955). Multiple range test and multiple *F*-tests. Biometrics

11:1-42.
Fancher BI, Jensen LS (1989). Influence on performance of three to six week old broilers of varying dietary protein contents with supplementation of essential amino acid requirements. Poult. Sci. 68:113-123.
Halder G, Roy B (2007). Effect of herbal or synthetic methionine on performance, cost benefit ratio, meat and feather quality of broiler chicken. Int. J. Agric. Res. 2: 987-996.
Igbasan FA. Ibrahim AM, Osho IB (2012). Comparative efficacy of herbal and synthetic methionine on performance, some haematological and biochemical parameters in domestic laying hens. Afr.J. Biotech. 11:10617-10625.
Itoe S Dafwang II Bawa GS (2010). Evaluation of Methiorep as a Substitute for Methionine in Broiler Diets. Int. J. Poult. Sci. 9: 809-812.
Kalbande VH, Ravikanth K, Maini S, Rekhe DS (2009). Methionine supplementation options in poultry. Int. J. Poult. Sci. 8: 588-591.
Lamb GN (1981). Manual of Veterinary Laboratory Technique, 5th ed. CIBA-GEIGY, Kenya. pp. 96-107.
Marjanovic D, Kozic L, Petrovic M, Polic T, Latricin V (1975). Serum transaminase activities in fowls. Poult. Sci. 35:550-551.
NRC National Research Council, Nutrient Requirements of Poultry (1994). 9th edn. National Academy Press, Washington DC.
Prabhakaran V, Chihravel V, Prabhakaran SK, Saravanan CS (1996). Haematological and biochemical profile of white leghorn chickens. Indian J. Anim. Health 35:11-15.
Rajurker S, Rekhe DS, Maini S, Ravikanth K (2009). Acute toxicity studies of poly herbal formulation (Methiorep premix). Vet. World 2: 58-59.
Rekhatel DH, Maini S, Pathak VP, Ther SV, Mahajan N, Datir DK, Kale M, Hardas RP (2010). Haemobiochemical profile study in commercial broilers fed on diet supplemented with herbal methionine (Methiorep) and DL-methionine. Indian J. Agric. Res. 44: 197-203.
SAS (2008). SAS User's Guide, 8th ed. SAS Institute Inc. Cary, NC, USA.

A study on evaluation and characterization of extruded product by using various by-products

Anjineyulu Kothakota[1], Navdeep Jindal[1] and Thimmaiah B.[2]

[1]Department of Food Engineering and Technology, Sant Longowal Institute of Engineering and Technology, Longowal-148106, Punjab, India.
[2]Department of Food Process Engineering, Sam Higginbottom Institute of Agriculture, Technology and Sciences, Allahabad- 211 007, UP, India.

Broken rice flour was added in proportions (75%) to equal amount of dehydrated pineapple waste pulp powder and red gram powder (12.5%) were extruded in a twin-screw extruder. The formulation was extruded at different moisture content (17-21%), screw speed (260-340 rpm) and die temperature (120-140°C). The lateral expansion, bulk density, water absorption index, water solubility index, hardness and sensory characteristics were measured as responses. In the experiments, increase in barrel temperature resulted in extrudate with higher expansion, higher hardness, lower bulk density, lower WSI and higher WAI. Increasing in screw speed resulted in higher expansion, lower bulk density, higher overall acceptability and lower hardness; whereas, increasing level of moisture resulted in lower hardness, lower expansion and minimum bulk density and higher overall acceptability. In the experiment, optimization studies resulted in 132.27°C of barrel temperature, 315 rpm of screw speed and 18.48% of feed moisture.

Key words: Twin screw extrusion, pineapple waste pulp, response surface methodology (RSM), broken rice flour, red gram powder.

INTRODUCTION

Fruit and vegetable wastes are inexpensive, available in large quantities, characterized by a high dietary fibre content resulting to high water binding capacity and relatively low enzyme digestible organic matter (Serena and Knudsen, 2007). A number of researchers have used fruits and vegetable by-products such as apple, pear, orange, peach, blackcurrant, cherry, artichoke, asparagus, onion, carrot pomace and pineapple waste pulp (Grigelmo-Miguel and Martin-Belloso, 1999; Ng et al., 1999; Nawirska and Kwasnievska, 2005) as sources of dietary fibre supplements in refined food. Cereal grains are generally used as major raw material for development of extruded snack foods due to their good expansion characteristics because of high starch content. The broken rice is a by-product of modern rice milling process. The rice portion can have varying percentages (5-7%) of broken kernels which contain nutritive value similar to whole rice and are available readily at relatively lower cost. Rice contains approximately 7.3% protein, 2.2% fat, 64.3% available carbohydrate, 0.8% fiber and 1.4% ash content (Zhoul et al., 2002).

Rice flour has become an attractive ingredient in the extrusion industry due to its bland taste, attractive white colour, hypoallergenicity and ease of digestion (Kadan et al., 2003). Cereal grains tend to be low in protein have a poor biological value due to their limited essential amino acid content and are usually fortified with lysine or pulse protein to produce nutritious snack foods. Among food legumes, pigeon pea is a valuable source of protein, minerals and vitamins and occupies a very important place in human nutrition in many developing countries (Singh, 1988). The method of dehulling of legumes significantly affects the formation of broken and powdered particles and in the case of pigeon pea it varies between 9 - 24.6% for broken and 5.5 - 6.1% for powder (Singh et al., 1992). Red gram powder is a by-product of milling process which has a high protein content (22%) similar to dhal and easily available at relatively lower cost as compared to whole pigeon pea dhal. Pineapple (*Ananas comoscus*) is sometimes called the 'King of Fruit,' one of the most popular of the non citrus tropical and subtropical fruit, though it is mainly processed into canned products and also processed into various food products such as jam, jelly, beverage and concentrate, which produce a large amount of solid waste such as skin, core. The production of canned pineapple was estimated at about 48 million standard cases as against 41 million standard case tones in 1996, an increase of almost 16% in the year 2008-2011.

Major world producers of canned pineapple are Thailand (40%), Philippines (25%), Indonesia (14%) and Kenya (9%) which altogether contribute to more than 80% of total world canned pineapple production in 2008-11. Malaysia's production amounting to 1,563,291 standard cases would be equivalent to 3.3% of total world production. In India, north east India, cover Tripura, Manipur, Nagaland, Meghalaya and Assam. There is a lot of solid pineapple waste by the canned industries every year; it includes cellulose, hemicelluloses and other carbohydrates. This waste cannot be used for further process and also causes lot of environmental problems and pollution problems. That is why we planned to recycle the pineapple pulp waste in the form of incorporation extruded product (Abdullah and Hanif, 2008).This particular study aims at utilization of various by-products (pineapple waste pulp, broken rice flour and red gram powder).

MATERIALS AND METHODS

Experimental design

Response surface methodology (RSM) was adopted in the design of experimental combinations (Altan et al., 2008a; Yagci and Gogus, 2008; Ding et al., 2005; Montgomery, 2001). The main advantage of RSM is the reduced number of experimental runs needed to provide sufficient information for statistically acceptable results. A three-variable (five levels of each variable) central composite rotatable experimental design was employed (Montgomery, 2001; Yagci and Gogus, 2008). The parameters and their levels were chosen based on the literature available on rice based extrudates (Yagci and Gogus, 2008; Ibanoglu et al., 2006; Ding et al., 2005; Upadhyay et al., 2008). The ingredients used for the preparation of ready-to-eat snack products were: pineapple waste pulp, broken rice flour, red gram powder (pigeon pea). The independent variables including the proportion of CPPP in rice flour (equal amount of pineapple waste pulp and red gram powder (75: 12.5+12.5%) : are maintained constant throughout the experiment, and moisture content (16-21%), screw speed (260 - 340 rpm) and die temperature (120-140°C) are variables. Response variables were expansion index, bulk density, water absorption and solubility index, hardness and sensory characteristics. The five levels of the process variables were coded as -1.68, -1, 0, +1, +1.68 (Montgomery, 2001) and design in coded (*x*) form and at the actual levels A, B, C, and D are given in Table 1.

Dry pineapple waste pulp powder preparation

Commercial variety (KEW) was procured from local market, Longowal, District. Sangrur, India. These were washed in running tap water two times and leafy top were removed by twisting off with hands. Then it was set aside, using a plane stainless steel knife, eyes were removed then the fruit and trimmed to remove extra hard material. The juice was extracted using a Juice Mixer Grinder cum Food Processor (Make: Supremo DLX, Maharaja appliance limited, New Delhi, India) to extract carrot juice (Goyal, 2004). The pineapple waste pulp was collected for further studies. A hot air oven (Make: Osaw Industrial Products Pvt. Ltd., Haryana, India) was used for drying pineapple waste pulp, which could regulate drying air temperature up to 250°C with ± 2°C accuracy.

The dryer consisted of a preheating and heating chamber with thermostat based control unit, an electrical fan, and measurement sensors. The samples were spread over the trays and the temperature of the dryer was set at 60°C. The drying procedure continued till the moisture content of the sample was reduced to about 5±1% (wet basis). The grinding was performed using the same food processor (Make: Supreme DLX, Maharaja appliance limited, New Delhi, India) with grinder attachment. The material was ground to pass through the sieve of 2 mm size. The pomace was stored in sealed polythene bag for further use.

Sample preparation

Ingredient formulations for extrusion products are: The rice flour : pineapple waste pulp powder and red gram powder were mixed in equal proportion in a food processor with mixer attachment. The moisture content of the formulation was estimated by hot air oven method (Ranganna, 1995). The moisture was adjusted by sprinkling distilled water in dry ingredients. All the ingredients were weighed and then mixed in the Supremo DLX food processor for 10 min based on preliminary study. The mixture was then passed through a 2 mm sieve to reduce the number of lumps formed due to addition of moisture. After mixing, samples were stored in polyethylene bags at room temperature for 24 h (Stojceska et al., 2008). The moisture content of all the samples were again determined by hot air oven method (Ranganna, 1995) prior to extrusion experiments.

Extrudates preparation

Extrusion of samples was performed using a co-rotating twin-screw extruder (Basic Technology Pvt. Ltd., Kolkata, India). The length-to-diameter (L/D) ratio of the extruder was 8:1. The main drive of extruder was provided with a 7.5 HP motor (400 V, 3 ph, 50 cycles). The output shaft of worm reduction gear was provided with a torque limiter coupling. The barrel of the extruder received the feed from a co-rotating variable speed feeder. The barrel was provided with two electric band heaters and two water cooling jackets. A temperature sensor was fitted on the front die plate, which was connected to temperature control unit placed on the control panel. The die was required to be fixed on the face of barrel by a screw nut tightened

Table 1. The central composite rotatable experimental design (in coded and uncoded levels of three variables and five levels) employed for development of pineapple waste pulp, broken rice flour and red gram powder based extrudates and the responses of developed extrudates.

	Independent variables						dependent variables					
S/N	Moisture content		Screw speed		temperature		LE (%)	BD g/cc	WAI (g/g)	WSI (%)	Force	Sensory
	Coded	Uncoded	Coded	Uncoded	Code	uncoded	Y_1	Y_2	Y_3	Y_4	Y_5	Y_6
1	-1	16	-1	120	-1	260	119.6	0.3202	4.115	16.2	11.50	5.21
2	1	21	-1	120	-1	260	112.2	0.3088	4.148	17.2	15.03	5.79
3	-1	16	1	140	-1	260	162.5	0.3503	4.001	18.8	7.803	6.15
4	1	21	1	140	-1	260	132.5	0.3125	4.132	16.2	14.17	5.89
5	-1	16	-1	120	1	340	95.2	0.2459	4.191	15.6	14.95	5.51
6	1	21	-1	120	1	340	140	0.2957	3.918	17.2	13.45	5.98
7	-1	16	1	140	1	340	165	0.3215	4.242	20.8	11.92	6.45
8	1	21	1	140	1	340	137.5	0.3483	4.484	14	13.40	5.89
9	-1.68	14.3	0	130	0	300	130	0.2087	4.801	16.4	11.12	5.22
10	1.68	22.7	0	130	0	300	122.5	0.2528	4.712	14.8	12.08	6.14
11	0	18.5	-1.68	113.8	0	300	140	0.2614	5.304	11.2	9.19	5.44
12	0	18.5	1.68	146.82	0	300	170.2	0.2521	5.128	14	10.89	6.2
13	0	18.5	0	130	-1.68	232.73	147.5	0.3138	4.046	18	12.08	6.85
14	0	18.5	0	130	1.68	367.27	160	0.1654	4.622	20.8	9.19	8.19
15	0	18.5	0	130	0	300	180	0.0202	5.997	11.2	10.96	6.67
16	0	18.5	0	130	0	300	180	0.0202	5.997	11.2	10.9	7.67
17	0	18.5	0	130	0	300	180	0.0202	5.997	11.2	10.9	7.67
18	0	18.5	0	130	0	300	180	0.0202	5.997	11.2	10.9	7.67
19	0	18.5	0	130	0	300	180	0.0202	5.997	11.2	10.9	7.67
20	0	18.5	0	130	0	300	180	0.0202	5.997	11.2	10.9	7.67

LE = Lateral expansion, BD = bulkdensity, WAI = water absorption index, WSI = water solubility index.

by a special wrench provided. The twin screw extruder was kept on for 30 min to stabilize the set temperatures and samples were then poured in to feed hopper and the feed rate was adjusted to 4 kg/h for easy and non-choking operation. The die diameter was selected at 4 mm as recommended by the manufacturer for such product and recommended by Stojceska et al. (2008). The product was collected at the die end and kept at 60 ± 0.5°C in an incubator (Orbital Incubator, Macro Scientific Works, New Delhi) for 12 h duration to remove extra moisture from the product. The samples were packed in polythene bags for further analysis.

Evaluation for lateral expansion of extrudates

The ratio of diameter of extrudates and the diameter of die was used to express the expansion of extrudates (Fan, 1996; Ibanoglu et al., 2006). Six lengths of extrudates (approximately 120 mm) were selected at random during collection of each of the extruded samples. The diameter of the extrudates was then measured at 10 different positions along the length of each of the six samples, using a vernier caliper. Lateral expansion (LE, %) was then calculated using the mean of the measured diameters:

$$LE = \frac{(diameter\ of\ product\ -\ diameter\ of\ hole)}{diameter\ of\ hole} X\ 100 \qquad (1)$$

Evaluation of bulk density of extrudates

Bulk density (BD, g/cm^3) was calculated using the following expression (Stojceska et al., 2008):

$$BD = \frac{4\ m}{\pi\ d^2\ L} \qquad (2)$$

Where m is mass (g) of length L (cm) of extrudate with diameter d (cm).

Evaluation of water absorption index (WAI) and water solubility index (WSI) of extrudates

WAI and WSI were determined according to the method developed for cereals (Stojceska et al., 2008). The ground extrudate was suspended in water at room temperature for 30 min, gently stirred during this period, and then centrifuged at 3000 g for 15 min. The supernatant was decanted into an evaporating dish of known weight. The WAI was the weight of gel obtained after removal of the supernatant per unit weight of original dry solids. The WSI was the weight of dry solids in the supernatant expressed as a percentage of the original weight of sample.

$$WAI\ (g/g) = \frac{Weight\ gain\ by\ gel}{Dry\ weight\ of\ extrudate} \qquad (3)$$

$$WSI\ (\%) = \frac{Weight\ of\ dry\ solid\ in\ supernatent}{Dry\ weight\ of\ extrudate} X\ 100 \qquad (4)$$

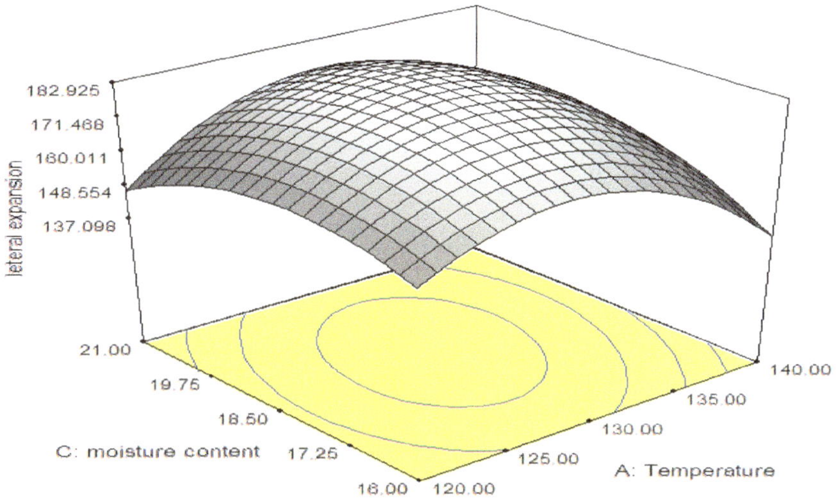

Figure 1A. Response surface plot for the lateral expansion of extrudate as a function of temperature and moisture content.

Evaluation for hardness of extrudates

Mechanical properties of the extrudates were determined by crushing method using a TA-XT2 texture analyzer (Stable Micro Systems Ltd., Godalming, UK) equipped with a 500 kg load cell. An extrudate of about 40 mm long, was compressed with a probe SMS P/75 - 75 mm diameter at a crosshead speed of 5 mm/s to 3 mm of 90% of diameter of the extrudate. The compression generated a curve with the force over distance. The highest first peak value was recorded as this value indicated the first rupture of snack at one point and this value of force was taken as a measurement for hardness (Stojceska et al., 2008).

Evaluation of sensory characteristics of extrudates

Sensory analysis was conducted for all the samples. Twelve panellists were asked to assess the expanded snacks and mark on a Hedonic Rating Test (1 - dislike extremely, 5 - neither like nor dislike and 9 - like extremely) in accordance with their opinion for taste, texture, color and overall acceptability.

Statistical analysis of responses

The responses (lateral expansion, bulk density, water absorption index, water solubility index, hardness and sensory evaluation of the extrudates) for different experimental conditions were related to coded variables (xi, I = 1, 2 and 3) by a second degree polynomial equation as given below:

$$y = b_0 + b_1x_1 + b_2x_2 + b_3x_3 + b_4x_4 + b_{12}x_1x_2 + b_{13}x_1x_3 + b_{14}x_1x_4 + b_{23}x_2x_3 + b_{24}x_2x_4 + b_{34}x_3x_4 + b_{11}x_1^2 + b_{22}x_2^2 + b_{33}x_3^2 + b_{44}x_4^2 + \varepsilon \quad (5)$$

Where, x_1, x_2 and x_3 are the coded values of temperature (°C), screw speed (rpm) and moisture content (%), respectively. The coefficients of the polynomial were represented by β_0 (constant); β_1, β_2, β_3 (linear effects); β_{12}, β_{13} $\beta_{,23}$, (interaction effects); β_{11} $\beta_{,22}$, β_{33} (quadratic effects); and ε (random errors). Data were modelled by multiple regression analysis and statistical significance of the terms was examined by analysis of variance for each response. The statistical analysis of the data of three dimensional plotting was performed using Design Expert Software (Statease 6.0). The adequacy of the regression model was checked by R^2 Adjusted R^2 Adequate precision and Ficher's F-test (Montgomery, 2001). The regression coefficients were then used to make statistical calculation to generate three dimensional plots for the regression model.

RESULTS AND DISCUSSION

Variation of responses (lateral expansion, bulk density, water absorption index, water solubility index, hardness and sensory evaluation) of extrudates with independent variables (feed proportion, moisture content, screw speed and temperature) are shown in Table 1. A complete second order model (Equation 5) was tested for its adequacy to decide the variation of responses with independent variables. To aid visualization of variation in responses with respect to processing variables, series of three dimensional response surfaces (Figures 1 to 6) were drawn using design expert software (Statease 6.0).

Lateral expansion

Lateral expansion of extrudates ranged from 95.25 and 190% with an average value of 110.76%. The coefficients of the model and other statistics are given in Table 2A. The model F value of 15.16 implies that the model is significant (P<0.0001). R^2 and Adjusted R^2 values of the model are 0.9379 and 0.7514, respectively. The Adequate precision value of 10.447 indicates that the model can be used to navigate the design space as it is greater than 4.0 (Montgomery, 2001). Considering these criteria, the following response model was selected for representing the variation of lateral expansion for further

Table 2A. Regression coefficients of second order polynomial and their significance for lateral expansion, bulk density and water absorption index.

Coefficient	Lateral expansion (%)			Bulk density (g/cc)			WAI (g/g)		
	Coefficient value	F	P	Coefficient value	F	P	Coefficient value	F	P
x_1	-2.40^{ns}	.79	0.3953	7.437E- 003^{ns}	0.20	0.6662	-1.22E-003^{ns}	1.538E-004	0.9903
x_2	13.27***	24.22	0.0006	0.001^{ns}	0.41	0.5363	0.014^{ns}	0.020	0.8899
x_3	2.34^{ns}	.75	0.4065	-0.024^{ns}	2.80	0.1794	0.10^{ns}	1.10	0.3199
x_1^2	-21.93***	69.79	0.0001	0.077***	22.57	0.0008	-0.51***	27.75	0.0004
x_2^2	-11.73**	19.97	0.0012	0.087***	28.25	0.0003	-0.34**	12.77	0.0051
x_3^2	-12.2***	21.63	0.0009	-0.081***	24.43	0.0006	-0.65***	46.60	0.0001
$x_1.x_2$	-11.86**	11.33	0.0072	-6.17E-003^{ns}	0.080	0.7834	0.077^{ns}	0.35	0.5647
$x_1.x_3$	6.84*	3.76	0.0810	0.016^{ns}	0.52	0.4855	-0.024^{ns}	0.036	0.8535
$x_2.x_3$	.51^{ns}	0.021	0.8873	0.012^{ns}	0.29	0.6012	0.093^{ns}	0.53	0.4847
R^2	0.9379			0. 8691			0.8840		
Adjusted R^2	0.8703			0.7514			0.7995		
Adequate precision	10.447			6.450			7.866		
Lack of fit	2.84			0.75			4.37		
F-value	9.44			7.38			8.46		

Significant at 10% (*), 5% (**), 1% (***), ns- non significant, E = 10^{-3}.

analysis.

$$LE = 182.81 - 2.40x_1 + 13.27x_2 + 2.34x_3 - 21.93x_1^2 - 11.73x_2^2 - 12.21x_3^2 - 11.86x_1x_2 + 6.84x_1x_3 + .51x_2x_3 \quad (6)$$

Where, x_1, x_2 and x_3 are the coded values of temperature (°C), screw speed (rpm) and moisture content (%), respectively. The following observations can be made from Equation 6 .The coefficients of x_1 are negative, but those of x_2 and x_3 are positive. Whereas, the coefficient of x_1^2, x_2^2 and x_3^2 is negative, F-values for square term of temperature, moisture content and screw speed (x_1^2, x_2^2 and x_3^2) were 69.79, 19.79 and 21.63 and p value was <0.0001, 0.0012 and 0.0009 ($P< 0.05$) respectively, validating that these terms are significant. From Table 2A, the coefficients of x_1^2 x_2^2 and x_3^2 are negative therefore they will show negative quadratic effect on lateral expansion ratio, among them, x_1^2 is more negative. A maximum expansion will occur in the range of temperature considered in the study.

It may be seen from Figure 1A that, there was slight expansion with the increase in moisture content, which may be due to gelatinization of starch, whereas further decrease in expansion with increase in moisture content may be attributed to the reduction of elasticity of dough through plasticization of melt as observed by Ding et al. (2005, 2006). The lateral expansion rapidly increased with increase in the die temperature which may be due to expandability of pulse powder at higher temperature, and further decrease in expansion with increase in the die temperature due to higher degree of superheating of water in the extruder encoun-tering the bubble formation (Ding et al., 2006). It may be observed from Figure 1B that the expan-sion increased with the increase in screw speed, which may be due to high mechanical shear resulting in higher expansion. Similar results have been reported by Ding et al. (2006).

Bulk density

Bulk density is a major physical property of the extrudate products. The bulk density, which

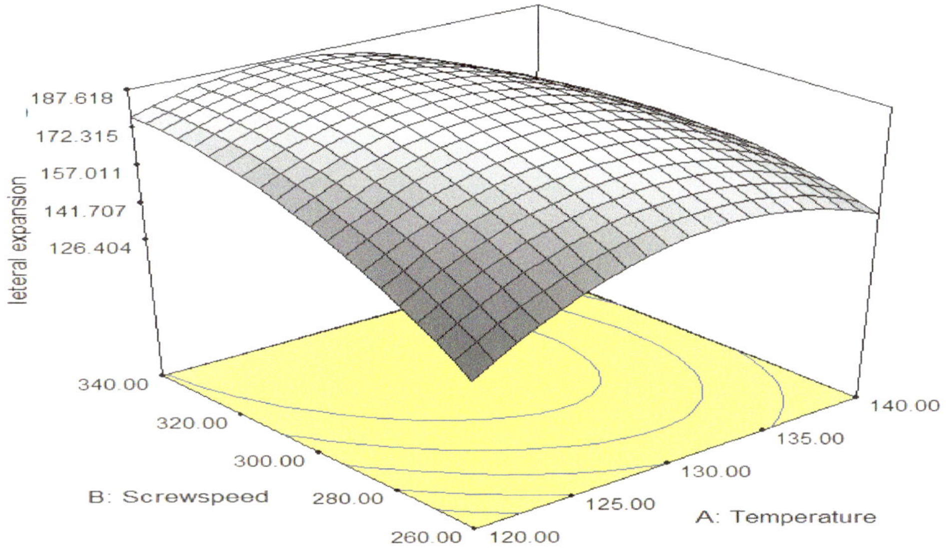

Figure 1B. Response surface plot for the lateral expansion of extrudate as a function of temperature and screw speed.

considers expansion in all direction, ranged from 0202 to 3503 g/cc for the rice flour and pulse powder, pineapple waste pulp powder extrudates. The coefficients of the model and other statistics are given in Table 2A. The Model F-value of 7.38 indicates that the model is significant ($P < 0.05$). R^2 (0.8691) and adjusted R^2 (0.7514) values are similar. The Adequate Precision (6.450) indicates that model can be used for prediction purposes. Considering these criteria, the following response model was selected for representing the variation bulk density for further analysis.

$$BD = 0.045 + 7.437E\text{-}003.x_1 + 0.011x_2 - 0.024x_3 + 0.077x_1^2 + 0.087x_2^2 + 0.081x_3 - 6.175E\text{-}003x_1x_2 + 0.016x_1x_3 + 0.012x_2x_3 \quad (7)$$

It is evident from Equation 7 that coefficients of x_1, x_2 are is positive, but that of x_3 is negative. Therefore, increase in temperature and screw speed may increase the bulk density, whereas increase in moisture content may reduce the bulk density of the product. Since coefficient of x_2^2 is higher than any other positive coefficients terms indicating that it is more significant than others, a minimum bulk density will occur in the range of screw speed selected for the study, x_1x_2 is negative, a maximum bulk density will be in the range of temperature considered in the study reported by Ding et al. (2006).

It is observed from Table 2A and Equation 7 that F values for x_1^2 x_2^2 and x_3^2 were 22.57, 28.25 and 24.43 and p value was 0.0008, 0.0003 and 0.0006, respectively, validating that these terms are significant, among x_2^2 high significance was observed in Table 2A.

It is perceived from Figure 2A that bulk density initially decreased with moisture content, which may be due to proper gelatinization and higher expansion, whereas further increase in bulk density may be because of reduction in elasticity of dough and lower expansion as reported by Ding et al. (2005, 2006). However, it decreased bulk density initially with screw speed , which may be attributed to lighter mass of the fiberous pomace in comparison to other constituents, whereas further increase in bulk density with screw speed may be because of more water binding property of non starch polysaccharides than protein and starch (Yagci and Gogus, 2008; Pansawat et al., 2008). The contour plot in Figure 2B demonstrate the initial increase in bulk density with the increase in temperature, which may be due to the presence of pineapple pulp in the feed formulations, whereas further decrease in bulk density with the increase in temperature may be attributed to higher expansion (Ding et al., 2006).

Water absorption index

Water absorption measures the amount of water absorbed by starch that can be used as an index of gelatinization (Anderson et al., 1969) and it is generally agreed that barrel temperature exert greatest effect on the extrudate by promoting gelatinization (Ding et al., 2005). The water absorption index of extrudates varied in the range of 3.91 to 5.99 g/g. The coefficients of the model and other statistics are given in Table 2A. The Model F-value of 8.46 signifies that the model is significant ($P < 0.05$). R^2 (0.8840) and Adjusted R^2 (0.7995) are in reasonable agreement. The Adequate Precision value of 7.866 implies that the model can be used for prediction purposes. Considering these criteria,

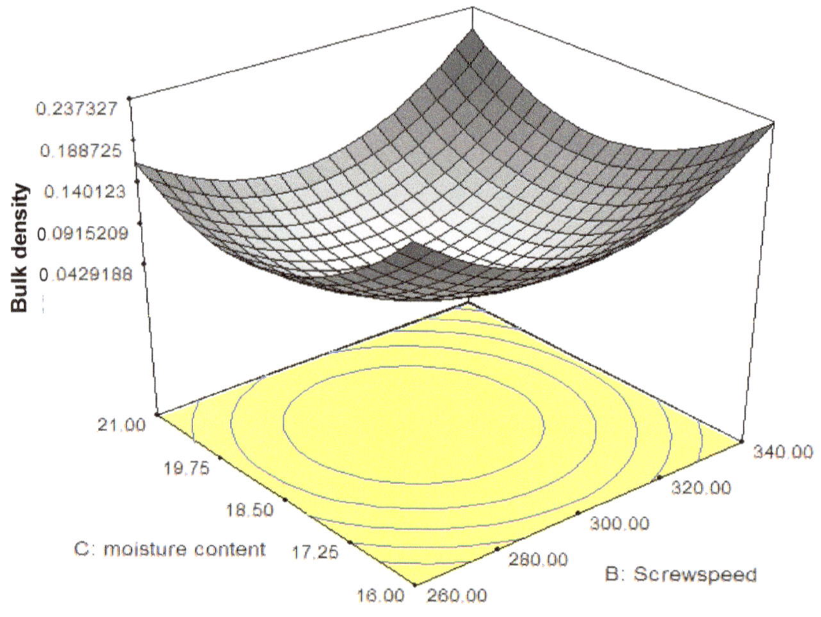

Figure 2A. Response surface plot for the bulk density of extrudate as a function of moisture content and screw speed.

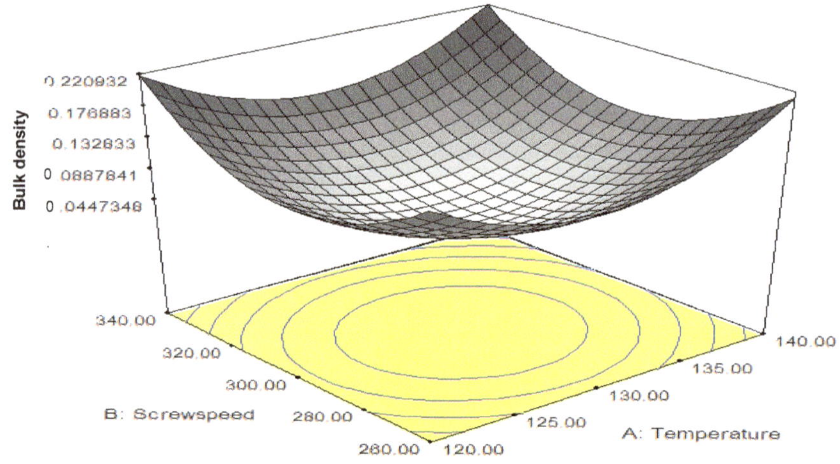

Figure 2B. Response surface plot for the bulk density of extrudate as a function of screw speed and temperature.

the following response model was selected for representing the variation water absorption of for further analysis:

$$WAI = 5.881\text{-}226E\text{-}003x_1 + 0.014x_2 + 0.10x_3 0.51x_1^2 - 0.34x_2^2 + 0.65x_3^2 + 0.077 + x_1x_2 - 0.024x_1x_3\text{-}0.093x_2x_3 \quad (8)$$

It can be seen from Equation 8 that the coefficients of x_2, x_3 are positive, but that of x_1 is negative; therefore increase in screw speed and moisture content may increase the water absorption index, whereas increase in temperature may reduce the water absorption index of the product. F-values for square term of temperature, screw speed and moisture content (x_1^2 x_2^2 and x_3^2) were 22.57, 28.25 and 24.43 and p value are (0.0008,0.0003and0.0006), respectively, validating that these terms are significant, among these x_2^2 highly significant was observed (Table 2A).

It was observed from Figure 3A that increase in feed moisture content resulted in quadratic decrease in water absorption index. Altan et al. (2008b) also reported the similar behaviour due to competition of absorption of water between pineapple pulp and available starch. It was observed from Figure 3B that increase in temperature

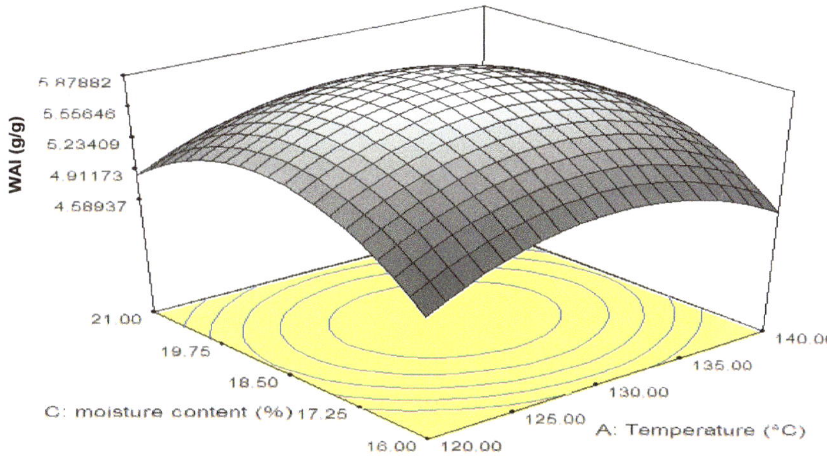

Figure 3a. Response surface plot for the variation of WAI of extrudate as a function of moisture content and temperature.

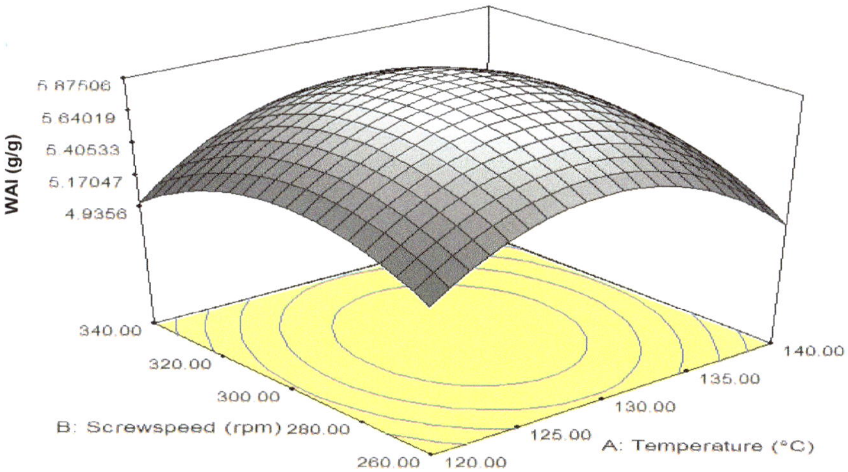

Figure 3b. Response surface plot for the variation of WAI of extrudate as a function of screw speed and temperature.

content resulted in quadratic decrease in water absorption index. Whereas, further increase water absorption index may be because dextrinization at higher temperature, similar observations were reported by Mercier and Feillet (1975). Water absorption index increased with increase in screw speed, may be attributed to high mechanical shear and higher expansion due to gelatinization. Whereas further decrease water absorption index may be because of plasticization of melt at higher moisture content (Ding et al., 2006).

Water solubility index

Water solubility index was used as an indicator of degradation of molecular components. It measures the amount of soluble polysaccharide released from the starch component after extrusion (Ding et al., 2005). Water solubility index of extrudates ranged from 11.2 to 20.2%. The coefficients of the model and other statistics are given in Table 2B. The Model F-value of 19.47 indicates that the model is significant ($P < 0.05$). R^2 (0.9460) and Adjusted R^2 (0.8974) are in reasonable agreement. Moreover, the adequate precision (12.217) is greater than 4, indicating a good fit of experimental data and the acceptability of the model for prediction purposes. Considering these criteria, the following response model was selected for representing the variation of lateral expansion for further analysis:

$$WSI = 11.59 - 0.065x_1 + 0.65x_2 + 0.33x_3 + 1.61x_1^2 + 0.54x_2^2 + 2.95x_3^2 - 1.59x_1x_2 - 0.53x_1x_3 - 0.025x_2x_3 \quad (9)$$

It was observed from Equation 9 that the coefficient of x_1

Table 2B. Regression coefficients of second order polynomial and their significance for water solubility index, hardness and sensory characteristics.

		Hardness (%)			WSI (N)			Sensory characteristics		
S/N	Coefficient	Coeff. value	F	P	Coeff. value	F	P	Coeff. value	F	P
1	x_1	0.84*	7.96	0.0181	-0.65*	5.46	0.0416	0.13 ^{ns}	1.27	0.2865
2	x_2	-1.11**	13.76	0.0040	0.65*	5.48	0.0413	0.23*	4.03	0.0726
3	x_3	0.17 ^{ns}	0.33	0.5764	0.33 ^{ns}	1.40	0.2635	0.22**	3.72	0.082
4	x_1^2	0.58*	4.00	0.0733	1.61***	35.03	0.0001	-0.72***	40.93	0.0001
5	x_2^2	1.53***	27.95	0.0004	0.54*	4.04	0.0723	-0.67***	35.50	0.0001
6	x_3^2	0.023 ^{ns}	9.19-E003	0.9255	2.95***	118.18	0.0001	-0.069 ^{ns}	0.38	0.5511
7	$x_1.x_2$	0.73*	3.49	0.0914	-1.59**	18.71	0.0015	-0.23 ^{ns}	2.40	0.1527
8	$x_1.x_3$	-1.24**	10.41	0.0098	-.53 ^{ns}	2.08	0.1799	-0.051 ^{ns}	0.12	0.7414
9	$x_2.x_3$	0.18 ^{ns}	0.22	0.6467	-0.025 ^{ns}	4.714E-003	0.9466	-0.024 ^{ns}	0.025	0.8782
10	R^2	0.8691			0. 9460			0.8905		
11	Adjusted R^2	0.7512			0. 8974			0.7920		
12	Adequate precision	10.689			12.217			8.382		
13	Lack of fit	3.48			2.18			1.43		
14	F-value	7.37			19.47			9.04		

Significant at 10% (*), 5% (**), 1% (***), ns - non significant, E = 10^{-3}.

is negative, but that of x_2 and x_3 are positive. Therefore, increase in temperature content may reduce the water solubility index, whereas increase in screw speed and moisture content may increase the water solubility index of the product. The coefficient of x_3^2 is the highest followed by the x_1^2 and x_2^2 in the positive direction, a minimum water solubility index will be in the range of moisture content and followed by screw speed. Among these x_2^2 had more significant maximum water solubility index which lie in the range of temperature considered in the study.

Analysis of variance of Equation 9 (Table 2B) show that F values for x_1, x_2 and x_3 were within 5.48 and p value more than 0.0413 (P < 0.05), indicating no direct significance on water solubility index. The F values of parameters x_1^2 x_2^2 and x_3^2 were below 118.08 and p values less than 0.0001 (P < 0.05), show the significant quadratic contribution.

It is evident from Figure 4A that water solubility index decreased initially and increased further with increase in temperature, which may be attributed to increased dextrinization at higher temperature and it causes more solubility due to change in the starch granule structure (Mercier and Feillet, 1975). Water solubility index increases with increase screw speed due to high mechanical shear exerted on extrudates. Similar observations were reported by Ding et al. (2005). It was observed from Figure 4B that water solubility index decreased initially with the increase in moisture content, which may be due to proper gelatinization and lateral expansion of the starch, whereas further increase with the increase in moisture content may be attributed to reduction in lateral expansion due to plasticization of melt as observed by Ding et al. (2005).

Hardness

The textural property of extrudate was determined by measuring the force required to break the extrudate (Singh et al., 1994). The higher the value of maximum peak force required in Newton, which means the more force required to breakdown

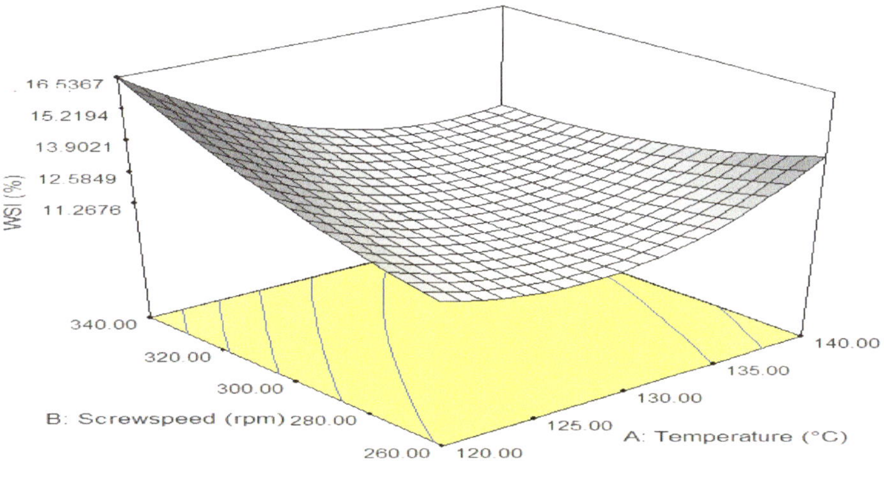

Figure 4A. Response surface plot for the variation of WSI of extrudate as a function of screw speed and temperature.

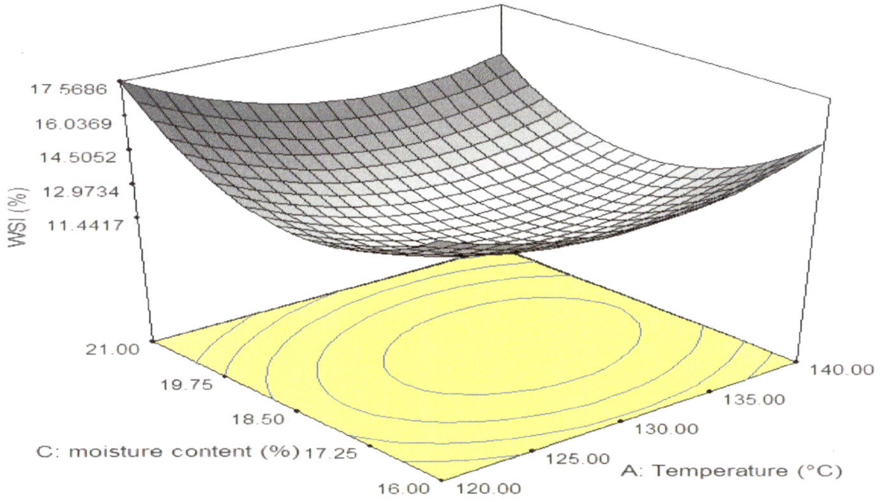

Figure 4B. Response surface plot for the variation of WSI of extrudate as a function of moisture content and temperature.

the sample, the higher the hardness of the sample to fracture (Li et al., 2005). Hardness of rice flour and pulse powder, pineapple pomace extrudate varied between 8.94 and 16.52 N. Table 2B shows the coefficient of the model and other statistical attributes of hardness. The Model F-value of 7.37 implies the model is significant. In this case x_1, x_2, x_2^2, $x_1.x_3$ are significant model terms at $P<0.05$. R^2 and Adjusted R^2 values of the model are 0.8691and 0.7512, respectively. Considering these criteria, the following response model was selected to represent the variation of lateral expansion for further analysis:

$$H = 10.36 + 0.84x_1 - 1.11\ x_2 + 0.17x_3 - 0.58x_1^2 - 1.53x_2^2 + 0.0286x_3^2 + 0.73x_1x_2 - 1.24x_1x_3 + 0.18x_2x_3 \quad (10)$$

The following observations can be made from Equation 10. The coefficient of x_1, x_3 is positive, but those of x_2 are negative. Therefore, increase in temperature and moisture content may increase the hardness whereas, increase in screw speed may reduce the hardness of the product. Since coefficient of x_2^2 is maximum, positive, minimum hardness will be in the range of screw speed considered in the study.

It is evident from Figure 5A that the hardness increase with the increase in moisture content, which may be due to increase expansion with the increase in moisture content. The increase in hardness with increase in temperature may be due to higher expansion at elevated temperatures and would decrease melt viscosity, but it also increases the vapor pressure of water. This favors

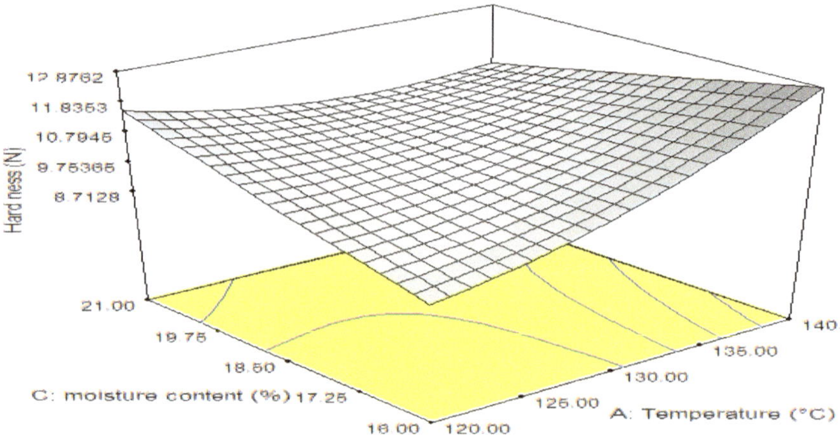

Figure 5A. Response surface plot for the variation of hardness of extrudate as a function of moisture content and temperature.

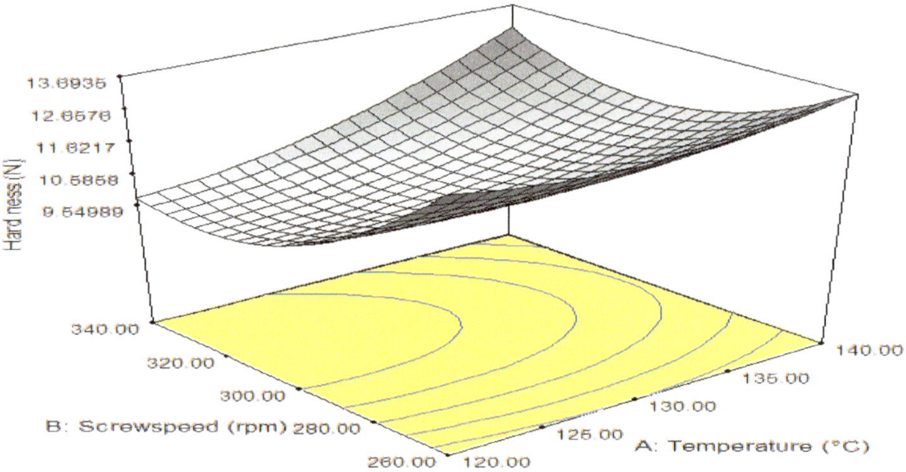

Figure 5B. Response surface plot for the variation of hardness of extrudate as a function of screw speed and temperature.

the bubble growth which is the driving force for expansion that produces low density products and thus increases hardness of extrudate. Similar findings have been reported by Altan et al. (2008b). It may be observed from Figure 5B that hardness decreased with the increase in screw speed. Similar decrease in hardness with increased screw speed due to lower melt density was observed by Ding et al. (2006).

Sensory characteristics

Sensory evaluation indicates the acceptability of the product. Hedonic scale is used to find the different aspect of sensory evaluation. The overall acceptability of the product ranges from 5.2 to 7.5 in the extrudates prepared from rice flour, red gram powder and pineapple waste pulp powder. The coefficients of the model and other statistics are given in Table 2B. The Model F-value of 9.04 indicates that the model is significant ($P < 0.0005$). R^2 (0.8509) and Adjusted R^2 (0.7920) values and Adequate Precision (8.352) indicates that the model can be used for prediction purposes. Considering these criteria, following response model was selected for representing the variation of lateral expansion for further analysis:

$$\text{Sensory evaluation} = 7.48 + 0.13x_1 + 0.23x_2 + 0.22x_3 - 0.72x_1^2 - .0.67x_2^2 - 0.069x_3^2 - 0.23x_1x_2 - 0.051x_1x_3 + 0.024x_2x_3 \quad (11)$$

It was observed from Equation 11 that coefficient of x_1, x_1

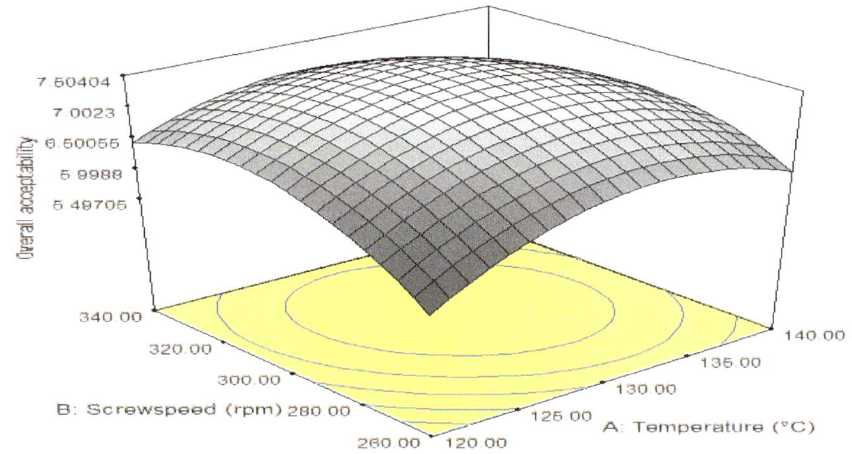

Figure 6A. Response surface plot for the variation of OA value of extrudate as a function of screw speed and temperature.

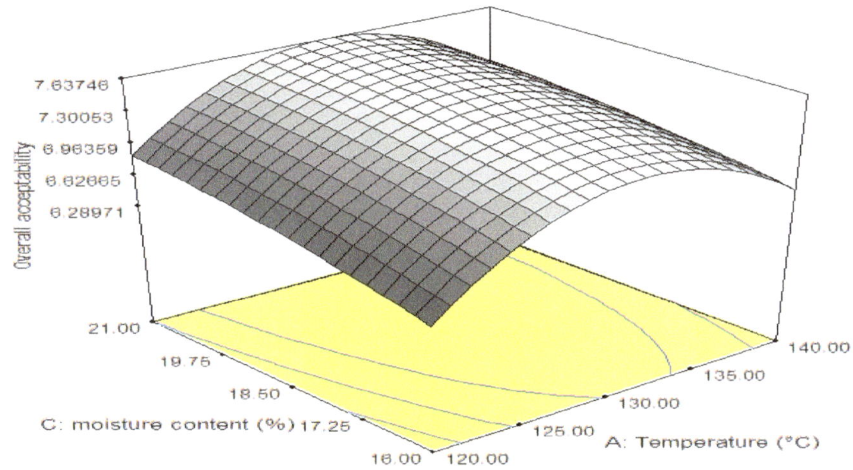

Figure 6B. Response surface plot for the variation of OA value of extrudate as a function of moisture content and temperature.

and x_3 are positive indicating that increases in temperature, screw speed and moisture content will increases the overall acceptability. All linear and interaction terms were not significant ($P > 0.05$). In this case only square terms x_1^2 and x_2^2 were significant model terms.

Analysis of variance of Equation 11 (Table 2B) show that F-values and p values for square term of temperature and screw speed were *40.93*, 0.0026 and 0.0001, 0.0001, respectively ($P < 0.0005$). Indicating that temperature and screw speed is significant for the overall acceptability of the product. Since coefficient of x_1^2, x_2^2 are negative; they will show negative effect on graph variation with the change in value of variables.

It is evident from Figure 6A that the overall acceptability value increased quadratically with increase in screw speed and temperature, and later decreased linearly with decreased screw speed and temperature, which may be attributed to the significant square terms of (x_1^2 *and* x_2^2). These coefficients of $x_1^2$and x_2^2 are negative therefore they will show negative quadratic effect on the overall acceptability value. It was also perceived from Figure 6B that the overall acceptability value increase with increase in moisture content. This may be due to effect of pineapple pomace. Similar results were reported by Upadhayay et al. (2008).

Compromised optimum condition

The compromised optimum condition for the development ready-to-eat extrudates was then determined using design expert software (Statease, DE 6). The final product would be considered optimum, if the sensory score, lateral expansion, water absorption index are as high as possible, whereas bulk density should be as low as possible. Therefore, compromised optimum condition criteria applied for numerical technique optimization were as follows: (1) maximum sensory score for product acceptability; (2) maximum lateral expansion for proper fluffiness; (3) maximum water absorption index for good digestibility and (4) minimum bulk density. The compromised optimum conditions obtained for the development of extrudates were: The product temperature, 132.27°C; moisture content, 18.48% and screw speed, 315 rpm.

Conclusion

From the present study, it was concluded that:

1. In this experiment, the product responses like lateral expansion bulk density were mostly affected by changes in temperature and moisture content.
2. Increasing in barrel temperature and moisture content resulted in maximum expansion, minimum bulk density was observed. WAI and WSI are mostly affected by changes in temperature and screw speed.
3. Increased barrel temperature and screw speed resulted in maximum WAI and minimum WSI was observed. The responses of hardness are mostly affected by changing temperature and moisture content.

REFERENCES

Abdullah A, Hanafi M (2008). Characterisation of solid and liquid pineapple waste, Reaktor, 12(1):48-52.

Altan A, McCarthy KL, Maskan M (2008a). Evaluation of snack food from barley-tomato blends by extrusion processing. J. Food Eng. 84:231-242.

Altan A, McCarthy KL, Maskan M (2008b). Twin-screw extrusion of barley-grape pomace plends: Extrudate characteristics and determination of optimum processing conditions. J. Food Eng. 89:24-12.

Anderson RA, Conway HF, Griffin EL (1969). Gelatinization of corn grits by roll and extrusion cooking. Cereal Sci. Today, 14:4-12.

Ding QB, Ainsworth P, Tucker G, Marson H (2005). The effect of extrusion conditions on the physicochemical properties and sensory characteristics of rice based expanded snacks, J. Food Eng. 66:283-289.

Ding QB, Ainsworth P, Plunkett A, Tucker G, Marson H (2006). The effect of extrusion conditions on the functional and physical properties of wheat based expanded snacks, J. Food Eng. 73:142-148.

Fan JM (1996). The effect of sugars on the extrusion of maize grits: I. The role of the glass transition in determining product density and shape. Int. J. Food Sci. Technol. 31:55-65.

Grigelmo-Miguel N, Martín-Belloso O (1999). Comparison of dietary fibre from by products of processing fruits and greens and from cereals. LWT- Food Sci. Technol. pp. 503-508.

Ibanoglu S, Ainsworth P, Ozer EA, Plunkett A (2006). Physical and sensory evaluation of a nutritionally balanced gluten-free extruded snack, J. Food Eng. 75:469-472.

Kadan RS, Bryant RJ, Pepper man AB (2003). Functional properties of extruded rice flours. J. Food Sci. 68:1669-1672.

Mercier C, Fillet P (1975). Modification of carbohydrate component of extrusion cooking of cereal product. Cereal Chemistry 52:283-297.

Montgomery DC (2001). Design and Analysis of Experiments, New York Wiley pp. 416-419.

Nawirska A, Kwasniewski M (2005). Dietary fibre fractions from fruit and vegetable processing waste, Food Chem. 91(2):221-225.

Ng A, Lecain S, Parker ML, Smith AC, Waldron KW (1999). Modification of cell wall polymers of onion waste. III. Effect of extrusion-cooking on cell wall material of outer fleshy tissue. Carbohydrate Polymers, 39:341-349.

Pansawat N, Jangchud K, Jangchud A, Wuttijumnong P, Saalia FK, Eitenmiller RR, Phillips RD (2008). Effect of extrusion conditions on secondary extrusion variables and physical properties of fish, rice-based snacks, LWT- Food Sci. Technol. 41:632-641.

Ranganna S (1995). Handbook of analysis and quality control for fruits and vegetable products, Tata Mc Graw Hill Publishing Company Limited, New Delhi.

Serena A, Knudsen B (2007). Chemical and physicochemical characterization of co-products from vegetable food and agro industries. Anim. Feed Sci. Technol. 139:109-124.

Singh U (1988). Antnutritional factors of chickpea and pigeon pea and their removal by processing. Plant Food Human Nutr. 38:251-261.

Singh U, Santosa BAS, Rao PV (1992). Effect of dehulling methods and physical characteristics of grains on dhal yield of pigeon pea(Cajanus cajan) genotypes. J. Food Sci. Technol. 29:350-353.

Stojceska V, Ainsworth P, Plunkett A, Ibanoglu E, Ibanoglu S (2008). Cauliflower by-products as a new source of dietary fibre, antioxidants and proteins in cereal based ready-to-eat expanded snacks. J. Food Eng. 87:554-563.

Upadhyay A, Sharma HK, Sarkar BC (2008). Characterization of dehydration kinetics of carrot pomace, The CIGR E J. Vol X (Impress).

Yagci S, Gogus F (2008). Response surface methodology for evaluation of physical and functional properties of extruded snack. Foods developed from food-by-products, J. Food Eng. 86:122-132.

Zhoul Z, Robards K, Helliwell S, Blanchard C (2002). Composition and functional properties of rice. Int. J. Food Sci. Technol. 37:849-868.

Proteolysis of milk heated at high temperatures by native enzymes analysed by trinitrobenzene sulphonic acid (TNBS) method

Lucy M. CHOVE[1] , Abdulsudi ISSA- ZACHARIA, Alistair S. GRANDISON[2] and Michael J. LEWIS[2]

[1]Department of Food Science and Technology, Sokoine University of Agriculture, P.O. Box 3006, Morogoro, Tanzania.
[2]Department of Food and Nutritional Sciences, University of Reading, P.O. Box 226, Reading RG6 6AP, UK.

Native enzymes play a significant role in proteolysis of milk during storage. This is significant for heat resistant native enzymes. Plasmin is one of the most heat resistant enzymes found in milk. It has been reported to survive several heat treatments, causing spoilage during storage. The aim of this study was to assess susceptibility of high temperature heated milk to proteolysis by native enzymes. The trinitrobenzene sulphonic acid (TNBS) method was used for this purpose. Raw milk was heated at 110, 120, 130,142°C for 2 s and 85°C for 15 s and milk processed at low temperature (85°C /15s) was selected to mimic pasteurisation. TNBS method confirmed that raw milk and milk processed at 85°C /15s were the most proteolysed, whereas treatment of milk at high temperatures (110, 120, 130 and 142°C for 2 s) inactivated the native enzymes. It may thus be concluded that high temperature processing positively affects proteolysis by lowering its susceptibility to spoilage during storage.

Key words: Plasmin, proteolysis, trinitrobenzene sulphonic acid (TNBS), heat treatment, isoelectric precipitation.

INTRODUCTION

Proteolysis in milk may be a positive or negative attribute depending on the processing purposes and conditions (Nielsen, 2002). It is important for cheese ripening through development of desirable changes in flavour and texture, but is undesirable when it results in gelation and bitterness as observed in ultra high temperature (UHT) milk (Datta and Deeth, 2001). It is caused by either bacterial enzymes or naturally occurring enzymes of which plasmin is significant (Grufferty and Fox, 1988). Plasmin has a high heat resistance (Metwali et al., 1998) and can also retain activity in UHT milk (Newstead et al., 2006). Active protease from either source can persist in UHT milk during storage, which can be up to 9 months at ambient temperature (Button et al., 2011). Even low residual or reactivated enzymatic activities can cause serious defects in UHT milk, which is approximately twice as sensitive as pasteurised milk to the action of proteolytic enzymes (McKeller, 1981). Desirable changes taking place during UHT processing of milk such as destruction of microorganisms and inactivation of enzymes occur, while undesirable effects such as browning, loss of nutrients, sedimentation, fat separation, cooked flavor also take place. Gelation of UHT milk during storage (age gelation) is a major factor limiting its shelf life (Chavan et al., 2011).

Ultra-high temperature processing refers to heating of a food product at a high temperature for a short time so as to extend its shelf life at room temperature. In milk, various temperature-time combinations are used to achieve this goal by heating either directly such as at 142°C for 4 s (Snoeren et al., 1979) or 138°C for 2 to 5 s (Samel et al., 1971), or directly and indirectly, respectively at 142°C for 5 s and 145°C for 3 s (Manji et al., 1986). These heating temperature and times have influence on casein stability, interactions with whey proteins and hence milk proteolysis. These heating temperature and times have influence on casein stability, interactions with whey proteins and hence milk proteolysis. Generally, heat labile enzymes are inactivated by unfolding followed by molecular scrambling, whereas heat resistant enzymes are inactivated by covalent modification such as cystine cross-links and deamination of asparagines and glutamine residues. Proteinases, lipases and phospholipase (from psychrophic bacteria, mostly *Pseudomonas*) are stable at room temperature but survive pasteurisation and UHT treatment (Chavan et al., 2011).

In a study to investigate the effect of raw milk quality and UHT temperature (145 or 150°C for 4 s) on proteolysis in UHT milk, it was found that proteolysis milk samples increased during storage, mostly from samples produced from low-quality milk at 145°C (Topcu et al., 2006). The same study also found that bitterness in UHT milk processed from low-quality milk at 145°C increased during storage whereas gelation occurred in that milk after 150 days. Bagliniere et al. (2013) found that milk proteolysis leading to milk destabilisation in UHT milk caused by *Pseudomonas fluorescens* was due to the activity of Apr X enzyme.

Homogenisation of milk prior to UHT treatment led to lower levels of enzymes that contributed to proteolysis during storage. However, the same study found that homogenisation induced modifications in proteins that may have played a role in their susceptibility to proteolytic attack- it brought about the attachment of caseins and whey proteins to the fat globule membrane and gave rise to smaller micellar particles not easily sedmentable by centrifugation (Garcla-Risco et al., 2001).

Other researchers have found that a high proportion of β -lactoglobulin and κ -casein (β-κ) complex present at the surface of casein micelles (favoured by high concentrations of denatured β-Lg) and by indirect heating, reduce the degree of proteolysis and gelation by inhibiting the access of proteases to caseins (Garcia-Risco et al., 1999, 2000; Enright and Kelly, 1999).

The aim of the current study was to investigate susceptibility of milk processed at various temperature-time profiles to proteolysis by native enzymes during storage, with raw milk as a control using TNBS method.

MATERIALS AND METHODS

Unless otherwise stated, all materials were from Fisher (Fisher Scientific UK Ltd, Leicestershire, UK.). Raw milk was obtained from the Centre of Dairy Research (CEDAR), University of Reading, UK. It was processed on an APV junior UHT plate heat exchanger (APV, Crawley, UK), with two stages of heating involving hot water (80°C) and steam (112 to 142°C) as described (Browning et al. 2001). A constant flow rate was used, giving a residence time of 2 s in the holding section at 110, 120, 130 and 142°C but 15 s at 85°C. Homogenisation took place between the heating stages at about 170 bars. These temperature-times combinations were selected based on studies of plasmin inactivation. The lowest (85°C/ 15 s) was chosen so as to mimic pasteurisation, whereas others were in a range where inactivation of plasmin could occur and therefore it would be interesting to monitor changes in proteolysis with time.

After cooling, the samples were stored at 2°C for 2 days. The six batches of milk samples were treated with sodium azide (0.05 %) to prevent bacterial contamination. They were then dispensed in sterile bottles in a laminar flow hood cabinet followed by incubation at 37°C for 28 days. Sampling for analysis was done on days 0, 3, 7, 14 and 28. Clarification to obtain pH 4.6 soluble extracts was carried out as detailed below. The soluble extracts were analysed by TNBS method. Clarification procedures were carried out after incubation for all milk samples studied. Prior to clarification, all milk samples were heated and held at 100°C for 10 min to denature the whey proteins. For isoelectric precipitation at pH 4.6, 50 mL warm water (40°C) was added to 5 mL of milk followed by 0.5 mL 10 % (w/v) acetic acid. After standing for 10 min, 0.5 mL 1 M sodium acetate was added and placed in cold water for 10 min before filtration through Whatman no. 41 filter paper and washing and making up to 100 mL. The clear extracts obtained were further filtered by 0.2 µm Millipore filter before being subjected to the TNBS method. The TNBS method was done as described by McKellar (1981). Triplicate samples of pH 4.6 soluble extracts (0.2 mL) were mixed with 2 mL 1 M-potassium borate buffer (pH 9.2) and 0.8 mL 5 mM-TNBS (Sigma-Aldrich, Gillingham, UK). After 30 min incubation in the dark at 25 °C, 0.8 mL freshly prepared 2 M-monobasic Na_2PO_4 [containing 18 mM-Na_2SO_3] and 5 mL distilled water were added. Absorbance was read at 420 nm by spectrophotometer (Cecil CE 1021 1000 series Cambridge, England).

Statistical analysis was carried out by using Statistical Package for Social Sciences (SPSS version 14). General Linear Model of analysis of variance (ANOVA) was used to determine statistical differences between means. Least square differences (LSD) and Duncan's multiple range tests were used to determine values that were statistically different (P <0.05). All analyses were carried out in triplicate, and results were expressed as mean ± standard deviation (SD).

RESULTS AND DISCUSSION

Effect of storage time (days) on proteolysis of milk heated at high temperatures

The present investigation aimed at examining the effect of various heat processes of milk on proteolysis during storage. Clarified samples of raw milk and milk heated at various temperatures and stored for 28 days were analysed. The results are shown in Figure 1. Generally, the absorbance of all milk samples increased with storage time (Figure 1) and had the significantly higher absorbance value (p<0.05) at day 28 indicating that proteolysis

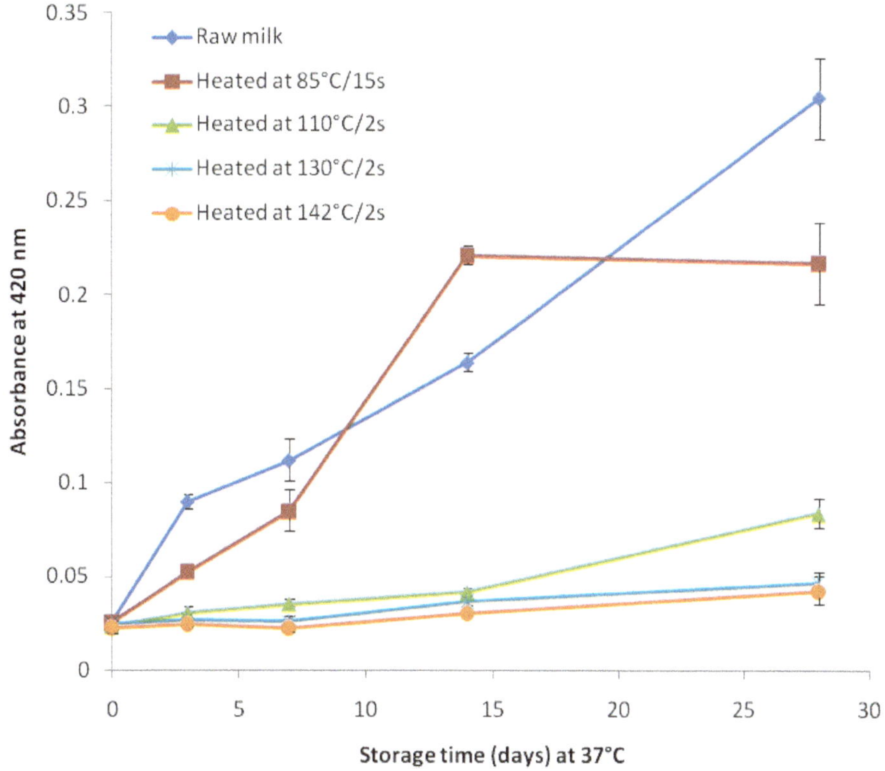

Figure 1. The effect of storage time (days) on proteolysis of raw milk and milk heated at 85°C/15s, 110, 120, 130 and 142°C for 2 s analysed by TNBS at 420 nm. The extent of proteolysis was represented by the absorbance of pH 4.6 soluble extracts at 420 nm measuring the breakdown products of raw milk and milk heated at various processing conditions during storage at 37°C for 28 days.

increased with storage time. Although proteolysis followed an increasing trend with time for all samples, the increasing rates were higher for raw milk and milk heated at 85°C than those of milk samples heated at 110, 120, 130 and 142°C for 2 s (Figure 1). The increase in percentage breakdown proteolysis from days 3 to 28 was as follows: raw milk (22.07%), processed at 85°C (16.84%), 110°C (5.43%), 120°C (3.38%), 130°C (2.05%) and 142°C (1.84%) which shows that it was more than 15% for the first two samples whereas it was less than 6% for the remaining four samples. For milk heated at 110, 120, 130 and 142°C for 2 s, small, steady but significant increase in proteolysis was observed. This is also evident from Table 1, where the increase in percentage protein breakdown decreased with the heating temperature. The increased absorbance with storage time was probably due to prolonged activities of either bacterial (released before incubation) or native enzymes or both. It has been reported that significant proteolysis may occur in the udder prior to milking leading to increased proteose peptones and γ-caseins as a result of plasmin activity on caseins (Schaar, 1985). These would account for proteolysis observed on day 0. Increased proteolysis of proteose peptones from 2.5 to 11.1 mg/ml in whey prepared from raw milk stored for 7 d at 37°C was also observed in previous study (Andrews and Alichanidis, 1983). The same study also revealed that hydrolysis of α_{s1} and β-casein occurred during storage, the peptide fragments of which were not precipitated at pH 4.6 but by TCA from whey where they contributed to the proteose peptones fractions.

It is known that plasmin inhibitors and plasminogen activator inhibitors are heat labile and therefore inactivated at lower temperatures as compared to the other components of the plasmin system (Richardson, 1983). About 81.1 and 35.8% of plasminogen activator inhibitors and plasmin inhibitors respectively were inactivated at 75°C for 15 s (Prado et al., 2006). This in turn increases plasmin activity as more plasminogen is also converted to plasmin (Richardson, 1983) and hence higher proteolysis. Although proteolysis was initially low, it increased significantly by day 28 which was a result of accumulation of breakdown products with storage time. These observations are consistent with the results reported by Gillis et al. (1985) who stated that although enzymatic activity may be decreased by UHT heat treatment, it is not

Table 1. Absorbance of pH 4.6 soluble extracts of raw milk and milk processed under various temperature-time conditions and incubated at 37°C for 28 days to examine the effect of proteolysis on storage time by the TNBS method at 420 nm.

Incubation time (days)	Treatments	Absorbance of pH 4.6 soluble extracts of UHT skim milk at 420 nm	Protein breakdown (%)
Day 0	Raw milk	0.026 ± 0.0023^{aA}	*
	Heated at 85°C	0.026 ± 0.0014^{aE}	*
	Heated at 110°C	0.024 ± 0.0008^{aI}	*
	Heated at 120°C	0.025 ± 0.0021^{aM}	*
	Heated at 130°C	0.025 ± 0.0020^{aU}	*
	Heated at 142°C	0.023 ± 0.0035^{aV}	*
Day 3	Raw milk	0.090 ± 0.0105^{bB}	6.57
	Heated at 85°C	0.053 ± 0.0040^{cF}	2.77
	Heated at 110°C	0.031 ± 0.0034^{dJ}	0.72
	Heated at 120°C	0.027 ± 0.0031^{dN}	0.21
	Heated at 130°C	0.027 ± 0.0015^{dU}	0.21
	Heated at 142°C	0.025 ± 0.0023^{dV}	0.20
Day 7	Raw milk	0.112 ± 0.0250^{eB}	8.83
	Heated at 85°C	0.085 ± 0.0111^{fF}	6.06
	Heated at 110°C	0.036 ± 0.0018^{gJK}	1.23
	Heated at 120°C	0.033 ± 0.0025^{gOP}	0.82
	Heated at 130°C	0.026 ± 0.0029^{gU}	0.10
	Heated at 142°C	0.023 ± 0.0028^{gV}	0.00
Day 14	Raw milk	0.164 ± 0.0060^{hC}	14.17
	Heated at 85°C	0.221 ± 0.0050^{iG}	20.02
	Heated at 110°C	0.042 ± 0.0017^{jK}	1.84
	Heated at 120°C	0.038 ± 0.0025^{kjP}	1.33
	Heated at 130°C	0.037 ± 0.0021^{kjS}	1.23
	Heated at 142°C	0.031 ± 0.0012^{kV}	0.82
Day 28	Raw milk	0.305 ± 0.0087^{lD}	28.64
	Heated at 85°C	0.217 ± 0.0216^{mH}	19.61
	Heated at 110°C	0.084 ± 0.0078^{nL}	6.15
	Heated at 120°C	0.060 ± 0.0091^{oQ}	3.59
	Heated at 130°C	0.047 ± 0.0056^{oT}	2.26
	Heated at 142°C	0.043 ± 0.0074^{oW}	2.05

Different lower case letters on the same column show significant differences ($p<0.05$) per treatment per day whereas different uppercase letters on the same column show significant differences ($p<0.05$) for the same treatment on different days; The experiment was replicated 3 times (N = 9); The pH 4.6 soluble extracts were diluted (20x). *Day 0 was used as a reference. Breakdown products (from days 3 to 28) were calculated with reference to this day (day 0).

inactivated.

The small but significant differences in each of the treatments could be due to inactivation of plasmin, plasminogen and plasminogen activators which are the main actors responsible for increased proteolysis in the samples. It may thus be concluded that lower activities in the heated samples was a result of higher temperature (>85°C) used which may have caused reduced activities of enzymes responsible for proteolysis. It has been reported that during severe heat treatments such as UHT processing, decreased plasmin activity is due to thiol-disulphide interactions between disulphide groups in plasmin and reactive SH groups of β-lactoglobulin during the unfolding and denaturation that occurs at high temperatures (Grufferty and Fox, 1986; Kelly and Foley, 1997). Other changes that were described in association with severe heat treatments include denaturation of whey proteins especially β-lactoglobulin, leading to the formation of β-lactoglobulin-κ-casein complex example through interaction with κ-casein (Datta and Deeth, 2003; McMahon, 1996).

Degradation of caseins throughout storage of direct and indirect UHT whole and skim milk was also observed in previous study where directly and indirectly processed whole and skim UHT milks were observed for 11 weeks (Lopez-Fandino et al., 1993). In their study, the researchers found that after 11 weeks, significant proteolysis occurred in skim milk as compared to whole milk as analysed by the TNBS method.

Effect of heating temperatures on the proteolysis in milk stored at 37°C

Table 1 shows the absorbance (at 420 nm) of soluble extracts of raw milk and milk processed under various temperature-time conditions. Samples with similar treatments were compared per day of storage as shown in Table 1. At day 0 of storage, the level of proteolysis of raw milk did not significantly differ ($p>0.05$) from milk samples heated at 85°C /15 s, 110, 120, 130 and 142°C for 2 s. This means that although there was some proteolysis probably occurring from udder, none of the samples had undergone any significant increase in proteolysis. On days 3, 7, 14 and 28 raw milk and milk heated at 85°C/15 s significantly differed ($p<0.05$) from the rest of the samples heated at 110, 120, 130 and 142°C for 2 s (Table 1). Although, a study by Andrews and Alichanidis (1983) concluded that proteolysis progressed faster in pasteurised milk than in raw milk, it was not exactly the case in the samples studied here. Raw milk showed higher proteolysis than milk heated at 85°C in the current study. Reasons for the difference in results is not clear as lower temperature was used in the previous study (72°C/18 s), as compared to 85°C/15 s in this study. It has been reported that plasmin levels in raw milk vary, so probably higher levels were already present in the raw milk samples used by the previous researchers, which increased further upon pasteurisation due to inactivation of inhibitors. Another possibility is the presence of bacterial enzymes in raw milk which could have resulted in increased proteolysis in raw milk. It was reported that a milk sample that has a high population of psychrotrophic bacteria could have a reduced shelf-life even after the viable bacteria have been inactivated (Adams et al., 1975). The former reason is more likely as the raw milk was of good bacterial quality (10^3 cfu/mL psychrotrophic count– results not shown). It is therefore suggested that higher proteolysis in raw milk was most likely due to native enzymes.

Moreover, these two samples (raw milk and milk heated at 85°C) showed the highest levels of proteolysis among all treatments analysed most probably due to activity of native enzymes. The rest of the heated samples (at 110, 120, 130 and 142°C) had very low proteolysis on days 3, 7, 14 and 28 as compared to raw milk and milk heated at 85°C for 15 s signifying possible inactivation of plasmin and its inhibitors. It is obvious from the trend that the increase in the rate of proteolysis was inversely proportional to the heat treatment implying that the lower the temperature employed, the higher the enzyme activity and the higher the proteolysis. It has been documented that variability of heat stability of UHT milk during storage results from multiple and complex reactions which occur in milk such as Maillard reactions, lactose degradation and other biochemical changes such as polymerisation by disulphide bridges and dephosphorylation (Gaucher et al., 2008). It is well established that radicals that are generated during the Maillard reaction can promote a selective attack on the protein backbone which would break specific bonds (Guinot-Thomas et al., 1995). However, it is unlikely that the heat treatment of the milk led to the Maillard reaction, as proteolysis was still low in the samples.

It may thus be concluded that raw milk and milk heated at 85°C have shown the highest proteolysis as compared to the rest of the samples due to the presence and activity of native enzymes. Lower proteolysis in milk processed at higher temperatures is associated with the destruction of these enzymes.

REFERENCES

Adams DM, Barach JT and Speck ML (1975) Heat resistant proteases produced in milk by psychrotrophic bacteria of dairy origin. J. Dairy Sci. 58: 828-834.

Andrews AT, Alichanidis E (1983). Proteolysis of caseins and the proteose-peptone fraction of bovine milk. J. Dairy Res. 50: 275-290.

Bagliniere F, Mateos A, Tanguy G, Jardin J, Briad-Bion V, Rousseau F, Robert B, Beaucher E, Gaillard JL, Amiel C, Humbert G, Dary A, Gaucheron F (2013) Proteolysis of Ultra High Temperature-treated casein micelle by AprX enzyme from *Pseudomonas fluorescens* F induces their destabilisation Int. Dairy J. 31 (2): 55-61

Browning E, Lewis MJ, Macdougall D (2001). Predicting safety and quality parameters for UHT -processed milks. Int. J. Dairy Technol. 54: 111-120.

Button PD, Roginski H, Deeth HC, Craven HM (2011) Improved shelf estimation of UHT milk by prediction of proteolysis. J. Food Qual. 34: 229–235

Chavan RS, Khedka CD, Jana AH (2011). UHT Milk Processing and Effect of Plasmin . Compr. Rev. Food Sci. Food Saf. 10 (5): 251-268.

Datta N, Deeth H C (2001) Age gelation of UHT milk. J. Food Process Eng. 79: 197-210.

Datta N, Deeth HC (2003). Diagnosing the cause of proteolysis in UHT milk. Lebensm Wiss Technol. 36: 173-182.

Enright E, Kelly A (1999). The influence of heat treatment of milk on sus-

ceptibility of casein to proteolytic attack by plasmin. Milchwisseschaft 54:491-493.
Garcia-Risco MR, Ramos M, Lopez-Fandino R (1999) Proteolysis, protein distribution and stability and stability of UHT milk during storage at room temperature. J. Sci. Food Agric. 79: 1171-1178
Garcia-Risco MR, Ramos M, Lopez-Fandino R (2000). Modifications in milk proteins induced by heat treatment and homogenisation and their influence on susceptibility to proteolysis. Int. Dairy J. 12: 679-688
Garcia-Risco MR, Villamiel M, Lopez-Fandino R (2001) Effect of homogenisation on protein distribution and proteolysis during storage of indirectly heated milk. Lait 82 (2002): 589-599.
Gaucher I, Mollé D, Gagnaire V, Gaucheron F (2008). Effects of storage temperature on physico-chemical characteristics of semi-skimmed UHT milk. Food Hydrocoll. 22: 130-143.
Gillis WT, Cartledge MF, Rodriguez IR, Suarez EJ (1985). Effect of raw milk quality on Ultra-High Temperature processed milk. J. Dairy Sci. 68: 2875-2879.
Grufferty MB, Fox PF (1986). Potassium iodate induced proteolysis in ultra high heat treated milk during storage: the role of B-lactoglobulin and plasmin. J. Dairy Res. 53: 601-613.
Grufferty MB, Fox PF (1988). Milk alkaline proteinase. J. Dairy Res. 55: 609-630.
Guinot-Thomas P, Al Ammoury M, Roux Y, Laurent F (1995) Study of proteolysis during storage of raw milk at 4°C. Effect of plasmin and microbial proteinases. Int. Dairy J. 5: 685-697.
Kelly AL, Foley J (1997). Proteolysis and storage stability of UHT milk as influenced by milk plasmin activity, plasmin/[beta]-lactoglobulin complexation, plasminogen activation and somatic cell count. Int. Dairy J. 7: 411-420.
Lopez-Fandino R, Olano A, Corzo N, Ramos M (1993). Proteolysis during storage of UHT milk: differences between whole and skim milk. J. Dairy Res. 60: 339-347.
Manji B, Kakuda Y, Arnott DR (1986). Effect of storage temperature on age gelation of ultra-high temperature milk processed by direct and indirect heating systems. J. Dairy Sci. 69: 2994-3001.
Mckellar RC (1981). Development of off-flavour in Ultra-High Temperature and Pasteurised Milk as a function of proteolysis. J. Dairy Sci. 64: 2138-2145.
Mcmahon DJ (1996). Age-gelation of UHT milk: changes that occur during storage, their effect on shelf life and the mechanism by which age-gelation occurs. Proceedings of the IDF Symposium held in Vienna (Austria), 6-8 September 1995, International Dairy Federation, Brussels (Belgium): International Dairy Federation. pp. 315-326.
Nielsen SS(2002) Plasmin system: characteristic, roles, and relationships. J. Agric. Food Chem. 50: 6628-6634.
Prado BM, Sombers SE, Ismail B, Hayes KD (2006). Effect of heat treatment on the activity of inhibitors of plasmin and plasminogen activators in milk. Int. Dairy J. 16: 593-599.
Richardson BC (1983). The proteinases of bovine milk and the effect of pasteurisation on their activity. New Zealand J. Dairy Sci. Technol. 18: 233-245.
Samel R, Weaver RWV, Gammack DB (1971). Changes on storage in milk processed by ultra-high temperature sterilisation. J. Dairy Res. 38: 323-332.
Schaar J (1985). Plasmin activity and protease-peptone content of individual milks. J. Dairy Res. 52: 369-378.
Snoeren THM, Speck ML, Dekker Both P (1979). Proteolysis during the storage of UHT-sterilised whole milk. 1. Experiments with milk heated by direct system for 4 seconds at 142ºC. Neth Milk Dairy J. 33: 31-39.
Topcu A., Numanoglu E, Saldamli I (2006) Proteolysis and storage stability of UHT milk produced in Turkey. Int. Dairy J. 16 (6): 633-638.

Thiaminase properties in the fillet and liver of *Tilapia zillii* from Osinmo Reservoir: A comparative study

Raphael Emuebie Okonji[1] , Komolafe Olaniyi Olusola[2], Leonard Ona Ehigie[3], Popoola Micheal Olaoluwa[2] and Kayode Samuel Bamitale[4]

[1]Department of Biochemistry, Obafemi Awolowo University, Ile-Ife, Nigeria.
[2]Department of Zoology, Obafemi Awolowo University, Ile-Ife, Nigeria.
[3]Department of Biochemistry, Ladoke Akintola University of Technology, Ogbomoso, Nigeria
[4]Department of Medical Pharmacology and Therapeutics, Obafemi Awolowo University, Ile-Ife, Nigeria.

The thiaminase enzymes were partially purified and characterized from *Tilapia zillii* fillet (flesh) and liver using ammonium sulphate precipitation. The enzyme showed specific activities of 5.20 and 1.17 umol/min/mg of protein respectively for *T. zillii* fillet and liver. The enzymes exhibited a maximal activity at pH 5.0 and 7.0 for the liver and fillet, respectively. The Michealis constant for thiamine as substrate for both tissues was of 0.4 mM but 0.5 mM and 22 mM were obtained for aniline as substrate in the liver and fillet respectively. The optimum temperature of *T. zillii* thiaminases were 50 and 85°C for fillet and liver, respectively. Amino acids of fillet did not show significant effect on the enzyme activity but in the liver, the amino acids showed great inhibition with lysine showing complete inhibition of the enzyme. The thiaminase activities of fillet and liver were inhibited by divalent metal ions (Zn^{2+}, Sn^{2+}, Mg^{2+} and Hg^{2+}) but the enzyme from the liver was completely inhibited by Mg^{2+}. The inhibitors (2-Mercaptoethanol, ethylenediaminetetraacetic acid (EDTA), reduced glutathione (GSH), citrate and ascorbic acid) also showed different inhibitory effects on the enzymes from both tissues. EDTA and ascorbic acid did not inhibit the fillet thiaminase enzyme.

Key words: *Tilapia zillii*, liver, fillet, thiaminase, characterization, kinetic properties.

INTRODUCTION

Although the existence of thiaminase enzymes has been known for decades, the physiological role of these enzymes remains unknown. Thiaminase is produced by several species of bacteria (Fujita, 1954; Boyd and Walton, 1977; Abe et al. 1987) and is also found in the tissues of a number of marine and freshwater fish and shellfish (Hilker and Peter, 1966; Greig and Gnaedinger, 1971; Hirn and Pekkanen, 1975; Tillitt et al., 2005), zooplankton (Zajicek et al., 2005), insects (Nishimune et al., 2000) and plants (e.g., brackens *Pteris aquiline*: Parker and McCreg 1965).

Thiaminases have been reported to degrade thiamine by replacing the thiazole moiety with a variety of nucleophiles. Thiamine deficiency complex causes mortality and sublethal effects in Great Lakes salmonines and results from low concentrations of egg thiamine that are thought to be caused by thiaminolytic enzymes (that is, thiaminase) present in the diet (Riley and Evans, 2008). Thiamine deficiency causes disease in fish (Fisher et al., 1996; Riley and Evans, 2008), foxes Vulpes spp. (Green and Evans, 1940), minks *Mustela vison* (Petrova et al., 2003), cats, *Felis silvestris* catus (Anderson and Morrow, 1987), dogs *Canis lupus familiaris*(Houston and Hulland, 1988); chickens,*Gallus gallus* (Ishihara et al.,

1974); gulls (Bartoli et al., 1997), ruminants (Roberts and Boyd, 1974), american alligators *Alligator mississippiensis* (Sepulveda et al. 2004), and humans (disease termed beriberi: Fujita, 1954; Luxemburger et al., 2003). In many cases, thiamine deficiency is associated with a diet of raw fish (Ishihara et al., 1974; Vimokesant et al., 1982; Rouvinen et al., 1997), some species of which may contain thiaminase.

Fishes are highly important in the development of Nigeria both economically and healthwise as source of protein with low cholesterol level in the diets of many populace as well as an intermediate host to some parasites. In sub-Saharan Africa, fish accounts for 10% of the animal protein consumed, and 98% of this is finfish (Delgado and McKenna, 1997). The average per capita consumption of fish in Africa in 1992 was about 8 Kg having increased from an average of 7 Kg per annum from 1969 to 1974 (Ahmed, 1997). Approximately 40% of fish consumed in Africa, south of the Sahara is freshwater fish as compared to the global average of 25% (Bonga, 1999).

Tilapia zillii considered a potential competitor with native fish for food and spawning areas (Molnar, 2008), is a voracious herbivore which constitutes a serious threat to native aquatic plants, and the organisms that depend on these. Osinmo reservoir is located in the South-Western region of Nigeria. The reservoir is home to diverse species of fishes with *Tilapia* species ranking tops. The reservoir serves to provide part of the region with inland water and fishing activities is the major occupation of the locals. Proteins, essentially the amino acids are required for foetal development and growth. In dietary protein, the amino acids are needed principally for growth, metabolism and maintenance, especially in the young ones. The high demand for fish and fish products call for studies into the nutritional status of the fish. Thiaminases are enzymes found in a few plants and the raw flesh and viscera of certain fish and shellfish (Fujita, 1954). Also, thiamine deficiency causes disease in fish and in human (Adamolekun et al., 1994; Fisher et al., 1996) and this implies that there is a substantial amount of thiaminase in fish. The present work, therefore, describes the comparative properties of thiaminases from two tissues of *T. zillii* (fillet and liver).

MATERIALS AND METHODS

Trizma-acids, Trizma-base, thiamine and EDTA, were purchased from Sigma Chemical Company, St. Louis, Mo, USA. Ascorbic acid, citrate, 2-mercaptoethanol and glutathione were purchased from Sigma Chemical Company, St. Louis, Mo, USA. The chlorides of magnesium, tin, mercury and zinc were used and purchased from Sigma Chemicals, USA. L-lysine, L-cysteine, L-serine, L-proline, and albumin (from bovine pancreas) were purchased from Sigma Chemical Company, St. Louis, Mo. USA. Dialysis tubing was also bought from Sigma Chemical Company. Ammonium sulphate (enzyme grade), aniline and hydrochloric acid were purchased from BDH Chemical Limited, Poole England. All other reagents were of analytical grade and were obtained from either Sigma or BDH. Apparatus used included, top load weighing balance (Mettler PN1210), pH meter (Mettler MP200), Cintra 101 double beam (UV/VIS) spectrophotometer.

Collection and extraction of samples

Osinmo reservoir lies between latitude 07° 52.8' N to 07° 53.2' N and Longitude 04° 21.2' E to 04°21.7' E, Osun state, Southwestern, Nigeria. The catchment area is about 102 Km^2. The surface area of the reservoir is about 0.78 Km^2 with a mean maximum depth of 3.2 m. Cast-net was used to collect fish samples. *T. zillii* fishes were stored in an ice-chest covered with ice before transporting to the laboratory where they were stored at temperature below 0°C until ready for use. They were identified and dissected at the Department of Zoology, Obafemi Awolowo University, Ile Ife. The fillet and liver of *T. zillii* were excised with the aid of sharpened knife wrapped with aluminum foil and kept in the refrigerator until it was required. A total of 60 g of the *T. zillii* liver and 121 g of fillet were extracted. The weighed sample were then minced and homogenized with the use of blender in five volumes (v/w) of the homogenization buffer containing 0.2 M sodium phosphate buffer, pH 6.5. The homogenized samples were filtered using double layered cheese cloth and the filtrates were stirred and centrifuged at 4000 rpm for 30 min. and the supernatants collected were used as the crude enzyme solution.

Thiaminase assay

Thiaminase I enzyme was measured according to the modification of Nishimune et al. (2000). Enzyme sample was incubated at 37°C in 0.1 M Tris-HCl buffer, pH 8.0 with 1.0 µ M thiamine and 0.4 mM aniline and incubated for 30 min. The remaining thiamine was oxidized with the addition of 1.0 ml of 20 % NaOH and assayed spectrophotometrically. Absorbance was taken at 411 nm. One unit of enzyme activity is that amount that catalyses the formation of one micromole of heteropyrithiamine in 30 min. The protein concentration was determined by the method of Bradford (1976). Absorbance was read at 595 nm. Protein concentration was extrapolated from a standard curve using bovine serum albumin (BSA) as standard.

Ammonium sulphate precipitation

The supernatant obtained from centrifugation was then precipitated with 70% (w/v) ammonium sulphate. This involved the dissolving of 27.132 g of ammonium sulphate in 60 ml of enzyme solution and stirred over a period of 1 h. This was allowed to stand for 18 h in the refrigerator. The resulting precipitate was collected by centrifugation at 4000 rpm at 25C for 30 min. The precipitated protein sample was subjected to dialysis in 5 mM Tris buffer, pH 7.5 for 8 h with hourly change of buffer. The dialysate was then centrifuged at 4000 rpm for 30 min. The precipitate was collected and reconstituted with small volume of buffer.

Determination of kinetic parameters

The kinetic parameters (V_{max} and K_m) of the enzyme were determined according to the modified method of Nishimune et al. (2000). The K_m of thiamine and aniline was determined by varying the concentrations of thiamine between 0.1 and 3 mM and that of aniline between 0.04 and 1.2 mM in 0.1 M Tris-HCl buffer, pH 8.0, respectively. The kinetic parameters were determined from the double reciprocal plot of Lineweaver and Burk (1934).

Table 1. Partial Purification profile of thiaminase activity in *T. zillii* Fillet and liver.

Fraction	Total protein (mg)	Total activity (µ)	Specific activity (µ/mg)	Purification fold	% Yield
Crude fillet	2725.3	1199.0	0.44	1.00	100.00
Live	926.00	776.1	0.84	1.00	100.00
70% ASP fillet	1701.8	5371.9	3.16	7.80	62.40
liver	354.8	416.0	1.172	1.40	38.00

ASP, Ammonium phosphate; U= unit of thiaminase enzyme in µmol/ml/min.

Table 2. Kinetic properties of *Tilapia zillii* fillet and liver Thiaminases.

Michealis constant	Thiamine (mM)		Aniline (mM)	
	Fillet	Liver	Fillet	Liver
K_m (mM)	0.40	0.40	0.02	0.50
V_{max} (U)	0.50	0.45	0.29	0.15

U, µmol/ml/min.

Effect of temperature on thiaminase activity

The enzyme was assayed at different temperatures between 30 and 100°C to investigate the effect of temperature on the activity of the enzyme and to determine the optimum temperature of the enzyme. The assayed mixture was first incubated at the indicated temperature for 30 min before initiating reaction by the addition of an aliquot of the enzyme which had been equilibrated. The residual enzyme was then assayed routinely at 30 min interval.

Effect of pH on thiaminase activity

The effect of pH on thiaminase activity was studied by assaying the enzyme at different pH values: citrate buffer pH (3.0 to 6.0), Tris buffer pH (7.0 to 8.0), borate buffer pH (9.0 to 10.0). A reaction mixture of 1 ml contained 0.4 ml of Tris buffer, 0.2 ml of thiamine, 0.2 ml of aniline, 0.1 ml of distilled water and 0.1 ml of enzyme solution. It was incubated for thirty minutes at 37°C and then 1 ml of 20% NaOH was added. Absorbance was read at 411nm

Effect of metal ions on thiaminase activity

The effects of metal ions on the activity of thiaminase from fillet and liver of *T. zillii* were investigated using the following metal ions: Mg^{2+}, Sn^{2+}, Hg^{2+} and Zn^{2+} at a concentration of 10 µM. In a typical assay, 0.05 ml of the metal ion was contained in the reaction mixture.

Effect of amino acids on thiaminase activity

The effects of amino acids on the activity of thiaminase were determined. The amino acids used were of L-proline, L-cysteine, L-serine and L-lysine. In a typical assay, 0.05 ml of the amino acid was contained in the reaction mixture.

Effect of inhibitors on thiaminase activity

The effects of inhibitors on thiaminase activity were determined. 0.01 mM of the inhibitors (ascorbate, citrate and glutathione (GSH), mercaptoethanol and ethyleneditetraacetic acid (EDTA)) were used in a typical enzyme assay. 0.05 ml of these inhibitors was contained in the reaction mixture.

RESULTS

Purification of thiaminase

Purification of thiaminase was carried out as described in the materials and methods. The results of the purification of thiaminase using ammonium sulphate precipitation and dialysis from the fillet and liver of *T. zillii* are summarized in the purification table (Table 1).

Kinetic parameters studies

The K_m and V_{max} values of arginine are presented in Table 2 while Figure 1A to D show the Lineweaver-Burk plot of the two substrates.

Effect of temperature on thiaminase activity

The activity of thiaminase was assayed at temperatures between 30 and 100°C and the optimum temperature of the enzyme was found to be 85°C as shown in Figure 2A and B.

Effect of pH on thiaminase activity

The assay mixture contained buffers with different pH values (Figure 3A and B).

Effects of Inhibitors

Tables 3 to 5 shows the different effects of metal ions, amino acids and inhibitors on fillet and liver thiaminase

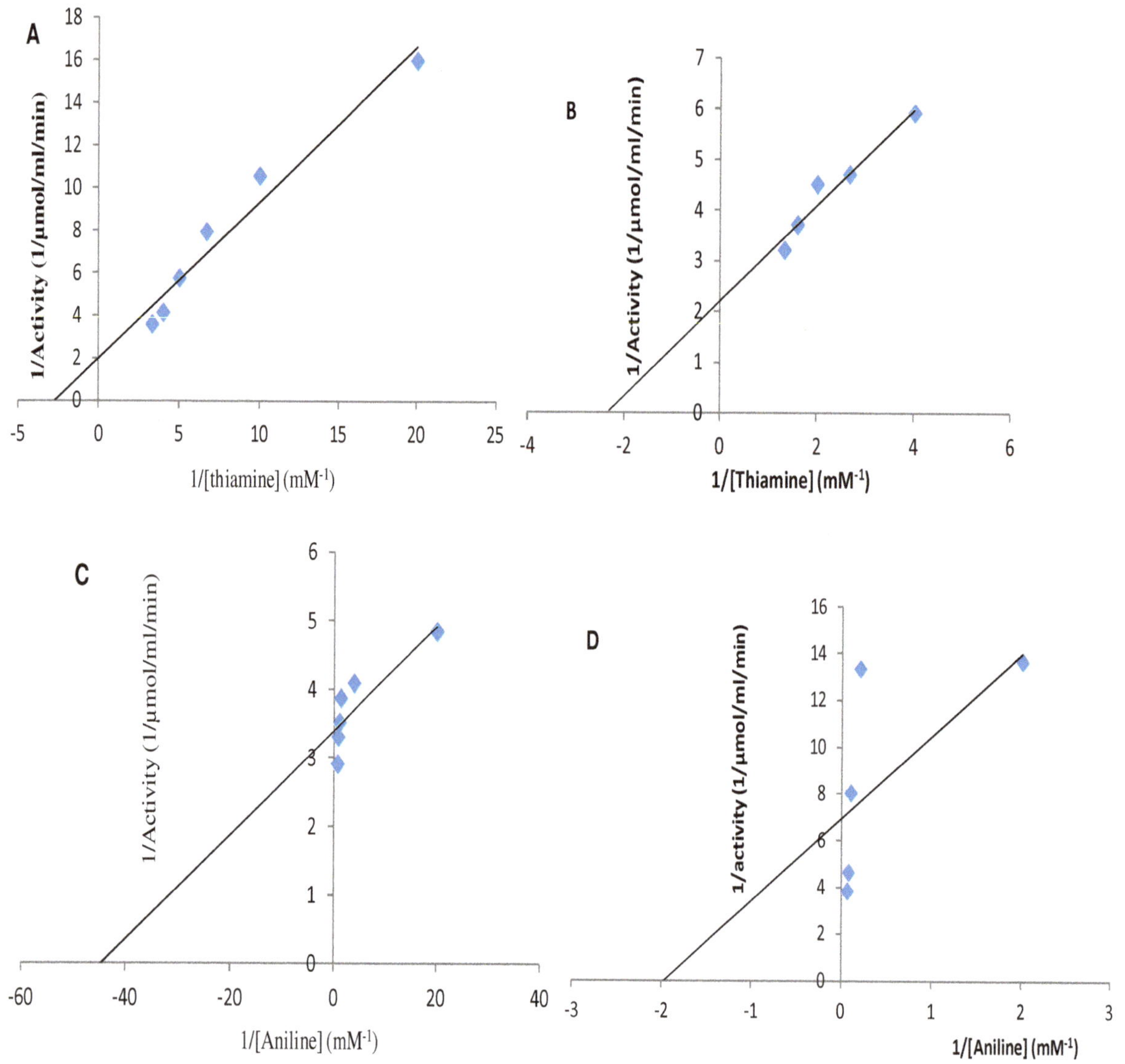

Figure 1. The Lineweaver-Burk plot showing the effect of varying concentrations of thiamine and aniline on the initial reaction velocity at pH 8.0. **(A)** Varying thiamine concentration for fillet thiaminase. **(B)** Varying thiamine concentration for liver thiaminase. **(C)** Varying aniline concentration for fillet thiaminase. **(D)** Varying aniline concentration for liver thiaminase.

activity expressed in percentage residual activities.

DISCUSSION

Fish and fish products provide as much as 17 to 63% protein intake of the large Nigerian populace (Abdullahi, 2001). The use of protein concentrates and fish oil in diets have been reported to reduce heart diseases, arthritis, arteriosclerosis, asthma, auto-immune disease, cancer, chronic infection, diabetes and multiple sclerosis (Cobiac et al., 1991; Gerhard et al., 1991; Adefemi, 2011).

The thiaminase activities of *T. zillii* fillet and liver tissues were partially purified using ammonium sulphate precipitation and dialysis. The specific activities of the preparations were found to be 1.172 and 5.2 U/mg of protein respectively. McCleary and Chick (1977) reported the thiaminase activity in Fronds of the fern nardoo *(Marsilea drummondii)* with a specific activity of 2.07 U/mg of protein.

The optimum temperature obtained for thiaminase in *T. zillii* fillet was 50°C while that for the liver thiaminase was 85°C. The difference in the optimum temperature could be due to the difference in the internal environment/compartmentalization of the tissues. It also possible that the temperature regulating nature of the fish could account for the difference in the optimum temperatures of the enzymes since fishes are poiklothermics (Hildebrand and Goslow, 2001). A much lower optimum temperature of 37°C was reported for extracellular thiaminase of *Bacillus thiaminolyticus* (Wittliff and Airth, 1968).

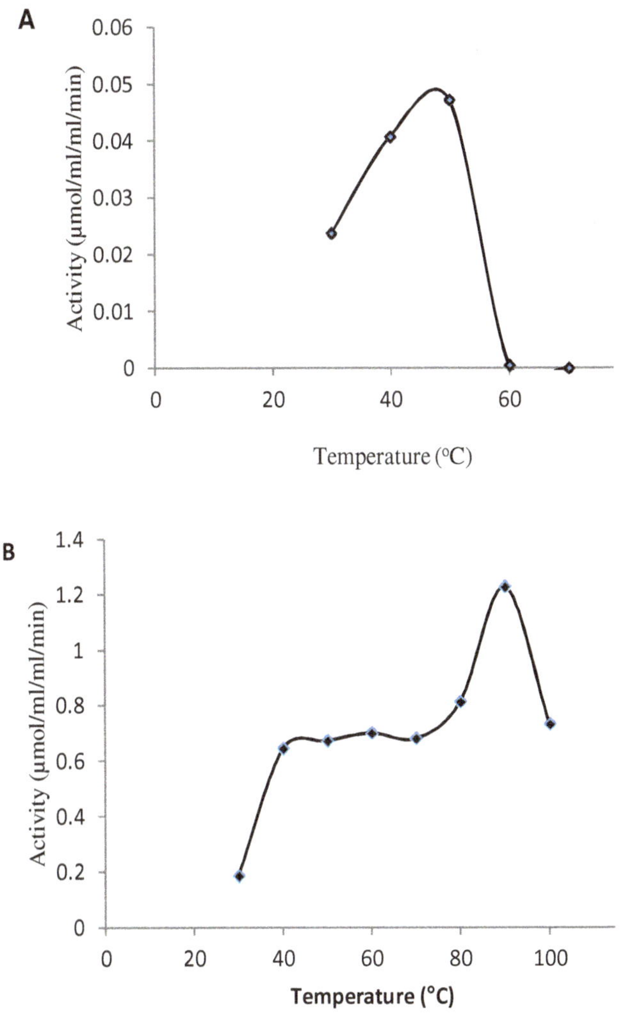

Figure 2. Effect of temperature on thiaminase activity of *T. zillii* Fillet (A) and Liver (B).

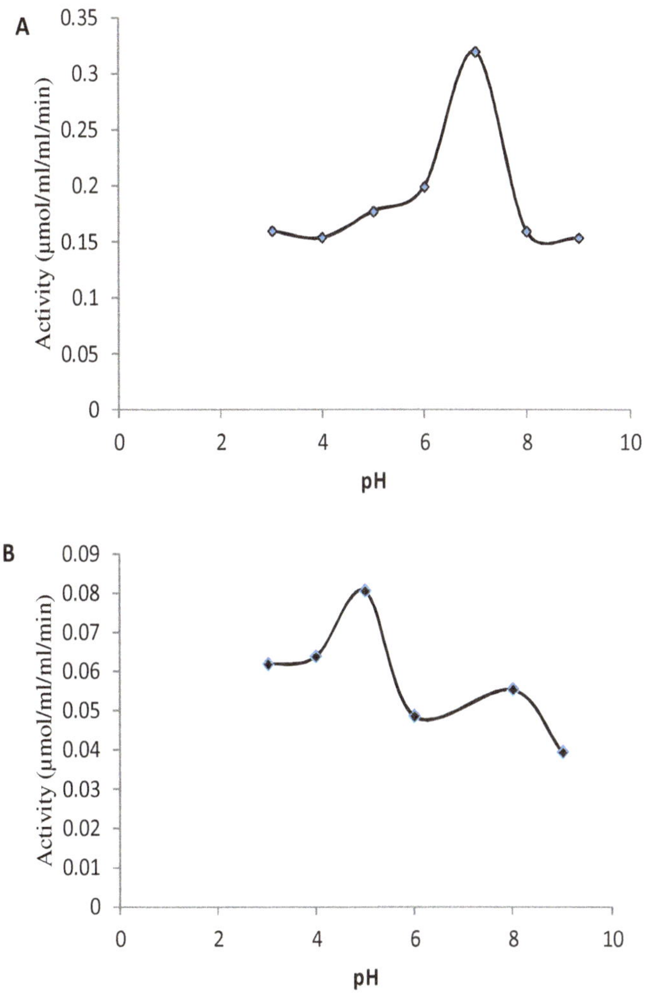

Figure 3. Effect of pH on thiaminase activity of *T. zillii* Fillet (A) and Liver (B).

Evans (1975) reported a range of temperature optimum between 37 to 40°C for *B. thiaminolyticus* enzyme; shell fish and fern enzymes, 55 to 66°C. An optimum temperature of 65°C was obtained for thiaminase enzyme from *Marsilea drummondii*, while the thiaminase in the buffer extract of *Anaphe* pupae was reported to have an optimal temperature of 70°C (Nishimune et al., 2000).

The K_m values for *T. zillii* fillet in Table 2 compares very well with previous reports from other sources. Kinetic studies of thiaminase I in extracts of ruminant faeces showed that the affinity for one substrate varied with the concentration of the other substrate in the manner of a two-step transfer mechanism (Ramos et al., 2006). Other compounds have been reported to act as substrates for thiaminase; such as pyridoxine, amino acids, glutathione, taurine and 4-aminopyridine (Bos and Kozik, 2000; Nishimune et al., 2000). Thiaminase has been reported to have a wide tolerance to the nature of the nucleophile (Leinhard, 1970). Thiamine functions as a precursor of thiamine diphosphate which serves as a coenzyme in a number of the major metabolic pathways (Friedrich, 1987). The ability of a wide range of aromatic amines, heterocyclic bases and sulphydryl compounds to participate in the hydrolysis of thiamine by thiaminase has been recognised for many years. Fujita (1954) showed that some bases were better co-substrates than others. Thiamine phosphate esters were inactive as substrates (Nishimune et al., 2000). Bos and Kozik (2000) also supported the use of numerous nucleophiles such as aniline, pyridine, and 2-mercaptoethanol by the enzyme as cosubstrates. Hanes et al. (2007) in their work on thiaminase assay in complex samples used 4-nitrothiophenolate as an alternative substrate. A wide range of heterocyclic bases, sulphydryl compounds and amines, including the non-aromatic amines 6-aminohexanoic acid and ethanolamine, act as co-substrates in the thiaminase I reaction; however, their effectiveness is dependent on both their degrees of basicity and to some extent, their stereochemistry (McCleary and Chick, 1977).

The optimum pH values of 5.0 and 7.0 reported for liver and fillet tissues, respectively compares very well with pH

Table 3. Effect of amino acids on thiaminase activity of *T. Zillii* flesh and liver.

Amino Acids (0.1 mM)	Residual activity (%)	
	Fillet	Liver
Lysine	92.73	0
Serine	97.63	48.05
Cysteine	83.38	9.87
Proline	99.11	6.75

The enzyme was assayed in a typical enzyme assay such that the assay mixtures contain 0.1 ml of the amino acids of 0.05 M concentration in 1.0 ml.

Table 4. Effect of metals on thiaminase activity of *T. zillii* fillet and liver.

Metals (0.01 mM)	Residual activity (%)	
	Fillet	Liver
Zinc chloride	73.15	35.84
Tin chloride	42.43	19.48
Magnesium chloride	74.18	0
Mercury chloride	100	19.74

The enzyme was assayed in a typical enzyme assay such that the assay mixtures contain 0.1 mM of the cation concentration in 1.0 ml.

Table 5. Effects of inhibitors on *T. zillii* Fillet and Liver thiaminases.

Chelating compound (0.01 mM)	Residual activity (%)	
	Fillet	Liver
2-Mecaptoethanol	19.59	59.13
EDTA	100	48.47
Glutathione	68.04	55.32
Citrate	0	64.56
Ascorbic acid	100	47.24

The enzyme was assayed in a typical enzyme assay such that the assay mixtures contain 1.0 mM of 2-mercaptoethanol, 1.0mM of Glutathione, 0.1 M ascorbic acid, 0.1M Citrate and 1.0 mM EDTA were used in the typical reactions.

values reported for other thiaminases. An optimum pH of 6.8 was reported for *B. thiaminolyticus* enzyme (Wittliff and Airth, 1968). An optimum pH of 7.0 to 8.0 was reported for bracken enzyme (Evans, 1975). Similarly pH values were reported for the enzymes in *Clostridium sporogenes* and *B. thiaminolyticus* at 5.2 and 5.6 respectively (Boyd and Walton, 1977) using citrate-phosphate buffer. Thiaminase activity was also found to have an optimum pH close to 8.5 in *Anaphe vernata* (Nishimune et al., 2000). Thiaminases from different sources exhibits varied responses to change in pH and to differing activators, cosubstrates and inhibitors (Edwin et al., 1982).

Table 3 shows the results obtained for the effect of various amino acids (lysine, serine, cysteine and proline) on the activity of *T. zillii* fillet and liver thiaminases. There was very little significant effect on the enzyme activity form fillet tissue by the presence of the amino acid compared to the effect seen on the liver enzyme, where the enzyme activity was greatly inhibit with lysine and proline completely inhibiting the enzyme. The effects of amino acids may not be uncommon as reports have shown that histidine residues and carboxyl groups may be essential for thiamine binding to the active site (Bos and Kozi, 2000).

The effect of metals on the fillet thiaminase activity showed slight inhibition as compared to the inhibition of the metals on the liver enzyme (Table 4). Magnesium ion completely inhibited the liver enzyme. The different effect of the metal on the tissue enzymes could be due to the compartmentization of the enzymes in the different tissues. It is possibly to attribute the effects of metal ions on the thiaminase of the fish to the anthropogenic activities around the reservoir (Atobatele, 2008; Komolafe and Arawomo, 2008). Some of these metals have also been reported to be abundant both on land and water environment (Eisler, 1991). Studies on the assessment of heavy metals and other pollutants in the Bompai-Jakara catchment basin of which Wasai reservoir is located has shown high amount of heavy metals contamination (Mustapha, 2008; Imam, 2010). The inhibition of bacterial thiaminase enzymes by various metal ions and by primary substrates such heteropyrithiamine has been reported (McCleary and Chick, 1977).

The inhibitory compounds (2-mercaptoethanol, ethylenediamine tetraacetic acid (EDTA), glutathione, citrate and ascorbate) showed varying degree of inhibition on the *T. zillii* fillet and liver thiaminase activities. EDTA and ascorbate activated the enzyme from fillet tissue, while glutathione showed slight inhibition on both tissue enzymes. 2-mercaptoethanol and citrate showed extensive to complete inhibition on the fillet enzyme but mildly inhibited the enzyme from liver tissue.

In conclusion, this study shows that the fillet and liver of *Tilapia zillii* contains thiaminases that are heat-resistant. Their physicochemical properties compares very well with reported thiaminases. Literatures have revealed that thiaminases are involved in disruption of energy production pathways and effective use of acetyl CoA in the body through the inactivation of the active form of thiamine and causing reduction of cellular thiamine concentration leading to impairment of energy production. These activities, therefore makes thiaminase an anti-nutritional enzyme especially as its present in fish (a major source of essential nutrient).

REFERENCES

Abdullahi SA (2001). Investigation of Nutritional status of *Chrysichthys nigrodigitatus*, *Bagrus filamentous* and *Auchenoglanis occidentalis*: Family Bagridae. J. Arid Zone Fish. 39-50.

Abe M, Ito S, Kimoto M, Hayashi R, Nishimune T (1987). Molecular studies on thiaminase I. Biochim. Biophys. Acta. 909:213-221.
Adamolekun B, Adamolekun WE, Sonibare AD, Sofowara G (1994). A doubleblind,placebo-controlled study of the efficacy of thiamin hydrochloride in a seasonal ataxia in Nigerians. Neurology 44:549-551.
Adefemi OS (2011). Chemical composition of Tilapia mosambis fish from major dams in Ekiti-State, Nigeria. Afr. J. Food Sci. 5(10): 550-554.
Ahmed M (1997). Fish for the poor under a rising global demand and changing fishery resources. NAGA: ICLARM Quaterly, July-December. pp. 73-76.
Anderson WI, Morrow LA (1987) Thiamine deficiency encephalopathy with concurrent myocardial degeneration and polyradiculoneuropathy in a cat. Cornell Vet. 77:251–257.
Atobatele OE (2008). Physicochemical parameters, plankton, macrozoobenthos and someaspects of the biology of two species of chrisichthys (Bagridae) of Aiba reservoir, Iwo, Osun State Nigeria. (PhD Thesis.) University of Ibadan. pp. 390
Bartoli P, Bourgeay-Causse M, Combes C (1997). Parasite transmission via a vitamin supplement. BioScience 47:251–253.
Bonga H (1999). Globalisation of trade relation and Africa trade for fish. Bimonthly Bullettin for the West Africa: programme on improvement for post-harvest utilization of artisanal fish catches. 45: 8-9.
Bos M, Kozik AJ (2000). Some molecular and enzymatic properties of a homogeneous preparation of thiaminase I purified from carp liver. J. Protein Chem. 19(2):75-84.
Boyd JW, Walton JR (1977). Cerebrocortical necrosis in ruminants attempt to identify source of thiaminase in affected animals. J. Comp. Path. 87:581–589.
Bradford MM (1976). A rapid and sensitive method for the quantification of microgramquantities of protein, utilizing the principle of protein dye binding. Anal. Biochem. 72, 248-254.
Cobiac L, Cliffton PM, Abbey M, Belling GB, Westel PJ (1991). Lipid, lipoprotein and lemostatic effects of fish is fish oil n-3 fatty acid is mildly perhipidenmic males. Am. J. Clin. Nutr. 53: 1210-1216
Delgado CL, McKenna AA (1997). Demand for fish in sub-Saharan Africa; NAGA: ICLARM Quaterly, July-December. pp. 79-82.
Edwin EE, Jackman R, Jones P (1982). Some properties of thiaminase associated with cerebrocortical necrosis. J. Agric. Sci. 99: 271-275.
Eisler R (1991). Cyanide hazard to fish, wildlife, and invertebrate. A synoptic review. US fish wildlife Service. Biol. Reports 85:1-23.
Evans WC (1975). Thiaminases and their effects on animals.Vit.and Horm., 33, 467-504. Intern. Biochem. Res. Rev. 2(1): 31-49, 2012 48
Fisher JP, Fitzsimons JD, Combs GF, Spitsbergen JM (1996). Naturally occurring thiamine deficiency causing reproductive failure in Finger Lakes Atlantic salmon and Great Lakes lake trout. Transact. Am. Fisheries Soc. 125:167–178.
Friedrich W (1987). Thiamin (Vitamin B1, aneurin). In: Handbuch der Vitamine. Munich: Urban and Schwartzenberg. 240–258.
Fujita A (1954). Thiaminase, Adv. Enzymol. Relat. Subj. Biochem. 15: 389–421.
Gerhard GT, Patton BD, Lindquist SA, Wander RC (1991). Comparison of three species of dietary fish: effects on serum concentrations of low-density lipoprotein cholesterol and apolipoprotein in normottingly ceridermic subjects. Am. J. Clin. Nutr. 4: 334-339
Green RG, Evans CA (1940). A deficiency disease of foxes. Science 92:154-155.
Greigh RA, Gnaedinger RH (1971). Occurrence of thiaminase in some common aquatic animals of the United States and Canada. NOAA Technical Report NMFS Special Scientific Report–Fisheries 631.
Hanes JW, Kraft CE, Begley TP (2007). An assay for thiaminase I in complex biological samples. Analytical Biochemistry 368: 33–38.
Hildebrand M, Goslow GE Jr. (2001). Principal ill. Viola Hildebrand. (2001). *Analysis of vertebrate structure*. New York: Wiley. pp. 429.
Hilker DM, Peter OF (1966). Antithiamine activity in Hawaii fish. J. Nutr. 89:419–421.
Hirn J, Pekkanen TJ (1975). Quantitative analysis of thiaminase activity in certain fish species. Nordisk Veterinaermedicin 27:646–648.
Houston DM, Hullard TJ (2007). Thiamine deficiency in a team of sled dogs. Can. Vet. J. 29: 383-385.
Imam TS (2010): Aspects of ecology and biomonitoring of heavy metals associated with industrial pollution in Bompai-Jakara catchment basin, Kano sate, Nigeria. A PhD Progress seminar paper presented at Biological Sciences Department, Bayero University Kano, 24th, February.
Ishihara T, Kinari H, Yasuda M (1974). Studies on thiaminase I in marine fish. 3. Vitamin B1 deficiency disease of chicken caused by inclusion of anchovy in diet. Bulletin of the Japanese Society of Scientific Fisheries 40:309–316.
Komolafe OO, Arawomo GAO (2008). Preliminary observations on fish species in a newly impounded Osinmo Reservoir. Turk. J. Fish. Aquat. Sci. 8: 288–292.
Leinhard G (1970). Kinetic evidence for a (4-amino-2-methyl-5-pyrimidinyl) methyl enzyme intermediate in the thiaminase I reaction. Biochem. 9: 3011–3020
Lineweaver H, Burk D (1934). The determination of enzyme dissociation constants, J. Am. Chem. Soc. 56: 658-666.
Luxemburger C, White NJ, Kuile F, Singh HM, Ailier-Frachon I, Ohn M, Chongsuphajaisiddhi T, Nosten F (2003). Beri-beri: the major cause of infant mortality inKaren refugees. Transact. Royal Soc. Trop. Med. Hyg. 97:251-255.
McCleary BV, Chick BF (1977). The purification and properties of a thiaminase I enzyme from Nardoo (*Marsilea drummondii*). Phytochem. 16:207-213.
Molnar J (2008). Assessing the global threat of invasive species to marine biodiversity. Frontiers Ecol. Environ. 6 (9): 485-492.
Mustapha A (2008). Environmental pollution in Kano: the contribution of wastewater discharge from kano old city and Bompai industrial estate to Jakara river basin system. J. Techno Africana 2(1): 83-88.
Nishimune T, Watanabe Y, Okazaki H, Akai H (2000). Thiamin is decomposed due to Anaphe spp. entomophagy in seasonal ataxia patients in Nigeria. J. Nutr. 130: 1625-1628.
Parker WH, McCreg CT (1965). Bracken (Pteris aquiline) poisoning of sheep in the North York moors. Vet. Record 77:861–865.
Petrova GG, Ilyina TN, Tyutyunnik NN, Chernikevich IP, Meldo HI (2003). Thiamine status in mink blood under alimentary hypovitaminosis B;1 after feeding thiaminase-containing fish. Science 26:106–111.
Ramos J, Marca C, Loste A, García de Jalón J, Fernández A, Cubel T (2006). Biochemical changes in apparently normal sheep from flocks affected bypolioencephalomalacia. Vet. Res. Comm. 27 (2): 111–124.
Riley SC, Evans AN (2008). Phylogenetic and Ecological Characteristics associatedwith Thiaminase Activity in Laurentian Great Lakes Fishes, Transact. Am. Fisheries Soc. 137(1):147-157.
Roberts GW, Boyd JW (1974). Cerebrocortical necrosis in ruminants occurrence of thiaminase in gut of normal and affected animals and its effect on thiamine status. J. Comp. Pathol. 84:365–374.
Rouvinen KI, Anderson DM, Alward SR (1997). Effects of high dietary levels of silver hake and Atlantic herring on growing-furring performance and blood clinical-chemistry of mink (Mustela vison). Can. J. Anim. Sci. 77:509–517.
Sepulveda F, Marın SL, Carvajal J (2004). Metazoan parasites in wild fish and farmed salmon from aquaculture sites in southern Chile. Aquacult. 235: 89–100
Tillitt DE, Zajicek JL, Brown SB, Fitzsimons ZD, Honeyfield DC, Holey ME, Wright GM (2005). Thiamine and thiaminase status in forage fish of salmonines from Lake Michigan. J. Aquat. Anim. Health 17:13–25.
Vimokesant S, Kunjara S, Rungruangsak K, Nakornchai S, Panijpan B (1982). Beriberi caused by antithiamin factors in food and its prevention. Annals of the New York Academy of Sciences 378:123–136.
Wittliff JL, Airth RL (1968). The extracellular thiaminase I of *Bacillus thiaminolyticus* I. Purification and physicochemical properties. Biochem. 7(2): 736-744.
Zajicek JL, Tillitt DE, Honeyfield DC, Brown SB, Fitzsimons JD (2005). A method for measuring total thiaminase activity in fish tissues. J. Aquat. Anim. Health 17:82–94.

Comparative evaluation of the nutritional and sensory quality of different brands of orange-juice in Nigerian market

Joel Ndife[1], Dehinde Awogbenja[2] and Umaru Zakari[1]

[1]Department of Food Technology, Kaduna Polytechnic, Kaduna State, Nigeria.
[2]Department of Home Science and Management, Nasarawa State University, Keffi, Nasarawa State, Nigeria.

The physico-chemical properties, vitamins, minerals, microbial and sensory qualities of four different brands of orange-juice (samples A, B, C and E) in the Nigerian market were evaluated to determine their overall quality. Results of the physico-chemical properties obtained show the following range of values for acidity (0.40 - 1.06%), total solids (5.50 - 11.80%), total sugar (9.15 - 13.39%) and fruit juice content (14.28 - 45.63%). The sweetness index for the juice samples A, B, C and D were 11.0, 9.66, 8.65 and 7.80, respectively. Sample D had the highest benzoic acid content of 217.21 ppm as preservative. The vitamin and mineral contents vary among the different orange juice samples. Samples D and B had the highest contents of vitamin A (140.12 IU) and C (42.57 mg), respectively. While samples A and C had the lowest values for potassium (119.10 mg) and sodium (5.65 mg) contents. The viable microbial counts for the juice samples were very low (1.0 x 10^2 to 5.2 x 10^2 cfu/100 ml). There were no coliform contaminations in all the samples. Sample D was adjudged the best in overall acceptance (8.45). Generally, the juice samples were within the regulatory specifications, and are fit for consumption

Key words: Orange, juice, vitamins, mineral, microbial and sensory quality.

INTRODUCTION

Fruits have been a part of human diet over the years. They are also considered as food supplements and are recommended internationally as essential to healthy nutrition, because they contain high quantity and quality of water, sugars, vitamins and minerals (Wardlaw, 2004; Potter and Hotchkiss, 2006). Fruit consumption has been reported to contribute to the prevention of degenerative processes, particularly lowering the incidence and mortality rate of cancer and cardio-cerebro-vascular diseases (Wenkam, 1990; Wardlaw, 2004; Mannay and Shadaksharaswany, 2005). They also contain phytochemicals which act against oxidative reactions in the human body (Vanamala et al., 2006; Okwu and Emenike, 2006). Orange fruits have been reported as a rich source of these phytochemicals such as flavonoids, especially flavanones, which have been shown to posses several physiological properties which can help inhibit cell proliferation and promote cell differentiation (Vanamala et al., 2006; Ndife and Abbo, 2009). However, due to the perishable nature of fruits and vegetables, high post harvest losses occur immediately after harvest, during distribution and marketing (Mannay and Shadaksharaswany, 2005; Potter and Hotchkiss, 2006), resulting from lack of cold storage facilities on the farms, improper handling and inadequate processing facilities (Alaka et al., 2003; Landon, 2007; Adubofuor et al., 2010). Reports show that the post harvest losses for oranges range between 31-50% (Alaka et al., 2003;

Landon, 2007). One of the ways of preserving these fruits and vegetables from deterioration and subsequent loss is to process them into fruit juices (Wenkam, 1990; Vanamala et al., 2006).

During the processing, a large part of the quality characteristics of the fresh fruits under-go remarkable changes which could reduce the nutritional value of the products (Wenkam, 1990; Landon, 2007). Moreover, the fruit juices may be stored for several months in unfavourable conditions before consump-tion, thus leading to undesirable quality changes due to the influence of temperature, time, oxygen content, light exposure and packaging material (Mannay and Shadaksharaswany, 2005; Landon, 2007; Averbeck and Schieberle, 2010).

The consumption of fruit juices is popular in Nigeria because of their health and invigorating benefits (Alaka et al., 2003; Okorie et al., 2009). Though some fruit juices are produced locally, most of the fruit juices and drinks found in the market are imported (Dosumu et al., 2009; Okorie et al., 2009).

Valid data are not available on the extent to which these commercial juice and drinks are either mislabeled, adulterated or of inferior quality (Dosumu et al., 2009). This is due to the fact that laboratories' testing capabilities are limited, and tests are expensive to conduct (Landon, 2007).

The objective of this study was therefore, to evaluate and provide information on the nutritional, microbial and sensory quality of the different brands of orange juice available in the Nigerian market with the aim of attracting the attention of the regulatory authorities and helping the un-informed consumers to make a healthful choice. The implications for the industry are also discussed.

MATERIALS AND METHODS

Four most popular brands of orange juice were purchased off the shelf from different supermarket in Kaduna town, of Northern Nigeria. Three of the different brands of juice were produced in Nigeria and were labeled A, B, C and D for the imported brand. All the juice samples were stored at ambient temperature, in sample bottles with tight-fitting lids, during the period of analytical investigation.

Physico-chemical analysis

The pH, brix (soluble solids) and specific gravity of the juice samples were measured using instrumental methods (Jacobs, 1999). The fruit juice content, benzoic acid content, acidity were determined by the method described by Ayo and Agu, (2012) with some modifications; while, the total sugar was determined by Lane Eynon method as described by Mehmoud et al. (2008). The sweetness and astringency indexes were calculated as the ratio of soluble solids to acidity and vice versa (Wardy et al., 2009).

Vitamin assay

The ascorbic acid (vitamin- C), B-carotene (vitamin -A), riboflavin (vitamin- B_2), vitamin-E and folic acid (niacin-vitamin-B_3) contents of the orange juice samples were determined by the method described by Nielsen (2003) with some modifications using UV-VIS spectrophotometer.

Mineral assay

The orange juice samples were digested by the wet ashing method prior to mineral content determination using atomic absorption spectrophotometer for Ca, Mg, and Fe and Corning 400 flame photometer for K and Na (Abulude et al., 2005). While the phosphorus content was determined colorimetrically with Jenway 6100 spectrophotometer using the method described by Nielson (2003).

Microbiological assay

The determination of the microbial quality (mesophilic aerobic bacteria, coliforms, yeasts and mold counts) of the products were performed by the method outlined in compendium of methods for the microbiological examination of foods (AMPH, 1992) with some modifications.

Sensory analysis

Sensory evaluation of the orange juice samples were carried out by 25 panelists on a 9 point hedonic scale for different parameters such as colour, aroma, taste, texture and overall acceptability as described by Iwe (2010).

Statistical analysis

The sensory evaluation data was statistically analyzed using the analysis of variance (ANOVA) and the Duncan Multiple range test with significance level at $p<0.05$ (Ihekoronye and Ngoddy, 1985).

RESULTS AND DISCUSSION

Physico-chemical analysis

The results obtained from the physico-chemical analysis of the different brands of orange juice samples are presented in Table 1.

The moisture content in the different brands of orange juice analyzed ranged from 88.20 to 94.50%. Sample D (imported brand) had the lowest value of 88.20%. The moisture content has an inverse relationship with the total fruit juice content.

The pH of the orange juice samples range from 3.40 to 4.08. Sample A had the highest pH value of 4.08. The reverse was observed in the results for acidity, in which sample D had the highest value of 1.06%, while samples A, B and C had 0.40, 0.68 and 0.85%, respectively. Kareem and Adebowale (2007) reported that the dominant acid in orange juice is citric acid.

Food acids dictate the dominant microflora in foods and to a large extent will determine the shelf stability of the juice (Ezeama, 2007). The more acidic the juice, the less susceptible to bacterial action but the more susceptible to the action of yeasts and moulds (Jay, 2000). Moreover

Table 1. Physico-chemical analysis of different brands of orange juice samples.

Parameter	Juice sample			
	A	B	C	D
Moisture (%)	94.50±3.01	91.83±3.11	90.22±3.05	88.20±3.03
pH	4.08±0.15	4.01±0.18	3.40±0.10	3.23±0.16
Acidity (%)	0.40±1.10	0.68±1.00	0.85±1.04	1.06±1.07
Specific gravity	1.03±0.50	1.02±0.55	1.04±0.53	1.08±0.55
Soluble solids (°Brix)	4.40±0.85	6.57±1.00	7.35±0.90	8.27±0.95
Total solids (%)	5.50±1.12	8.17±1.20	9.78±1.24	11.80±1.20
Total sugar (%)	13.39±2.04	14.25±1.95	9.15±2.06	10.76±1.98
Fruit juice content (%)	14.2±2.25	22.6±2.16	37.60±2.32	45.63±2.30
Sweetness Index	11.00±0.55	9.66±0.50	8.65±0.48	7.80±0.50
Astringency Index	0.09±0.01	0.10±0.03	0.12±0.01	0.13±0.02
Benzoic acid (ppm)	185.41±3.30	170.80±3.10	195.60±3.21	217.21±3.16

[a]Data are mean values of duplicate determinations ± standard deviation.

Anvoh et al. (2009) reported that fruit acids influence colour, flavour and gustative characteristics of the juice products.

The specific gravity for the orange juice sample ranges from 1.02 to 1.08, with sample B recording the lowest value of 1.02. While the Brix values ranges from 4.40 to 8.27°. Sample A had the lowest Brix of 4.40°. The Brix values of orange juices should be between 4 - 9° (Kareem and Adebowale, 2007). Also, the total solids content of the fruit juices were: 5.50, 9.60, 1.70 and 11.10% for samples A, B, C and D, respectively. The total (soluble and non-soluble) solids are used as indicators of the fruit juice content which ranged from 14.2 to 45.63% in the orange juice samples. The total solids and juice content are used in characterizing the quality of juice and other beverage products (Egbekun and Akubor, 2007; Adubofuor et al., 2010).

The total sugar was highest in sample B (14.25%) and lowest in sample C (9.15%). The total (reducing and non-reducing) sugars to a large extent determine the sweetness of juices and beverages. It could be used for masking the astringency derived from organic acids (Anvoh et al., 2009; Adeola and Aworh, 2010). When compared with the recommended dietary allowances (RDA) of 130 g/day (El-Sheikha et al., 2010), for total sugars, the orange juice samples will contribute to the average 15.40%. Wardlaw (2004) reported that frequent consumption of sugar sweetened beverages may be associated with larger weight gain and increased risk of type 2 diabetes.

The sweetness index for samples A, B, C and D were 11.0, 9.66, 8.65 and 7.80, respectively. Sweetness index (SI) and the astringency index (AI) are used for the prediction of flavours in juices (Wardy et al., 2009; Adeola and Aworh, 2010). The ratio of sugars to acids and vice-versa gives an accurate prediction of the tartness and sweetness of acid foods which affects organoleptic perception (Wardy et al., 2009; Averbeck and Schieberle, 2010). Fruit juices with sweetness index greater than 19 are regarded as sweet and with less acid by taste (Wardy et al., 2009).

Benzoic acid was detected in all the brands of orange juice, with values that ranged from 170.80 to 217.21 ppm. Benzoic acid is particularly used as a preservative against yeasts and moulds for long term storage (Mehmood et al., 2008).

Table 2 shows the result of the analysis of the vitamin content of the orange juice samples. It was observed that the vitamin contents vary among the different brands of orange juices. Sample D had the highest vitamin A (140.12 IU) while sample B had the highest content of vitamin C (42.57 mg). The vitamin B_1 (Thiamine) and vitamin B_2 (Riboflavin) contents were lowest in sample C with values of 0.90 and 0.43 mg, respectively. Sample A had the highest value of vitamin B_3 (Niacin) content (7.23 mg). Fruit juices are important in the delivery of body fluid and essential micro-nutrients such as vitamins (Landon, 2007) and the nutritional significance of food nutrients is related to their contribution to the recommended dietary allowance (RDA) (Wardlaw, 2004). Ascorbic acid content of fruit juices is the most prominent quality index of fruit juices due to its health significance as a vitamin and cellular antioxidant (Landon, 2007). The fruit juices will contribute on the average, 40% to the recommended dietary intake of vitamin C. More so, when compared when the recommended dietary allowances (RDA), the orange juice samples A, B, C and D will contribute 44.75, 51.30, 27.72 and 32.90% to vitamin A requirements. Thus, confirming that the different brands of orange juice are good sources of essential vitamins.

The mineral compositions of the orange juice samples are shown in Table 3. Potassium is the most abundant element in all the orange juice samples, with an average value of 143.28 mg, followed by calcium (14.12 mg) and

Table 2. Vitamin analysis of different brands of orange juice samples.

Parameter	Juice sample (mg/100 g)				
	A	B	C	D	RDA (mg/day)
Vitamin-A (IU)	91.23±3.10	108.34±3.16	132.55±3.28	140.12±3.10	750-900
Vitamin-C	37.14±2.10	42.57±2.15	23.01±2.18	27.28±2.25	75-90
Vitamin-B_1	1.04±0.11	1.10±0.15	0.90±0.18	1.03±0.15	1.1-1.2
Vitamin-B_2	0.86±0.05	1.0±0.03	0.43±0.05	0.65±0.04	1.1-1.3
Vitamin-B_3	7.23±1.05	11.14±1.10	4.38±1.14	5.15±1.08	14-16

*RDA- Recommended dietary allowance; [a]data are mean values of duplicate determinations ± standard deviation.

Table 3. Mineral analysis of different brands of orange juice samples.

Parameter	Juice sample (mg/100 g)				
	A	B	C	D	RDA (mg/day)
Sodium	10.35±1.50	13.82±1.36	5.65±2.10	8.47±1.85	1000-1500
Potassium	119.10±2.00	128.31±1.53	155.28±1.91	170.42±1.64	4000-4700
Calcium	16.81±0.42	18.54±0.25	10.10±0.22	11.01±0.36	800-1100
Magnesium	13.32±0.10	15.36±0.14	10.01±0.20	11.63±0.16	110-300
Iron	1.22±0.04	3.43±0.08	1.70±0.06	3.50±0.08	8-22
Phosphorus	11.83±2.11	9.41±0.18	7.32±2.00	9.35±1.17	380-1055
Ash (%)	0.98±0.10	1.32±0.11	0.64±0.10	0.75±0.12	-

*RDA- Recommended dietary allowance; [a]data are mean values of duplicate determinations ± standard deviation.

Table 4. Microbial analysis of different brands of orange juice samples.

Parameter	Juice sample (cfu/100 ml)				
	A	B	C	D	*SON
Total aerobic counts	$5.2±0.25x10^2$	$3.6±0.20x10^1$	$1.5±0.14x10^2$	$1.0±0.12x10^2$	$<2.0x10^3$
Total coliform counts	Nil	Nil	NIL	NIL	3
Total yeast and mould counts	$1.1±0.13x10^1$	$2.3±0.15x10^1$	$4.7±0.20x10^1$	$3.5±0.18x10^1$	50

[a]SON-Standards Organization of Nigeria maximum allowable counts; [b]Data are mean values of duplicate determinations ± standard deviation.

magnesium (12.58 mg), while iron is present in trace amounts (2.46 mg). This is in agreement with the result reported by Dosumu et al. (2009). Natural fruits and vegetables are good sources of potassium and are low in sodium, an advantage reported to protect against arterial hypertension (Nnam and Njoku, 2005). Inadequate intakes of micronutrients (minerals) have been associated with severe malnutrition, increased disease conditions and mental impairment (Mannay and Shadaksharaswany, 2005; Dosumu et al., 2009).

In comparison with the RDA (Wardlaw, 2004), the orange juice samples would contribute on the average: 0.96, 3.58, 1.80, 11.44, 30.75 and 2.50% for Na, K, Ca, Mg, Fe, and P, respectively. The results from this study show that the orange juice samples would need to be supplemented with other mineral sources in the diet to meet up with the RDA requirements for a healthy nutrition.

Microbial analysis

The results of microbial analysis of the different orange juice samples are presented in Table 4. The results obtained for total aerobic counts were low in all the orange juice samples ($<1.5 \times 10^2$ cfu/100 ml). They were also below the maximum limit in the SON (2008) specification for commercial fruit juices ($<2.0 \times 10^2$ cfu/100

Table 5. Microbial analysis of different brands of orange juice samples.

Parameter	Juice sample			
	A	B	C	D
Appearance	5.83[b]	5.14[b]	7.65[a]	8.35[a]
Flavour	6.75[b]	6.15[b]	8.50[a]	7.72[a]
Residual after taste	7.55[a]	8.31[a]	5.68[a]	5.23[a]
Aroma	5.61[b]	5.45[b]	7.89[a]	8.51[a]
Overall acceptability	6.44[b]	6.21[b]	7.90[a]	8.45[a]

*Means within a row with different letters are significantly different at P<0.05.

ml). However, the aerobic plate counts were found to be higher in the sample C (1.2 x 10^2 cfu/100 ml) than in the other orange juice samples. The acid and sugar content, are critical to the survival of microbes and will ultimately affect the shelf stability and sensory quality of the orange juice samples (Jay, 2000; Ezeama, 2007).

There were no observable coliform growths from all the orange juice samples. This eliminates the possibility of faecal contamination in the different brands of orange juice samples, which is a pointer to good manufacturing and handling practice (Ezeama, 2007). This also conforms to the regulatory specification of <3 cfu/ml for total coliform counts in fruit juices (SON, 2008). The total mould and yeasts counts in the orange juice samples range from 1.1 x 10^1 to 4.7 x 10^1 cfu/100 ml. While the highest mould and yeasts counts (4.7 x 10^1 cfu/100 ml) were observed in sample C. This is slightly lower than the standard specification of 5.0 x 10^1 cfu/100 ml for fruit juices (SON, 2008). The low microbial counts in orange juice samples could also be due to the use of benzoic acid as a preservative. Benzoic acid is known to have antimicrobial properties due to the low pH. The results show that the different brands of orange juices are safe for human consumption as the range is below the upper limit of 10^3 (cfu/100 ml) of total bacteria counts, considered safe for foods (ICMSF, 1998).

Sensory analysis

Table 5 summarizes the results for the sensory evaluation and overall acceptability of the different brands of orange juice samples. The statistical analysis revealed that there were significant difference (p<0.05) among the orange juice samples in the sensory attributes observed. Sample D (imported brand) had the highest score (8.35), while sample B had the lowest score (5.14) for appearance. The appearance was based on how the colours appeal to the panelists. Browning in the beverages could have been due to Maillard-type reactions (Potter and Hotchkiss, 2006) resulting from the presence of reducing sugars, proteins and amino acids, also, due to the effect of severe heating during processing on the quality attributes (Mannay and Shadaksharaswany,2005).

Similar trends were observed for the sensory ratings of flavour and aroma in the orange juice samples, sample B had the lowest scores (6.15 and 5.45) followed by sample A (6.75 and 5.61). Acidity contributes to the development of flavour through a proper sugar-acid ratio thereby modifying the sweetness of sugar (Adeola and Aworh, 2010).

The residual after taste was characterized by perceived bitterness after swallowing the orange juice samples by the panelists. This bitter perception was prominent in samples A (7.55) and B (8.31). The panelists affirmed that this was responsible for their low scores on flavour and aroma. Mannay and Shadaksharaswany (2005) reported that the inclusion of additives could impact on the organoleptic qualities of food products. Orange samples C and D had the best overall acceptability ratings of 7.90 and 8.45, respectively.

Conclusion and recommendation

The difference in the quality attributes of the different brands of orange juice may be attributed to the different processing procedures employed and storage conditions. Consumption of these beverages is desirable as they would serve as good sources of vitamins and body electrolytes. The results from this study show that the orange juice samples would need to be supplemented with other nutrient sources in the diet to meet up with the RDA requirements for a healthy nutrition. The results also show that the different brands of orange juices are safe for human consumption considering their low microbial content. The orange samples C and D had the best overall acceptability.

Therefore, the regulatory authorities such as Standards Organization of Nigeria (SON) and National Authority for Food Drugs Administration and Control (NAFDAC) and Consumers Protection Agency (CPA) should ensure that foods are safe, wholesome, and honestly labeled, by periodically inspecting production facilities and occasionally sampling and testing products on the shelf, to ensure they conform to specified standards.

REFERENCES

Abulude FO (2005). Distribution of selected minerals in some nigerian white bread. Nig. Food J. 23:139-147.

Adeola AA, Aworh OC (2010). Development and sensory evaluation of an improved beverage from Nigeria's tamarind (tamarindus indica l.) fruit. J. Food Agric. Nutr. Dev. 10(9):4079-4093.

Adubofuor J, Amankwah E, Arthur B, Appiah F (2010).Comparative study related to physico-chemical properties and sensory qualities of tomato juice and cocktail juice produced from oranges, tomatoes and carrots. Afr. J. Food Sci. 4(7):427-433.

Alaka O, Aina J, Falade K (2003). Effect of storage conditions on the chemical attributes of Ogbomoso mango juice. Eur. Food Res. Technol. 218:79-82.

AMPH (1992). Compendium of methods for the microbiological examination of foods. Washington DC: American Public Health Association.

Anvoh K, Zoro-Bi A, Gnakin D (2009).Production and characterization of juice from mucilage of cocoa beans and its transformation to marmalade. Pak. J. Nutr. 8(2):129-133.

Averbeck M, Schieberle P (2010). Influence of different storage conditions on changes in the key aroma compounds of orange juice reconstituted from concentrate. Eur. Food Res. Technol. 217:1366-1380.

Ayo JA, Agu HO (2012). Simplified Manual of Food Analysis, Vol.1. Amana Printing and Advertising Ltd, Kaduna. pp. 113-120

Dosumu O, Oluwaniyi O, Awolola GV, Okunola MO (2009). Stability studies and mineral concentration of some Nigerian packed fruit juices, concentrate and local beverages. Afr. J. Food Sci. 3(3):082-085.

Egbekun MK, Akubor PI (2007). Chemical composition and sensory properties of melon seed- orange juice beverage. Nig. Food J. 24(1):42-45.

El-Sheikha A, Zaki M, Bakr A, El-Habashy M, Montet D (2010). Biochemical and sensory quality of physalis (physalis pubescens l.) juice. J. Food Process. Preserv. 34:541-555.

Ezeama CF(2007). Food Microbiology: Fundamentals and Applications. Natural Prints Ltd. Lagos.

ICMSF (1998). International Commission on Microbiologcal Specifications for Foods In: Microbiological Ecology of Food Commodities. Academic Press, New York. pp. 25-71.

Ihekoronye AI, Ngoddy PO (1985). Integrated Food Science and Technology for the Tropics. (2nd ed.) Macmillan Publishers Ltd. London.

Iwe MO (2010). Handbook of Sensory Methods and Analysis. Rojoint Communication Services Ltd. Enugu. pp. 75-78.

Jacobs BM (1999). The Chemical Analysis of Foods and Food Products. (3rd ed.) CBS Publishers and Distributors. New Delhi, India.

Jay MJ (2000). Modern Food Microbiology. (6th ed.) Aspen Publishers Inc.,Gaithersburg, Maryland.

Kareem SO, Adebowale AA (2007). Clarification of orange juice by crude fungal pectinase from citrus peel. Nig. Food J. 25(1):130-137.

Landon S (2007). Fruit juice nutrition and health (Review). Food Australia 59 (11):533-538.

Mannay S, Shadaksharaswany CM (2005). Foods: Facts and Principles. (2nd ed.). New Age International Ltd. Publishers. New Delhi, India.

Mehmood Z, Zeb A, Ayub M, Bibi A, Badshah A, Ihsanullah (2008).Effect of pasteurization and chemical preservatives on the quality and shelf stability of apple juice. Am. J. Food Technol. 3(2):147-153.

Ndife J, Abbo E (2009). Functional foods: prospects and challenges in Nigeria. J. Sci. Technol. 1(5):1-6.

Nielsen SS (2003). Food analysis laboratory manual. (3rd ed.). Kluwer Academic Plenum Publishers, New York.

Nnam NM, Njoku IE (2005). Production and Evaluation of Nutrient and Sensory Properties of Juices Made from Citrus Fruits. Nig. J. Nutr. Sci. 26(2): 62-66.

Okorie O, Enwere NJ, Udensi EA (2009). Effect of ambient storage conditions on pH and vitamin-C content of selected tetra-pak packaged fruit juice marketed in Nigeria. Nig. Food J. 27:4-10.

Okwu DE, Emenike IN (2006). Evaluation of the phytonutrients and vitamin content of citrus fruits. Int. J. Mol. Med. Adv. Sci. 2:1- 6.

Potter H, Hotchkiss I (2006). Food Science. (5th ed.). CBS Publishers and Distributors. New Delhi, India.

SON (2008). Standard for fruit juice. Standards Organization of Nigeria. ICS 67.160.20. NIS 235.

Vanamala J, Reddivari L, Sun-Yoo K, Pike L, Patil B (2006). Variation in the content of bioactive flavonoids in different brands of orange and grapefruit juices. J. Food Compost. Anal. 19:157-166.

Wardlaw GM (2004). Perspectives in Nutrition. (6th ed.). McGram Hill Companies, New York, U.S.A.

Wardy W, Saalia F, Steiner-Asiedu M, Budu A, Sefa-Dedeh S (2009). A comparison of some physical, chemical and sensory attributes of three pineapple (Ananas comosus) varieties grown in Ghana. Afr. J. Food Sci. 3(1):022-025.

Wenkam A (1990). Utilization and Processing of Fruits. Macmillan Press, London. pp. 388 -506.

Genotoxicity and anti-genotoxicity of fennel plant (*Foeniculum vulgare* Mill) fruit extracts using the somatic mutation and recombination test (SMART)

Sanae Amkiss[1] , Abdelkrim Dallouh[1] and Mohamed Idaomar[1], Bouchra AMKISS[2]

[1]Biology and Health Laboratory, Department of Biology, Faculty of Sciences, Abdelmalek Essaâdi University, BP 2121, 93002, Tétouan, Morocco.
[2]Chemical department, Faculty of Sciences, Abdelmalek University, BP 2121,93002, Morocco.

The antigenotoxic action of fennel (*Foeniculum vulgare*) were assessed using the eye (w/w^+) somatic mutation and recombination test (SMART) assay of *Drosophila melanogaster*. Fennel is used for various medicinal purposes. Methyl methanesulfonate (MMS) was used as the positive control. Fruit extracts of fennel did not show genotoxicity at the doses used. In the co-treatment, a dose-dependent decrease in mutation frequency was observed for this plant. The fennel at 8% (w/v) demonstrated a marked decrease in MMS with an inhibition rate of 41.16%. The results demonstrate that the fennel exerted a significant and potent anti-mutagenic activity in eyes spot of *D. melanogaster*.

Key words: *Drosophila melanogaster*, somatic mutation and recombination test (SMART), antigenotoxicity, *Foeniculum vulgare*, methyl methanesulfonate (MMS).

INTRODUCTION

Recently, there has been considerable interest in the mutagenicity and antimutagenicity of medicinal plants (Romero-Jimenez et al., 2005; Patenkovic et al., 2009; Ribeiro and Salvadori, 2003). It is sometimes argued that frequently used plants in traditional medicine are safe, due to their long-term use and are considered to have no side effects because they are natural (Van den Berg et al., 2011). This concept is largely circumstantial and it is important to determine toxicology of plant extracts, especially those that are used frequently over long periods.

Foeniculum vulgare Mill. (Apiaceae family) is commonly known as fennel. The plant is native to the Mediterranean region, and temperate regions of Asia. It was introduced and distributed to subtropical regions, Europe and North America and also cultivated worldwide (He and Huang, 2011). The fruits, leaves and roots can be used, but the fruits are most active medicinally and are the part normally used (EFSA, 2009).

Numerous compounds including *trans*-anethole, estragole, fenchone, sesquiterpenoids, coumarins and polyphenolics were isolated from this plant, most of which exhibited significant bioactivities. Fennel is a commonly used household remedy, having been claimed to be useful in the treatment of a variety of complaints, especially those of the digestive system (EFSA, 2009). The plant is said to be analgesic, anti-inflammatory, antispasmodic, aromatic, carminative, diuretic, emmenagogous, expectorant, galactogogous, hallucinogenic, laxative, stimulant and stomachic (El Bardai et al., 2001; Boskabady et al., 2004; Raffo et al., 2011)

The somatic mutation and recombination tests (SMART) system has been used for over 20 years to study anti-genotoxic effects of many mutagens and carcinogens (Zimmering et al., 1990). A wide variety of compounds and complex mixtures have been assayed in this test, such as food additives, beverages and insecticides (Osaba et al., 2002). These one generation tests make use of the wing or eye imaginal disc cells in larvae and have proven to be very efficient and sensitive. They are based on the principle that the loss of heterozygosity of suitable recessive markers can lead to the formation of mutant clones of cells that are then expressed as spots on the wings or eyes of the adult flies (Graf et al., 1998).

The purpose of this work was to study the genotoxic activity of fennel and to evaluate its potential anti-genotoxic effect using methyl methanesulfonate (MMS) as a mutagen

MATERIALS AND METHODS

Preparation of the aqueous extract

The fresh fruit of fennel was purchased at the local market in the city of Tetouan, Morocco. Plant material was prepared according to the traditional method used in Morocco. Ten grams of dried, ground plant material were extracted in 100 ml of distilled water and allowed for over-night shaking. After filtering, solutions were used immediately in treatments.

Negative and positive controls

Pure distilled water served as a negative control. A 1 mM aqueous solution of MMS [66-27-3] (Sigma) were used as the positive control. MMS was purchased from Sigma and was reagent grade.

Eye spot test

Markets and strains

The following *Drosophila melanogaster* strains were used: *Ok-white*, a strain carrying the X-linked eye color marker *white (w)* and *Ok-yellow*.

Treatment procedure

Plant extract and mutagen were administered by chronic co-treatment feeding exposure from the egg stage. The *Oregon k (Ok)* was used. Twenty *Oregon K- yellow-* virgin female flies were mated with *w* males (20) flies for three days and then transferred to bottles (200 mL) containing 3 g of instant Drosophila medium (Carolina Biological Supply, Burlington, USA) dissolved with equal volumes of the test plant extract and MMS.

Drosophila larvae were exposed to fennel extract at three concentrations: 2, 4 and 8%. This was done by adding the solutions to the flies' medium. The solutions were always freshly prepared, immediately before use. The flies were permitted to lay eggs for three days. Newly hatched flies were counted and the females were transferred to fresh medium; two to six days later, their eyes were inspected for the presence of white spots. All experiments were carried out at 25°C and 65% relative humidity.

Scoring of eyes

The scoring of the etherized flies was carried out in a liquid consistin g of 90 parts ethanol, one part Tween-80 and nine parts water (Smijs et al., 2004). Because of the fast evaporation of this solution, a second solution (90% water + 10% first solution) was added when necessary (Gaivao and Comendador, 1996).

Inspection of the eyes for the presence of white spots was performed using stereomicroscope at a magnification of 80x using a light source that was provided by a glass fiber illumination of 20 W. Large white spots are seen as white parts in the red eyes and small white spots are dark ommatidia in red eye. Spots separated from each other by at least four nonmutated ommatidia were counted as independent events, and the smallest size of white clone expected to be counted accurately is 2 (Vogel and Nivard, 1993). An increase in white clone frequency was only accepted as a positive response if it was significantly higher than that of both the concurrent and pooled experimental controls.

Statistical test

The statistical significance of the differences between spot frequencies in the experimental groups and control was calculated using the Chi-square test for proportions as described by Frei and Wurgler (1988). The percentage of genotoxicity inhibition was calculated based on the spot frequencies according to the formula: [(mutagen alone-mutagen + plant)/mutagen alone] x 100 (Abraham, 1994).

RESULTS

The fennel fruits extract was first analyzed for the evaluation of its mutagenic activity. The numbers of small spots, large spots with the total number of spots are given in Table 1. Fennel extract was not toxic in the chronic feeding. At the concentrations tested, it did not show any effect on the frequencies of somatic mutations when compared with their respective negative controls. The data obtained with this approach showed that the fennel extracts does not induce mutagenicity at the selected doses.

MMS showed to be clearly mutagenic (Figure 1). A high value of genotoxic activity with a high significant response ($p < 0.001$) was detected for all the spot categories: small spot, large spots as well as total spots for all the concentration used when compared with the negative control. MMS at 1 mM showed a highly significant increase in induced somatic mutations (53.03 spot per 100 eyes).

The anti-genotoxic effect was established by means of a chronic co-administration of fruits water extract of fennel with MMS to larvae of *D. melanogaster*. Our result showed a dose-dependent inhibition of genotoxicity in larvae (Table 2). The overall inhibition activity is 31.38, 34.83 and 41.16% for 2, 4 and 8% concentration of fennel, respectively. The chronic co-administration of fennel was effective in significantly reducing the frequencies of large and totals spots induced by MMS.

DISCUSSION

The use of *D. melanogaster* in the evaluation of genotoxicity has been well established as a test-system. Due

Table 1. Mutagenic effect of fennel fruit extract in *Drosophila* eye spot test (w/w+).

Compound concentration (%)	Number of eyes	Distribution of spots size (ommatidia)				Spots per 100 eyes size classes(-)		
		2-4	4-32	>32	T	S	L	T
0	560	40	30	0	70	7.14[b]	5.36[a]	12.5[a]
2	434	47	24	1	72	10.83[a]	5.76[a]	16.59[a]
4	458	37	08	1	46	8.78[a]	1.96[b]	10,04[a]
8	640	43	11	2	56	6.72[a]	2.03[b]	8.75[b]

(-) Size classes: S, small 1-4 ommatidia affected; L, large >4 ommatidia; T, total spots, a: significantly not different from positive control, p>0.05; b, significantly different from control at p<0.05.

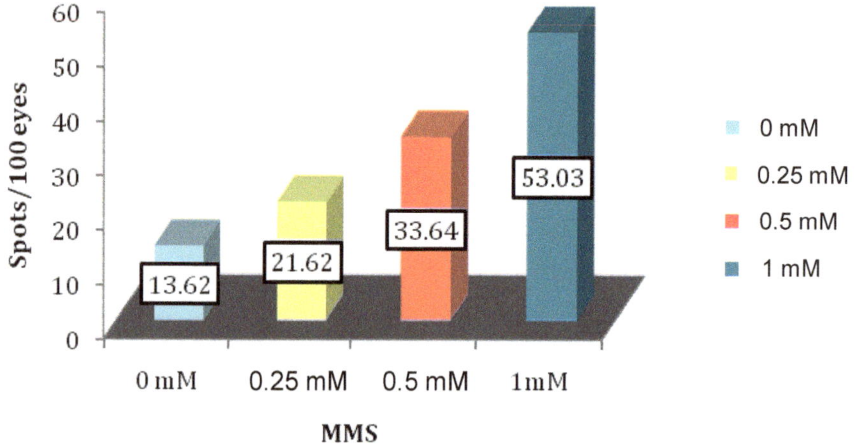

Figure 1. Mutagenic effect of MMS in eye spot test of *D. melanogaster*.

Table 2. Antimutagenic effect of fennel fruit extract on MMS that induced mutagenicity in *Drosophila* eye spot test (w/w+).

	Number of eyes	Distribution of spots size (ommatidia)				Spots per 100 eyes size classes(-)			Inhibition (%)
		2-4	4-32	>32	T	S	L	T	
MMS 1 mM	340	70	48	7	125	20.6	16.17	36.77	-
MMS and 2% Fen	214	16	35	3	54	**7.47[d]**	17.75[a]	**25.23[b]**	31.38
MMS and 4% Fen	242	12	36	10	58	**4.95[d]**	19.00[a]	**23.96[c]**	34.83
MMS and 8% Fen	298	9	46	9	64	**3.02[d]**	18.45[a]	**21.47[c]**	41.16

(-) Size classes: S, small 1-4 ommatidia affected; L, large >4 ommatidia; T, total spots; MMS: methyl methanesulfonate; a: significantly not different from positive control, p>0.05; b, c, d: Significantly different from positive control at p<0.05, 0.01 and 0.001, respectively.

to a genome similarity as compared to mammals and easy maintenance in the laboratory, these flies represent an appropriate organism to run *in vivo* short-term tests (Graf et al., 1996; Vogel et al., 1999). SMART is a simple and fast short-term assay when compared with other *in vivo* tests. Through the use of these test-systems, it is possible to evaluate the genotoxic activity of a single compound as well as complex mixtures (Patenkovic et al., 2009: Schneider et al., 2009; Lehmann et al., 2010).

The somatic assays take advantage of the possibility to expose such large populations of mitotically growing cells in the imaginal discs of larvae. If a genetic alteration occurs in one of these imaginal disc cells, this alteration will be present in all the descendant cells and will form a clone of mutant cells. If the alteration causes a visible change in the phenotype, the mutant cell clone can be detected as a spot of mutant cells on the body surface of the adult flies. Thus, due to its capabilities, SMART was chosen to evaluate the anti-genotoxic effects of the aqueous fennel extract.

The fennel was first analyzed for the evaluation of its genotoxic/mutagenic activity. Three concentrations of the fennel were tested. The concentrations chosen are not toxic according to progeny mortality relative to the control. Fennel, at the concentrations tested, did not show any effect on the frequencies of somatic mutations when compared with their respective negative controls.

The evaluation of the antimutagenic potential of the aqueous extract from fennel was established by means of a co-treatment protocol in which the extract was administered simultaneously with the mutagens used, MMS a direct-acting genotoxin. MMS is a monofunctional alkylating agent known for its ability to interact directly with DNA, and produces genotoxic damage in different models *in vitro* and *in vivo* (Arnaiz et al., 1996). Under our experimental conditions, MMS showed to be clearly mutagenic. The aqueous extract from fennel displayed significant antimutagenic effect on the somatic mutations induced by MMS.

The mutational spectra induced in *Drosophila* by MMS suggest the involvement of apurinic sites as mutagenic lesions (Vogel et al., 1990). In addition, a clear relationship exists between the extent of the DNA *N*-alkylation and the efficiency of the MMS to induce mitotic recombination in the *Drosophila* wing-spot test (Arnaiz et al., 1996). Against this direct acting mutagen, the fennel used in this study showed a significant antimutagenic activity. Considering the genotoxin co-administered with the plant, there is possibility that the antimutagens from fennel exert their protective effect by interacting with MMS in desmutagenic manner without affecting the genetic material directly (Kuroda et al., 1992). MMS does not require metabolic activation; therefore, the natural compounds present in this plant may interact directly with the methyl radical groups of MMS and inactivate them by chemical reaction. It is also possible that these compounds compete to interact with the nucleophilic sites in DNA, thus altering the binding of the mutagen to these sites.

The inhibitory effect detected in this study can be attributed to wide range of constituents of plant including *trans*-anethole, estragole, fenchone, sesquiterpenoids, coumarins and polyphenolics studied. Aqueous extract of fennel fruits contains rich phenolic compounds. Many of them have antioxidant activities, such as 3-caffeoylquinic acid, 4-caffeoylquinic acid, 1.5-Odicaffeoylquinicacid, rosmarinic acid, eriodictyol-7-Orutinoside, quercetin-3-O-galactoside, kaempferol-3-Orutinoside and kaempferol-3-O-glucoside. Besides these compounds, fennel was reported to contain hydroxylcinnamic acid derivatives, flavonoid glycosides and flavonoid aglycones (Parejo et al., 2004). Naturally-occurring antioxidants can be used to protect gainst oxidative stress damage (Scalbert et al., 2005). Fennel was known as excellent source of nature antioxidants and contributed to the daily antioxidant diet (Shahat et al., 2011).

The mutagenicity, antimutagenicity and anticarcinogenicity of fennel and some of its components were subjected to several investigations (Ebeed et al., 2010; Hassan et al., 2011) and indicated that trans-anethole, the main component of fennel oil, does not increase the mutant frequency in the Salmonella/microsome test and it did not induce chromosome aberrations in Chinese hamster ovary cells. In addition, pretreatment with trans-anethole and eugenol led to significant antigenotoxic effects against ethyl methane sulfonate (EMS), cyclophosphamide (CPH), procarbazine (PCB), N-methyl-N'-nitro-N-nitrosoguanidine (MNNG) and urethane (URE). Both trans-anethole and eugenol exerted dose-related antigenotoxic effects against PCB and URE. There was no significant increase in genotoxicity of trans-anethole and eugenol even when administered at high doses (Abraham, 2001).

Conclusion

This research indicates that fennel may serve as potential dietary sources of natural anti-mutagen for improving human nutrition and health. Purification and identification of the active compounds in these herbs is required for a better understanding of the protective mechanisms involved and for the possible application in the food industry and in medicine.

REFERENCES

Abraham S (2001). Anti-genotoxicity of trans-anethole and eugenol in mice. Food Chem. Toxicol. 39:493-498.

Abraham SK (1994). Antigenotoxicity of coffee in the Drosophila assay for somatic mutation and recombination. Mutagenesis 9: 383-386

Arnaiz R, Soto POJ, Oyarzun C, Graf U (1996). Analysis of mitotic recombination induced by several mono and bifunctional alkylating agents in the Drosophila wing-spot test. Mutat. Res. 351:133-145.

Boskabady MH, Khatami A, Nazari A (2004). Possible mechanism for relaxant effects of *foeniculum vulgare* on guinea pig tracheal chains. Pharmazie. 59:561-564.

Ebeed NM, Abdou HS, Booles HF, Salah SH, Ahmed ES, Fahmy Kh (2010). Antimutagenic and Chemoprevention Potentialities of Sweet Fennel (*Foeniculum vulgare* Mill.) Hot Water Crude Extract. J. Am. Sci. 6:831-842

EFSA (2009). Advice on the EFSA guidance document for the safety assessment of botanicals and botanical preparations intended for use as food supplements, based on real case studies. EFSA Journal. 7:281-280.

El Bardai S, Lyoussi B, Wibo M, Morel N (2001). Pharmacological evidence of hypotensive activity of Marrubium vulgare and Foeniculum vulgare in spontaneously hypertensive rat. Clin. Exp. Hypertens. 23:329-343

Frei H, Würgler FE (1988). Statistical methods to decide whether mutagenicity test data from Drosophila assays indicate a positive, negative, or inconclusive result. Mutat. Res. 203:297-308.

Gaivao I, Comendador MA (1996). The *w/w+* somatic mutation and recombination test (SMART) of Drosophila melanogaster for detecting reactive oxygen species: characterization of 6 strains. Mutat. Res. 360: 145-151

Graf U, Abraham SK, Rincon JG, Würgler F (1998). Antigenotoxicity studies in Drosophila melanogaster. Mutat. Res. 402: 203-209.

Graf U, Spano MA, Rincon JG, Abraham SK (1996). The wing somatic mutation and recombination test (SMART) in *Drosophila melanogaster*. An efficient tool for the detection of genotoxic activity of pure compounds or complex mixtures as well as for studies of antigenotoxicity. Afr. Newslett Occup. Heal. Safe. 6(9):1-6.

Hassan EA, Abdel Aziz KB, El-Awadi ME, Hanaa MR (2011). Anti-mutagenic effects of fennel plant (Foeniculum vulgare Mill) seeds and pure anethole: An in vitro test on mice. Nat. Sci. 9:21-26.

He W, Huang B (2011). A review of chemistry and bioactivities of a medicinal spice: Foeniculum vulgare. J. Med. Plants Res. 5:3595-3600.

Kuroda Y, Jain AK, Tezuka H, Kada T (1992). Antimutagenicity in cultured mammalian cells. Mutat. Res. 267:201-209.

Lehmann M, Danesi CC, Bellagamb BC, Dihla RR, de Andradea HHR, Cunha KS, Spano MA, Reguly ML (2010). Mutagenic evaluation of combined paclitaxel and cisplatin treatment in somatic cells of Drosophila melanogaster. Mutat. Res. 696: 139-143.

Osaba L, Rey MJ, Aguirre A (2002). Evaluation of genotoxicity of captan, maneb and zineb in the wing spot test of Drosophila melanogaster: role of nitrosation. Mutat. Res. 518: 95-106.

Parejo I, Viladomat F, Bastida J, Hirschman GS, Burillo J, Codina C (2004). Bioguided isolation and identification of the nonvolatile antioxidant compounds from fennel (F. vulgare Mill.) waste. J. Agric. Food Chem. 52: 1890-1897.

Patenkovic A, Radak MS, Banjanac T, Andjelkovic M (2009). Antimutagenic effect of sage tea in the wing spot test of Drosophila melanogaster. Food Chem.Toxicol. 47:180-183.

Raffo A, Nicoli S, Leclerc C (2011). Quantification of estragole in fennel herbal teas: Implications on the assessment of dietary exposure to estragole. Food Chem. Toxicol. 49:370-375.

Ribeiro LR, Salvadori DMF (2003). Dietary components may prevent mutation related diseases in humans. Mutat. Res. 544:195-201.

Romero-Jimenez M, Compos-Sanchez J, Analla M, Munoz-Serrano A, Alonso-Moraga A (2005). Genotoxicity and anti-genotoxicity of some traditional medicinal herbs. Mutat. Res: 585:147-155.

Scalbert A, Manach C, Morand C, Remesy C (2005). Dietary of polyphenols and the prevention of diseases. Crit. Rev. Food Sci. 45:287-306.

Schneider LE, do Amaral VS, Dihl RR, Lehmann M, Reguly ML, de Andrade HHR (2009). Assessment of genotoxicity of Lidocaine®, Prilonest® and Septanest® in the drosophila wing-spot test. Food Chem. Toxicol. 47:205-208.

Shahat AA, Ibrahim AY, Hendawy SF, Omer EA, Hammouda FM, Abdel- Rahman FH, Saleh MA (2011). Chemical composition, antimicrobial and antioxidant activities of essential oils from organically cultivated fennel cultivars. Molecules 16:1366-1377.

Smijs TGM, Nivard MJM, Schuitmaker HJ (2004). Development of a Test System for Mutagenicity of Photosensitizers Using Drosophila melanogaster. Photochem. Photobiol.79:332-333.

Van den Berg SJPL, Restani P, Boersma MG, Delmulle L, Rietjens IMCM (2011). Levels of Genotoxic and Carcinogenic Compounds in Plant Food Supplements and Associated Risk Assessment. Food Nutr. Sci. 2:989-1010.

Vogel EW, Barbin A, Nivard MJM, Bartsch H (1990). Nucleophilic selectivity of alkylating agents and their hypermutability in Drosophila as predictors of carcinogenic potency in rodents. Carcinogenesis 11:2211-2217.

Vogel EW, Graf U, Frei HJ, Nivard MJM (1999). The results of assays in Drosophila as indicators of exposure to carcinogens. IARC Sci. Publ. 146:427-470.

Vogel EW, Nivard MJM (1993). Performance of 181 chemicals in a Drosophila assay predominantly monitoring interchro-mosomal mitotic recombination. Mutagenesis 8:57-81.

Zimmering S, Olvera O, Hernandez ME (1990). Evidence for a radioprotective effect of chlorophilin in Drosophila. Mutat. Res. 245:47-49.

Nutritional qualities and trace metals concentration of six fish species from Oba reservoir, Ogbomoso, Nigeria

Fawole, O. O., Yekeen, T. A., Adewoye, S. O., Ogundiran, M. A., Ajayi, O. E. and Nwaiya, M. N.

Department of Pure and Applied Biology, Ladoke Akintola University of Technology, P.M. B. 4000, Ogbomoso, Nigeria.

The nutritional qualities and concentration of trace metals (Pb, Fe, Cu, Zn, and Cd) in six fishes (*Hemichromis fasciatus, Siluranodon auritus, Sarotherodon galileaus, Tilapia zilli, Chrysichthys nigrodigitatus,* and *Hepsetus odoe*) obtained from Oba reservoir in Ogbomoso, Nigeria were evaluated. The value ranges of 55.95 to 68.80, 16.91 to 26.18, 5.46 to 18.18, 0.51 to 9.05, 7.67 to 10.14 and 0.97 to 1.41% were obtained for moisture, protein, fat, carbohydrate, ash and fibre contents, respectively. Highest protein and fat contents were obtained from *H. odoe.* The trace metals in fishes were found at various concentrations (P≤0.05). High levels of essential metals; Fe, Cu and Zn, were found in the fishes. The concentrations of Pb and Cd were also high in the fishes, but not detected in the water sample from the reservoir. A positive correlation was observed between fat and fibre (P≤0.01) as well as Zn and Pb (P≤0.05) while fat and iron were negatively correlated. The results show that the fishes evaluated are rich in essential nutrients. However, prior exposure of the fishes to pollutants was not unlikely as shown by Pb and Cd levels. There is the need to initiate further study to identify sources of these pollutants and their effects on biota from Oba reservoir.

Key words: Fish, nutrient, environment, pollutant, reservoir.

INTRODUCTION

Fish is widely acceptable in global menu due to its palatability, low cholesterol level and tenderness of its flesh (Eyo, 2001). It also receives an increased level of attention because of its potential source of animal protein and some essential nutrients for human diets (Zenebe et al., 1998). It is the cheapest source of animal protein and other nutrient required in human diets. Fish meal contains significantly low lipids and higher water than beef and chickens and it is greatly favoured over all white and red meats (Nestel, 2000). The nature and quality of nutrients in most animals depends largely on the type of food consume. The feeding habit of an individual fish species greatly affects its body nutrient composition.

The measurement of some proximate and nutrient profiles such as protein contents, lipid, ash content, nitrogen free extract and crude fibre including some essential heavy metals, such as copper, zinc and iron to mention a few, are very important to know the nutritional quality of the fish. Proximate composition is a good indicator of physiology which is needed for routine analysis of fisheries (Cui and Wootton, 1988). Evaluation of fishes to determine the concentration of many other heavy metals, that are non-essential and may be body burden to human due to accumulation from fish consumption, is often necessary to ensure that they meet the dietary requirements and commercial specifications (Watermann, 2000).

Report has shown that heavy metals are major pollutants that pose serious health risk and environmental concerns (Onyia et al., 2007). Pollution of the aquatic environment with heavy metals has become a worldwide problem because they are indestructible and most of

Figure 1. Some fresh water fishes of commercial importance obtained from Oba reservoir, Ogbomoso. **A.** *Hemichromis fasciatus*. **B.** *Siluranodon auritus*. **C.** *Sarotherodon galileaus*. **D.** *Tilapia zilli*. **E.** *Chrysichthys nigrodigitatus*. **F.** *Hepsetus odoe*.

them have toxic effects on aquatic organisms (Sen et al., 2011). Among environmental pollutants, metals are of particular interest, due to their potential toxic effects and ability to bioaccumulates in aquatic ecosystems (Gledhill et al., 1997). Transport of metals in fish occurs through the blood where the ions are usually bound to protein (Kaoud and El-Dahshan, 2010). Some animals have ability to accumulate heavy metals to levels higher than the ambient metal concentrations without any detectable adverse effects on their normal physiological functions (Otitoloju et al., 2007). High concentrations of heavy metals in human body often result in severe damage to vital organs and systems. Reported damages include enhanced lipid peroxidation, DNA damage, enzyme inactivity and the oxidation of protein sulfydryl groups (Taiz and Zeiger, 1998). Toxic heavy metal can also cause dermatological diseases, skin cancer and internal organs cancers, cardiovascular disease, diabetes and anaemia, as well as reproductive, developmental, immunological and neurological effects in the human body (Kaoud and El-Dahshan, 2010). The toxicity of these elements is due to their ability to cause oxidative damages to living tissues (Kaou and El-Dahshan, 2010)

Information on the chemical composition of freshwater fishes is invaluable in different aspect of food science. The nutritional components of the freshwater fishes differ greatly between species, sexes, seasons and geographical localities (Zenebe et al., 1998) and also found to be influenced by post-harvest processing (Clement and Lovell, 1994). However, nutritional content of fish is ultimately a function of the food that is available for the fish to feed on and the assimilative capacity of individual fish. Fresh water fishes from Oba reservoir in Ogbomoso (N8° 7' 59. 9442; E4° 14' 59. 9238) are being sold to the populace. Despite the long existence of Oba reservoir, only few studies have reported the aspect of fish biology (Fawole and Adewoye, 2004; Fawole et al., 2003) while information on the nutritional quality of some fishes in the area is yet to be reported. The fact that the reservoir receives water from different tributaries, majorly from Afon and Asa rivers, makes it vulnerable to pollution especially through the water sources as well as run off. In view of this, we assessed six selected fish species (*Hemichromis fasciatus*, *Siluranodon auritus*, *Sarotherodon galileaus*, *Tilapia zilli*, *Chrysichthys nigrodigitatus* and *Hepsetus odoe*) of commercial importance from Oba reservoir to determine their nutritional qualities and possible bioaccumulation of heavy metals.

MATERIALS AND METHODS

Collection of water and fish samples

Water sample from Oba reservoir used for physicochemical analysis was collected into pre-cleaned plastic container (200 ml). Six fishes (n=6) were sampled with each species collected separately into polythene bag from Oba reservoir located in Ikose, Orire Local Government area of Ogbomoso. The fish species (Figure 1 A-F) were identified as *Hemichromis fasciatus* (Peters 1858), *Siluranodon auritus* (Geoffroy St. Hiaire, 1808-09), *Sarotherodon galileaus* (Linnaeu, 1758), *Tilapia zilli* (Gervais, 1848), *Chrysichthys nigrodigitatus* (Laceépède, 1803) and *Hepsetus odoe* (Bloch, 1794).

Preparation and digestion of fish samples

The fish samples were blotted dry and weighed, then each fish wrapped in foil and oven dry at 105°C until constant weight was

Table 1. Physicochemical properties of water from Oba reservoir.

Parameter	Oba reservoir	Aqua culture standard[b]
pH	6.8	6.7-8.5
Total hardness	14.0	-
Dissolved Oxygen	11.05	5-10
BOD	6.20	≤10
Alkalinity	325	50- 3000
Total dissolved solid	67.0	≤500
Total suspended solid	12.0	-
Total solid	79.0	-
Salinity	0.06	-
Sulphate	2.06	-
Nitrates	3.947	-
Fe	0.89	-
Pb	ND	-
Cu	ND	-
Cd	ND	-

Unit: mg/L except pH. [b]Chaudhary et al., (2004).

obtained. The fish were homogenized using blender (MC- 320B). Sample digestion was as described by Kumar et al. (2010), with slight modification. Briefly, 1.0 g of homogenized powdered of fish sample was placed in a Teflon beaker and digested with few drops of sodium chloride solution (30%) and a 10 ml mixture (1:5) of concentrate nitric acid (65%) and concentrated perchloric acid (70%). The free chlorine developed loosen the chemical bonds in organic compounds after gentle heating (at 70±5°C) in a water bath for 12 h and destroys the organic matter in order to transfer the metals into the solution. The digested samples were centrifuged and the supernatant was analyzed for lead, cadmium, copper, zinc and iron using atomic absorption spectrophotometer (Buck 210 AAS, 2005, USA).

Proximate analysis

The proximate components were analyzed using AOAC (1990). Fish samples were analyzed for the following.

Moisture content: The fish moisture content was determined by drying the sample in a hot air oven at 105°C until constant weight was obtained. The percentage difference due to loss of water was then calculated.

Protein content: It was determined by Kjeldahl method which involves digestion of organic nitrogen of 1 g of grand fish sample using concentrated teraoxosulphate (VI) acid which converts it to ammonium teraoxosulphate VI. It was then diluted and made alkaline with sodium hydroxide, followed by distillation. The liberated ammonia was collected in a boric acid solution and was determined via titration. The percentage concentration of protein in the sample was then calculated.

Ash content: It was determined using drying ashing procedure by burning off 1 g of homogenized dried sample in a muffle furnace at 550 to 600°C, until the residue become completely white. The percent of ash was calculated as:

Percentage (%) of ash = (Weight of ash / Weight of Sample) × 100

Fat content: This was estimated using 1.0 g of completely dried fish sample, placed in the Soxhlet extractor and extracted with a nonpolar solvent, ethyl ether. After extraction, the solvent was evaporated and the extracted materials were weighed. The percentage of fat content was calculated as follows:

Percentage (%) of fat = (Weight of extract/Weight of sample) × 100

Fibre content: A 1.0 g of dried sample was digested with 0.128 M H_2SO_4 with two drops of octanol to prevent foaming. The content was boiled for 30 min, and then filtered and washed to remove acid. The residue was boiled with 0.223 M KOH for 30 min, and then washed in boiling water and acetone. The residue was dried and ignited in muffle furnace. The loss of weight represents the crude fibre.

Carbohydrate content: This was determined by the percentage difference of other contents.

Physicochemical parameters

The following physicochemical parameters were determined from water samples collected; pH, total hardness, dissolved oxygen, alkalinity, total dissolved solid, salinity, sulphate, nitrate, total suspended solid. All parameter were measured in mg/L except pH and salinity (psu) using standard method (APHA, 1995). Trace metals: Zn, Fe, Pb, Cu and Cd were also analyzed.

Statistical analysis

Data collected were subjected to statistical analysis using One-way analysis of variance (ANOVA) at 0.05 level of significant. Further analysis was done using Duncan Multiple Range Test. Spearman correlation was used to evaluate the relationships between the proximate parameters and heavy metals analyzed. The P≤0.05 and P≤0.01 were considered to indicate statistical significance. SPSS version 15 software was employed.

RESULTS

The result of the physical and chemical parameters of water sample from Oba reservoir is shown in Table 1. The non-essential heavy metals; cadmium and lead were not detected. The proximate composition of the fishes analyzed is shown in Table 2. The moisture content values obtained for the fishes ranged from 55.95 to 68.80%. The least moisture content value was obtained in *S. galilaeus* while highest in *C. nigrodigitatus*. The protein values obtained for the fishes showed that *H. odoe* and *H. fasciatus* had similar values which were significantly higher than in *S. auriatus, T. zilli, S. auriatus and C. nigrodigitatus* but similar to *S. galilaeus*. Protein value range was between 16.9 and 26.18%. The values of fat content of all the six species were in the order of *H. odoe* > *H. fasciatus* > *S. auritus* > *S. galilaeus* > *C. nigrodigitatus* > *T. zilli*, with significant difference among the values except those obtained for *C. nigrodigitatus, S. galilaeus* and *T. zilli*. The carbohydrate content of the fishes range from 0.51 to 9.05% with the highest value obtained for *S. auriatus* significantly different from other values, while the values obtained for *H. fasciatus, T. zilli* and *C.*

Table 2. Proximate analysis of selected fish species from Oba reservoir.

Fish sample	Moisture	Protein	Fat	Carbohydrate	Ash	Fibre
Hemichromis fasciatus	59.79 ±2.61^{b}	25.42 ±0.80^{a}	9.78 ±0.41^{b}	0.51 ±0.08^{d}	10.14±0.47^{a}	1.41 ±0.01^{a}
Siluranodon auriatus	64.12 ±1.52^{a}	16.91c±120^{c}	7.81 ±0.06^{c}	9.05 ±2.14^{a}	9.89 ±0.69^{a}	1.19±0.03^{a}
Sarotherodon galilaeus	55.95 ±1.20^{c}	23.05 ±1.26^{ab}	6.72 ±0.31^{cd}	3.79 ±1.56^{c}	9.43±0.25^{a}	1.06 ±0.08^{ab}
Tilapia zilli	61.95 ±1.22^{ab}	21.31 ±1.55^{b}	5.46 ±0.12^{d}	1.13 ±0.03^{d}	9.18 ±0.41^{a}	0.97 ±0.12^{b}
C. nigrodigitatus	64.80 ±0.41^{a}	18.32 ±0.65^{c}	5.72 ±0.18^{a}	0.53 ±0.12^{d}	9.15 ±0.04^{a}	1.03 ±0.02 ^{ab}
Hepsetus odoe	59.53 ±1.11^{b}	26.18 ±0.18^{a}	18.18 ±0.18^{a}	7.52 ±0.36^{b}	7.67 ±0.77^{b}	1.28 ±0.02 ^{a}

Measurements were done in percentage. Different alphabets along column indicate significant difference using ANOVA and Duncan's multiple range tests (P≤0.05).

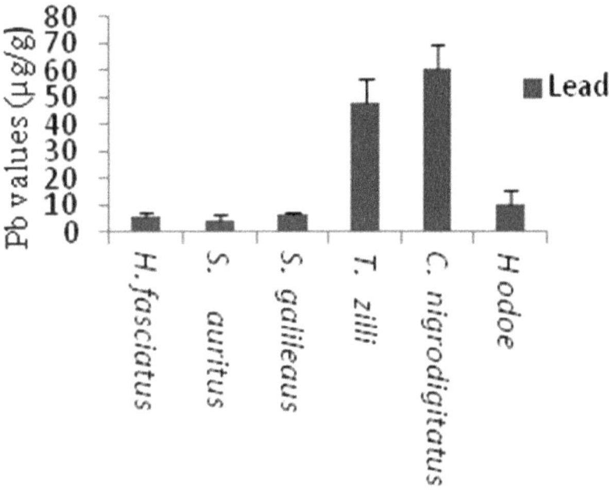

Figure 2. Lead concentrations obtained from selected fishes from Oba reservoir.

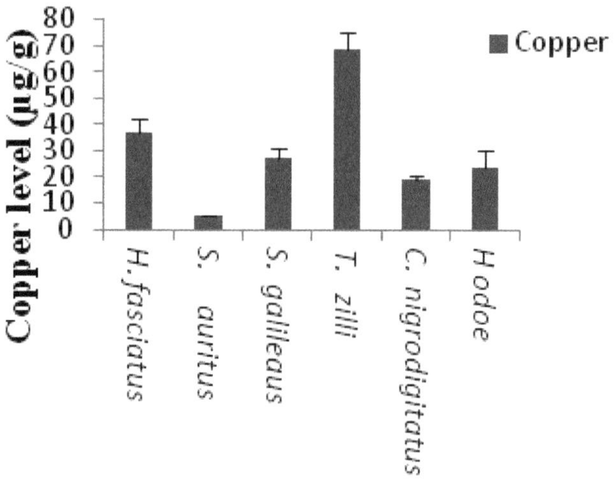

Figure 4. Zinc concentrations obtained from selected fishes from Oba reservoir.

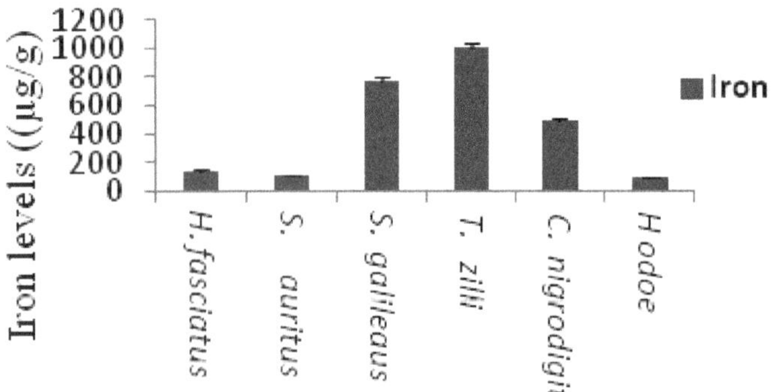

Figure 3. Iron concentrations obtained from selected fishes from Oba reservoir.

nigrodigitatus were similar (Table 1). Similarity was observed in the values obtained for the ash content of the fishes without significant difference except only in *H. odoe* with lowest value compared to other fishes. The highest ash content was seen in *H. fasciatus*. The fibre content was within the range of 0.97 to 1.41% with the lowest value obtained for *T. zilli* but similar to *S. galilaeus* and *C. nigrodigitatus*.

The concentrations of the selected metals obtained in different fish species are shown in Figure 2 to 6. Lead values (Figure 2) ranged from 4.42 to 60.35 µg/g with the highest obtained in *C. nigrodigitatus*. There was significant difference in the Pb values among the fishes with similarity in the values of *H. fasciatus, S. auriatus* and *S. galilaeus*. The values of Fe (Figure 3) and Zn (Figure 4) were significantly higher in *T. zilli, C. nigrodigitatus* and *S. galilaeus* compared to other three *fishes* (P≤0.05). The highest value of Cu (Figure 5) was obtained in *T. zilli* while the least value was obtained in *S. auriatus* and the range was 5.04 to 68.29 µg/g. Compared to all other metals evaluated, the values obtained for Cd (Figure 6) were the least in all fish species. Mean values of Cd showed variation among the fish species with values ranging from 0.07 to 2.85 µg/g.

Table 2 shows the Spearman correlation between the proximate parameters and the heavy metals evaluated. Significant positive correlation was found between fat and

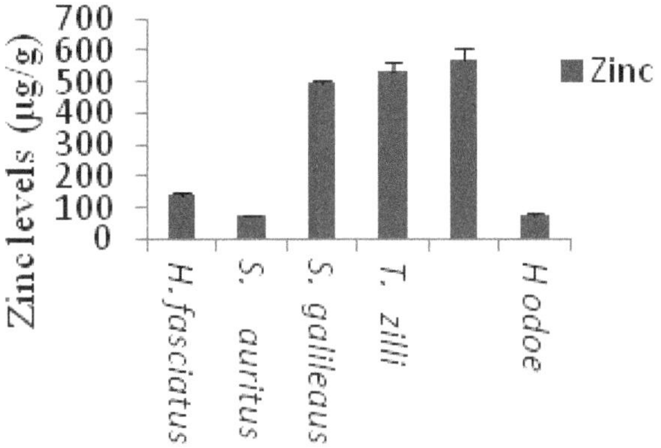

Figure 5. Copper concentrations obtained from selected fishes from Oba reservoir.

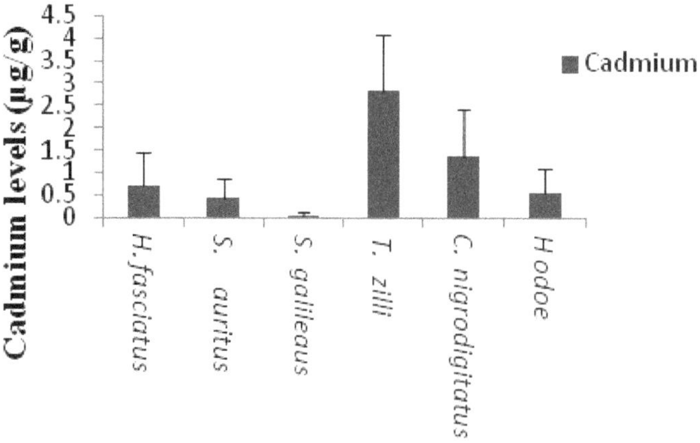

Figure 6. Cadmium concentrations obtained from selected fishes from Oba reservoir.

fibre (P≤0.01), Zn and Pb (P≤0.05) while fat and Iron were negatively correlated (P≤ 0.05)

DISCUSSION

Fish is widely acceptable for consumption because of its high palatability, low cholesterol and tender flesh (Eyo, 2001) and also a cheap source of protein. The demand for fresh water fishes can still be improved on, if information dissemination on the valuable nutritional composition of fishes reaches the grass route. Besides this, there is the need to raise consumer awareness to ensure that contaminated fishes are avoided. This will relieve the body of the burden of these toxic chemicals. Although physicochemical properties of water from Oba reservoir favours production of fishes which conforms with standard for aquaculture (Chaudhary et al., 2004) and with heavy metals like Pb and Cd not detected. However the high concentration of non essential heavy metals in some of the fishes evaluated (*T. zilli, C. nigrodigitatus and H. odoe*) is a source of concern. This trend could probably be due to the prior exposure of these fishes to toxicants from polluted water body before swimming down to the reservoir since they are pelagic fishes. Mersch et al. (1993) linked the accumulation of the toxicants in the fish to the possibilities of the fish to accumulate metals and thus could serve as indicators of pollution.

Report has shown that Fe is one of the most abundant metals in the earth crust (Ibrahim and Tayel, 2005). The high value of Fe observed in this study compared to other metals may be due to increase in total dissolved iron in the Oba reservoir. The values of Fe were in the order: *T. zilli* > *S. galilaeus* > *C. nigrodigitatus*. High concentration of Fe was reported in fish organs (El-Naggar et al., 2009) and was reported as the most abundant in *C. nigrodigitatus* compared to other fishes evaluated (Asuquo et al., 2004). However, the values obtained for Fe in this study were lower compared to freshwater fishes in China (Du et al., 2012). Copper, in its many forms is regarded as the third most common minerals in human body. It is a fundamental micronutrient to all forms of life in enzyme activity or random rearrangement of natural protein (Bower, 1979). High values were observed in all fish evaluated with the least value in *S. auriatus*. The high value obtained could be due to elevated metal– binding protein synthesis. Similar observation of high value of copper in fishes attributed to elevated metal binding protein synthesis has also been reported (Yacoub, 2007).

Zinc is known to be involved in most metabolic pathways in humans and zinc deficiency can lead to loss of appetite, growth retardation, skin changes and immunological abnormalities (Malakootian et al., 2011). Fish takes up Zn directly from water, especially via mucous and gills. The maximum zinc level permitted for fish is 50 mg/kg according to Food Codex. The values of Zn obtained were significantly high in *T. zilli*, *S. galilaeus and C. nigrodigitatus*, compared to the other fishes but lower compared to Zn values reported for freshwater fishes in China (Du et al., 2012). United States environmental protection agency and the European Commission (US-EPA and EC) have not considered any standards or limits for the zinc concentrations in fish(WHO, 2008).

Lead is non-essential element that constitutes body burden and a great threat to life if present in substantial quantity. It is toxic even at low concentrations and has no known function in biochemical processes (Burden et al., 1998). The values of Pb were very high in the fishes. The standard level Pb was reported to be 0.5 mg/kg dry weight (FAO, 1983). Similar to Pb, Cadmium was not detectable in the water but fairly high value observed in the fishes and highest in *T. zilli*. The standard permissible level of 0.05 to 5.5 mg/kg fish dry weight was reported for Cd (FAO, 1983). None detection of Pb and Cd in the water body and high values of these metals in the fishes call for further research on the reservoir in order to determine the source(s) of these pollutants.

Copper is an essential element in the body and is

Table 3. Relationships between proximate analysis and heavy metals evaluated from six selected fishes in Oba reservoir, Ogbomoso.

Item	Moisture	Protein	Fat	Carbohydrate	Ash	Fibre	Lead	Iron	Copper	Zinc	Cadmium	Weight
Moisture	1											
Protein	-0.771	1										
Fat	-0.486	0.600	1									
Carbohydrate	0.257	-0.200	0.257	1								
Ash	-0.429	-0.200	0.086	-0.200	1							
Fibre	-0.600	0.543	0.943**	0.029	0.371	1						
Lead	0.429	0.029	-0.543	-0.371	-0.771	-0.657	1					
Iron	0.029	-0.257	-0.886*	-0.429	0.086	-0.771	0.429	1				
Copper	-0.600	0.486	-0.257	-0.543	0.143	0.143	-0.200	0.600	1			
Zinc	0.257	-0.143	-0.771	-0.657	-0.314	-0.714	0.829*	0.771	0.371	1		
Cadmium	0.377	-0.116	-0.464	-0.638	0.290	-0.406	0.696	0.319	0.319	0.638	1	
Weight	0.029	0.257	-0.086	0.340	-0.657	-0.314	0.429	0.143	-0.086	0.257	-0.348	1

Spearman's correlation: **Correlation is significant at the 0.01 level (2-tailed).*Correlation is significant at the 0.05 level (2-tailed).

required for different functions. Standard level of 20 mg/kg dw (dry weight) has been reported (TFC, 2002). However, after new regulation in 2008, no maximum level was specified for Cu in the fish species and sea foods (TFC, 2008). Also, there is no guideline on acceptable levels of Cu in fish suggested by European Economic Community or FAO/WHO (Mol et al., 2010). The level of Cu in the evaluated fishes is considerably high.

Body composition is a good indicator for the physiological condition of fish. The percentage of water is a good indicator of its relative content of energy, proteins and lipids. The lower the percentage of water, the greater the lipid and protein contents, and the higher the energy density of the fish (Dempson et al., 2004; Ali et al., 2005).This pattern was observed in this study although not consistent for all the fishes evaluated. Protein and fat are the major nutrient in fish and their level help to define the nutritional status of a particular organism (Aberoumad and Pourshafi, 2010). The chemical composition of fish varies from one individual to another depending on age, sex, environment and season with protein level ranging from 16 to 21%, lipid from 0.1 to 25 %, ash 0.4 to 1.5%, moisture content of 60 to 80 % with extreme of 96% having been reported (Muraleedharan et al., 1996). Also, moisture content range of 65.88 to 78.62 (Mazumder et al., 2008), 77.93 to 82.7% (Ravichandran et al., 2011) and 68.6 to 77.1% (Aberoumand and Pourshafi, 2010) were reported for fishes. The values we obtained for the proximate composition where within the the reported ranges with higher percentage in the case of fibre. The result obtained for percentage of crude protein in this study shows that the selected fishes are good source of protein.

In view of the importance of fish in human diet, it is necessary that biological monitoring of the fish meant for consumption should be carried out regularly to ensure safety. WHO (1995) reported that heavy metals must be controlled in food sources in order to assure public safety. Excessive concentration of food heavy metals is associated with the etiology of a number of diseases, especially cardiovascular, renal, neurological and bone diseases (Chailapakul et al., 2008). A major reason to monitor levels of toxic metals in foods follows from the fact that contamination of the general environment has increased (Heidarieh et al., 2013). Information on the proximate and heavy metal composition of fresh water fishes is of particular importance for the determination of their nutritive values as well as for investigating the level of pollution or contamination. Although, high levels of some of the selected metals were observed in the fishes, the water analysis indicates that Oba reservoir might not be polluted. Further study is however required for verification.

REFERENCES

Aberoumad A, Pourshafi K (2010). Chemical and proximate composition properties of different fish species obtained from Iran. World J. Fish Mar. Sci. 2 (3): 237-239.

Ali M, Iqbal F, Salam A, Iram S, Athar M (2005). Comparative study of body composition of different fish species from brackish water pond. Inter. J. Environ. Sci. Technol. 2(3): 229-232

AOAC (1990). *Official Methods of Analysis of the Association of Official Analytical Chemists* (W. Horwitz, Ed.), 15th edn., Methods: 981.10, 925.45, and 945.16. Association of Officiall Analytical Chemists, Washington, DC.

APHA (American Public Health Association) (1995). Standard methods for the examination of water and wastewater 19th ed. OPH Washington D.C.

Asuquo FE, Ewa-Oboho I, Asuquo EF, Udo PJ (2004). Fish Species Used as Biomarker for Heavy Metal and Hydrocarbon Contamination for Cross River, Nigeria. Environmentalist. 24: 29–37

Bower JJM (1979). Environmental chemistry of the element. Academic Press Landon.

Burden VM, Sandheinrich MB, Caldwell CA (1998). Effects of lead on the growth and alpha aminolevulinic acid dehydrates activity of juvenile rainbow trout, *Oncorhynchus mykiss*. Environ. Poll. 101: 285-289.

Chailapakul O, Korsrisakul S, Siangproh W, Grudpan K (2008). Fast and simultaneous detection of heavy metals using a simple and reliable microchip-electrochemistry route: An alternative approach to food analysis. Talanta 74:83–689.

Chaudhary S, Nuradha A, Sastry KV (2004). Pollution of River Yamuna at Faridabad (Haryana). Pollu. Res. 23(4): 749-756.

Clement S, Lovell RT (1994). Comparison of processing yield and nutrient composition of Nile tilapia and catfish. Aquaculture 119: 299-310.

Cui Y, Wootton RJ (1988). Bioenergetics of growth of a cyprinid, *Phoxinus phoxinus* (L.): the effect of ration and temperature on growth rate and efficiency. J. Fish Biol. 33 (5): 763–773.

Dempson IB, Schwarz CJ, Shears M, Furey G (2004). Comparative proximate body composition of atlantic salmon with emphasis on parr from fluvial and lacustrine habitats. J. Fish Biol. 64: 1257-1271

Du Z, Zhang J, Wang C, Li L, Man Q, Lundebye A, Frøyland L (2012). Risk–benefit evaluation of fish from Chinese markets: Nutrients and contaminants in 24 fish species from five big cities and related assessment for human health. Sci. Total Environ. 416: 187–199

El-Naggar AM, Mahmoud SA, Tayel SI (2009). Bioaccumulation of Some Heavy Metals and Histopathological Alterations in Liver of Oreochromis niloticus in Relation to Water Quality at Different Localities along the River Nile, Egypt. World J. Fish Mar. Sci. 1 (2): 105-114

Eyo AA (2001). Fish Processing Technology in the tropics. University of Ilorin, Nigeria. pp. 403.

FAO (1983). Compilation of legal limits for hazardous substances in fish and fishery products, Food and Agricultural Organization, Fishery circular No. 464, pp. 5–100.

Fawole OO, Adewoye SO (2004). Aspects of the Biology of *Clarias gariepinus* BURCHELL, 1822) in Oba reservoir, Ogbomoso, Nigeria. Sci. Focus 8: 105 – 109.

Fawole OO, Adewoye SO, Yekeen TA, Kehinde AT (2003). Studies on the meristic and morphometric characteristics of *Clarias gariepinus* (Burchell:1822) from two freshwater environments. Sci. Focus 3: 81-83.

Gledhill M, Nimmo M, Hill SJ, Brown MT (1997). The toxicity of copper (II) species to marine algae, with particular reference to macroalgae. J. Phycol. 33: 2-11.

Heidarieh M, Maragheh MG, Shamami MA, Behgar M, Ziaei F and Akbari Z (2013). Evaluate of heavy metal concentration in shrimp (Penaeus semisulcatus) and crab (Portunus pelagicus) with INAA method. SpringerPlus 2:72. doi:10.1186/2193-1801-2-72

Ibrahim SA, Tayel SI (2005). Effect of heavy metals on gills of *Tilapia zillii* inhabiting the River Nile water (Damietta branch) and El-Rahawy drain. Egypt. J. Aquat. Biol. Fish. 9: 111-128.

Kaoud HA, El-Dahshan AR (2010). Bioaccumulation and histopathological alterations of the heavy metals in *Oreochromis niloticus* fish. Nat. Sci. 8(4): 147-156

Kumar B, Senthilkumar K, Priya M, Mukhopadhyaya DP, Saha R (2010). Distribution, partitioning, bioaccumulation of trace elements in water, sediment and fish from sewage fed fish ponds in eastern Kolkata, India. Toxicol. Environ. Chem. 92(2): 243-260.

Malakootian M, Tahergorabi M, Daneshpajooh M, Amirtaheri K (2011). Determination of Pb, Cd, Ni, and Zn Concentrations in Canned Fish in Southern Iran. Sacha. J. Environ. Stud. 1(1): 94-100

Mazumder MSA, Rahman MM, Ahmed ATA, Begum M, Hossain MA. (2008). Proximate composition of some small indigenous fish species (sis) in Bangladesh. Int. J. Sustain. Crop Prod. 3(4):18-23

Mersch J, Dubost N, Pihan J (1993). Comparison of several inert and biological substrates to assess the trace metals concentration in the reservoir of the nuclear power plant in catlenom. France. 29:325-33

Mol S, Ozden O, Oymak SA (2010). Trace metal contents in fish species from Atatürk Dam Lake (Euphrates, Turkey). Turk. J. Fish. Aquat. Sci. 10: 209-213.

Muraleedharan V, Antony KP, Perigreen PA, Gopakumar K (1996). Utilization of unconventional fish resources for surimi preparation. Proceedings of the second workshop on scientific results of FORV SAGAR. Sampada. Dept. of Ocean development. New Delhi, (India). pp. 539-543.

Nestel PJN (2000). Fish oil and cardiovascular disease: lipids and arterial function. Am. J. Clin. Nutr. 71: 228-231.

Onyia LU, Milam C, Ese DS (2007). Investigation of heavy metals in four commercial fishes along upper River Benue, Yola, Adamawa State. J. Phys. Sci. 2(2):32-33.

Otitoloju AA, Don-Pedro NK, Oyewo OE (2007). Assessment of potential ecological disruption based on heavy metal toxicity, accumulation and distribution in media of the Lagos Lagoon. Afr. J. Ecol. 45: 454–463.

Ravichandran S, Kumaravel K, Pamela FE (2011). Nutritive composition of some edible fin fishes. Int. J. Zool. Res. 7(3): 241-251

Sen I, Shandil A, Shrivastava VS (2011). Study for Determination of Heavy Metals in Fish Species of the River Yamuna (Delhi) by Inductively Coupled Plasma-Optical Emission Spectroscopy (ICP-OES). Adv. Appl. Sci. Res. 2 (2): 161-166

Taiz L, Zeiger E (1998). Plant defenses: surface protectants and secondary metabolites.In:Taiz L, Zeiger E. (eds.). Plant physiology. Sinauer Associates, Massachusetts. pp. 347-377

Turkish Food Codex (TFC) (2002). The communique on maximum limits of contaminants in foodstuffs. (Communique No:2002/63).

Turkish Food Codex (TFC) (2008). The communique on maximum limits of contaminants in foodstuffs. (Communique No: 2008/26).

Watermann JJ (2000). Composition and Quality, fish, Edinburgh, Torry Research Station.

WHO (2008). Guidelines for Drinking-water Quality Third Edition Incorporating the First and Second Addenda. Volume 1. Recommendations. Geneva

WHO (1995). Lead Environmental Health Criteria. WHO, Geneva

Yacoub AM (2007). Study on some heavy metals accumulated in some organs of three River Nile fishes from Cairo and Kalubia governorates. Afr. J. Biol. Sci. 3: 9-21.

Zenebe T, Ahigren G, Gustafsson B, Boberg M (1998). Fatty acid and lipid content of *Oreochromis niloticus* L. in Ethiopian lakes. Dietary effects of phytoplankton. Ecol. Freshwater Fish. 7: 146-158.

Proximate composition and microbiological study of five marine fish species consumed in Gabon

Alain Serges ONDO-AZI[1] , Brice Serge KUMULUNGUI[1], Ludovic MEWONO[2] and Armel MBINA KOUMBA[1] and Crépin ELLA MISSANG[1,3]

[1]Unité de Recherche Agrobiologie, Université des Sciences et Techniques de Masuku, B.P. 941 Masuku Franceville Gabon.
[2]Laboratoire d'Immunologie, de Microbiologie appliquée et d'Hygiène, Ecole Normale Supérieure, B.P. 440 Libreville Gabon.
[3]Institut de Recherche Technologique, CENAREST, B.P. 14.070, Libreville, Gabon.

The aim of the current study was to determine proximate composition (protein, lipid, ash and moisture), energy and microbiological spoilage in five fish species. The chemical analysis revealed that the protein content of *Pseudotolithus typus* was 19.19%, *Pomadasys peroteti* 16.05%, *Galeoides decadactylus* 17.15%, *Carassius auratus* 20.78% and *Pimelodus grosskopfii* 18.57% of the fresh weight, respectively. The lipid content, the ash, and carbohydrate ranged from 0.46 to 1.76, 1.01 to 1.40 and 0.20 to 0.49% of the fresh weight correspondingly. Microbiological analysis showed that the Total Viable Count (TVC) varied between 5.20 and 8.42 LogUFC/g. The results obtained on TVC were not within the acceptable limits set by the International Commission on Microbiological Specification for Foods. *Staphylococcus* sp. bacteria were observed in all five species and fecal coliforms in *Carassius auratus* and *Pimelodus grosskopfii* only. The bacteria isolated on each of the five fish species included *Staphylococcus aureus*, *Staphylococcus xylosus*, *Enterobacter agglomerans* and one unidentified species. However, fungi were not found in these five samples.

Key words: Franceville, Gabon, proximate composition, microbiological quality, fish species.

INTRODUCTION

Fish is known to be highly nutritious and excellent source of animal proteins which is consumed by a larger percentage of populace of Gabon because of its availability and palatability (Foran et al., 2005; Sutharshiny and Sivashanthini, 2011a). Fish is also, a vitamin and mineral rich food for young and old consumers (Edem, 2009; Koffi-Nevry et al., 2011). The nutritional characteristics of fish and fishery products are of vital interest to consumers. However, the consumption of fish has been linked to health benefits such as reduced risk of coronary heart disease and maintenance of a healthy body (Arannilewa et al., 2005). Polyunsaturatted fatty acids from fish have been reported to have preventive and/or curative effects for several diseases including arterial hypertension (Turkmen et al., 2005), cancers and inflammatory diseases (Marchioli, 2001).

Fish meat contains significantly low lipids and high water compared to that of beef or chicken and is favoured over other white or red meats (Nestel, 2000). Fish products contain most important nutritional components

Table 1. List of fish samples collected in Franceville for our study.

Local name	Common name	Scientific name	Family
Capitaine	Croaker	*Pseudotolithus typus*	Moronidae, Polynemidae
Dorade grise	Black sea bream	*Pomadasys peroteti*	Sparidae, Haemulidae
Bar	Threadfins (Sea bass)	*Galeoides decadactylus*	Lethrinidae, Sciaenidae
Poisson rouge	Goldfish	*Carassius auratus*	Cyprinidae
Mâchoiron		*Pimelodus grosskopfii*	Bragidae, Pimelodidae

and serves as source of energy for human beings (Koffi-Nevry, 2009; Surtharshiny and Sivashanthini, 2011a). However, these values vary considerably within and between species, size, sexual condition, feeding season and physical activity (Chandrashekar and Deosthale, 1993). Moreover, measurements of the some proximate profiles such as protein, lipids and moisture contents are often necessary to ensure that they meet the requirements of food regulations and commercial specifications (Waterman, 2000; Tawfik, 2009; Surtharshiny and Sivashanthini, 2011a). However, all these positive interests can be decrease by microbial activities.

In fact, tropical areas are not suitable for conservation methods and fish consumption is a problem since it is a highly perishable product. In urban areas, freezing is the most used method but the cold supply chain often not guaranteed. This study was carried out to determine proximate composition (including protein, lipid, ash, carbohydrate and moisture), the caloric value and microbial impact on five fish species collected from Franceville market.

MATERIALS AND METHODS

Source of samples

All fishes were purchased from the local market in the Franceville city, South-East of Gabon. Five fish species were included in this study (Table 1) during 2010. These species are among the most commonly consummated in Gabon. Five kilograms (12 to 18 entire fishes) of each fish species collected were immediately dipped in ice, kept and transported in sterile polystyrene boxes to sustain freshness. Then, samples convoyed to the laboratory for analysis.

Proximate analysis

The protein content was determined by estimation of the total Nitrogen by Kjeldahl method and protein content was calculated by multiplying total nitrogen by 6.25 factor (AOAC, 2000) while the moisture content of the pulp was determined after drying at 103°C for 24 h. The percentage of ash was determined by ignition in a muffle furnace at 550°C as described by Pearson (1976), while lipid content in pulp was extracted using hexane in a soxhlet extractor as described by AOAC (2000). The method of Dubois et al. (1956) was used to determine the carbohydrate content of the dry sample. The energetic density was calculated using the specific coefficients of Atwater and Rosa (1899). All the experients were made in triplicate, the results were expressed on fresh weight and mean values were given.

Microbiological analysis

The muscles of the fishes were cut aseptically using sterile forceps and scalpels and five grams (5g) aseptically weighted into conical flask containing 45 ml of sterile deionised water. The contents were then transferred into a sterile mortar and homogenate into a watery paste all within 40 min. Serial dilutions of the samples were used for microbial enumeration with the following media: Standard Plate count Agar for Total Viable Counts (TVC), Sabouraud Chloramphenicol (for fungi), Chapman (for staphylococci) and Mac Conkey (for total and fecal coliforms) medium for bacteria specifications. Proportions (0.1 ml) of appropriate dilution were spread plated. The standard plate count method described by Coulin et al. (2006) was used to determine the total viable cell count. On each Petri dish, colonies were inventoried before culturing. Pure cultures of the isolates were Gram-stained, while the cell morphology was examined by phase contrast microscopy. Fungi were isolated on the sabouraud chloramphenicol (SC) and potatoes dextrin agar (PDA) media. After incubation at 37°C, the Petri dishes were examined in the laboratory.

Statistical analyses

The Statistical Package for Social Science (SPSS version 16.0) was used. Data analysis involved one-way Analysis of Variance (ANOVA). The mean differences were determined using Duncan's Multiple Range Test. A significant difference was established at $p < 0.05$.

RESULTS

Proximate composition of the samples

Proximate composition included moisture, lipid, protein, ash and carbohydrate contents, and caloric value of the samples is presented in Table 2. The moisture content ranged from 74.46 to 78.50% for *C. auratus* and *G. decadactylus*, respectively. Crude ash ranged from 1.01% in *P. peroteti* to 1.40% in *C. auratus*. Carbohydrates composition estimated in the samples analysed varied between 0.20% (*P. typus*) and 0.49% (*P. grosskopfii*). The crude protein content ranged from

Table 2. Chemical and nutritional parameters of fishes collected from Gabon (mean percentage ± SD).

Species	Parameter**					
	Moisture	Lipid	Crude protein	Ash	Carbohydrate	Energy
Pseudotolithus typus	77.39±0.16^{b}	0.47±0.00^{a}	19.19±1.19^{d}	1.15±0.15^{ab}	0.20±0.00^{a}	81.79±0.64^{b}
Pomadasys peroteti	78.50±0.14^{c}	0.67±0.03^{b}	16.05±0.20^{a}	1.01±0.14^{a}	0.44±0.02^{b}	71.99±1.15^{a}
Galeoides decadactylus	78.46±0.20^{c}	1.50±0.17^{c}	17.15±0.19^{b}	1.10±0.07^{ab}	0.29±0.04^{a}	83.26±2.45^{b}
Carassius auratus	74.36±0.23^{a}	1.76±0.05^{d}	20.78±0.37^{e}	1.40±0.07^{c}	0.41±0.10^{b}	100.6±1.35^{c}
Pimelodus grosskopfii	77.29±0.33^{b}	0.46±0.03^{a}	18.57±0.20^{c}	1.23±0.08^{bc}	0.49±0.07^{b}	80.38±1.35^{b}

*Means within a row followed by different letters are significantly different ($p<0.05$); **expressed in kcal and other parameters in g.

Table 3. Bacterial burden of fish samples collected from Gabon.

Species	TVC (Log CFU)	Coliforms** (Log CFU)	Staphylococcus (Log CFU)
Pseudotolithus typus	8.42^{b}	-	5.33^{b}
Pomadasys peroteti	5.64^{a}	-	5.43^{b}
Galeoides decadactylus	7.54^{b}	-	4.90^{ab}
Carassius auratus	5.20^{a}	3.08^{a}	4.20^{a}
Pimelodus grosskopfii	5.23^{a}	4.15^{b}	4.51^{a}

*Means within a row followed by different letters are significantly different ($p<0.05$); **: - no growth.

Table 4. Microorganisms isolated from some fish species collected from Gabon.

Organism	Characteristics	Fish species				
		P. typus	*P. peroteti*	*G. decadactylus*	*C. auratus*	*P. grosskopfii*
S. aureus	Raised, white, cocci, G+, regular	×	×	×	×	×
S. xylosus	Raised, yellow, cocci, G+, regular, opaque	×	×	×	×	×
E. agglomerans	Raised, white, stick, G-, irregular	-	-	-	×	×
Not indentified species	Raised, white, stick, G-, regular	-	-	-	×	×

Note: ×, Presence; -, absence; G+, Gram+; G-, Gram-.

16.05% (*P. peroteti*) to 20.78% (*C. auratus*). Caloric value varied from 71.99 (*P. peroteti*) to 100.6 kcal/100 g (*C. auratus*). The lipid content was found to be between 0.46 to 1.76% in this study.

Microbiological analysis

The microbiological status of the samples is summarized in Table 3. This table shows that total viable count (TVC) of the five fish species varied from 5.20 to 8.42 Log CFU/g. In three species (*P. typus*, *P. peroteti* and *G. decadactylus*), coliforms were absent but found in the two others species as 3.08 Log CFU/g for *C. auratus* and 4.15 Log CFU/g for *P. grosskopfii*. On fungal specific media, none microbial growth was observed for the five samples. Table 4 shows the microbial isolates obtained from several fish species samples analysed. *Staphylococcus aureus* and *Staphylococcus xylosus* were found on all fish samples; *Enterobacter agglomerans* and an unidentified species on both of *C. auratus* and *P. Grosskopfii*.

DISCUSSION

The growing urban population in South Sahara Africa countries calls for increasing attention on food safety and quality. In Gabon, fish is among the most preferred food product. In the present study, we have investigated the five most consumed fish species in Franceville, Gabon. The proximate composition, caloric value and microbial safety were examined. The values obtained on proximate composition (moisture, protein, lipid, ash and carbohydrate contents) of the five fish samples analysed in this study were in-line with those reported elsewhere (Tawfi, 2009; Pirestani, 2010; Nurnadia et al., 2011). For moisture content, the highest contents were found on *P. peroteti* (78.50%) and *G. decadactylus* (78.46%) and the lowest value was obtained with *C. auratus* (74.48%). Based on the lipid content, *P. typus*, *P. peroteti* and *P. grosskopfii* were distinguished as lean fish, where the lipid content was lower than 1% for fresh weight (Tawfik, 2009; Sutharshiny and Sivashanthini, 2011b), while *G. decadactylus* and *C. auratus* were classified as semi-lean (half-lean) fish. In addition, we found that the species (*P. peroteti*, *P. typus* and *P. grosskopfii*) which had low fat content had the highest water content as reported by Osman et al. (2001) As a consequence, their flesh is withering in colour. On the five fish species, crude proteins had high values. This result is in-line with the reports of other authors (Stefens, 2006; Elagba et al., 2010), that protein forms the largest quantity of dry matter in fish sample. Likewise, these high values obtained may explain the important living resources of dietary protein as other sea fish (Zuraini et al., 2006; Elagba et al., 2010).

Crude ash content varied little between species and was low in comparison with the finding of Pirestani (2010) and Nurnadia et al. (2011), who found values of up to 3.3%. Caloric value calculated for the five fish species was dominated by lipid and protein. The lipid content showed a large variation between lean and semi-lean fishes. The protein content showed low variation between species. The nutritional proprieties of the five fish samples are acceptable, showed significant differences (P > 0.05) between fish species. Regarding proximate composition, values presented in this study are near to those found by others (Tawfik, 2009; Nurnadia et al., 2011; Barua et al., 2012) and specially, for specie *P. typus* (Osibona, 2011). Microbiological count of the *C. auratus*, *P. grossopfii* and *P. peroteti* species was lower than for *G. decadactylus* and *P. typus*. Results recorded on the three firsts species indicated values nearer to the findings of Ouattara (1986) who observed 6.11 Log CFU/g. The higher values observed on *G. decadactylus* and *P. typus* was similar to results reported by Kovacevic (1970). All the values obtained in microbiological analysis of the fish samples were higher than the acceptable International Commission on Microbiological Specification for Foods (ICMSF) and French Norms (FN) limits. In fact, ICMSF and FN have a limit of 4.70 Log CFU/g for frozen fishes. The analysis on specific medium showed that the predominant microorganisms were staphylococci and coliforms; fungi were not detected. The coliforms were found on *C. auratus* and *P. peroteti*; Seydi et al. (1992) observed 2.8 Log CFU/g on fish sample. Staphylococci were present on all the fish samples with values ranging from 4.20 to 5.43 Log CFU/g. These values were higher than findings from other authors (Bernadac et al., 1985; Ouattara, 1986; Ndiaye, 1998) and opposite to Bernadac et al. (1985) who reported that fish samples were free from staphylococci.

The inventory of microorganisms found on the five samples reveals the presence of four bacterial species: *S. aureus*, *S. xylosus*, *E. agglomerans* and one unidentified species. So, some strains of *S. aureus* are capable of producing a highly heat-stable protein toxin that causes illness in humans and *S. xylosus* is implicated in fish degradation (Jeantet et al., 2006).

E. agglomerans belongs in the family Enterobacteraceae. They have been associated with nosocomial outbreaks, and are considered opportunistic pathogens. This study shows that the preferential fish species consumed in Franceville presented a good nutritional composition. The high proteins content confirms that these species are good animal proteins sources; they are particularly consigned for the children growing stages. The caloric value confirms their energetic plus value. However, microbial results show large count; these microbial qualities are not acceptable. The bacterial microorganisms isolated and identified can disrupt human health.

REFERENCES

AOAC (2000). Official Methods of Analysis 17th ed. W. Hortuntwed (Ed), Association of Official Analytical Chemists. Washington.

Arannilewa ST, Salawu SO, Sorungbe AA, Aubrey GD (2005). Effect of frozen period on the chemical, microbiological and sensory quality of frozen tilapia fish (Sarotherodun galiaenus). Afr. J. Biotechnol. 4:852-855.

Atwater W, Rosa EA (1899). A new respiratory calorimeter and experiments on the conservation of energy in human body. II. Phys. Rev. 9:214-251.

Bernadac M, Scheid A, Hugon B (1985). Aptitude à la conservation et contrôle microbiologique de poisons congelés, conditionnés sous pellicule plastique en atmosphère compensée. R.T.V.A. 208:25-34.

Chandrashekar K, Deasthale YG (1993). Proximate composition, amino acid, mineral, and trace element content of the edible muscle of 20 Indian fish species. J. Food Compost. Anal. 6:195-200.

Coulin P, Faraz Z, Assanvo J, Spillmann H, Puhan Z (2006). Characterisation of the microflora of attieke, a fermented cassava product during traditional small-scale preparation. Int. J. Food Microbiol. 106:131-136.

Dubois M, Gilles KA, Hamilton JK, Rebers PA, Smith F (1956). Colorimetric method for determination of sugars and related substances. Anal. Chem. 28:350-356.

Edem DO (2009). Vitamin A: A review. Asian J. Clin. Nutr. 1: 65-82.

Foran JA, Carpenter DO, Hamilton MC, Knuth BA, Schwager SJ (2005). Risk- based consumption advice for farmed Atlantic and wild pacific salmon contaminated with dioxins and dioxin- like compounds. Environ. Health Perspect. 33:552-556.

Jeantet R, Croguennec T, Schuck P, Brilé G (2006). Science des aliments. Stabilisation biologique et physicochimie. Tec & Doc (Ed). Lavoisier. Paris France. pp. 59-94.

Koffi-Nevry R, Ouina TST, Koussemon M, Brou BK (2011). Chemical composition and lactic microflora of Adjuevan, a traditional Ivorian

fermented fish condition. Pak. J. Nutr. 10:332-337.
Kovacevic M (1970). Quelques renseignements sur la microbiologie alimentaire : Objectifs et Méthodes. PNUD/FAO. 1-12.
Marchioli R (2001). Efficacy of n-3 polyunsaturated fatty acids after myocardial infarction: results of GISSI-Prevenzione trial. Gruppo Italiano per lo Studio della Sopravvivenza nell'Infarto Miocardico. Lipids 36 Suppl:S119-26.
Ndiaye A (1998). Contribution à l'évolution de la qualité bactériologique des produits de la pêche destinés à l'exportation en 1996 et 1997. Thèse de Médecine Vétérinaire, Dakar. pp. 73.
Nestel PJN (2000). Fish oil and cardiovascular disease: lipids and arterial function. Am. J. Clin. Nutr. 71:228-231.
Nurnadia AA, Azrina A, Amin I (2011). Proximate composition and energetic value of selected marine fish and shellfish from the west coast of Peninsular Malaysia. Int. Food Res. J. 18:137-148.
Osman H, Suriah AR, Law EC (2001). Fatty composition and cholesterol content of selected marine fish in Malaysian water. Food Chem. 73:55-60.
Ouattara P (1986). Etude de la qualité microbiologique des poissons congelés. Thèse de Médecine Vétérinaire Dakar. 20.
Pearson AD (1976). The Chemical analysis of food. 7th Ed. Edinburg: Churchill, Livingstone. pp. 422-511.
Pirestani S (2010). Lipid, cholesterol and fatty acid profile of some commercially important fish species from South Caspian. J. Food Biochem. 34(4):886-895.
Seydi M, Pangui A, Azibe B (1992). Qualité hygiénique des filets de poissons congelés produits au Sénégal. Rev. Microbiol. Hyg. Aliment. 9(4):12-17.
Sutharshiny S, Sivashanthini K (2011a). Total lipid and cholesterol content in the flesh of the five important commercial fishes from around Jaffna Peninsula, Sri Lanka. Int. J. Biol. Chem. 6: 161-169.
Sutharshiny S, Sivashanthini K (2011b). Proximate composition of three species of *Scomberoides* fish from Sri Lankan Waters. Asian J. Clin. Nutr. 3(3):103-111.
Tawfik MS (2009). Proximate composition and fatty acids profiles in most common available fish species in Saudi market. Asian J. Clin. Nutr. 1(1):50-57.
Turkmen A, Aro T, Nurmi T, Kallio H (2005). Heavy metals in three commercially valuable fish species from Iskenderun Bay. Northern East Mediterranean Sea, Turkey. Food Chem. 91:167-172.
Waterman JJ (2000). Composition and Quality of Fish. Torry Research Statation. Edinburgh.

Effect of moisture content on selected physical properties of shea kernel of varying slice thickness

Divine N. BUP[1,3] , Charles F. ABI[2], Dzudie TENIN[3] Cesar KAPSEU[3] and Clergé TCHIEGANG[4]

[1]Higher Institute of the Sahel, University of Maroua, Cameroon.
[2]Department of Chemistry, Higher Teachers' Training College, University of Yaounde 1, PO Box 47, Yaounde, Cameroon.
[3]Department of Process Engineering, ENSAI, University of Ngaoundere, P.O. Box 455 Ngaoundere, Cameroon.
[4]Department of Food Science and Nutrition, ENSAI, University of Ngaoundere, P.O. Box 455 Ngaoundere, Cameroon.

Bulk density, kernel density, particle porosity and shrinkage of sheanut krnel slices were determined as a function of moisture content and slice thickness in order to establish preliminary design parameters for drying in an indirect solar dryer. Solid density increased linearly while bulk density decreased non-linearly during drying as the moisture content decreased from 150 to 5 %d.b. The porosities of the kernels ranged from 60.12 -67.91, 57.05-64.87 and 56.88-62.49%, respectively, for the 5, 10 and 15 mm slices. The volumetric shrinkage of the shea slices reduced with decrease in moisture content for all the particle sizes evaluated in this study. Among the shrinkage models tested, the empirical quadratic model gave the best fit of the experimental data. It is proposed that the mechanism of shrinkage in sheanut kernel slices is linked to the removal of water that leads to structural collapse.

Key words: Density, porosity, volumetric shrinkage, sheanut kernel, slices

INTRODUCTION

The shea tree has long been reported as important livelihood tree (Maranz et al., 2003). This is because in addition to the medicinal value its leaves and roots has, the tree produces fruits with high oil content (45-50%) usually referred to as shea butter. In the processing of shea fruits to obtain butter, drying of the kernels after depulping and dehusking has been described as one of the key steps indispensable in the production process (Womeni et al., 2004). This is because sheanut kernels at harvest contain about 45-60% moisture content on a wet mass bases and are therefore prone to deterioration if not properly preserved. The kernel size varies from one locality to the other. For example, it measures 28.3 by 20.6 mm with respect to length and diameter in Chad (Mohagir, 2010), 36.63 by 27.93 mm in Nigeria (Olajide et al., 2000). In Cameroon, it measures averagely 45 by 30 mm (Bup Nde et al., 2012). Sheanut kernel is therefore one of the biggest oilseed in the world. Traditional drying of the whole kernel is carried out either under the sun or in solar dryers. This may take up to 20 days which therefore prolongs the processing time of these kernels. Drying experiments of whole kernels carried out in an electric dryer at 45°C reduced drying times to about 12 days but allowed the development of a surface coat which hindered uniform drying of the kernels (Kapseu et al., 2007). Drying for more than 12 days may equally lead to the development of moulds which contain lipases that catalyse hydrolyses thereby leading to rancidity in the resultant oil (Kapseu et al., 2007). To reduce this drying time it is advisable to dry these kernel as thin slices in solar dryers which can be cheap and easily affordable to local African women, the main processors of shea fruits to butter. The understanding and modelling of the drying behaviour of sheanut kernel slices and the development of a solar dryer for sheanut kernel slices requires the mastery of its physical properties such as densities, porosities and shrinkage. Although some of these physical properties have been reported in the literature for shea fruits, nuts and kernels (Olajide et al., 2000; Aviara et al., 2005; Bup Nde et al., 2012) those of sheanut kernels slices remain very scarce.

An important aspect of the study of shrinkage involves modelling so that they can be incorporated into drying models to study the effect of shrinkage on the drying behaviour of the product. Many of such models have been reported in the literature and one needs to make a judicious choice of which one best suits the product under study. The aims of this work were therefore to:

1. Determine the bulk density, solid density, porosity and volumetric shrinkage of sheanut kernel slices as a function of moisture content and slice thickness and
2. Model the volumetric shrinkage phenomenon in sheanut kernel slices.

MATERIALS AND METHODS

Sampling and sample treatment

Sheanuts obtained from Tchabal Ngaoundere were cooked in water using a laboratory oven at 80°C for 120 min (Bup Nde et al., 2012). They were then withdrawn from water, allowed to cool on a laboratory bench overnight, cracked and the kernels were cut into 5, 10 and 15 mm thick slabs using a Tommy Slicer (model Siemens, Erlangen, Germany). The sheanut kernel slices were then dried in an indirect solar dryer (drying air temperature 45 ± 5°C; airspeed 1.4 m/s, air realative humidity 55 ± 5%) for predefined periods of time and the resulting moisture contents was measured by the oven method. Results of the physical properties were then expressed as a function of moisture content.

Determination of bulk density of the kernels

The bulk density was determined using the AOAC (1980) method. This involved the filling of a 500 ml cylinder having an inner diameter of 49.54 mm with sheanut kernel slices to a height of 15 cm and weighing the contents. The bulk density ρ_b in kg/m^3 was given by:

$$\rho_b = \frac{m_b}{v_b} \quad [1]$$

where V_b is the bulk volume. These experiments were carried out for the 5, 10 and 15 mm thick slices. Each experiment was replicated four times at each particle size.

Determination of solid density of sheanut kernel slices

To determine the true or solid density of the shea kernel slices, an analytical balance (model Scout Pro SPU402, OHAUS, USA) adapted for this purpose was used. The balance was set to the specific gravity mode. A spring was attached to the balance from which the sample tied to a string of negligible weight was hung. The weight of the sample was taken in air. The sample was then immersed into a beaker of water placed on the balance and the new weight was taken. The specific gravity of the sample was then determined from the relation (as indicated in the User's Manual of the balance):

$$Specific\ gravity = \frac{weight\ in\ air}{weight\ in\ air - weight\ in\ water}$$

This was then converted to solid density given that the density of water at 22°C is 1 g/cm^3. At each moisture content, five kernel slices were used for each determination. The experiment was replicated thrice at each moisture content. These studies were carried out on 5, 10 and 15 mm thick sheanut kernel slices.

Determination of porosity

The porosity was calculated from the solid and bulk densities using the relationship given by Pabis et al. (1988). The porosity, ε, was given by the equation:

$$\varepsilon = 100\left(1 - \frac{\rho_b}{\rho_p}\right) \quad [2]$$

Determination of volumetric shrinkage of sheanut kernel slices

Volumetric shrinkage ratio was determined by measuring the volume of the kernels at each moisture content and defining the ratio V/V_0 as volumetric shrinkage ratio. V_0 is the corresponding volume of the kernels at the initial moisture content X_0, V is the corresponding volume of the kernels at moisture content X. Both empirical and fundamental shrinkage models with and without porosity terms were tested and some selected to describe the shrinkage behaviour in sheanut kernel slices. The first group of these models included empirical linear (Equation 3) and quadratic models (Equation 4):

$$\frac{V}{V_0} = a + b\frac{X}{X_0} \quad [3]$$

$$\frac{V}{V_0} = c + d\frac{X}{X_0} + e\left(\frac{X}{X_0}\right)^2 \quad [4]$$

The second group included the Vacarezza (1975) (Equation 5) and Suzuki et al. (1976) (Equation 6) models which are fundamental models without porosity:

$$\frac{V}{V_0} = q_0 + q_1\frac{X}{X_0}; \quad q_0 = \frac{1}{X_0\left(\rho_S/\rho_w\right)+1}; \quad and \quad q_1 = \frac{X_0\left(\rho_S/\rho_w\right)}{X_0\left(\rho_S/\rho_w + 1\right)} \quad [5]$$

$$\frac{V}{V_0} = q_2 + q_3 X$$

$$q_2 = \frac{1 - q_5}{X_0 - X_e - q_4\left(q_5X_0 - X_e + q_5 - 1\right)} \quad q_3 = \frac{q_5X_0 - X_e - q_4\left(p_5X_0 - X_e + q_5 - 1\right)}{X_0 - X_e - q_4\left(q_5X_0 - X_e + q_5 - 1\right)}$$

$$q_4 = \frac{\rho_e - (1 - X)\rho_e}{\rho_e} \quad \text{and} \quad q_5 = \frac{(1 + X_e)\rho_0}{(1 + X_e)\rho_e} \quad [6]$$

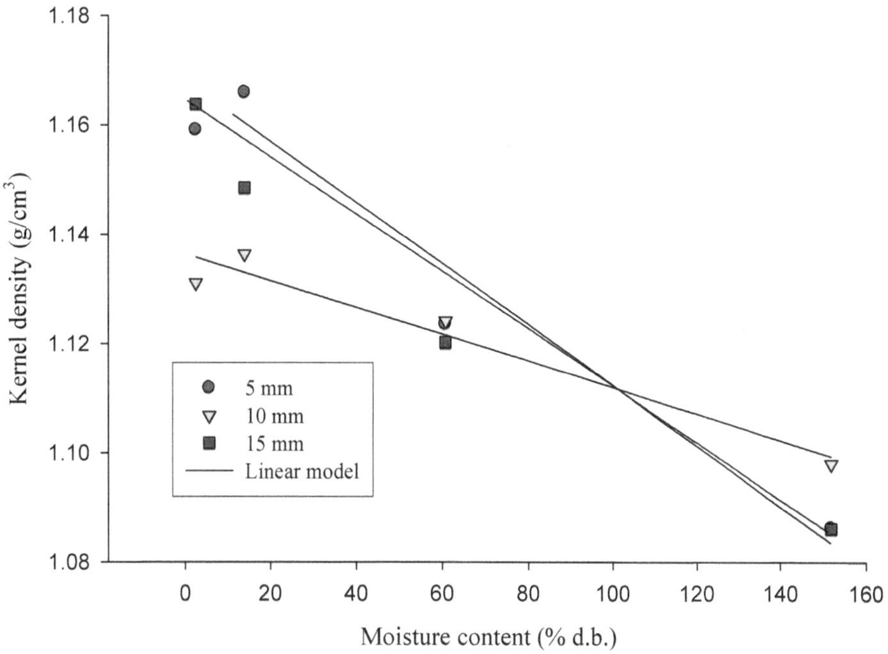

Figure 1. Influence of moisture content on the particle density of sheanut kernel slices.

The third group was the modified Perez and Calvedo model (Mayor and Sereno, 2004) (Equation 7) which included a porosity term, that is:

$$\frac{V}{V_0} = \frac{1}{(1-\varepsilon)}\left[1 + \frac{\rho_0(X - X_0)}{\rho_w(1 + X_0)} - \varepsilon_0\right] \quad [7]$$

a, b, c, d, e are constants of the empirical models, q_i are parameters of fundamental models (variable), X_e is the equilibrium moisture content and p_e is the equilibrium density, p_0 is density at X_0, p_s and p_w are the densities of the dry solid and water respectively. q_5 is a constant when the equilibrium density (p_e) and the equilibrium moisture content (X_e) are known. ε and ε_0 are the porosities at moisture contents X and X_0.

Validation of models

The criteria for evaluating the reliability of the simulations were the correlation coefficients and/or the standard relative error of deviation observed on the moisture content between the experimental and theoretical results. The standard relative error (SRE) of deviation of theoretical from experimental results was determined from Equation 8.

$$SRE(\%) = \frac{100}{p}\sum_{i=1}^{p}\left|\frac{Y_{exp} - Y_{mod}}{Y_{exp}}\right| \quad [8]$$

where Y_{exp} and Y_{mod} are the values obtained from experiments and from the model respectively. P is the number of points at which measurements were carried out.

RESULTS AND DISCUSSION

Solid density of sheanut kernel slices

The solid density increased linearly during the drying period as the moisture content decreased from 150 to 5% d.b. (Figure 1). The values ranged from 1.086-1.160, 1.098-1.131 and 1.086-1.163 g/cm^3 for the 5, 10 and 15 mm slices, respectively, in the moisture range studied. Solid density is simulated with linear equations with regression coefficients of 0.956, 0.951 and 0.963 for the 5, 10 and 15 mm thick kernels, respectively. The increase in solid density with decrease in moisture content could be due to the fact that, during the drying process, the kernels lost water and became more compact and were therefore capable of packing more regularly and closely in the measuring cylinder. This regular and closed packing could have been acompanied by a more rapid decrease in volume as compared to its mass and led to increased solid density which obviously increased with decrease in moisture content. The solid density for all the slices at all moisture contents was greater than unity, implying that the slices are heavier than water. This property can be very useful in the design of cleaning and separation equipments for the slices. No clear relation was established between solid density and particle size as the solid density was higher for the 10 mm thick slices at higher moisture content (150-100% d.b.) as compared to the 5 and 15 mm thick slices. The trends were reversed as the moisture content decreased below 100% d.b. The variation of solid density with moisture content was

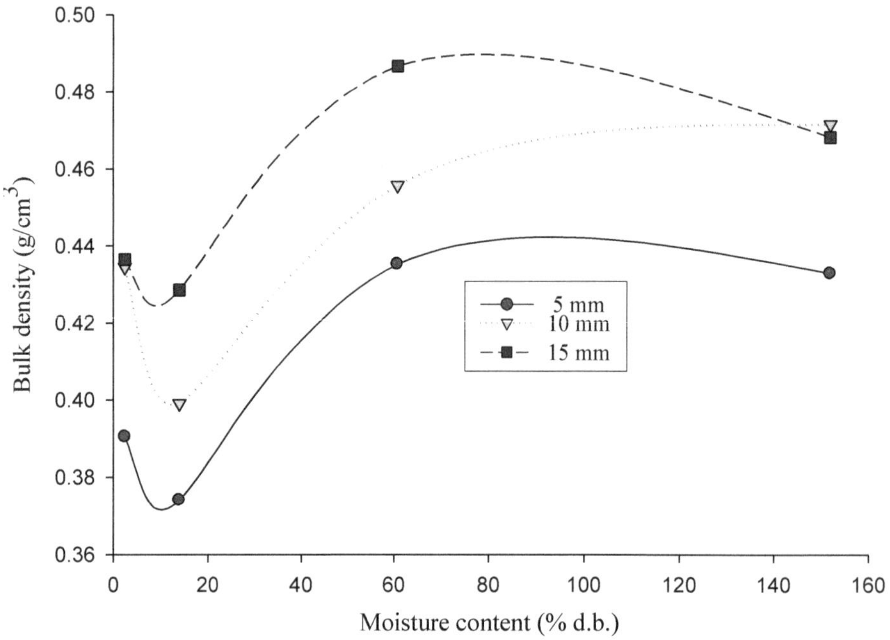

Figure 2. Influence of moisture content on the bulk density of sheanut kernel slices.

modelled with a linear equation with the regression coefficients greater than 0.95% and the SRE less than 5%.

Bulk density of sheanut kernel slices

Figure 2 shows the influence of moisture content on the bulk density of sheanut kernels slices. The bulk density of the sheanut kernel slices decreased non-linearly as the moisture content decreased from 150 to 5% d.b. In the final stages of drying, the bulk density slightly increased. The non-uniform decrease of bulk density with moisture content could be due to the fact that the bulk volume and mass of the kernels might not have changed uniformly in the course of drying. The increase in bulk density in the final stages of drying could be attributed to the fact that, at that stage, the mass of the product became constant due to the limitation of the movement of tightly bound water as observed in the drying of sheanut kernel slices (Kapseu et al., 2007). However, its volume continued to decrease due to the unbalanced pressure between the inner portion of the product and the outer environment thereby leading to an increase in the mass to volume ratio (bulk density). The bulk density decreased from 0.435 to 0.374, 0.472 to 0.399 and 0.468 to 0.429 g/cm^3 for the 5, 10 and 15 mm slices, respectively, in the moisture range studied. From Figure 2, it was equally observed that the bulk density was generally dependent on the particle size, increasing as the particle size increased.

Porosity of sheanut kernel slices

The porosity of the kernel slices increased non-uniformly with moisture content in the range 150-5% d.b. (Figure 3). The porosities of the kernels ranged from 60.12 - 67.91, 57.05-64.87 and 56.88-62.49%, respectively, for the 5, 10 and 15 mm slices. The porosities decreased with an increase in kernel size at all moisture contents. The slight decrease in porosity observed at the final stages of drying was properly due to the decrease in bulk volume at that stage as explained for bulk density above. The variation of porosity with moisture content of sheanut kernel slices observed in this work were different from trends reported by Aviara et al. (2005) for sheanuts in which porosity of the nuts increased with an increase in moisture content to a maximum value and then decreased sharply afterwards.

Volumetric shrinkage of sheanut kernel slices

The volumetric shrinkage of the shea slices reduced with decrease in moisture content for all the particle sizes evaluated in this study. It is clearly observed from Figure 4 that shrinkage of the kernel slices decreased non-linearly as the moisture content of the kernels decreased. The higher decrease in volume observed at higher moisture contents could be linked to the high elastic behaviour of the material still completely in its rubbery state. Under these conditions, they were sufficiently elastic to shrink into the space left by the evaporated water. This elasticity reduced as more and more water was evaporated from the product. The material became more rigid and led to a reduction in the rate at which the product shrunk. This suggested that shrinkage was probably caused by the volume of water removed.

As earlier mentioned, five models were tested to fit the data for shrinkage (Vacarezza, 1975; Suzuki et al., 1976); empirical linear and quadratic models and the modified

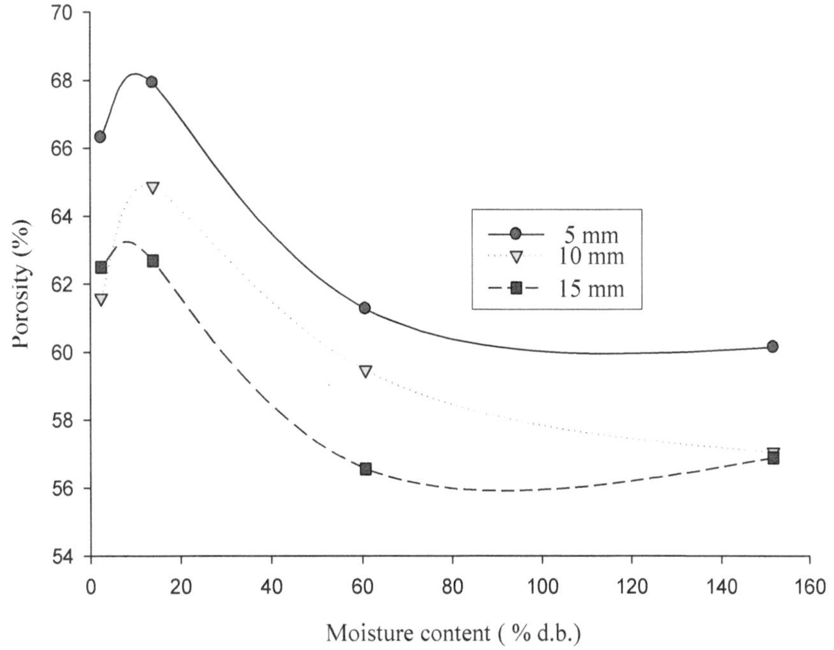

Figure 3. Influence of moisture content on the porosity of sheanut kernel slices.

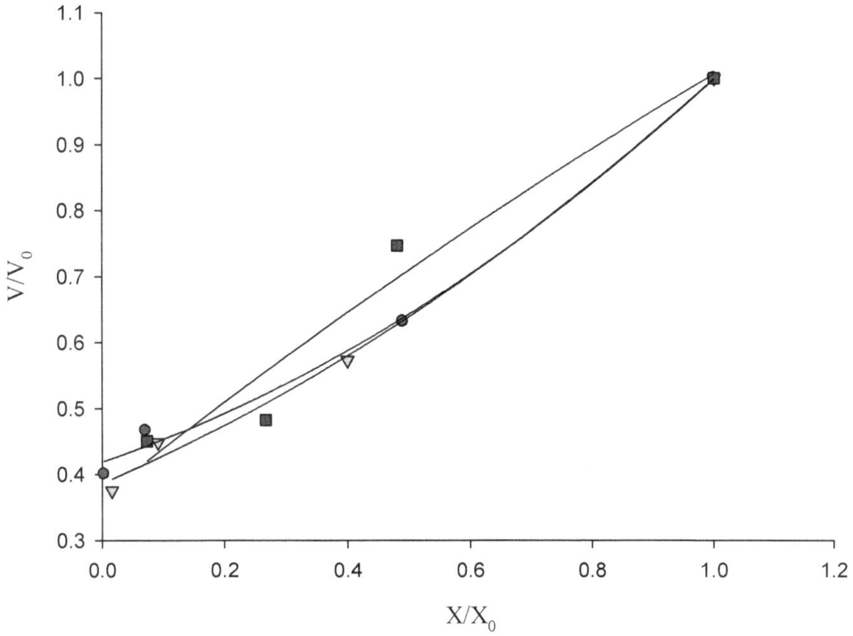

Figure 4. Influence of moisture content on the volumetric shrinkage of sheanut kernel slices.

Perez and Calvedo model (Mayor and Sereno, 2004). Of these models, the empirical quadratic model gave the best fit of the experimental data as indicated by its very high R^2 and very low SRE (Table 1). This model was therefore used to describe the shrinkage of the sheanut kernel slices. The modified Perez and Calvedo model gave very low R^2 and high SRE values indicating that it could not be used to describe shrinkage in sheanut kernel slices. The values of its constants are therefore not reported. The failure of the modified Perez and Calvado model could be attributed to the fact that, the porosity considered in this study was external porosity (voidage) instead of intrinsic porosity of the kernels. The non linear variation of volumetric shrinkage with moisture

Table 1. Model constants and R^2 and SRE values for equations used in modelling shrinkage of sheanut kernel slices.

Model / Particle size (mm)	Linear empirical				Vicarezza (1975)				Suzuki et al. (1976)		
	A	B	R^2	SRE	Q_0	Q_1	R^2	SRE	Q_5	R^2	SRE
5	0.3978	0.0038	0.982	5.59	0.4996	0.5004	0.954	12.8	0.4827	0.977	5.03
10	0.3649	0.0041	0.989	3.00	0.4996	0.5004	0.989	19.8	0.5021	0.988	17.90
15	0.384	0.0041	0.954	5.94	0.4996	0.5004	0.982	14.50	0.4847	0.999	9.75
						Quadratic					
				C	D	E	R^2	SRE			
5				0.418	0.319	0.261	0.996	2.76			
10				0.386	0.399	0.214	0.982	2.82			
15				0.367	0.735	-0.095	0.982	7.29			

content in the investigated moisture content range indicated that, during the drying process shrinkage is not essentially uniform. This assertion was buttressed by the fact that more than 60% of the samples were visually cracked after the drying experiments. Mayor and Sereno (2004) stated that surface cracking occurs during drying when shrinkage is not uniform leading to the formation of unbalanced stresses and failure of the material. A uniform shrinkage of the kernel slices could be linked to the fact that phase transition (change from a rubbery to the glassy state) in the material through out the drying process may have been highly limited or negligible. That is, if the material remained mostly in the rubbery state through out the drying process, then shrinkage almost entirely compensated for moisture lost and the volume of the material decreased linearly with moisture content. However, if the material passed from a rubbery to the glassy state (phase transition), the mobility of the solid matrix reduced and the rate and extent of shrinkage decreased significantly leading to deviations from linearity. This last phenomenon was observed for the sheanut kernel slices investigated in this work. Deviations of shrinkage from linearity especially at low moisture contents have been reported for garlic, potato and sweet potato (Ratti, 1994) and for apple and potato (Wang and Brennan, 1995). So in this work, it is proposed that the mechanism of shrinkage in sheanut kernel slices is linked to the removal of water that leads to structural collapse. Water removal becomes highly limited in the later periods of drying. This culminates into deviations from linearity which could be attributed to phase transitions. Further work is needed on the glass transition temperatures of sheanut kernel slices to verify this assertion.

Conclusion

This work reported some of the physical properties of sheanut kernel slices as a function of moisture content. Apart from the solid density whose variation was linear, bulk density and porosity varied non lineraly with moisture content and differed from those of whole large kernels in the mass range 21-28 g. Shrinkage was modelled by an empirical quadratic equation and it is proposed that the mechanism in sheanut kernel slices is linked to the removal of water that leads to structural collapse. The results of these physical properties can be employed in modelling drying processes for an eventual design of a dryer for sheanut slices.

ACKNOWLEDGEMENTS

This research was supported by the International Foundation for Science, Stockholm, Sweden and the Conseil Ouest et Centre Africain pour la recherche et le Development Agricoles–West African Council for Agricultural Research and Development (CORAF/WECARD), Dakar, Senegal, through a grant to Bup Nde Divine.

REFERENCES

Aviara NA, Oluwole FA, Haque MA (2005). Effect of moisture content on some physical properties of sheanut (*Butyrospernum paradoxum*). Int. Agrophys. 19: 193-198.

Bup Nde D, FON ABI C, Tenin D, Kapdeu C, Tchiegang C (2012). Optimisation of the cooking process of sheanut kernels (vitellaria paradoxa gaertn.) Using the Doehlert experimental design. Food Bioprocess Technol. 5:108–117. DOI: 10.1007/s11947-009-0274-z.

Kapseu C, Bup Nde D, Tchiegang C, Fon Abi C, Broto F, Parmentier M (2007). Effect of particle size and drying temperature on drying rate and oil extracted yields of *Buccholzia Coriacea* (MVAN) and *Butyrospermum Parkii* ENGL. Int. J. Food Sci. Technol. 42: 573-578.

Maranz S, Wiesman Z, Garti N (2003). Phenolic constituents of shea kernels (*Vitellaria paradoxa*). J. Agric. Food Chem. 51:21: 6268-6273.

Mayor L, Sereno AM (2004). Modelling shrinkage during convective drying of food materials: a review. J. Food Eng. 61: 373–386.

Mohagir MA (2010). Optimisation of press extraction and decolourisation of shea butter (*Vitellaria paradoxa* Gaertner F.). Ph.D thesis University of Ngaoundere, Cameroon.

Olajide O, Ade-Omowaye, BIO, Otunola ET (2000). Some Physical Properties Of Shea Kernel. J. Agric. Eng. Res. 76(4): 419-421.

Pabis S, Jayas DS, Cenkowski S (1998). Grain Drying: Theory and Practice, John Wiley and Sons, New York, USA. p. 303.

Ratti C (1994). Shrinkage during drying of foodstuffs. J Food Eng. 23(1): 91–105.

Suzuki K, Kubota K, Hasegawa T, Hosaka H (1976). Shrinkage in dehydration of root vegetable. J. Food Sci. 41 :189–1193.
Vacarezza L (1975). Cinetica y mecanismo de transporte de agua durante la deshidrataci on de la remolacha azucarera. Ph.D. Thesis. Universidad de Buenos Aires, In: Mayor L, Sereno AM (2004). Modelling shrinkage during convective drying of food materials: a review. J. Food Engr. 61: 373–386.
Wang N, Brennan JG (1995). Changes in structure, density and porosity of potato during dehydration. J. Food Eng. 24 (1): 61–76.
Womeni HM, Ndjouenkeu R, Kapseu C, Parmentier M (2004). Application du Procédé Séchage-Friture aux Amandes de Karité: Cinétique du séchage. OCL. 11(6):457-463.

Comparative evaluation of physico-chemical properties of two varieties of pomogranate fruits-Ganesh and Arakta

Patil Priyanka, Sayed H. M., A. A. Joshi and Jadhav B. A.

Department of Food Chemistry and Nutrition, College of Food Technology, Vasantrao Naik Marathwada Krishi Vidyapeeth, Parbhani-431402, MH, India.

The present investigation was carried out to analyze different physico-chemical characteristics of two different varieties of pomogranate fruits viz. Ganesh and Arakta. The study reveals that the morphological characteristics of pomegranate, that is, colour of fruit of var. Ganesh and Arakta was yellowish with pink patches and dark red, respectively. The average weight was 210.90 and 254.20 g for Ganesh and Arakta cultivars. The edible index and waste index were found to be 66.55 and 33.45% for Ganesh, and 65.89 and 34.11% for Arakta cultivar. The chemical parameters of fruit determined, total soluble solid (TSS), acidity, pH, moisture, ash content, carbohydrate, protein, fat and ascorbic acid were quite similar for both varieties. The dark red colour of Arakta cultivar had eye appeal effect to the consumers. With the various nutritional benefits, the fruits could be recommended for commercial exploitation and preparation of different value added products.

Key words: Pomogranate, anardana, physical properties, processing, chemical properties.

INTRODUCTION

The pomegranate is a non-climacteric many-seeded berry. The fruits are round, oblate or obovate in shape and vary in weight and size (8 to 12 cm diameter). On an average, edible portion fruit represents about 52% of total fruit weight comprising about 78% juice and 22% seeds (Berry, 2005).

The fruit skin may be thick or thin but smooth, leathery and tough with colour varying from pale-yellow to crimson-red. Inside, the fruit is filled with numerous seeds coated with juicy pulp shining like pearls or jewels. These are called arils, which are the edible portion. The pulp in fruits of superior quality is thick, soft, fleshy and dripping juicy. The colour of the pulp varies from light pink to crimson-red. The taste of the pulp also varies from sweet and aromatic to sour and insipid. The seeds may be hard or soft but edible. The softer seeded varieties are designated seedless (Bose et al., 2002).

The fruit is a good source of sugars and vitamin C and a fair source of iron and poor in calcium. The juice contains amino acids such as glutamic acid, aspartic acid, tryptophan, methionine, etc. It is rich in phosphorus. The seeds are rich source of lipids, protein, crude fibre and ash, and also contain sugars and pectin. The seeds yield oil which is composed of several fatty acids. The rind contains polyphenols and mineral matter, which is composed of potassium, calcium, iron and copper. The water and solvent extracts of the rind contains gallo-catechin, quercetin, gallic acid, catechin, etc. which possess strong antioxidant properties.

Pomegranate though minor fruit is included under

Table 1. Physical properties of pomegranate fruit.

Variety	Physical properties						
	Colour of fruit	Colour of arils	Weight of fruit (g)	Weight of arils (g)	Weight of peel (g)	Edible index (%)	Waste Index (%)
Ganesh	Yellowish with pink patches	Pinkish	210.90	140.35	70.55	66.55	33.45
Arakta	Dark red	reddish	254.20	167.50	86.70	65.89	34.11
S.E.	-	-	1.74	1.74	0.72	0.83	0.18
C.D.	-	-	0.80	0.83	0.84	0.28	0.74

medicinal plants. Since ancient time it has been commonly used in folk medicine in the Middle East, Iran and India as medication for antiviral, antifungal, antibacterial and antimicrobial activity. The juice from fresh pomegranate fruit is an excellent cooling beverage for alleviating thirst in cases of fever and sickness. The extract of pomegranate may be used as an antifungal agent for the treatment of candidosis associated with denture stomatitis (Vasconcelos et al., 2003). In addition, it is widely used in therapeutic formula, cosmetics, and food seasonings. Since ancient times, the pomegranate has been regarded as a "healing food" with numerous beneficial effects in several diseases (Vidal et al., 2003).

Knowing the importance of health benefits of pomegranate and all related aspects, the present investigation entitled "Comparative evaluation of physico-chemical properties of two varieties of pomogranate fruits-Ganesh and Arakta" was carried out to analyze different physico-chemical characteristics of different varieties of pomogranate fruits, that is, Ganesh and Arakta cultivars and to explore these fruits for commercial processing.

MATERIALS AND METHODS

Pomegranate fruits cultivars Ganesh and Arakta were selected from local market from Parbhaani, Maharashtra. The fruits of uniform size, colour and maturity were selected by visual observation and used as experimental material.

Morphological features

Morphological features like colour, weight of fruit, percent edible portion, percent waste, percent of juice were determined.

Proximate analysis of pomogranate fruits

Pomogranate fruits were analyzed for its total soluble solid (TSS) (°Bx), acidity, pH, moisture, ash, fat, protein, carbohydrate and ascorbic acid content by standard methods (AOAC, 1990).

Statistical analysis

The analysis of variance of the data obtained was done by using completely randomized design (CRD) for different treatments as per the methods given by Panse and Sukhatme (1967). The analysis of variance revealed at significance of $P < 0.05$ level, S.E. and C.D. at 5% level is mentioned wherever required.

RESULTS AND DISCUSSION

The data on physical characteristics of pomegranate (Table 1) shows that the colour of fruits of Ganesh cultivar was yellowish with pink patches and that of Arakta cultivar was dark red. The colour of arils in Ganesh cultivar was pinkish whereas it was blood red in Arakta cultivar. The average weight of fruit of Ganesh cultivar was 210.90 g while weight of Arakta was 254.20 g.

The edible portion and waste index in Ganesh cultivar were 66.55 and 33.45%, respectively, while in Arakta cultivar these were 65.89 and 34.11% respectively. The edible portion was found to be more in Ganesh cultivar than in Arakta cultivar. The results of present study were in good accordance with results reported by Jagtap et al. (1992).

It is observed from Table 2 that the TSS of juice from Ganesh (15°Bx) was slightly higher than that of Arakta (14°Bx). The acidity in juice of Ganesh and Arakta cultivar was 0.3 and 0.35%, respectively. The pH of juice of Ganesh and Arakta fruit was 3.2 and 3.1, respectively. The moisture content of Ganesh cultivar (80.50%) was slightly higher than that of Arakta (79.80%). The ash content of Ganesh and Arakta was 0.8 and 0.9%, respectively.

The fat content of Ganesh and Arakta was 0.2 and 0.25%, respectively. The protein content found in Ganesh was 1.6 and 1.5% in Arakta cultivar. The carbohydrates are important from nutritional point of view as it provides energy to the body. The carbohydrate content of Ganesh was 14%, but 14.5% carbohydrate was found in Arakta. The amount of ascorbic acid present in Ganesh was 9 mg/100 g and 9.5 mg/100 g in Arakta cultivar. Similar results were reported by Fadavi et al. (2006) for different cultivars of pomegranate.

Conclusion

Nevertheless, the pomegranate has commercial potential for production of health related food products, but the

Table 2. Chemical characteristics of pomegranate fruit.

Variety	Parameter								
	TSS (^{0}Bx)	Acidity (%)	pH	Moisture (%)	Ash (%)	Fat (%)	Protein (%)	Carbohydrate (%)	Ascorbic acid (mg/100 g)
Ganesh	15	0.30	3.2	80.50	0.8	0.2	1.6	14.0	9.0
Arakta	14	0.35	3.1	79.80	0.9	0.25	1.5	14.5	9.5
S.E.	0.37	0.083	0.091	0.38	0.091	0.045	0.115	0.420	0.238
C.D.	1.46	0.32	0.35	1.50	0.35	0.17	0.452	1.64	0.933

systematic and organized approach should be followed with other sophisticated methods for retention of other bio-active components, storage life, clinical studies and packaging requirements.

It was concluded that the chemical parameters of fruit determined were TSS, acidity, pH, moisture, ash content, carbohydrate, protein, fat and ascorbic acid which were quite similar for both varieties. The dark red colour of Arakta cultivar had eye appeal effect to the consumers. With the various nutritional benefits, the fruits could be recommended for commercial exploitation and preparation of different value added products.

REFERENCES

AOAC (1990). Official methods of analysis of the association of official analytical chemists (15th edition), Washington D.C. pp. 992-995.

Berry SK (2005) Pomegranate: A Miracle Fruit. Review article. Indian Food Packer. Nov.-Dec. pp. 189-194.

Bose TK, Mitra SK, Sanyal D (2002) Fruits: tropical and subtropical, vol. 2,3rd revised edition, Naya Udyog, Kolkata, India. pp.125-162.

Fadavi A, Berzegar M, Azizi MH (2006) Determination of fatty acids and total lipid content in seeds of 25 pomegranate varieties grown in Iran. J. Food Comp. Anal. 19:676-680

Jagtap DB, Desai UT, Kale P N (1992) Chemical composition of some indigenous and exotic cultivars of pomegranate. Maharashtra J. Hort. 6(1):10-12.

Panse VS, Sukhatme PV (1967). Statistical Methods for Agricultural Workers. Indian Council of Agricultural Research. New Delhi. pp. 70-72.

Vasconcelos LC, Sampaio MC, Sampaio FC, Higino J S (2003) Use of *Punica granatum* as an antifungal agent against candidosis associated with denture stomatitis. Mycoses 46(5-6):192-196.

Vidal A, Fallarero A, Pena BR, Medina ME, Gra B, Rivera F, Gutierrez Y, Vuorela PM (2003) Studies on the toxicity of *Punica granatum* L. (Punicaceae) whole fruit extracts. J. Ethnopharmocol. 89:295-300.

Effect of different types of honey on the microbial shelf stability of cassava-wheat composite bread

Adeboye, A. S.[1] , Babajide J. M.[2] and Fayemi O. E. [3]

[1]Department of Food Technology, Moshood Abiola Polytechnic, Abeokuta, Ogun State, Nigeria.
[2]Department of Food Science and Technology, University of Agriculture, Abeokuta, Ogun State, Nigeria.
[3]Department of Microbiology, Federal University of Technology, Akure, P.M.B. 704, Ondo State, Nigeria.

The effect of physicochemical properties (moisture content, total acidity, total phenolic content and viscosity) of honey on the microbial shelf stability of cassava-wheat composite bread was investigated. The percentage weights of the sugar (sucrose) required in the formulated recipe was substituted with different types of honey at the same level (70% sucrose: 30% honey). Freshly baked and cooled cassava-wheat composite bread loaves were stored on shelf at ambient temperature and the total aerobic bacteria and mould counts were determined after 4 and 6 days. The physicochemical analyses revealed that the total acidity, total phenolic content, moisture and viscosity of the different types of honey used varied. The physicochemical properties of the various honey used influenced the microbial shelf stability of the cassava-wheat composite bread. Highest total aerobic bacteria counts of 0.25 × 104 cfu/g was recorded for the cassava-wheat composite bread baked with Hamba honey 2 which had significant ($p \leq 0.05$) lowest total acidity (42.41 mEq/kg) and total phenolic contents (48.97 GAE/100 g) as well as significant ($p \leq 0.05$) highest % moisture after 4 days of storage. However, the incorporation of dark and golden honey with higher total acidity and total phenolic contents compared with other types of honey used, resulted in the least mould counts of 0.42 × 104 and 0.45 × 104 cfu/g, respectively, after 6 days of storage. The physicochemical properties of honey could enhance the microbial shelf stability of baked bread if the minimum inhibitory concentration of the honey after baking is employed.

Key words: Honey, cassava-wheat composite bread, microbial shelf stability.

INTRODUCTION

Bread has become the second most widely consumed non-indigenous food products in Nigeria. To cut the nation's expense on wheat importation and find wider utilization for the increasingly produced cassava roots in Nigeria, the Federal Government mandated the use of composite cassava wheat flour for baking by adding minimum of 10% cassava flour to wheat for a start (Shittu et al., 2007). According to Shittu et al. (2007), fresh crumb moisture, density, porosity and softness as well as the dried crumb hardness were significantly affected by both the baking temperature and time. The studies of Defloor et al. (1993, 1994, 1995) and Khalil et al. (2000)

established that 10% substitution of wheat with cassava flour gave bread with quality not significantly different from 100% wheat bread. Comparative studies have shown that honey has less impact on blood sugar level because it offers low glycemic index (GI) response (Foster, 2008). Beyond many health claims and ability to mask any taste deficiency that may have resulted from ingredient interactions, inclusion of honey into bread formulation is said to offer functional benefits (Foster, 2008). Baking technology has been evolving continuously as new materials; equipment and processes are being developed (Selomulyo and Zhou, 2007). The impacts of various ingredients on sensory and nutritional quality of bread have been widely studied (Barcenas and Rosell, 2005; Plessas et al., 2005).

Pure honey has been shown to be bactericidal to many pathogenic microorganisms and the antimicrobial activity has been reported to be as a result of the presence of osmotic effect, acidity and hydrogen peroxide (Radwan et al., 1984; Jeddar, et al., 1985). The pH of honey is reported to be low enough to slow down or prevent the growth of many species of bacteria. The high sugar content of honey makes the water unavailable for microorganisms: no bacteria or fungi can grow in fully ripened honey, but the more diluted honey becomes, the more species can grow in it (Molan, 1992). It was also reported that glucose oxidase enzyme activated by dilutions of honey generates hydrogen peroxide which generally is the major antibacterial factor in honey. This enzyme is inactivated by heating honey, and by exposure to light in some honeys which contain a sensitizing factor. Some honeys also contain substances which destroy the hydrogen peroxide generated by the enzyme (Molan, 1992).

According to Beatriz et al. (2011), the water content is important for honey stability, while the acidity content of honey is a function of honey fermentation. Studies have shown that honeys from the tropics with high water content tend to ferment readily and the free acidity is increased (Sibel et al., 2010; Beatriz et al., 2011) and CA (2001) prescribes a maximum value of 50 milliequivalents of free acids per kilogram of honey. Antimicrobial effects of honey against microorganism associated with disease or infection, food spoilage, including foodborne pathogens and yeasts, have been reported and honey biological activity has been attributed not only to the high sugar concentration and hydrogen peroxide production but also to different compounds such as acids, phenolics, proteins and carbohydrates (Gheldof et al., 2002; Olasupo et al., 2003; Mundo et al., 2004; Guerrini et al., 2009; Gomes et al., 2010). However, the findings of Olaitan et al. (2007) revealed that microorganisms that survive in honey are those that withstand the concentrated sugar, acidity and other antimicrobial characteristics of honey. Stefan et al. (2008) reported that honey from different sources differs in appearance, sensory perception and composition. This suggests that their antimicrobial properties/potential may vary. Although, information is still scarce on utilization of honey in bread formulation developed from cassava wheat composite flour, previous study on the influence of a single type of honey on microbiological shelf stability of cassava-wheat composite bread revealed that the honey used extended the shelf life of the baked bread loaves (Adeboye et al., 2013). Therefore, this current study therefore aims at investi-gating the microbial shelf life stability of cassava-wheat composite bread baked with different types of honeys obtained from different locations in South Western Nigeria.

MATERIALS AND METHODS

The sweet cassava variety (*Manihot esculenta Crantz*) tubers used for the production of cassava flour in this study was harvested from the farm of Moshood Abiola Polytechnic, Abeokuta, Nigeria and the white wheat flour was obtained from honey well flour mill, Lagos, Nigeria. Other ingredients used were granulated sucrose sugar (Dangote groups (Nig) Ltd. Nigeria), Fermipon baking yeast (DSM bakery ingredient, Dordrecht -Holland) and baking fat (Pt Intibuca Sehtera, Jakarta, Indonesia). The various honey used were obtained from local bee keeping and honey production farms (Ibadan, Nigeria). The composite flour was made by mixing 10 part cassava flour with 90 part wheat flour.

Determination of the physicochemical properties of honey types

Total phenolic content

The procedure for the determination of total phenolic content was adapted from Zalibera et al. (2008) with some modifications. 50 µL of the honey or methanolic extract was added to 125 µL of Folin-Ciocalteu reagent. The mixture was sonicated for 5 min after which 625 µL of sodium carbonate was added and the absorbance was determined after 2 h at 760 nm. The results were expressed as milligram of gallic acid equivalents per kilogram of honey (mg GAE/kg).

Moisture content, total acidity and viscosity

Moisture content, total acidity and viscosity of the honey types were determined according to methods described by the AOAC (2000) (Official Methods 979.20, 969.38, 962.19).

Preparation of cassava flour

The matured cassava tubers were peeled, washed and grated in a mechanical grater and the pulp obtained was dewatered in a 'muslin' cloth placed in between a screw press. The pulverized cassava mash obtained was then dried in cabinet dryer (Lukas Engineering Nig. Ltd) at 70°C to a constant weight to give 4% moisture content. The dried cassava mash was milled in a hammer mill (Lukas Engr. Ltd. Lagos) and sieved with a mesh of 0.5 mm pore size and fine cassava flour obtained was stored in an airtight container.

Preparation of the recipe

The calculated weight of the sugar (sucrose) in the recipe used was substituted with the different honey types at the optimum

Table 1. The physicochemical properties of the different types of honey used for the baking of the cassava-wheat composite bread.

Sample	Total acidity (mEq/kg)	Total phenolic content (GAE/100 g)	Moisture (%)	Viscosity (cP)
Dark honey	63.99^{d} (0.02)	66.03^{c} (0.007)	16.03^{a} (0.028)	1.75^{a} (0.014)
Hamba Honey1	62.93^{c} (0.02)	67.25^{d} (0.007)	16.26^{a} (0.03)	3.55^{a} (0.014)
Hamba Honey 2	42.41^{a} (0.02)	48.97^{a} (0.127)	22.60^{c} (0.007)	5.00^{b} (0.14)
Light Hamba Honey	49.75^{b} (0.01)	56.40^{b} (0.084)	20.41^{e} (0.014)	3.40^{ab} (0.021)
Golden honey	64.41^{e} (0.01)	68.29^{e} (0.021)	19.86^{b} (0.014)	2.50^{a} (0.021)

Values are the means and standard deviations of three replicate experiments (n =3). Means with different superscripts in the same column are significantly different ($P \leq 0.05$).

substitution ratio of 30:70 of honey (H) to sugar (S). The six bread samples were one control with 100% sugar and five with different honey types. The ratio of honey and sugar was 7.56: 17.64 g in all samples.

Production and storage of honey-cassava wheat composite bread

The honey-cassava wheat bread was prepared using the procedure described by Shittu et al. (2007) with slight modification. The ingredients (yeast, water (at 40°C), honey and butter) were combined in a large liquid measuring cup and stirred until the yeast has dissolved and the baking fat has melted. The sugar, composite flour and salt were dry mixed in a large bowl. The yeast mixture was thoroughly incorporated into the mixture of dry ingredients. The dough obtained was then transferred into a lightly floured work surface of the kneading machine and kneaded for about 15-20 min to form smooth and elastic dough. The dough was cut into uniform sizes (300 g), placed in lightly greased pan and proofed at 30°C and 78-80% RH for 2 h then baked in an oven at 220°C for 30 min. Samples from each honey-cassava wheat bread were stored at ambient temperature in nylon 6-guage of polythene bag for 6 days during which the total viable bacteria and mould growth for each sample type were monitored

Microbiological analysis

The microbial profile of the total aerobic bacteria and yeasts and moulds of the composite bread was determined using nutrient agar (NA) (Oxoid, Basingstoke, Hampshire, England, UK) and potatoes dextrose agar (PDA) (Merck, Darmstadt, Germany) (supplemented with 50 mg/litre of streptomycin), respectively. Ten gram (10 g) of each of the baked bread sample was aseptically homogenized with 90 mL sterile 0.1% buffered peptone water (BPW) (Merck) solution. After serial dilutions of all the samples, the appropriate dilution was spread plated and the NA (Oxoid) agar plates were incubated at 37°C for 24 h while yeast and mould plates were incubated at 25°C for 3-5 days. Samples were analysed after 4 and 6 days of storage.

Statistical analysis

Analysis of variance (ANOVA) was used to determine if physicochemical properties of honey affected the microbial shelf stability of honey cassava-wheat bread significantly at 95% confidence levels. Each experiment was repeated in triplicate and data were analysed using the Statistical Package for the Social Sciences (SPSS) 21.0 (IBM SPSS Inc., Chicago, IL) and mean separation was carried out with Duncan's new multiple range test.

RESULTS AND DISCUSSION

Physico-chemical properties of honey

There was a significant difference ($P \leq 0.05$) in all the physicochemical properties of the honey samples. The total acidity of the honey types used for the cassava-wheat composite bread ranged between 42.41 and 64.41 mEq/kg with the golden honey having the highest mean value while the lowest value was recorded for the Hamba honey 2 (Table 1). Similarly, Golden honey had the highest total phenolic content while Hamba honey 2 had the least total phenolic content of 68.29 mEq/kg and 48.97 GAE/100 g, respectively.

The variation in phenolic contents of the different types of honey used in this study is in accordance with the findings of Estevinho et al. (2003) and Alzahrani et al. (2012) who reported that the differences between honey samples in terms of antibacterial and antioxidant compounds could be attributed to the natural variations in floral sources of nectar and the different locations. Studies have also shown that honey phenolic compounds composition and consequently antioxidant capacity depends on their floral sources used to collect the honey predominance of which dependents on seasonal and environmental factors (Al-Mamary et al., 2002; Yao et al., 2003).

There was a positive correlation (Figure 2) between total acidity and total phenolic content of honey types used in this study. Moisture and viscosity of the Dark, Hamba 1, Hamba 2, Light hamba and Golden honey were; 63.99, 62.93, 42.41, 49.75 and 64.41 $mEqkg^{-1}$; 66.03, 67.25, 48.97, 56.40 and 68.29 GAE/100 g, 16.03, 16.26, 22.60, 20.41 and 19.86%; 1.75, 3.55, 5.00, 3.40 and 2.50. Hamba honey 2 had the highest moisture and viscosity value (22.60% and 5.00P) while Dark honey had the least moisture and viscosity value (16.03% and 1.75P).

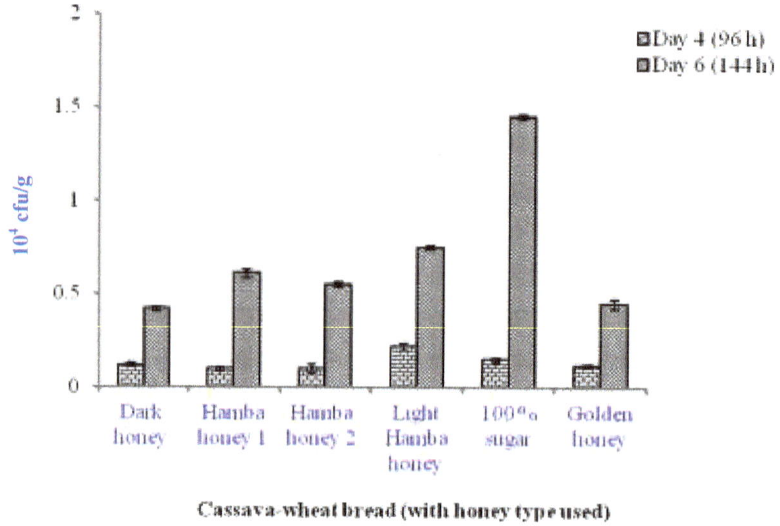

Figure 1. Microbial counts of moulds during the storage of cassava-wheat composite bread baked made with different types of pure honey for 4 and 6 days at 30°C (n = 3). Error bars = ± standard deviation.

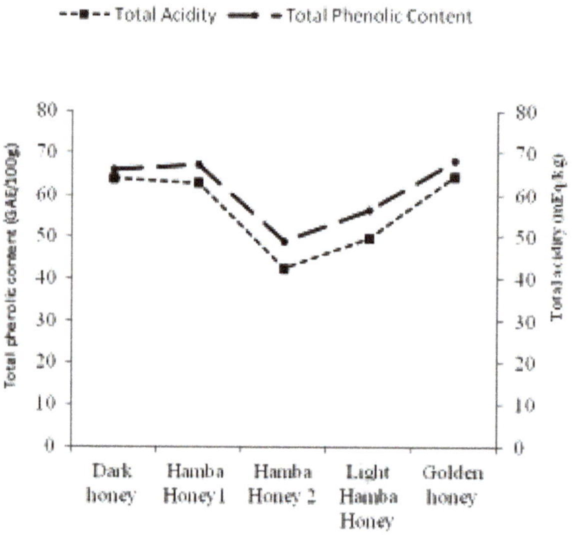

Figure 2. The positive correlation between total acidity and total phenolic content of different types of pure honey incorporated in the cassava-wheat composite bread.

Microbial shelf stability

The addition of the different types of pure honey used had a significant ($p \leq 0.05$) effect on the total aerobic bacteria and mould counts in all the cassava-wheat composite bread. Bacteria and moulds were not detected within the first 3 days of storage at 30°C. This can be attributed to the antimicrobial potency of the physico-chemical properties of honey used for baking against the bacteria and moulds growth. Studies have reported that most bacteria and other microbes are dormant in honey and therefore cannot grow due to the antibacterial activity of honey (Al Somai et al., 1994; Olaitan et al., 2007). However, total aerobic counts (TAC) were detected after 4 days of storage with highest significant ($p \leq 0.05$) counts of 0.25×10^4 cfu/g in the cassava-wheat bread incorporated with Hamba honey 2 (Table 2) while the cassava-wheat composite bread baked with Light hamba honey has highest mould counts (Figure 1). The level of TAC and moulds in the cassava- wheat bread incorporated

Table 2. Total viable bacteria counts in cassava-wheat composite bread baked with different types of pure honey after 4 and 6 days of storage at 30°C.

Sample	Day 4 (96 h) h (10^4) cfu/g	Day 6 (144 h) h (10^4) cfu/g
Golden Honey	0.1^a (0.0)	0.65^a (0.0)
Dark Honey	0.1^a (0.0)	0.65^a (0.0)
Hamba Honey1	0.1^a (0.0)	0.80^a (0.0)
Hamba Honey 2	0.25^b (0.01)	0.13^a (0.01)
Light Hamba Honey	0.1^a (0.0)	0.65^a (0.0)
0H:100S	0.15^a (0.0)	1.15^b (0.0)

Values are the means and standard deviations of three replicate experiments (n =3); Means with different superscripts in the same column are significantly different (P ≤ 0.05); H = Honey, S = Sugar (Sucrose); 0H:100S = Cassava-wheat composite bread baked with 100% honey (Control); Dark honey = Cassava-wheat composite bread baked with 30% Dark honey; Hamba honey 1 = Cassava-wheat composite bread baked with 30% Hamba honey 1; Hamba honey 2 = Cassava-wheat composite bread baked with 30% Hamba honey 2; Light Hamba honey = Cassava-wheat composite bread baked with 30% Light Hamba honey; Golden honey = Cassava-wheat composite bread baked with 30% Golden honey.

with Hamba honey 2 and Light Hamba honey could be attributed to the significant (p ≤ 0.05) lower total acidity and total phenolic contents in these two honey types in comparison with Dark, Hamba honey 1 and Golden honey used for the baking of the cassava-wheat bread (Table 1). This is in accordance with the findings of Weston et al. (1999) who revealed that antimicrobial activity of honey depends on the aromatic acids or phenolic contents derived from the honey. Furthermore, the significant (P ≤ 0.05) higher percentage moisture content in the Hamba honey 2 and Light hamba honey could have possibly influenced the water activity of the cassava-wheat composite bread and consequently enhanced the bacteria and moulds growth during storage at ambient temperature (Cooper et al., 1999). There was no significant (p > 0.05) difference in the TAC in all the cassava-wheat bread baked with different types of honey after 6 days of storage despite the significant (P ≤ 0.05) variations in their physicochemical properties (Table 2). This could possibly be as a result of the baking temperature which might have resulted in a reduction in the antimicrobial potency of the different types of honey used against some species of bacteria. Molan (1992) reported reduction in all the antimicrobial activities of honey such as total acidity and phenolic contents after exposure to 100°C for 15 min. Furthermore, exposure of honey to 100°C for 10 min has been reported to cause a complete loss of activity against different species of bacteria, but only partial loss of activity against *Bacillus pumilus* and *Streptomyces*, and no loss of activity against *Bacillus subtilis* and *Sarcina lute* (Molan 1992). Our previous study also reported a significant reduction for mould and total viable bacteria counts in the cassava-wheat composite bread baked with a single type of honey ((Adeboye et al., 2013). However, in this current study, lowest mould counts of 0.42 and 0.45 × 10^4 cfu/g were recorded for the cassava-wheat composite bread baked with the Dark honey and Golden honey, respectively, after 6 days of storage. This can be attributed to the higher significant (p ≤ 0.05) level of total acidity and phenolic contents in the Dark and Golden honey which resulted in partial inhibitory effect on the moulds. Studies have also shown that the phenolic compounds in honey have growth inhibition on a wide range of Gram-positive and negative bacteria (Davidson et al., 2005; Estevinho et al., 2008). However, the TAC and mould counts were significantly (P ≤ 0.05) higher in the cassava-wheat composite bread baked with sugar (control sample) than the in the cassava-wheat composited bread baked with different types of honey after 6 days of storage. This could be as a result of the fact that the physicochemical properties of honey which are responsible for its antimicrobial potency were not completely loss during baking thereby enhancing partial inhibition of the bacteria and moulds. Partial loss of antibacterial properties of honey after heating has been reported (Molan, 1992). Furthermore, the increase in the TAC and mould counts in all the cassava-wheat composite bread between 4th and 6th day of storage could possibly be attributed to the concentration of the honey incorporated into the bread which was not up to the minimum inhibitory concentration needed after baking to prevent the growth of bacteria and moulds. Study has shown that the antimicrobial potency of honey depends on the minimum inhibitory concentration (MIC) (Cooper et al., 1999) and the MIC of honey has been found to be eight times higher after exposure of honey to 55°C for 8 h (Wooton et al., 1978).

Conclusions

This study has demonstrated that the physicochemical properties of honey could influence the microbial shelf stability of cassava-wheat composite bread. However, it may be necessary to increase the ratio of partial substitution with concentration of the honey used in this current study in order to determine the minimum inhibitory concentration of honey in bread and to circumvent the effect of the baking temperature on the antimicrobial potential of the honeys.

REFERENCES

Adeboye AS, Babajide JM, Shittu TA, Omemu, AM, Oluwatola OJ (2013). Effect of honey as partial sugar substitute on pasting properties, consumer preference and shelf stability of cassava-wheat composite bread. NIFOJ 31(1):13-22.

Al Somai N, Coley KE, Molan PC, Hancock BM (1994). Susceptibility of Helicobacter pylori to the antibacterial activity of Manuka honey. J. Royal Soc. Medicine. 87:9-12.

Al-Mamary M, Al-Meeri A, Al-Habori M (2002). Antioxidant activities and total phenolics of different types of honey. Nutrition

Research. 22:1041-1047.

Alzahrani HA, Alsabehi R, Boukraâ L, Abdellah F, Bellik Y, Bakhotmah BA (2012). Antibacterial and antioxidant potency of floral honeys from different botanical and geographical origins. Molecules. 17:10540-10549.

AOAC (2000). Association of Official Analytical Chemist. Methods 979.20, 969.38, 962.19, approved methods for neutraceutical analyses. Official methods of analysis of AOAC International, (17th ed.). Gaithersburg, MD, USA: AOAC.

Barcenas ME, Rosell CM (2005). Effect of HPMC on the microstructure, quality and aging of wheat bread. *Fd Hydrocolloids.* 19:1037−1043.

Beatriz A, Sandra M, Montserrat H, Eduardo C (2011). Quality parameters and antioxidant and antibacterial properties of some Mexican honeys. J.Fd. Sc. 71: 1

CA (2001).Codex Alimentarius Commission: Codex standard 12, revised Codex standard for honey standards and standard methods, Vol. 11. Sperimentale per la Zoologia, Rome. Available from: http://www.codexalimentarius.net/ Accessed July 15, 2013.

Cooper RA, Molan PC, Harding KG (1999). Antibacterial activity of honey against strains of *Staphylococcus aureus* from infected wounds. J.of the Royal Soc. Med. 92: 283-285.

Davidson PM, Sofos JN, Brenem AL (2005). Antimicrobials in foods, third ed. Marcel Dekker Inc., New York. pp. 291–306.

Defloor I, De Geest C, Schellekens M, Martens A, Delcour JA (1994). Impact of genotype and cropdage on the breadmaking and physiochemical properties of flour produced from cassava (*Manihot esculenta Crantz*) planted in the dry season. J. Sc. Fd Agr. 66:193-202.

Defloor I, Leijskens R, Bokanga M, Delcour JA (1995). Impact of genotype, crop age and planting season on the breadmaking and gelatinization properties of flour produced from cassava (*Manihot esculenta Crantz*). J. Sc. Fd Agr. 68: 167–174.

Defloor I, Nys M, Delcour JA (1993). Wheat starch, cassava starch, and cassava flour impairment of the bread making potential of wheat flour. Cereal Chem. 78 :525–530

Estevinho L, Pereira AP, Moreira L, Dias LG, Pereira E (2008). Antioxidant and antimicrobial effects of phenolic compounds extracts of Northeast Portugal honey. Food Chem. Toxicol. 46:3774-3779.

Foster RJ (2008). Fortification on the rise. Bulletin of Food Product Design, official media site of supply side innovative ingredient international. Retrieved on 22/05/10 7:00am from http// www.foodproductdesign.com

Gheldof N, Wang X, Engeseth NJ (2002). Identification and quantification of antioxidant components of honey from various floral sources. J Agric Food Chem 50:5870-5877.

Gomes S, Dias LG, Moreira LL, Rodrigues P, Estevinho L (2010). Physicochemical, microbiological and antimicrobial properties of commercial honey from Portugal. Food Chem. Toxicol. 48:544–548.

Guerrini A, Bruni R, Poli F, Paganetto G, Muzzoli M, Scalvenzi L, Sacchetti G (2009). Ecuadorian stingless bee (*Meliponinae*) honey: a chemical and functional profile of an ancient health product. Food Chem. 114:1413-1420.

Jeddar A, Kharsany A, Ramsaroop UG, Bhamjee A, Haffejee IE, Moosa A (1985). The antibacterial action of honey. An in vitro study. SA Med. J. 67(7):257-258.

Khalil AH, Mansour EH, Dawood FM (2000). Influence of malt on rheological and baking properties of wheat-cassava composite flours. J. Fd Sc. Tech. 33: 159–164.

Molan PC (1992). The antibacterial activity of honey. The nature of the antibacterial activity. Bee World 73(1):5-28.

Mundo MA, Padilla-Zakour OI, Worobo RW (2004). Growth inhibition of foodborne pathogens and food spoilage organisms by select raw honeys. Int. J. Food Microbiol. 97:1-8.

Olaitan PB, Adeleke OE, Iyabo OO (2007). Honey: a reservoir for microorganisms and an inhibitory agent for microbes. J. Afr. Health Sc. 7(3):159-165.

Olasupo NA, Fitzgerald DJ, Gasson MJ, Narbad A (2003). Activity of natural antimicrobial compounds against *Escherichia coli* and *Salmonella enterica* serovar Typhimurium. Lett App Micro. 37 (6): 446-451.

Radwan SS, El-Essawy AA, Sarhan MM (1984). Experimental evidence for the occurrence in honey of specific substances active against microorganisms. Zentralblattfür Mikrobiologie. 139(4): 249-255.

Selomulyo VO, Zhou WB (2007). Frozen bread dough: Effects of freezing storage and dough improvers. J. Cereal Sc. 45:1-17.

Shittu TA, Raji AO, Sanni LO (2007). Bread from composite cassava-wheat flour: Effect of baking time and temperature on some physical properties of bread loaf. Fd. Res. Int. 40:280-290.

Sibel S, Osman S, Lutfiye E (2010). Total phenolic content, antiradical, antioxidant and antimicrobial activities of Rhododendron honeys. Food Chemistry 121:238243.

Stefan B, Tomislav J, Robert S, and Peter G (2008). Honey for Nutrition and Health: a Review. American J. Coll. of Nutr. 27:677-689.

Weston RJ, Mitchell KR, Allen KL (1999). Antibacterial phenolic components of New Zealand manuka honey. Food chemistry, 64:295-301.

Wooton M, Edwards RA, Rows EA (1978): Antibacterial properties of some Australian honeys. Food Technology in Australia (May): 175-176.

Yao L, Datta N, Tomás-Barberán FA, Ferreres F, Martos I, Singanusong R (2003). Flavonoids, phenolic acids and abscisic acid in Australian and New Zealand Leptospermum honeys. Food Chemistry, 81:159-168.

Zalibera M, Stasko A, Slebodav´a A, Jancovicov´a V, Cerm´akov´a T, Brezov´a V (2008). Antioxidant and radical-scavenging activities of Slovak honeys: an electron paramagnetic resonance study. Food Chem. 110:512-521.

Permissions

All chapters in this book were first published in AJFS, by Academic Journals; hereby published with permission under the Creative Commons Attribution License or equivalent. Every chapter published in this book has been scrutinized by our experts. Their significance has been extensively debated. The topics covered herein carry significant findings which will fuel the growth of the discipline. They may even be implemented as practical applications or may be referred to as a beginning point for another development.

The contributors of this book come from diverse backgrounds, making this book a truly international effort. This book will bring forth new frontiers with its revolutionizing research information and detailed analysis of the nascent developments around the world.

We would like to thank all the contributing authors for lending their expertise to make the book truly unique. They have played a crucial role in the development of this book. Without their invaluable contributions this book wouldn't have been possible. They have made vital efforts to compile up to date information on the varied aspects of this subject to make this book a valuable addition to the collection of many professionals and students.

This book was conceptualized with the vision of imparting up-to-date information and advanced data in this field. To ensure the same, a matchless editorial board was set up. Every individual on the board went through rigorous rounds of assessment to prove their worth. After which they invested a large part of their time researching and compiling the most relevant data for our readers.

The editorial board has been involved in producing this book since its inception. They have spent rigorous hours researching and exploring the diverse topics which have resulted in the successful publishing of this book. They have passed on their knowledge of decades through this book. To expedite this challenging task, the publisher supported the team at every step. A small team of assistant editors was also appointed to further simplify the editing procedure and attain best results for the readers.

Apart from the editorial board, the designing team has also invested a significant amount of their time in understanding the subject and creating the most relevant covers. They scrutinized every image to scout for the most suitable representation of the subject and create an appropriate cover for the book.

The publishing team has been an ardent support to the editorial, designing and production team. Their endless efforts to recruit the best for this project, has resulted in the accomplishment of this book. They are a veteran in the field of academics and their pool of knowledge is as vast as their experience in printing. Their expertise and guidance has proved useful at every step. Their uncompromising quality standards have made this book an exceptional effort. Their encouragement from time to time has been an inspiration for everyone.

The publisher and the editorial board hope that this book will prove to be a valuable piece of knowledge for researchers, students, practitioners and scholars across the globe.

List of Contributors

A. Abedin
Faculty of Agriculture, University for Development Studies, Tamale, Ghana

A. H. Abubakari
Faculty of Agriculture, University for Development Studies, Tamale, Ghana

A. K. Quainoo
Faculty of Agriculture, University for Development Studies, Tamale, Ghana

F. A. Chimsah
Faculty of Agriculture, University for Development Studies, Tamale, Ghana

G Nyarko
Faculty of Agriculture, University for Development Studies, Tamale, Ghana

G. K. Mahunu
Faculty of Agriculture, University for Development Studies, Tamale, Ghana

F. K. Abagale
Faculty of Agriculture, University for Development Studies, Tamale, Ghana

G. Addai
Faculty of Agriculture, University for Development Studies, Tamale, Ghana

S. A. O Adeyeye
Department of Food Science and Technology, Federal University of Agriculture, Abeokuta Ogun State, Nigeria

O. B. Oyewole
Department of Food Science and Technology, Federal University of Agriculture, Abeokuta, Ogun State, Nigeria

A. O Obadina
Department of Food Science and Technology, Federal University of Agriculture, Abeokuta, Ogun State, Nigeria

A. M. Omemu
Department of Hospitality and Tourism Management, Federal University of Agriculture, Abeokuta, Ogun State, Nigeria

H. A. Oyedele
Department of Food Science and Technology, Federal University of Agriculture, Abeokuta, Ogun State, Nigeria

S. O. Adeogun
Department of Agricultural Administration, Federal University of Agriculture, Abeokuta, Ogun State, Nigeria

Chanida Saenkod
National Research Center of Engineering and Technology for Utilization of Botanical Functional Ingredients, Changsha, Hunan 410128, People's Republic of China

Zhonghua Liu
Key Laboratory of Tea Science of Ministry of Education and Hunan Province and Key Laboratory of Crop Germplasm Innovation and Utilization, Hunan Agricultural, University, Changsha, Hunan 410128, People's Republic of China
National Research Center of Engineering and Technology for Utilization of Botanical Functional Ingredients, Changsha, Hunan 410128, People's Republic of China

Jianan Huang
Key Laboratory of Tea Science of Ministry of Education and Hunan Province and Key Laboratory of Crop Germplasm Innovation and Utilization, Hunan Agricultural, University, Changsha, Hunan 410128, People's Republic of China
National Research Center of Engineering and Technology for Utilization of Botanical Functional Ingredients, Changsha, Hunan 410128, People's Republic of China

Yushun Gong
Key Laboratory of Tea Science of Ministry of Education and Hunan Province and Key Laboratory of Crop Germplasm Innovation and Utilization, Hunan Agricultural, University, Changsha, Hunan 410128, People's Republic of China
National Research Center of Engineering and Technology for Utilization of Botanical Functional Ingredients, Changsha, Hunan 410128, People's Republic of China

T. O. Adejumo
Department of Microbiology, Adekunle Ajasin University, P.M.B. 01, Akungba-Akoko, Ondo State, Nigeria

Kra Kouassi Aboutou Séverin
Laboratoire de Biotechnologies, Filière Biochimie-Microbiologie de l'Unité de Formation et de Recherche en Biosciences de l'Université de Cocody, 22 BP 582 Abidjan 22, Côte d'Ivoire

Akoa Edwige
Laboratoire de Biotechnologies, Filière Biochimie-Microbiologie de l'Unité de Formation et de Recherche en Biosciences de l'Université de Cocody, 22 BP 582 Abidjan 22, Côte d'Ivoire

Mégnanou Rose-Monde
Laboratoire de Biotechnologies, Filière Biochimie-Microbiologie de l'Unité de Formation et de Recherche en Biosciences de l'Université de Cocody, 22 BP 582 Abidjan 22, Côte d'Ivoire

Yéboué Koffi
Laboratoire de Biotechnologies, Filière Biochimie-Microbiologie de l'Unité de Formation et de Recherche en Biosciences de l'Université de Cocody, 22 BP 582 Abidjan 22, Côte d'Ivoire

Akpa Eric Essoh
Laboratoire de Biotechnologies, Filière Biochimie-Microbiologie de l'Unité de Formation et de Recherche en Biosciences de l'Université de Cocody, 22 BP 582 Abidjan 22, Côte d'Ivoire

Niamké L. Sébastien
Laboratoire de Biotechnologies, Filière Biochimie-Microbiologie de l'Unité de Formation et de Recherche en Biosciences de l'Université de Cocody, 22 BP 582 Abidjan 22, Côte d'Ivoire

Danielle Cristina Guimarães da Silva
UFLA, Lavras, MG, Brasil

Danilo Florisvaldo Brugnera
UFLA, Lavras, MG, Brasil

Luiz Ronaldo de Abreu
UFLA, Lavras, MG, Brasil

Kuku Adenike
Department of Biochemistry, Obafemi Awolowo University, Ile-Ife, Nigeria

Alabi Mercy Olufunmilayo
Department of Biochemistry, Obafemi Awolowo University, Ile-Ife, Nigeria

Ibironke Samson Ishola
Department of Food Science and Technology, Obafemi Awolowo University, Ile-Ife, Nigeria

H. O. Agu
Department of Food Science and Technology, Federal Polytechnic Bauchi, P.M.B. 0231 Bauchi, Nigeria

M. H. Badau
Department of Food Science and Technology, University of Maiduguri, P.M.B. 1069 Maiduguri Nigeria

M. U. Abubakar
Department of Food Science and Technology, University of Maiduguri, P.M.B. 1069 Maiduguri, Nigeria

V. A. Jideani
Department of Food Science and Technology, Cape Peninsula University of Technology, Cape Town 8000, South Africa

Guy Eshun
Department of Home Economics Education, Faculty of Science Education, University of Education, Winneba, Ghana

Emmanuel Adu Amankwah
Department of Home Economics Education, Faculty of Science Education, University of Education, Winneba, Ghana

John Barimah
Department of Food Science and Technology, Faculty of Biosciences, Kwame Nkrumah University of Science and Technology, Kumasi, Ghana

Jeremia N. Moyane
Department of Food Science and Technology, School of Agriculture, University of Venda, Private Bag X5050, Thohoyandou 0950, Limpopo Province, South Africa
Agricultural Research Council, Animal Production Institute, Gastrointestinal Tract Microbiology and Biotechnology Unit, Private Bag X2, Irene, 0062 South Africa

Afam I. O. Jideani
Department of Food Science and Technology, School of Agriculture, University of Venda, Private Bag X5050, Thohoyandou 0950, Limpopo Province, South Africa

E. K. Ajani
Department of Wildlife and Fisheries Management, University of Ibadan, Oyo State, Nigeria

S. B. Setufe
Department of Wildlife and Fisheries Management, University of Ibadan, Oyo State, Nigeria

O. O. Oyebola
Olabisi Onabanjo University, Ogun State, Nigeria

Akinwunmi Kemi Feyisayo
Department of Biochemistry, Faculty of Science, Obafemi Awolowo University, Ile-Ife, Osun State, Nigeria

Oyedapo Oluboade Oluokun
Department of Biochemistry, Faculty of Science, Obafemi Awolowo University, Ile-Ife, Osun State, Nigeria

Miao Yu
Department of Food Science and Engineering, Jinan University, Guangzhou 510632, China

Shiyi Ou
Department of Food Science and Engineering, Jinan University, Guangzhou 510632, China

Deng Liumengzi
Department of Food Science and Engineering, Jinan University, Guangzhou 510632, China

Caihuan Huang
Department of Food Science and Engineering, Jinan University, Guangzhou 510632, China

Guangwen Zhang
Department of Food Science and Engineering, Jinan University, Guangzhou 510632, China

Misheck Chandiposha
Department of Agronomy, Midlands State University, P. Bag 9055, Gweru, Zimbabwe

Chagonda Ignatius
Department of Agronomy, Midlands State University, P. Bag 9055, Gweru, Zimbabwe

Makuvaro Veronica
Department of Agronomy, Midlands State University, P. Bag 9055, Gweru, Zimbabwe

E. Oladeji Alamu
International Institute of Tropical Agriculture, Ibadan, Oyo State, Nigeria

Bussie Maziya-Dixon
International Institute of Tropical Agriculture, Ibadan, Oyo State, Nigeria

C. Christian Okonkwo
International Fertilizer Development, Centre, Nigeria

Asiedu Roberts
International Institute of Tropical Agriculture, Ibadan, Oyo State, Nigeria

Méndez-Iturbide Daniel
Laboratorio de Química de Alimentos, Escuela de Nutrición, Facultad de Ciencias Médicas, Universidad Autónoma de Tlaxcala. Av. Xicohtencatl No. 466, Barrio de San Nicolás, San Pablo del Monte 90920, Tlaxcala, México

Banderas-Tarabay José Antonio
Laboratorio de Química de Alimentos, Escuela de Nutrición, Facultad de Ciencias Médicas, Universidad Autónoma de Tlaxcala. Av. Xicohtencatl No. 466, Barrio de San Nicolás, San Pablo del Monte 90920, Tlaxcala, México

Nieto-Camacho Antonio
Instituto de Química, Universidad Nacional Autónoma de México, Circuito Exterior s/n, Ciudad Universitaria, Coyoacán, México, DF, C.P. 04510, México

Rojas-Chávez Agustina
Centro de Investigación en Biotecnología Aplicada, Instituto Politécnico Nacional, Ex-Hacienda San Juan Molino Carretera Estatal Tecuexcomac-Tepetitla Km 1.5, Tlaxcala C.P. 90700, México

García-Meza Maria
Centro de Investigación en Biotecnología Aplicada, Instituto Politécnico Nacional, Ex-Hacienda San Juan Molino Carretera Estatal Tecuexcomac-Tepetitla Km 1.5, Tlaxcala C.P. 90700, México

Ochulayi P. Ora
Department of Biochemistry, National Research Institute for Chemical Technology, Zaria Nigeria

G. E. Anekwe
Department of Biochemistry, University of Jos, Plateau state, Nigeria

ANDABATI Brain
School of Food Technology, Nutrition and Bioengineering, Makerere University, P.O. Box 7062, Kampala, Uganda

MUYONGA John
School of Food Technology, Nutrition and Bioengineering, Makerere University, P.O. Box 7062, Kampala, Uganda

A. A. Dada
Department of Fisheries and Aquaculture Technology, Federal University of Technology, Akure, Ondo State, Nigeria

B. D. Olugbemi
Department of Fisheries and Aquaculture Technology, Federal University of Technology, Akure, Ondo State, Nigeria

E. U. Onyeka
Federal University of Technology, Owerri, Imo State, Nigeria

O. S. Obeleagu
Federal University of Technology, Owerri, Imo State, Nigeria

J. A Folayan
Department of Agricultural Economics and Extension Service, Joseph Ayo Babalola University, Ikeji-Arakeji, P.M.B. 5006 Ilesa, Osun State, Nigeria

Israel Olusegun Otemuyiwa
Department of Chemistry,Obafemi Awolowo University,Ile-Ife, Nigeria

Steve Adeniyi Adewusi
Department of Chemistry,Obafemi Awolowo University,Ile-Ife, Nigeria

Khouloud LAKRARI
Laboratoire de Chimie Physique Générale, Département de Chimie, Université Mohammed V-Agdal, Faculté des Sciences, Avenue Ibn Batouta, BP 1014 Rabat

Mouloud EL MOUDANE
Laboratoire de Matériaux, Nanotechnologies et Environnement (LMNE), Département de Chimie, Université Mohammed V-Agdal, Faculté des Sciences, Avenue Ibn Batouta, BP 1014 Rabat

Imane HASSANAIN
Laboratoire de Chimie Physique Générale, Département de Chimie, Université Mohammed V-Agdal, Faculté des Sciences, Avenue Ibn Batouta, BP 1014 Rabat

Imane ELLOUZI
Laboratoire de Chimie Physique Générale, Département de Chimie, Université Mohammed V-Agdal, Faculté des Sciences, Avenue Ibn Batouta, BP 1014 Rabat

Said KITANE
Laboratoires de chimie appliquée à l'Ecole Nationale de l'Industrie Minérale, Avenue Hadj Ahmed Cherkaoui BP 753 Agdal Rabat

Mohamed ALAOUI El BELGHITI
Laboratoires de chimie appliquée à l'Ecole Nationale de l'Industrie Minérale, Avenue Hadj Ahmed Cherkaoui BP 753 Agdal Rabat

Sekar Ashok kumar
International Maritime Academy, Chennai, India

Packyam Mayavu
Faculty of Marine Sciences, Annamalai University, Parangipettai,India

ASSA Rebecca Rachel
Unité de Formation et de Recherche (UFR) Biosciences, Laboratoire de Biochimie et Sciences des Aliments; Université Félix Houphouët Boigny-Abidjan, (*) 21 BP 1126 Abidjan 21, Côte d'Ivoire
CNRA- Station Marc Delorme pour la recherche sur le cocotier de Port- Boüet, 07 BP 13 Abidjan 07, Côte d'Ivoire

PRADES Alexia
Centre de Coopération Internationale en Recherche Agronomique pour le Développement (CIRAD), Département PERSYST "Performances des systèmes de production et de transformation tropicaux", Unité Mixte de Recherche "QUALISUD", TA B 95/16. 73, rue JF Breton 34398 Montpellier Cedex 5 France

KONAN Amoin Georgette
Unité de Formation et de Recherche (UFR) Biosciences, Laboratoire de Biochimie et Sciences des Aliments; Université Félix Houphouët Boigny-Abidjan, (*) 21 BP 1126 Abidjan 21, Côte d'Ivoire

NEMLIN Jean
Centre National de Recherche Agronomique (CNRA) - Station sur la Recherche Technologique, 08 BP 881 Abidjan 08, Côte d'Ivoire

KONAN Jean-Louis
CNRA- Station Marc Delorme pour la recherche sur le cocotier de Port- Boüet, 07 BP 13 Abidjan 07, Côte d'Ivoire

J. C. Ani
Department of Food Science and Technology, University of Nigeria, Nsukka, Enugu State, Nigeria

U. E. Inyang
Department of Food Science and Technology, University of Uyo, Uyo, Akwa Ibom State, Nigeria

I. Udoidem
Department of Food Science and Technology, University of Uyo, Uyo, Akwa Ibom State,Nigeria

P. G. Udofia
Department of Hotel and Catering Management, Akwa Ibom State Polytechnic, Ikot Osurua, Akwa Ibom State, Nigeria

P. J. Udoudo
Department of Hotel and Catering Management, Akwa Ibom State Polytechnic, Ikot Osurua, Akwa Ibom State, Nigeria

N. O. Eyen
Department of Hotel and Catering Management, Akwa Ibom State Polytechnic, Ikot Osurua, Akwa Ibom State, Nigeria

F. A. Igbasan
Department of Animal and Health, School of Agriculture and Agricultural Technology, Federal University of Technology, P. M. B. 704, Akure, Ondo State, Nigeria

O. A. Olugosi
Department of Animal and Health, School of Agriculture and Agricultural Technology, Federal University of Technology, P. M. B. 704, Akure, Ondo State, Nigeria

Anjineyulu Kothakota
Department of Food Engineering and Technology, Sant Longowal Institute of Engineering and Technology, Longowal-148106, Punjab, India

Navdeep Jindal
Department of Food Engineering and Technology, Sant Longowal Institute of Engineering and Technology, Longowal-148106, Punjab, India

B. Thimmaiah
Department of Food Process Engineering, Sam Higginbottom Institute of Agriculture, Technology and Sciences, Allahabad- 211 007, UP, India

Lucy M. CHOVE
Department of Food Science and Technology, Sokoine University of Agriculture, P.O. Box 3006, Morogoro, Tanzania

Abdulsudi ISSA- ZACHARIA
Department of Food and Nutritional Sciences, University of Reading, P.O. Box 226, Reading RG6 6AP, UK.

Alistair S. GRANDISON
Department of Food and Nutritional Sciences, University of Reading, P.O. Box 226, Reading RG6 6AP, UK.

Michael J. LEWIS
Department of Food and Nutritional Sciences, University of Reading, P.O. Box 226, Reading RG6 6AP, UK

Raphael Emuebie Okonji
Department of Biochemistry, Obafemi Awolowo University, Ile-Ife, Nigeria

Komolafe Olaniyi Olusola
Department of Zoology, Obafemi Awolowo University, Ile-Ife, Nigeria

Leonard Ona Ehigie
Department of Biochemistry, Ladoke Akintola University of Technology, Ogbomoso, Nigeria

Popoola Micheal Olaoluwa
Department of Zoology, Obafemi Awolowo University, Ile-Ife, Nigeria

Kayode Samuel Bamitale
Department of Medical Pharmacology and Therapeutics, Obafemi Awolowo University, Ile-Ife, Nigeria

Joel Ndife
Department of Food Technology, Kaduna Polytechnic, Kaduna State, Nigeria

Dehinde Awogbenja
Department of Home Science and Management, Nasarawa State University, Keffi, Nasarawa State, Nigeria

Umaru Zakari
Department of Food Technology, Kaduna Polytechnic, Kaduna State, Nigeria

Sanae Amkiss
Biology and Health Laboratory, Department of Biology, Faculty of Sciences, Abdelmalek Essaâdi University, BP 2121, 93002, Tétouan, Morocco

Abdelkrim Dallouh
Biology and Health Laboratory, Department of Biology, Faculty of Sciences, Abdelmalek Essaâdi University, BP 2121, 93002, Tétouan, Morocco

Mohamed Idaomar
Biology and Health Laboratory, Department of Biology, Faculty of Sciences, Abdelmalek Essaâdi University, BP 2121, 93002, Tétouan, Morocco

Bouchra AMKISS
Chemical department, Faculty of Sciences, Abdelmalek University, BP 2121,93002, Morocco

O. O. Fawole
Department of Pure and Applied Biology, Ladoke Akintola University of Technology, P.M. B. 4000, Ogbomoso, Nigeria

T. A. Yekeen
Department of Pure and Applied Biology, Ladoke Akintola University of Technology, P.M. B. 4000, Ogbomoso, Nigeria

S. O. Adewoye
Department of Pure and Applied Biology, Ladoke Akintola University of Technology, P.M. B. 4000, Ogbomoso, Nigeria

M. A. Ogundiran
Department of Pure and Applied Biology, Ladoke Akintola University of Technology, P.M. B. 4000, Ogbomoso, Nigeria

O. E. Ajayi
Department of Pure and Applied Biology, Ladoke Akintola University of Technology, P.M. B. 4000, Ogbomoso, Nigeria

M. N. Nwaiya
Department of Pure and Applied Biology, Ladoke Akintola University of Technology, P.M. B. 4000, Ogbomoso, Nigeria

Alain Serges ONDO-AZI
Unité de Recherche Agrobiologie, Université des Sciences et Techniques de Masuku, B.P. 941 Masuku Franceville Gabon

Brice Serge KUMULUNGUI
Unité de Recherche Agrobiologie, Université des Sciences et Techniques de Masuku, B.P. 941 Masuku Franceville Gabon

Ludovic MEWONO
Laboratoire d'Immunologie, de Microbiologie appliquée et d'Hygiène, Ecole Normale Supérieure, B.P. 440 Libreville Gabon

Armel MBINA KOUMBA
Unité de Recherche Agrobiologie, Université des Sciences et Techniques de Masuku, B.P. 941 Masuku Franceville Gabon

Crépin ELLA MISSANG
Unité de Recherche Agrobiologie, Université des Sciences et Techniques de Masuku, B.P. 941 Masuku Franceville Gabon
Institut de Recherche Technologique, CENAREST, B.P. 14.070, Libreville, Gabon

Divine N. BUP
Higher Institute of the Sahel, University of Maroua, Cameroon
Department of Process Engineering, ENSAI, University of Ngaoundere, P.O. Box 455 Ngaoundere, Cameroon

Charles F. ABI
Department of Chemistry, Higher Teachers' Training College, University of Yaounde 1, PO Box 47, Yaounde, Cameroon

Dzudie TENIN
Department of Process Engineering, ENSAI, University of Ngaoundere, P.O. Box 455 Ngaoundere, Cameroon

Cesar KAPSEU
Department of Process Engineering, ENSAI, University of Ngaoundere, P.O. Box 455 Ngaoundere, Cameroon

Clergé TCHIEGANG
Department of Food Science and Nutrition, ENSAI, University of Ngaoundere, P.O. Box 455 Ngaoundere, Cameroon

Patil Priyanka
Department of Food Chemistry and Nutrition, College of Food Technology, Vasantrao Naik Marathwada Krishi Vidyapeeth, Parbhani-431402, MH, India

H. M. Sayed
Department of Food Chemistry and Nutrition, College of Food Technology, Vasantrao Naik Marathwada Krishi Vidyapeeth, Parbhani-431402, MH, India

A. A. Joshi
Department of Food Chemistry and Nutrition, College of Food Technology, Vasantrao Naik Marathwada Krishi Vidyapeeth, Parbhani-431402, MH, India

B. A. Jadhav
Department of Food Chemistry and Nutrition, College of Food Technology, Vasantrao Naik Marathwada Krishi Vidyapeeth, Parbhani-431402, MH, India

A. S. Adeboye
Department of Food Technology, Moshood Abiola Polytechnic, Abeokuta, Ogun State, Nigeria

J. M. Babajide
Department of Food Science and Technology, University of Agriculture, Abeokuta, Ogun State, Nigeria

O. E. Fayemi
Department of Microbiology, Federal University of Technology, Akure, P.M.B. 704, Ondo State, Nigeria